Bui's Maths Book

A Compendium

of Mathematical Invention

Volume 2

R H S White

The left hand circle on the front cover and on the back cover is a picture of the Kalachakra Sand Mandala painted by the Tibetan monks of Namgyal Monastery, Dharamsala, India, when they visited the Royal Academy of Arts in London. Just before they returned to India the sand painting was poured into the river Thames.

The right hand circle on the front cover is a picture of the E8 symmetry group, the most complicated structure ever studied by mankind, that hit the media headlines in March 2007. E8 lives in a 248 dimensional space. The circle is the map of its projection onto two dimensions.

The number, written in various scripts, is 8128, the fourth perfect number, the only perfect number with four digits.

Order this book online at www.trafford.com
or email orders@trafford.com

Most Trafford titles are also available at major online book retailers.

Note for Librarians: A cataloguing record for this book is available from Library and Archives Canada at www.collectionscanada.ca/amicus/index-e.html

Printed in Victoria, BC, Canada.

ISBN: 978-1-4269-1412-6

We at Trafford believe that it is the responsibility of us all, as both individuals and corporations, to make choices that are environmentally and socially sound. You, in turn, are supporting this responsible conduct each time you purchase a Trafford book, or make use of our publishing services. To find out how you are helping, please visit www.trafford.com/responsiblepublishing.html

Our mission is to efficiently provide the world's finest, most comprehensive book publishing service, enabling every author to experience success. To find out how to publish your book, your way, and have it available worldwide, visit us online at www.trafford.com

www.trafford.com

North America & international
toll-free: 1 888 232 4444 (USA & Canada)
phone: 250 383 6864 • fax: 812 355 4082

Preface

On contemplating his life as a mathematician, Sir Isaac Newton said that at times he felt like a small boy playing on the seashore and occasionally finding an interesting pebble while the vast ocean of mathematics lay before him.

Most of us enjoy the tales of Shakespeare or Charles Dickens and feel good if we manage to solve a crossword. We can also enjoy the subtlties of logic, symmetries and patterns, the rediscovery of a theorem that we had learnt at school but had since forgotten or uncovering the reasoning in a proof that we had not seen before.

Mathematics is a neglected part of British culture. In England, if you watch the quiz shows that are a popular diversion on British TV you will find plenty of evidence. On 'The Weakest Link' the host, Ann Robinson, thinks that the height of mathematical achievement is to be able to add together a pair of two digit numbers. Jeremy Paxman, the quizmaster for 'University Challenge' will sneer at the student who thinks that 'The Jew of Malta' was written by Shakespeare and not Christopher Marlowe while happily professing an ignorance of mathematics. When interviewing Professor Brian Cox, John Humphrys freely admitted to 'knowing nothing about science'. It seems that the celebrities of today may reasonably be expected to have some knowledge of poetry, literature, history, the cinema and pop music but to be asked a question concerning mathematics would be deemed to be unfair (unless it was adding two numbers together).

If the ideas that we learn at school are not continually used or refreshed then they will recede into the dusty corners of our minds and slowly fade away. Children study Shakespeare and Dickens at school and enjoy the stories on film or TV in later life but most children will study mathematics to the standard of O level and from that point onwards lose what interest they had. TV and radio are very good at refreshing our love of the Arts but very poor at shoring up the mathematical confidence of their audiences. Without such stimulation one's mathematical confidence will diminish and not grow.

Part of the function of this book is to show how a set of simple axioms and definitions can quickly build up into a solid theory by means of a few theorems and that the results we obtain are often evident in or relevant to the real world around us.

As with the chapters of volume 1, many of the chapters in volume 2 can be read independently of the others. Chapter 16, Towers of Numbers, examines the basic operations of addition, multiplication and raising to a power showing that each operation is a repeated application of the previous simpler operation and extends the idea to further towers of numbers finally discovering a relation between towers of numbers and Ackerman's function. Chapter 17 introduces the arithmetic and geometric sequences and solves the paradox of Achilles and the tortoise. Chapter 18 examines the link between the Fibonacci sequence, the golden ratio, the pentagram and Morse codes. Chapter 19 concerns the basic trig ratios of sine, cosine and tangent but also poses the question, can you do trig without right angle triangles? The answer is yes. Chapter 20 questions things that most of us take for granted like the area of a rectangle or the area of a circle but extends the discussion as far as the derivation of formulae for the area of a ball and the volume of a doughnut. Chapter 21 proves a number of theorems for figures made up of straight lines: Ceva's theorem, Menelaus' theorem, Pascal's theorem, Desargues theorem and ends with the surprising result that the tri-sectors of the angles of any triangle produce an equilateral triangle - Morley's triangle. Chapter 22 treats the circle theorems including the seven circles theorem and the nine circles theorem. Chapter 23 is on special relativity and shows how it is that time has to slow down and lengths contract and why you cannot travel on a beam of light. Chapter 24 is devoted to vectors. Chapter 25 is concerned with the development of different number systems including the complex number system and chapter 26 introduces calculus. Chapters 25, 26, 27 and 28 are related in that they are concerned with how the standard functions of mathematics, the trig functions, powers and logs, that are initially defined only for positive real numbers, have their definitions extended to include negative numbers and complex numbers to form a more complete theory. Chapter 28 also introduces the conic sections showing how a gardener can construct a flower bed and why a car headlamp is the shape it is.

This is not a mathematics textbook although there are many solved examples and exercises, all with answers, and it is hoped that it will appeal both to the general reader and to the mathematics specialist.

Bui's Maths Book is also an appeal to TV and radio programme makers to bring mathematics into mainstream programming and not to hide it away at 5 o'clock in the morning, but the treatment needs to be more than that: we need programmes that are there for the general public interest and not just for those studying for an examination in mathematics. It is not only those studying for an examination in English Literature who would watch "The Merchant of Venice" on TV. In the age of the Internet and space travel people want to see the equations and how they are solved. The theorems that we learned at school should be recalled and built upon, not forgotten. The cynical publishers dictum that "each equation will halve the number of readers" should not hold any more than an equation on the TV screen would halve the number of viewers.

However, the main motivation behind Bui's Maths Book is simply a passion for a subject that I feel should be enjoyed and better understood by many more people.

"Now we can see what makes mathematics unique. Once the Greeks had developed the deductive method they were correct in what they did, correct for all time. Euclid was incomplete and his work has been extended enormously but it has not had to be corrected. His theorems are, every one of them, valid to this day" —Isaac Asimov

"It isn't right to admit that you know nothing about the Arts so why should it be acceptable for science?"
—Professor Brian Cox commenting after John Humphrys said he knew nothing about science.

I would like to express my appreciation and gratitude to Dr Guy Vine, Jeff Terrell and Gerry Craddock for their encouragement and for reading (and criticizing) large chunks of the text.

Any errors belong to me and if you find any I would be grateful if you would tell me where they may be located.

Lastly, of course, I express my gratitude to my wife Phillipa for her encouragement, her penetrating criticisms of some of my verbiage and grammar and for her help with the design of the cover.

The Introduction to Volume 1

Friday afternoon in mid-summer is not the most favourable time to teach a class of electronics students in their late teens but that was the situation in which I found myself when I took up a temporary teaching post at Croydon College in the third term of the 1998-99 academic year. Most of the students were anxious to get away early and only a handful stayed on till 5pm to catch up with assignments. Two students, however, known to me as Miah and Bui would always wait till the end of the session when all the other students had left and then start plying me with all kinds of questions on mathematics.

Friday afternoons always meant an extra half an hour or so at the board "doing maths".

Realistically, as long as you can read the gas meter and check the speedo on your car, that's about all the maths the average citizen needs to know to survive in the modern world. Once you can press the right buttons on the calculator, the arithmetic skills needed today are minimal. So, why do maths? But we could also ask why read poetry? Why listen to music? Why look at paintings? Why do anything that is not really necessary for your survival as an intelligent animal in a competitive world? Perhaps the answer is that you are not just an animal trying to survive. You have in your head, the most intricate and complicated computing machine known to man. A dog has a brain but a dog cannot read poetry or gain pleasure from hearing music or looking at paintings. You have an enquiring mind that appreciates patterns, symmetries, logical argument and paradoxes. This is why we 'do maths', not to survive (unless you happen to be an engineer or a maths teacher) but because you can gain some pleasure and satisfaction from recognizing patterns, symmetries and subtle arguments. Mathematics is not just another subject on the school syllabus, it is a way of looking at the world and a way of thinking about things. I recall sixty maths students packed into the Mill Lane lecture theatre, in Cambridge, many years ago. The lecture on mathematical topology was coming to a close. The 20 foot blackboard was covered wall to wall with symbols, without a word of English and I, like most of those present, had drifted into a bemused soporific torpor having been completely lost after the first ten minutes. But the lecturer concluded with the words "and there you have it ladies and gentlemen - the Jam Sandwich Theorem" - there was a rustle and stirring among the students - here at last was something we could recognize. The lecturer proceeded to explain that he had proved that any three point sets in three dimensional space could "each and severally" be divided into equal halves by a single flat plane. In other words, given a hunk of bread (two slices if you like), a lump of butter and a blob of jam, you can always cut the jam sandwich exactly in half so that each half of the sandwich has half the bread, half the butter and half the jam. It was most refreshing to realize that such a mass of mathematical symbols and notation could produce such an easily understood result. Maybe this is the problem that many people have with mathematics, being able to cut through the mass of notation and equations that are so often encountered, to see some relevance to the real world or simply stated ideas.

This is a book for mathematics enthusiasts. Not necessarily competent mathematicians but those who have the curiosity to want to know how it works and what it is all about. I hope that it will appeal to both beginners, engineers, scientists, teachers and those of a less scientific persuasion. No attempt has been made to avoid equations and symbols, after all, that is the language of mathematics, but I hope that each chapter will impart some of the wonder and thrill that we experienced on that day in Mill Lane. It is dedicated to Bui Van Dong who once playfully suggested that I should write a book on "all that stuff" and to those students who have provided me with many of those golden moments when the spirit of youthful enquiry awakens the interest of a teacher to realize that his passion for mathematics is something worth sharing.

R.H.S.White
Surrey, England
2008

Volume 2

Contents

Chapter 21

Chapter 22

Circles 219

Chapter 23

Chapter 24

Chapter 28

==============================

CHAPTER 16

Towers of Numbers and Ackerman's Function

Definition: recursion..........(see recursion)

Recursive Definitions

We have seen in chapter 15 that the number of arrangements of n objects can be written

$$\text{perm}(n) = n \times (n-1) \times (n-2) \times .. 3 \times 2 \times 1$$

We now describe a neat alternative method of expressing this result. The method is variously called inductive definition, a recurrence relation, a reduction formula or recursive definition. We choose to call it an example of recursive definition:

We define

$$\text{perm}(n) = n \times \text{perm}(n-1) \quad \text{for all } n>1$$

and $\text{perm}(1) = 1$

Thus we can calculate perm(n) for any whole number n, for example,

$$\text{perm}(3) = 3 \times \text{perm}(2) = 3 \times 2 \times \text{perm}(1) = 3\times2\times1 = 6$$

Perm(n) is also called factorial n and written n! or $\underline{|n}$

We have also seen, in chapter 15, that there is good reason for defining perm(0) to be 1. The number of ways of selecting r things from a group of n things is ${}^nCr = \dfrac{n!}{r!.\,(n-r)!}$ and if we select all n things, which can clearly be done in just one way, then this formula gives $\dfrac{n!}{n!.(0)!}$ and only gives the correct answer ifwe define (0)! = 1.

Thus we decide that there is just one way of arranging nothing in a row: perm(0) = 1.

Examples

[1] Factorial n may be defined recursively by

n! = n x (n-1)!

0! = 1

[2] If n is a positive number then a↑n may be recursively defined by

a↑n = nx a↑(n-1) for n>0

a↑0 = 1

================================

A note on precedence:

Usually, in arithmetic, if we have expressions using operators of equal precedence, then the rule is to evaluate from the left, thus

6 – 4 –1 = (6-4) – 1 = 2 – 1 = 1

7 – 3 + 1 = (7-3) + 1 = 4 + 1 = 5

However, "raising to a power" has been an exception to this rule:

Before computers were commonplace, powers were always written using smaller and smaller suffices, for example,

$$2^{3^{2}} = 2^{9} = 512$$

$$(2^{3})^{2} = 8^{2} = 64$$

However, with the advent of the up arrow ↑ on calculators, things have changed around.

If we enter 2 ↑ 3 ↑ 2 we find that it is evaluated from the left as 8 ↑ 2 = 64
In what follows, we will often find it convenient to evaluate from the right.

Recursive definition of Arithmetic Functions

A Hierarchy of Functions on the natural Numbers

[1] Definition of inc

We define function on the natural numbers including zero, called **inc.** **inc** returns the next integer thus

inc(n) = n+1 for all n>=0.

Thus inc(2) = 3 and inc(4) = 5

Next we define recursively, a **plus** function in terms of **inc:**

[2] Definition of plus

a plus n = inc(a plus(n-1)) for n>=1

a plus 0 = a

Example: 2 plus 3 = inc(2 plus 2)

= inc(inc(2 plus 1))

= inc(inc(inc(2 plus 0)))

= inc(inc(inc(2)))

= inc(inc(3)) = inc(4) = 5

We can easily prove that **plus** is the usual arithmetic addition operator using induction on n.

Theorem 1 **a plus n = a+n** **for a,n>=0**

Proof

Suppose that **a plus k = a+k** for some value k

Then **a plus(k+1) = inc(a plus k) = a+k+1**

Therefore, if true for k, then the theorem is true for (k+1)

Now **a plus 0 = a+0**, by initial definition, so theorem 1 is true for **k=0**

therefore **a plus n = a+n** for all **a and n>=0** ■

Next we define a **times** function in terms of **plus:**

[3] Definition of times

a times n = a plus (a times(n-1)) for n>=1

a times 0 = 0

Example

2 times 3 = 2 plus (2 times(2))

= 2 plus (2 plus (2 times 1)

= 2 plus (2 plus (2 plus (2 times 0)))

= 2 plus (2 plus (2 plus 0))

= 2 plus (2 plus 2) = 2 plus 4 = 6

Again, we appeal to proof by induction to prove that **a times n = a x n:**

Theorem 2 **a times n = a x n**

Proof

Suppose that **a times k = a x k** for some integer k

Then **a times (k+1) = a plus(a times k)**

= a plus a x k

= a + ak = a x (k+1)

therefore, if the theorem is true for the number k, it is also true for (k+1)

but **a times 0 = 0 = a x 0** so the theorem is true for **k=0**

therefore, by the principle of induction **a times n = a x n** for all **n >=0**

■

We now define a **power** function in terms of **times**

[4] Definition of power

a power n = a times(a power (n-1)) **for n>=1**

a power 0 = 1

Example

2 power 3 = 2 times (2 power 2)

= 2 times (2 times (2 power 1))

= 2 times (2 times (2 times (2 power 0)))

= 2 times (2 times (2 times 1))

= 2 times (2 times 2) = 2x2x2 = 2↑3

==============================

Theorem 3 **a power n = a ↑ n** **for all n>=0**

Proof

Suppose that **a power k = a ↑ k** for some integer k

Then **a power (k+1) = a times (a power k)**

= a times a ↑ k

= a x a ↑ k = a ↑ (k+1)

therefore, if the theorem is true for some integer k, the theorem is also true for (k+1)

but the theorem is true for **k=0**, since **a power 0 = 1 = a ↑ 0**

Therefore, by the principle of induction, **a power n = a ↑ n** fo all **n >=0**

■

Towers of Numbers

A tower of four twos is called "two to the tower 4" and is written

$$2^{2^{2^{2}}} = 2\,T\,4$$

[5] Definition of aTn

a T n = a↑ (a T (n-1)) **for n>=1**

a T 0 = 1

Examples

2T1 = 2↑(2T0) = 2↑1 = 2

2T2 = 2↑(2T1) = 2↑2

2T3 = 2↑(2T2) = 2↑(2↑2) = 2↑2↑2

2T4 = 2↑(2T3) = 2↑2↑2↑2

etc

We note that the up arrows are evaluated from the right.

Theorem 4 **a T n is a tower of a's of height n** **for all n>=1**

Proof

Suppose that **a T k** is a tower of height **k,** for some integer **k**

Then **a T (k+1) = a ↑ (a T k)**

= **a** ↑ (a↑a↑.....↑a)

which is a tower of height **k+1**

Now **a T 1 = a↑(aT0) = a↑1 = a** which is a tower of height 1

Therefore, by the principle of induction, **aTn** is a tower of height n for all **n>=1**

■

The next tower function will be written **aTTn**

[6] Definition of aTTn

aTTn = aT(aTT(n-1)) **for n>=1**

aTT0 = 1

Similarly we have

aTTTn =aTT(aTTT(n-1))

aTTT0 = 1

and higher order tower functions are defined in the same way, with the same initial values;

aT0 = aTT0 =aTTT0 = aTTTT0 = ... = 1

We can then prove that

aT1 = aTT1 = aTTT1 = aTTTT1 = ... = a

For example, **aTTTT1 = aTTT(aTTTT0)**

= aTTT1 = aTT(aTTT0)

= aTT1 = aT(aTT0)

= aT1 = a↑(aT0) = a↑1 = a

so that we could, if we wished, write these tower functions in the same way that we write powers, thus

aTTT4 = aTT(aTTT3) = aT(aTT(aTTT2)) = aTT(aTT(aTT(aTTT1)))

= aTT(aTT(aTT(aTTa)))

which we could write as

$$aTTT4 = aTT^{aTT^{aTT^{aTTa}}}$$

$$aTT4 = aT^{aT^{aT^{aTa}}}$$

$$aT3 = a^{a^{a}}$$

Higher orders of towers quickly get very large, for example:

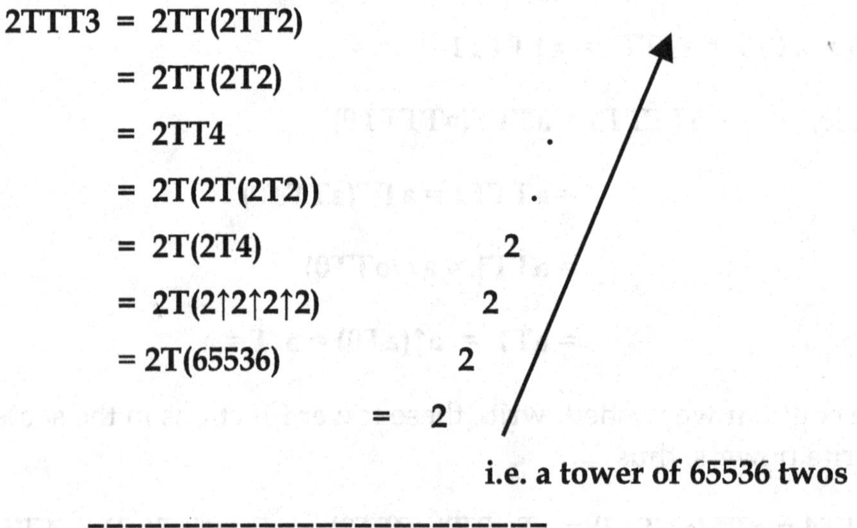
2TTT3 = 2TT(2TT2)
= 2TT(2T2)
= 2TT4
= 2T(2T(2T2))
= 2T(2T4)
= 2T(2↑2↑2↑2)
= 2T(65536)
= 2
2
2
2
.
.
i.e. a tower of 65536 twos

========================

We now introduce a more general notation that allows us to write the familiar arithmetic operations, plus, times and power, using the same recursive definition.

Definitions

The symbol $a \overset{m}{T} n$ that we may describe as "a to the mth tower of n" is defined by

Tower Def 1

$$a \overset{m}{T} n = a \overset{m-1}{T} (a \overset{m}{T}(n-1)) \qquad \text{for } m,n > 0$$

with initial values:

Tower Def 2

$$a \overset{0}{T} n = 1+n \qquad \text{for all } n \text{ and } a >= 0$$

Tower Def 3

$$a \overset{1}{T} 0 = a \qquad \text{for all } a>=0$$

Tower Def 4

$$a \overset{2}{T} 0 = 0 \qquad \text{for all } a>=0$$

Tower Def 5

$$a \overset{m}{T} 0 = 1 \qquad \text{for } m>2 \text{ and } a>0$$

$a \overset{m}{T} n$ is not defined for **a=0**

Thus

$a \overset{0}{T} n$ is the same as the inc function, where inc(n) returns the next integer.

First Tower Theorem

$$a \overset{1}{T} n = a+n \qquad \text{for all } a,n>=0$$

Proof

Suppose that $a \overset{1}{T} k = a+k$ for some integer k

$$a \overset{1}{T} (k+1) = a \overset{0}{T} (a \overset{1}{T} k) \qquad \text{(First tower def)}$$

$$= a \overset{0}{T} (a+k) \qquad \text{supposition}$$

$$= (a+k+1) \quad \text{since } a \overset{0}{T} n \text{ is the inc function.}$$

Therefore, if the theorem is true for k, then it is also true for (k+1)

But the theorem is true for k=0 since $a \overset{1}{T} 0 = a = a+0$ Tower def 3

Therefore, by the principle of mathematical induction,

$$a \overset{1}{T} n = a+n \qquad \text{for all } n>=0 \text{ and all } a \qquad \blacksquare$$

======================

Second Tower Theorem

$$a \overset{2}{T} n = a.n \qquad \text{for all } a ,n>=0$$

Proof Suppose that $a \overset{2}{T} k = a.k$ for some integer k

$$a \overset{2}{T} (k+1) = a \overset{1}{T} (a \overset{2}{T} k) \qquad \text{Tower def 1}$$

$$= a \overset{1}{T} (a.k) \qquad \text{supposition}$$

$$= a+a.k \qquad \text{First Tower Theorem}$$

$$= a.(k+1)$$

Therefore, if the theorem is true for k, then it is also true for (k+1)

But the theorem is true for k=0 since $a \overset{2}{T} 0 = 0 = a.0$ Tower def 4

Therefore, by the principle of mathematical inductin

$$a \overset{2}{T} n = a.n \qquad \text{for all } a,n>=0 \qquad \blacksquare$$

Third Tower Theorem

$$a \overset{3}{T} n = a^n \qquad \text{for all } n>=0 \text{ and } a>0$$

Proof

Suppose that $a \overset{3}{T} k = a^k$ for some integer k

$$a \overset{3}{T} (k+1) = a \overset{2}{T} (a \overset{3}{T} k) \qquad \text{Tower def 1}$$

$$= a \overset{2}{T} (a^k) \qquad \text{supposition}$$

$$= a.a^k \qquad \text{Second tower theorem}$$

$$= a^{k+1}$$

Therefore, if the theorem is true for k, then it is also true for (k+1)

But the theorem is true for k=0 since $a \overset{3}{T} 0 = 1 = a^0$ Tower def 5

Therefore, by the principle of mathematical induction

$$a \overset{3}{T} n = a^n \qquad \text{for all } n>=0 \qquad \blacksquare$$

====================

Fourth Tower Theorem

$$a \overset{4}{T} n = a^{Tn} \qquad \text{for all } n>=0$$

Proof

Suppose that $a \overset{4}{T} k = a^{Tk}$ for some integer k

then $$a \overset{4}{T} (k+1) = a \overset{3}{T} (a \overset{4}{T} k) \qquad \text{Tower def 1}$$

$$= a \overset{3}{T} (a^{Tk}) \qquad \text{supposition}$$

$$= a \uparrow a^k \qquad \text{by theorem 3}$$

$$= a^{T(k+1)}$$

Therefore, if the theorem is true for k, then it is also true for (k+1)

But the theorem is true for k=0 since $a \overset{4}{T} 0 = 1 = a^0$ (Tower def 5)

Therefore, by the principle of mathematical induction,

$$a \overset{4}{T} n = a^{Tn} \qquad \text{for all } n>=0 \text{ and } a>0 \qquad \blacksquare$$

Fifth Tower Theorem

$$a \overset{m}{T} 1 = a \qquad \text{for all } m>=2$$

Proof

Suppose that $a \overset{k}{T} 1 = a$ for some integer $k>=2$

then $a \overset{k+1}{T} 1 = a \overset{k}{T} (a \overset{k+1}{T} 0)$ Tower def 1

$= a \overset{k}{T} 1$ Tower def 5 (k+1>2)

$= a$ the supposition

Therefore, if the theorem is true for k, then it is also true for (k+1)

But it is true for k=2 since $a \overset{2}{T} 1 = a.1$ second tower theorem

$= a$

Therefore, by the principle of mathematical induction,

$$a \overset{m}{T} 1 = a \qquad \text{for all } m>=2$$ ■

====================

Sixth Tower Theorem

$$2 \overset{m}{T} 2 = 4 \qquad \text{for } m>0$$

Proof

Suppose that $2 \overset{k}{T} 2 = 4$ for some integer k

Then $2 \overset{k+1}{T} 2 = 2 \overset{k}{T} (2 \overset{k+1}{T} 1)$

$= 2 \overset{k}{T} 2$ theorem 5

$= 4$ supposition

Therefore, if the theorem is true for **k**, then it will be true for **k+1**

But, the theorem is true for **m=1** since $2 \overset{1}{T} 2 = 2+2 = 4$

Therefore, by the principle of mathematical induction, theorem 6 is true for all **m>0** ■

Ackermann's Function

It is a fact that nowadays, computers are rarely used to compute. During the second world war, while most able-bodied men were scattered abroad fighting, any calculations that were required by the armed forces, for example the firing range of a new tank, were carried out by women using mechanical calculators that had to be turned by hand. If the range of a new gun was required, the figures for angles of elevation and the muzzle velocities would be sent to the calculators, the ladies in the calculating office, who would apply formulae derived from Newton's laws of motion to give the generals the results they wanted.
Nowadays, the computer takes on the functions of the library, the record player and the post office. Rarely is it asked to do any arithmetic.

Ackermann's function was designed to test the limits of a computer's calculating powers.

It is defined recursively in computing books by

Def Ack1 (m,n) = (m-1, (m, n-1)) for m≠0 , n≠0

Def Ack2 (m,0) = (m-1, 1)

Def Ack3 (0, n) = n+1

Theorem Ack1

(1,n) = n+2

Proof

Suppose that (1,k) = k +2 for some integer k

then	**(1,k+1) = (0, (1,k))**	**def Ack1**
	= (0, k+2)	**supposition**
	= k+2 + 1	**def Ack3**
	= (k+1) + 2	

Therefore, if true for integer k, then theorem Ack7 is also true for integer k+1

However

(1,0) = (0,1) = 2 using def Ack2 and def Ack3

and (1,1) = (0, (1,0)) using def Ack1

= (0,2) = 3 using the above and def Ack3

Therefore, theorem 7 is certainly true for n=0 and n=1

Therefore, by the principle of induction, theorem Ack1 is true for all n>=0 ■

========================

Theorem Ack2

(2,n) = 2n+3

Proof

Suppose that (2,k) = 2k +3 for some integer k

Then	(2,k+1) = (1, (2,k))	def Ack1
	= (0, 2k+3)	supposition
	= 2k+5	Theorem Ack1
	= 2(k+1) + 3	

Therefore, if true for integer k, then theorem Ack2 is also true for integer k+1

However

(2,0) = (1,1) = 3 using def Ack2 and theorem Ack1

= 2x0 + 3

Therefore, theorem Ack2 is certainly true for n=0

Therefore, by the principle of mathematical induction, theorem Ack2 is true for all n>=0 ■

Theorem Ack3

$$(3,n) = 2^{n+3} - 3$$

Proof

Suppose that $(3,k) = 2^{k+3} - 3$ for some integer k

Then	$(3,k+1) = (2, (3,k))$	def Ack1
	$= (2, 2^{k+3} - 3)$	supposition
	$= 2\{2^{k+3} - 3\} + 3$	Theorem Ack2
	$= 2^{k+4} - 3$	
	$= 2^{(k+1)+3} - 3$	

Therefore, if true for integer k, then theorem Ack3 is also true for integer k+1

However

$(3,0) = (2,1) = 5$ using def Ack2 and theorem Ack2

$= 2^{0+3} - 3$

Therefore, theorem Ack3 is certainly true for n=0

Therefore, by the principle of induction, theorem Ack3 is true for all n>=0 ■

==============================

Theorem Ack4

$$(4,n) = 2T(n+3) - 3$$

Proof

Suppose that	$(4,k) = 2T(k+3) - 3$	for some integer k
Then	$(4,k+1) = (3, (4,k))$	def Ack1
	$= (3, 2T(k+3) - 3)$	supposition
	$= 2^{2T(k+3) - 3 + 3} - 3$	Theorem Ack3

$$= 2^{2T(k+3)} - 3$$

$$= 2T(k+4) - 3$$

Therefore, if true for integer k, then theorem Ack4 is also true for integer k+1

However

(4,0) = (3,1)		using def Ack2
	$= 2^4 - 3$	theorem Ack3
	$= 2^{T2} - 3 = 2T3 - 3$	

Therefore, theorem 10 is certainly true for n=0

Therefore, by the principle of induction, theorem Ack4 is true for all n>=0 ■

==================================

Gathering these results and using the more general notation $a\overset{m}{T}n$, we have

(0,n) = n+1 (def)	$(0,n) = 2\overset{0}{T}(n+3) - 3$
(1,n) = 2+(n+3) – 3	$(1,n) = 2\overset{1}{T}(n+3) - 3$
(2,n) = 2(n+3) – 3	$(2,n) = 2\overset{2}{T}(n+3) - 3$
(3,n) = 2n+3 – 3	$(3,n) = 2\overset{3}{T}(n+3) - 3$
(4,n) = 2T(n+3) – 3	$(4,n) = 2\overset{4}{T}(n+3) - 3$

The next theorem encompasses all of theorems Ack1 to Ack4.

The Ackerman Tower Theorem

$$(m,n) = 2\overset{m}{T}(n+3) - 3$$

Proof

Let the truth set for this statement be T, that is, all pairs (m,n) that satisfy the theorem.

Suppose that (m,k) ε T , m>0, (supposition 1)

and that the complete row m-1 ε T (supposition 2)

If we display all the pairs (m,n) in an array:

n ⟶

m ↓

(0,0) (0,1) (0,2) (0,3) (0,4) (0,5) ....
(1,0) (1,1) (1,2) (1,3) (1,4) (1,5).....
(2,0) (2,1) (2,2) (2,3) (2,4) (2,5) ......
(3,0) (3,1) (3,2) (3,3) (3,4) (3,5).....
.
.
(m-1,0) (m-1,1) ....(m-1,k)
(m,0) ....(m,k)

part 1

We are supposing that there is some **(m,k)** ε **T** and that all pairs in the row above also ε T.

Now,

(m,k+1) = (m-1,(m,k)) Ackermann def 1

and **(m-1,(m,k))** ε **T** supposition 2

therefore (m,k+1) ε **T**

hence, (m,k) ε **T ⇒ (m,k+1)** ε **T**

part 2

further, since $(m\text{-}1,1)\ \varepsilon\ T$ supposition 2

we have $(m\text{-}1,1) = 2 \overset{m-1}{T} 4 - 3$

but $(m,0) = (m\text{-}1,1)$ Ackermann def 2

so that $(m,0) = 2 \overset{m-1}{T} 4 - 3$

now $2 \overset{m}{T} 3 - 3 = 2 \overset{m-1}{T} (2 \overset{m}{T} 2) - 3$ tower definitions

$= 2 \overset{m-1}{T} 4 - 3$ second theorem

therefore $(m,0) = 2 \overset{m}{T} 3 - 3$

hence $(m,0)\ \varepsilon\ T$

Thus, using part 1, by the principle of induction, the whole of row m ε T

We have shown therefore, that if row K (=m-1) ε T then row K+1 (=m) ε T
But rows 0,1,2,3 and 4 have all been shown ε T .
Therefore, by the principle of induction, all rows ε T and this proves the theorem. ■

===

CHAPTER 17

Sequences and Series

The messenger burst into the court of the Maharajah. "Sire, I have an urgent message from my master the Prince. He says that he cannot sleep and cannot concentrate on the affairs of his estate because his head is filled with thoughts of your beautiful daughter the Princess. He begs you to grant him an audience to discuss his plight."

"Very well, " said the Maharajah, "tell him I will see him at dawn tomorrow".

The messenger left and the next day the Prince arrived at the Maharajah's palace.

The Prince said, "Your majesty, I toss and turn all night thinking of the Princess. I have no appetite and I cannot keep my mind on important affairs of state. All I can see in my mind is the fair face of your beautiful daughter and if I cannot find peace in my mind I will be ruined. I beg you, please may I have the hand of your daughter in marriage?"

The Maharajah loved his daughter but his estate was in dire straits and for the present, the Prince was rich. He thought for a while and then asked a servant to bring him a chessboard.

"I will give you a test to see if you are worthy of her love. There are sixty-four squares on this chessboard. Come tomorrow with one grain of rice to place on the first square of the board. Come the next day with two grains for the second square, the next day bring four grains for the third square. Come each day with twice the number of grains of rice for the next square on the board and on the sixty fourth day the Princess will be yours."

The Prince was overjoyed. The next day he came to the palace and placed one grain of rice on the first square of the board. The following day he placed two grains on the next square and on the third day he came with four grains of rice. The Prince did not win his Princess but the Maharajah soon became rich from selling rice.

The number of grains of rice that the Prince brought to the Maharajah each day form a sequence:

$$1, \quad 2, \quad 4, \quad 8, \quad 16, \quad 32, \quad 64, \quad 128, \ldots..$$

Adding the grains together forms a series:

After day 1 the Maharajah has 1 grain of rice in his pile
After day 2 the Maharajah has 1+2 = 3 grains of rice in his pile
After day 3 the Maharajah has 1+2+4 = 7 grains of rice in his pile
After day 4 the Maharajah has 1+2+4+8 = 15 grains of rice in his pile
After day 5 the Maharajah has 1+2+4+8+16 = 31 grains of rice in his pile
After day 6 the Maharajah has 1+2+4+8+16+32 = 63 grains of rice in his pile
After day 7 the Maharajah has 1+2+4+8+16+32+64 = 127 grains of rice.

You may notice that there is something special about the number of grains of rice in the Maharajah's pile at the end of each day. It involves powers of 2:

$$1 = 2^1 - 1,\ 3 = 2^2 - 1,\ 7 = 2^3 - 1,\ 15 = 2^4 - 1,\ 31 = 2^5 - 1,\ 63 = 2^6 - 1,\ 127 = 2^7 - 1$$

It appears that after day number n, the total number of grains of rice in the Maharajah's pile is $2^n - 1$.
A short argument proves this:

If we list the number of grains added on each day, we have:

Day	1	2	3	4	5	6 .... n
Grains added	1	2	4	8	16	32 .,....... ...$2^{(n-1)}$

On day n the Prince adds 2^{n-1} grains of rice to the pile, and on day n+1, the Prince adds twice as many which is 2^n grains to the pile.

If on day n, there are 2^n -1 grains in the pile then, on the next day, day n+1, there will be $2^n-1 + 2^n = 2x2^n-1 = 2^{n+1}-1$ grains.
Hence, if the formula 2^n -1 is correct after day n, then the same formula will be correct after day n+1.
Now the formula is correct for day 1, (2^1-1 = 1 grain) and so, by the principle of mathematical induction, the formula is correct for any day.

On my kitchen scales, the weight of 1000 grains of Basmati rice is about 15 grams so I estimate that the weight of one grain of rice to be about 0.015 grams.
On the tenth day, the prince has handed over a total of 2^{10}-1 = 1023 grains of rice weighing about 15 grams.
On the 20th day, the Maharajah's pile has 2^{20}-1 grains weighing about 16 kilograms.

On the 30th day, the weight of the pile is about 16000 kg. A fair size sack of rice might weigh about 40kg so after the 30th day, the Prince has given the Maharajah about 400 sacks of rice. At this stage the Prince would have been starting to get worried. The next day he has to hand over an extra 2^{30} grains which is another 400 sacks or so.
By day 40 the Prince would have handed over about four hundred thousand 40 kg sacks of rice.
After day 50 the Prince should have handed over a total of about 400 million sacks of rice.
After day 64, the Prince should have handed over a total of about 7million million (7×10^{12}) sacks of rice which represents a total of 2.8×10^{14} kilograms. Now in 1995, the entire output of rice from India was about 9.1×10^{10} kilograms so the Maharajah was asking for something like 3000 times the total annual production of the country. If the Prince had done some arithmetic that morning when he first went to visit the Maharajah perhaps he would not have been so happy.

Arithmetic Sequences

Can you find the next three numbers in this sequence?

5, 8, 11, 14, 17 ...

Sometimes it is easy spot a rule for getting the next number in a sequence like this but it is not always so obvious. In this case the rule is "add three" and we express this in the equation

$$U_{n+1} = U_n + 3$$

The symbol U_n is used to represent the nth term of the sequence so that U_1 is the first term, U_2 is the second term and so on.
(This form of definition is called a recurrence relation.)

Here we have, $U_1 = 5$, $U_2 = 8$, $U_3 = 11$.

What are U_6, U_7 and U_8? What is U_{50}?

No one would expect you to work your way through the sequence all the way to term number 50 but we can often find a formula for the n th term of such sequences. In this case the formula for the nth term is

$$U_n = 3n + 2$$

We can then immediately say that $U_{50} = 3x50 + 2 = 152$

The formula for the n th term of a sequence, if there is one, is called the general term of the sequence.

Example 1

Even numbers

The sequence of even numbers can be defined by the rule

$$U_1 = 2,$$
$$U_{n+1} = U_n + 2$$

or alternatively, we can define the sequence of even numbers by the formula for the nth term:

$$U_n = 2n$$

Example 2

Odd numbers

The sequence of odd number can be defined using the recurrence relation:

$$U_1 = 1$$
$$U_{n+1} = U_n + 2$$

or using the formula

$$U_n = 2n - 1$$

We cannot always find rules or formulae for important sequences, for example,

2, 3, 5, 7, 11, 13, 17, 19,

is the sequence of prime numbers that have no factors apart from 1 and the number itself. Every integer greater than 1 can be written uniquely as a product of prime numbers (sometimes only one prime number!) thus:

$2=2$, $3=3$, $4=2^2$, $5=5$, $6=2\text{x}3$, $7=7$, $8=2^3$, $9=3^2$, $10=2\text{x}5$,

(note that 1 cannot be a prime number for this to work because for example, $6=1\text{x}2\text{x}3=1^2\text{x}2\text{x}3=1^3\text{x}2\text{x}3=1^4\text{x}2\text{x}3$ etc and we would not get unique factorization into primes)

No one has yet found a formula for the n th prime number P_n, nor is there a formula for finding P_{n+1} from P_n.

Exercise 1

1. Given the sequence 1, 3, 7, 15, 31 ...

   Write down the rule for finding U_{n+1} from U_n .

2. Given the sequence 2, 5, 8, 11, 14, 17, 20, ...

   Write down the formula for U_n

Example 3 A sequence is defined by the recurrence formula

$$U_{n+1} = U_n + 3$$

$$U_n = 4$$

The sequence is 4, 7, 10, 13, 16, ... and we can see that the terms increase by a constant value 3. If we plot the terms of this sequence on a graph the plotted points lie on the straight line $y = 3x + 1$

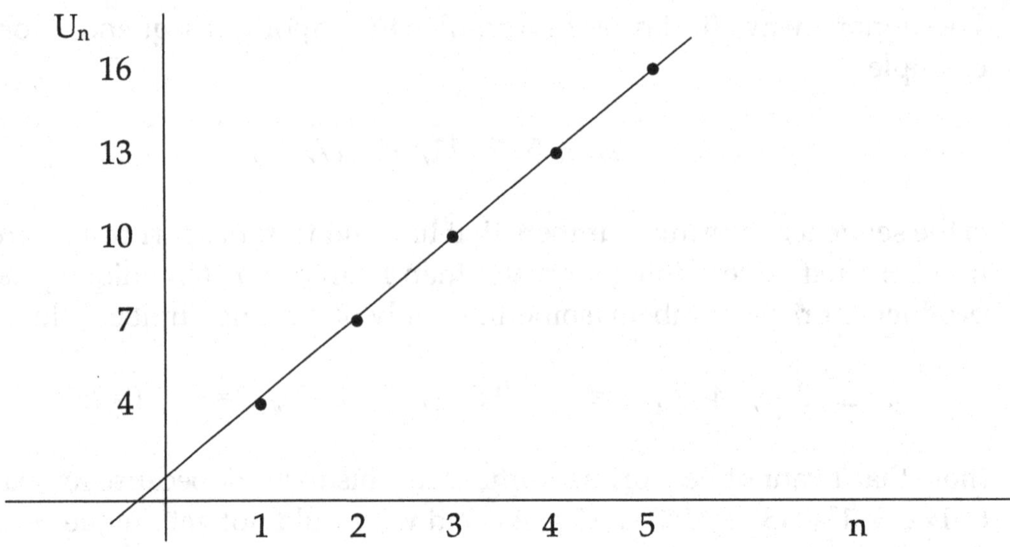

This sequence has the recurrence relation

$$\mathbf{U_{n+1} = U_n + 3}$$
$$\mathbf{U_1 = 4}$$

Note the following method for deriving the formula for U_n. This method is referred to a the method of differences:
We have

$$\mathbf{U_n = U_{n-1} + 3}$$
$$\mathbf{U_{n-1} = U_{n-2} + 3}$$
$$\mathbf{U_{n-2} = U_{n-1} + 3} \quad \text{and so on.}$$

Transposing from right side to left side now gives

$$\mathbf{U_n - U_{n-1} = 3}$$
$$\mathbf{U_{n-1} - U_{n-2} = 3}$$
$$\mathbf{U_{n-2} - U_{n-3} = 3}$$
$$. \quad \text{and so on}$$
$$.$$
$$\mathbf{U_3 - U_2 = 3}$$
$$\mathbf{U_2 - U_1 = 3}$$
$$\mathbf{U_1 = 3+1}$$

Adding these n equations together, everything except U_n cancells out on the left hand side and there are n threes on the right, giving

$$\mathbf{U_n = 3n + 1}$$

==========

Example 4

Find the general term for the sequence

7, 11, 15, 19, …

Solution

We note the constant difference 4 and so our graph of U_n against n gives a straight line graph with gradient 4. (one step along gives four steps up)

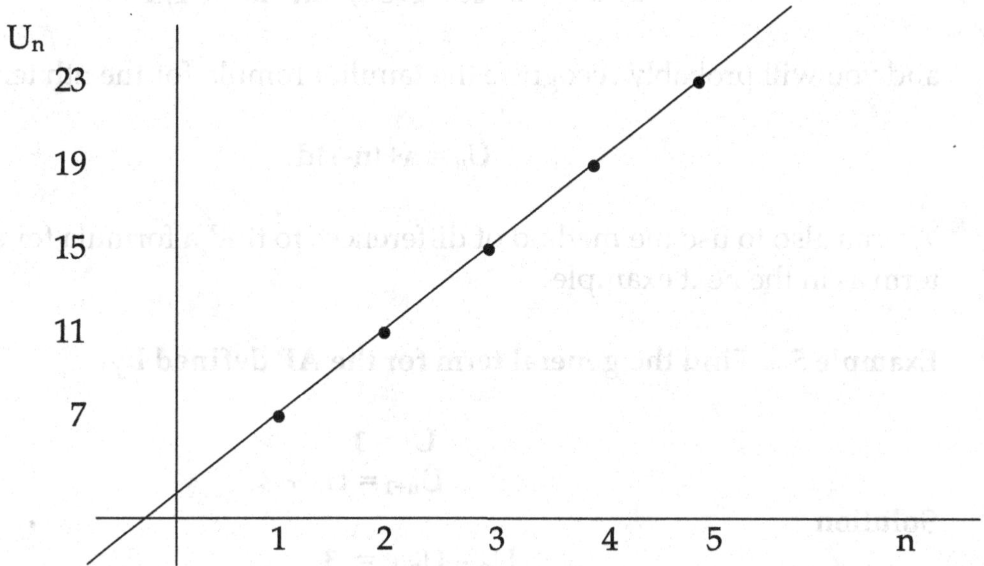

The line has the form **$y=4x + c$**

and our sequence has the formula **$U_n = 4n + c$**

Checking with U_1 gives c=3

Hence

$$U_n = 4n + 3$$

APs

A sequence that increases (or decreases) by a fixed amount from one term to the next is called an arithmetic sequence or arithmetic progression (A.P.). An arithmetic progression with first term **a** that increases by the fixed amount (common difference) **d** is defined by

$$U_1 = a$$
$$U_n = U_{n-1} + d$$

This AP is given by the sequence

$$a,\ a+d,\ a+2d,\ a+3d,\ \ldots.\ a+(n-1)d$$

and you will probably recognise the familiar fomula for the nth term as

$$U_n = a+(n-1)d$$

We can also to use the method of differences to find a formula for a general term as in the next example.

Example 5 Find the general term for the AP defined by

$$U_1 = 1$$
$$U_{n+1} = U_n + 3$$

Solution

$$U_n - U_{n-1} = 3$$
$$U_{n-1} - U_{n-2} = 3$$
$$U_{n-2} - U_{n-3} = 3$$
$$\vdots$$
$$U_2 - U_1 = 3$$
$$U_1 = 3\text{-}2$$

Adding these n equations, all terms except U_n cancel on the left hand side and the right hand side is 3xn – 2 so that we have the formula

$$U_n = 3n - 2$$

Alternatively, since the common difference is 3, we may try

$$U_n = 3n + c$$

And find c from $U_1 = 1 = 3x1 - 2$ giving $c = -2$

=======================

Series and Sigma Notation

A sequence or progression is simple a row of terms for example, the set of even numbers forms the following sequence:

2, 4, 6, 8, 10, 12, 14, 16, 18, ...

A sequence become a series when the terms are added together. The series of even numbers, starting at 2 is formed by adding up the even numbers:

2 + 4 + 6 + 8 + 10 + 12 + 14 + 16 + 18 + ...

The sum of n terms of a series is denoted by S_n thus

$$S_n = U_1 + U_2 + U_3 + U_4 + ... + U_{n-1} + U_n$$

Using sigma notation this would be written

$$S_n = \sum_{r=1}^{r=n} U_r$$

for example,

$$S_n = \sum_{r=1}^{r=5} 2r = 2+4=6+8+10$$

When the limits of the summation are clear, it is not necessary to put them in thus we could write

$$S_n = \Sigma U_r$$

===========

Exercise 2 Calculate

[1] $\sum_{r=1}^{r=10} r$

[2] $\sum_{21}^{30} r$

[3] $\sum_{1001}^{1010} r$

[4] $\sum_{1}^{5} (2r - 1)$

[5] $\sum_{1}^{6} (2r-1)$

[6] $\sum_{1}^{10} (2r)$

[7] $\sum_{1}^{10} (r-1)$

[8] $\sum_{n=1}^{n=5} \left(\sum_{1}^{n} (2r-1) \right)$

[9] $\sum_{r=1}^{5} r(r+1)$

[10] $\sum_{r=1}^{r=6} r(r+1)$

The Sum of an AP

Traditional textbooks usually give formulae for the sum of n terms of an Arithmetic Progression. Suppose the first term is a and that the common difference is d. Let the sum of n terms of the AP be S_n and suppose that the last term is l.
Then

$$S_n = a + (a+d) + (a+2d) + (a+3d) + ... + (l - 2d) + (l - d) + l$$

Reverse the series and we have

$$S_n = l + (l - d) + (l - 2d) + (l - 3d) + ... + (a+2d) + (a+d) + a$$

Adding these two equations, all of the d's cancel out giving us n brackets, each bracket being (a+l):

$$2S_n = (a+l) + (a+l) + (a+l) + (a+l) + ... + (a+l) + (a+l) + (a+l)$$

Thus $2S_n = n(a+l)$

Hence $$\mathbf{S_n = \frac{n}{2}(a+l)}$$

since **l= a+(n-1)d**

this gives $$\mathbf{S_n = \frac{n}{2}(2a+(n-1)d)}$$

=====================================

Example 5

Find the sum of n terms of the AP defined by

$$\mathbf{U_1 = 5}$$
$$\mathbf{U_{n+1} = U_n + 6}$$

Solution

This is an AP with a=5 and d=6

$$\mathbf{S_n = \frac{n}{2}(2a+(n-1)d)}$$

$$= \frac{n}{2}(10 + (n-1)\times 6)$$

$$= n\ (5 + (n-1)\times 3)$$

$$= n\ (5 + 3n - 3)$$

$$S_n = 3n^2 + 2n$$

====================================

Exercise 3

[1] The sixth term of an A.P. is 19 and the 30th term is 91. What is the sum of the first ten terms?

[2] The tenth term of an A.P. is equal to the sum of the third and fourth terms. The third term and the sixth term add up to 60. Find the sum of the first three terms.

[3] The sum of the first three terms of an A.P. is 30 and the sum of the next two terms is 40. Find the sum of the first eight terms.

[4] The sum of the first and the last terms of an A.P. is 100. The sum of all the terms is 1000. How many terms are there?

[5] A sequence is defined by the recurrence relation

$$U_n = U_{n-1} + U_{n-2} \qquad U_1=0,\ U_2=1$$

Prove that

$$\sum_{1}^{n} U_r = U_{n+2} - 1$$

==

Geometric Series

The Geometric Sequence or geometric progression with first term **a** and commom ratio **r** is

$$a,\ ar,\ ar^2,\ ar^3,\ ar^4,\ \ldots\ ar^{n-1},\ \ldots$$

The sequence can be defined using the recurrence relation

$$U_1 = a$$
$$U_{n+1} = r.U_n$$

The first term is **a** and we always get the next term by multiplying by the common ratio **r**.

The general term of the GP is

$$Un = a.r^{n-1}$$

The sum of n terms of a GP

The traditional method for deriving the formula for the sum of a GP is as follows:
Let

$$S_n = a + ar + ar^2 + ar^3 + \ldots + ar^{n-2} + ar^{n-1}$$

Multiply through by r

$$r.S_n = ar + ar^2 + ar^3 + ar^4 + \ldots + ar^{n-1} + ar^n$$

and subtract

$$S_n - r.S_n = a - ar^n$$

$$S_n(1-r) = a(1 - r^n)$$

$$\therefore \quad S_n = \frac{a(1 - r^n)}{(1 - r)}$$

============================

Alternatively we could proceed as follows:

$$U_n = r.U_{n-1}$$
$$U_{n-1} = r.U_{n-2}$$
$$U_{n-2} = r.U_{n-3}$$
$$\vdots$$
$$U_2 = r.U_1$$
$$U_1 = a$$

Multiplying these equations together and cancelling gives $U_n = a.r^{n-1}$

Adding gives $$S_n = a + r(S_n - U_n)$$

$$S_n(1 - r) = a - r.U_n$$

$$S_n(1 - r) = a - r.ar^{n-1}$$

$$S_n(1 - r) = a(1 - r^n)$$

$$S_n = \frac{a(1 - r^n)}{(1 - r)}$$

==============================

Exercise 4
Show that

1. $$1 + x + x^2 + x^3 + x^4 \ldots + x^n = \frac{1 - x^{n+1}}{1 - x}$$

2. $$1 + 2 + 4 + 8 + 16 + \ldots + 2^n = 2^{n+1} - 1$$

3. $$1 + 10 + 100 + 1000 + 10000 = (100000-1)/9$$

4. $$x^7 - 1 = (x - 1)(x^6 + x^5 + x^4 + x^3 + x^2 + x + 1)$$

5. $$0.99999 = 0.9 + 0.009 + 0.0009 + 0.00009 + 0.000009 = (1 - 0.1^5)$$

==============================

Example 7

The sixth term of a GP is 64 and the ninth term is 512. Find the sequence.

Solution

The ninth term is 512 hence $ar^8 = 512$
The sixth term is 64, hence $ar^5 = 64$

Divide to get $r^3 = 8$

Hence $r = 2$
Giving $a = 2$

The sequence is therefore **2, 4, 8, 16, 32, 64, 128, 256, 512...**

==================================

Exercise 5

[1] The fifth term of a GP is 162 and the third term is 18. Find the first five terms of the sequence.

[2] The seventh term of a GP is 192 and the fourth term is 24. Find the first five terms of the sequence.

[3] The sum of the first and second terms of a GP is 27. The sum of the fourth and fifth terms is 216. Find the first five terms of the sequence.

[4] The sum of the fourth and fifth terms of a GP is 216.
The sum of the second and the fourth terms is 60.
Prove that

$$5r^3 - 13r^2 - 18 = 0$$

and deduce that $r = 3$.

Arithmetic-Geometric Sequences

A typical arithmetic-geometric sequence is defined by the recurrence relation

$$\mathbf{U_1 = 0}$$
$$\mathbf{U_{n+1} = 2U_n + 3}$$

This recurrence relation is made up of an arithmetic part $\mathbf{U_{n+1} = T_n + 3}$ and a geometric part $\mathbf{T_n = 2U_n}$. We can build up the sequence starting from U_1 as far as we like by multiplying by 2 and then adding 3:

0, 3, 9, 21, 45, 93, 189, ...

but can we find a formula for the nth term U_n or for the sum S_n of n terms of the sequence?

In order to answer these questions we use differences.

For any give sequence, we call $U_n - U_{n-1}$ the sequence of first order differences. From the first sequence, we produce another sequence by writing down these differences:

	0		3		9		21		45		93		189 ...
1 st diff		3		6		12		24		48		96 ...	

If we define $\quad \mathbf{d_n = U_{n+1} - U_n}$ for all n>0, $d_1 = 3$, $d_2 = 6$...

Since
$$\mathbf{U_{n+1} = 2U_n + 3}$$
$$\mathbf{U_n = 2U_{n-1} + 3}$$

Subtracting gives
$$\mathbf{U_{n+1} - U_n = 2(U_n - U_{n-1})}$$

So that $\quad \mathbf{d_n = 2d_{n-1}}$ for all values of n>1

Thus
$$\mathbf{d_n = 2d_{n-1} = 2^2d_{n-2} = 2^3d_{n-3}}$$
$$\mathbf{= 2^{n-1}d_{n-(n-1)}}$$
$$\mathbf{= 2^{n-1}d_1 = 3x2^{n-1}}$$

so that

$$d_{n-1} = U_n - U_{n-1} = 3x2^{n-2}$$
$$d_{n-2} = U_{n-1} - U_{n-2} = 3x2^{n-3}$$
$$d_{n-3} = U_{n-2} - U_{n-3} = 3x2^{n-4}$$

.
.
.

$$d_2 = U_3 - U_2 = 3x2^1$$
$$d_1 = U_2 - U_1 = 3x2^0$$

Adding

$$U_n - 0 = 3(2^0 + 2^1 + 2^2 + \dots + 2^{n-2})$$

$$= 3(2^{n-1} - 1)$$

$$U_n = 3x2^{n-1} - 3$$

Having found the formula for the general term of the sequence, we can now find the sum S_n of the arithmetic-geometric series:

$$S_n = \sum_{r=1}^{r=n} U_r = \Sigma\,(3x2^{r-1} - 3)$$

$$= 3(2^0 + 2^1 + 2^2 + \dots + 2^{n-1}) - 3n$$

$$= 3(2^n - 1) - 3n$$

$$S_n = 3x2^n - 3 - 3n$$

========================

(Note that in this example we have $S_n = U_{n+1} - 3n$)

Exercise 6

[1] A sequence is defined by the recurrence relation

$$U_n = 2U_{n-1} - 1 \qquad U_1 = 1$$

Prove that $S_n = n$

[2] A sequence is defined by the recurrence relation

$$U_n = 3U_{n-1} - 1 \qquad U_1 = 1$$

Prove that $U_n = \frac{1}{2}(3^{n-1} + 1)$

and hence deduce that $S_n = \frac{3^n - 1}{4} + \frac{n}{2}$

[3] A sequence is defined by the recurrence relation

$$U_n = 4U_{n-1} - 1 \qquad U_1 = 1$$

Prove that $U_n = \frac{2 \times 4^{n-1} + 1}{3}$

and hence deduce that

$$S_n = \frac{2(4^n - 1)}{9} + \frac{n}{3}$$

[4] A sequence is defined by the recurrence relation

$$U_n = 5U_{n-1} - 1 \qquad U_1 = 1$$

Prove that $U_n = \frac{3.5^{n-1} + 1}{4}$

and hence deduce that

$$S_n = \frac{3(5^n - 1)}{16} + \frac{n}{4}$$

==================================

The General Arithmetic-Geometric Sequence

This sequence is defined by

$$U_1 = 0$$
$$U_n = aU_{n-1} + b$$

The general term is found using the method of differences. The sequence is:

$$0 \quad b \quad ab+b \quad a^2b + ab + b \quad a^3b + a^2b + ab + b \ldots$$

We have

$$U_n = aU_{n-1} + b$$

and

$$U_{n-1} = aU_{n-2} + b$$

subtracting gives

$$U_n - U_{n-1} = a(U_{n-1} - U_{n-2})$$

Using d_{n-1} for the difference U_n-U_{n-1} this means that $d_{n+1} = a\, d_n$ for all n>0

therefore we have

$$d_{n-1} = a.d_{n-2} = a^2\, d_{n-3} = a^3\, d_{n-4} = \ldots = a^{n-2}d_1 = a^{n-2}b$$

hence

$$U_n - U_{n-1} = a^{n-2}b$$

$$U_{n-1} - U_{n-2} = a^{n-3}b$$

$$U_{n-2} - U_{n-3} = a^{n-4}b$$

....

$$U_3 - U_2 = a^1b$$

$$U_2 - U_1 = a^0b$$

Adding gives

$$U_n = b(a^0 + a^1 + a^2 + \ldots. + a^{n-2})$$

$$U_n = \frac{b\,(a^{n-1} - 1)}{a-1}$$

The sum of the general arithmetic-geometric series is therefore given by

$$S_n = [\ b(\ a^0 + a^1 + a^2 + ... + a^{n-1}) \ - \ nb\ ] / (a-1)$$

$$S_n = \frac{b}{(a-1)}\left(\frac{a^n - 1}{a - 1}\right) - \frac{bn}{(a-1)}$$

which can be written $S_n = \dfrac{[\,U_{n+1} - bn\,]}{a - 1}$

==================================

Exercise 7

1. A sequence U_1 U_2 U_3 ... satisfies the recurrence relation

$$U_{r+1} = rU_r - r^2 - r + 3$$

$$U_1 = 3$$

Find U_2, U_3, U_4 and suggest a formula for U_r in terms of r.
Verify that your formula satisfies the given relation.

2. The sequence U_1 U_2 U_3 ... satisfies the relation

$$U_{r+1} = rU_r - r^2 + 2$$

$$U_1 = 2$$

Find U_2, U_3 and U_4. Suggest a formula for U_r in terms of r and verify that your formula satisfies the relation.

==

There is a leak in the roof of the Restaurant at the End of the Universe and to avoid a gigantic puddle, the waiter had placed a bucket under the leak to catch the drips.
"How long has that been there?", said Zaphod.
"Its been there ever since we opened " the waiter said, "it drips into the bucket every minute but it doesn't bother us so we just leave it there."
"Well it must bother your sooner or later.", said Zapod.
"No " said the waiter, "its no problem, it never gets full."
"How can it never get full if water drips into the bucket every minute of every day of every year from now till the end of time, ad infinitum to eternity?" said Zaphod.
"Well the first drip was pretty big," said the waiter, " in fact it was half a bucket full, but the drips are getting smaller. Each drip is only half as big as the one before so there always seems to be some space left"

Zaphod was puzzled. He took a piece of paper and wrote out:

Drip	1	2	3	4	5	6	7	8 ……
Water in Bucket	½	¼	1/8	1/16	1/32	1/64	1/128	1/256 …

Zaphod's coffee was getting cold. He took a sip and then tipped the dregs into the bucket.
The next day, the restaurant was closed due to flooding.

==============================

Zaphod realized that the drips form a G.P. After n drips, the amount of water in the bucket is the sum of n terms of the G.P:

$$\frac{a(1 - r^n)}{1 - r} = \frac{\frac{1}{2}(1 - (\frac{1}{2})^n)}{1 - \frac{1}{2}} = 1 - (\tfrac{1}{2})^n$$

The next drip will be half of $(\frac{1}{2})^n = (\frac{1}{2})^{n+1}$

So the bucket will now have $1 - (\frac{1}{2})^n + (\frac{1}{2})^{n+1} = 1 - (\frac{1}{2})^n(1 - \frac{1}{2})$

$= 1 - (\frac{1}{2})^{n+1}$ buckets of water

As the number of drips into the bucket goes to infinity, or, as $n \to \infty$
Then $(1/2)^n \to 0$

The water in the bucket $\to 1$ but the bucket never quite gets to 1 bucket full.
(At least, until Zaphod finished his cup of coffee).

==============================

Sum to Infinity

The sum of n terms of a GP is

$$S_n = \frac{a(1 - r^n)}{(1 - r)}$$

If $|r| < 1$, that is $-1 < r < +1$ then $r^n \to 0$ so that $S_n \to \frac{a}{(1 - r)}$

This is called "the sum to infinity" and we write $S_\infty = \frac{a}{(1 - r)}$

Example 8

Find the sum to infinity of the GP **9/10 + 9/100 + 9/1000 + ...**

Solution

Here a=9/10 and r=1/10

Hence $S_\infty = \frac{9/10}{1-1/10} = \frac{9/10}{9/10} = 1$

indicating that we can write 0.9999.... = 1

====================================

Example 9

Write the recurring decimal 0.123123123123.... as a rational number

Solution

$$\frac{123}{1000} + \frac{123}{1000000} + \frac{123}{1000000000} + \frac{123}{1000000000000} + \ldots..$$

is a GP with first term $\frac{123}{1000}$ and common ratio $\frac{1}{1000}$

therefore, the sum to infinity is $\frac{123}{1000} \Big/ \left(1 - \frac{1}{1000}\right)$

$$= \frac{123}{999}$$

==============================

Achilles was fast! He could cover 100 metres in 10 seconds! Tortoise was slow, in fact Achilles could cover tens times the distance that Tortoise could run in the same time. The philosopher Zeno was surprised to hear that Tortoise had placed a bet with Achilles that if he was given a 100 metres start, then Achilles would never catch him.

After all, when Achilles had covered the 100 metres, Tortoise would have moved on 10 metres ahead.

When Achilles had caught up that 10 metres, Tortoise would have moved one metre ahead. Whenever Achilles had caught up the lead, so Tortoise reasoned, he would have moved on bit further. Therefore Achilles would never catch him!

The tortoise had drawn up a table:

	Distance from the start	
Time	**Achilles**	**Tortoise**
0	0	100
10	100	10
11	110	111
11.1	111	111.1
11.11	111.1	111.11
11.111	111.11	111.111

"There you are " said Tortoise to Zeno, "I'm always ahead so I'm bound to win!"

Question: If Achilles runs at 10 metres per second, how long would it take him to cover $111\frac{1}{9}$ metres?

(and if the tortoise runs at 1 metre per second, how long does it take the tortoise to cover $11\frac{1}{9}$ metres?)

Exercise 8

[1] Write the recurring decimal 11.1111.... as a fraction.

[2] Write 369.369369369369...... as the sum of a G.P.
Use the sum to infinity of your G.P. to write the number as a fraction.

[3]

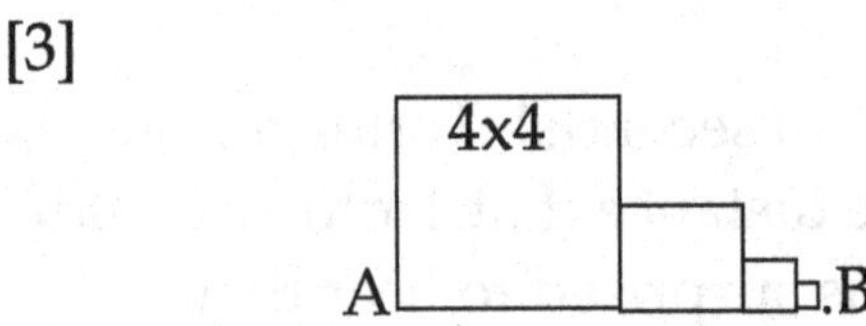

The first square is 4x4 and the lengths of the sides of the following squares are halved to produce the infinite sequence of squares 4x4, 2x2, 1x1, ½ x ½ ...... as shown in the figure above.

What is the sum total area of all the squares?
What is the length AB, the sum of the lengths of the sides?

[4] Small squares are continually added standing at the points of trisection of the sides.

The largest is a unit square.
What is the total area?

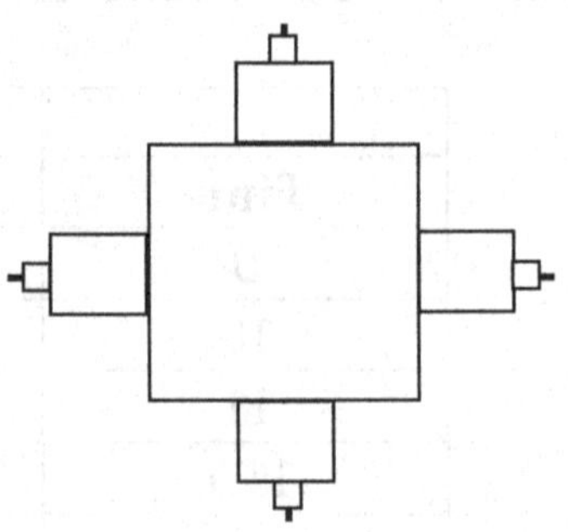

==============================

Answers: Question on arithmetic sequences: 19, 22, 25

Exercise 1

[1] $U_{n+1} = U_n + 2n$
[2] $U_n = 3n-1$
[3] 4,7,10,13,16

Exercise 2

[1]. 55, [2]. 255, [3]. 10055, [4]. 25, [5]. 36

[6]. 1010, [7]. 45, [8]. 55, [9]. 70, [10]. 112

Exercise 3

[1]. 175, [2]. 60, [3]. 160, [4]. 20

[5] this is the Fibonacci sequence. (You do not need $U_1 = 0$)

Exercise 5

[1]. 2, 6, 18, 54, 162 or 2, -6, 18, -54, 162

[2]. 3, 6, 12, 24, 48

[3]. 9, 18, 36, 72, 144

Exercise 7

[1]. 4, 5, 6, $U_r = r+2$ [2]. $U_r = r+1$

Exercise 8

[1] $11\frac{1}{9}$

[2] $369 + 369(10^{-3}) + 369(10^{-6}) + 369(10^{-9}) + \ldots = 369\frac{41}{111}$

[3] area = 64/3, AB = 8 [4] area = 1½

====================================

CHAPTER 18

Fibonacci's Rabbits and the Golden Section

Signor Bonaccio was a Venetian shipping merchant who travelled the Mediterranean during the early part of the twelfth century. In 1170 his wife gave birth to a son in the city of Pisa. Thus was born the son of Bonaccio, "filius Bonaccio" or as we call him, Fibonacci or Leonardo of Pisa. As a young boy Fibonacci travelled with his father to Bejaia near Algiers and while there he saw how much better the Hindu-Arabic decimal number system was than the clumsy Roman number system that was then in use in Western Europe. This kindled in him a keen interest in mathematics and as a result he later toured the great Mediterranean cities collecting the writings of the Arabic scholars. By 1202 he had compiled his discoveries into a book that he called "Liber Abaci" the book of calculations, in which he explained the use of the Hindu-Arabic number system. In 1220 he published a book on geometry and trigonometry, "Practica Geometriae" and in 1225 he published a book on algebra, "Liber Quadratorum". Fibonacci became one of the most important mathematicians of his time and was instrumental in the introduction of the Hindu decimal number system into Europe.

His most famous problem endures to this day and concerns breeding rabbits. It leads to what is probably the most well known sequence of numbers in mathematics, the Fibonacci Sequence.

Each pair of rabbits begets a new pair each month but they only become fertile from the second month of their birth. We start with no rabbits at all, but after a dull and boring month we decide to introduce just one pair of newborn rabbits. After two months, they bear another pair of rabbits.

Fibonacci asks, how many pairs of rabbits will there be in a few years time?

The analysis goes like this:

End of Month	1 month old pairs	breeding pairs	Newborn pairs	Total
1	0	0	0	0
2	0	0	1	1
3	1	0	0	1
4	0	1	1	2
5	1	1	1	3
6	1	2	2	5
7	2	3	3	8

The changes in the table take place as follows:

Month	1 month old pairs	Begetting pairs	Newborn pairs	Total
n	a	b	b	a+2b
n+1	b	a+b	a+b	2a+3b
n+2	a+b	a+2b	a+2b	3a+5b

The number of newborn pairs, column 4, is always the same as the number of begetting pairs, column 3.
The number of begetting pairs, column 3, is the sum of the previous columns 2 and 3.
The number of 1 month old pairs, column 2, is the same as the previous year's column 4.
We notice that the total for month n+2 is the sum of the two totals for months n and n+1 but of course, in mathematics we require a more rigorous proof:

Theorem 1
If the number of pairs in month n is P and the number of pairs in month n+1 is Q then the numbers of pairs in month n+2 will be P+Q

Proof
In month n we have P pairs, and in month n+1 we have Q pairs.

Month	Number of Pairs
n	P
n+1	Q
n+2	?

In the next month, n+2, each of the P pairs of month n will produce an extra pair of rabbits giving P+P pairs. The number of newborn pairs in month n+1 is Q-P but none of these newborn pairs will produce anything for the next month.
Therefore, the number of pairs in month n+2 is P+P +(Q-P) = P+Q.
Therefore, the number of pairs in month n+2 will be Q+P. ■

Notation
Let the number of pairs of rabbits in month n be denoted by Un (=P), then the number of pairs of rabbits in month n+1 will be denoted by Un+1 (=Q). The number of pairs of rabbits in month n+2, we have proved, is equal to P+Q thus we have the recurrence relation for Fibonaccis' rabbits:

$$U_{n+2} = U_n + U_{n+1}$$

The Fibonacci Sequence giving the number of pairs of rabbits in each month (=Un) can therefore be defined by the recurrence relation

$$U_1 = 0 \quad U_2 = 1$$

$$U_{n+2} = U_n + U_{n+1}$$

giving the Fibonacci sequence:

0, 1, 1, 2, 3, 5, 8, 13, 21, 34, 55, 89, 144, 233, 377, 610, ……..

The Golden Section

Psychologists tell us that there is a certain size for a picture frame that makes it most pleasing to the eye. The average person does not like pictures to be in a frame that is too long and narrow but on the other hand , a frame that is exactly square does not seem to satisfying. Somewhere between the long thin rectangle and the square one, there is a 'perfect rectangle'.

Greek architects and mathematicians determined that for this perfect rectangle, it should be possible to divide it into a square and a smaller rectangle of the same proportions:

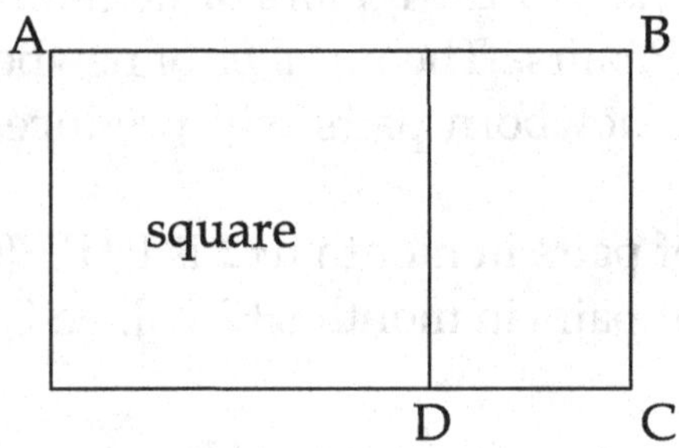

For the perfect rectangle $\frac{AB}{BC} = \frac{BC}{CD}$ and the **golden section** is the ratio of the sides AB/BC.

Finding the Golden Section

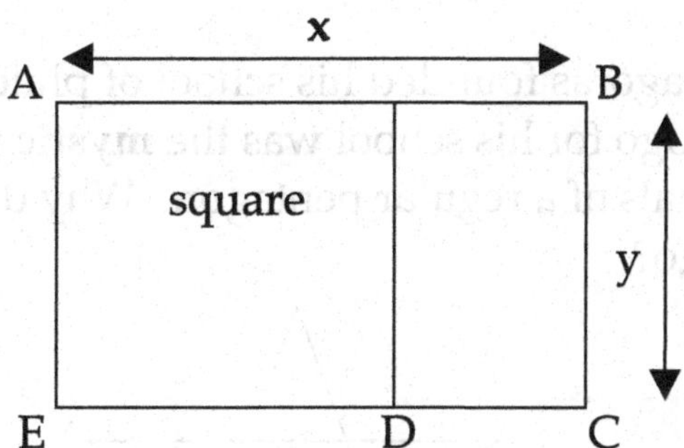

Let AB = x and BC = y so that DC = x-y (because the side DE = y).

Then since AB/BC = BC/CD we have

$$\frac{x}{y} = \frac{y}{x-y}$$

this gives the equation

$$x^2 - xy - y^2 = 0$$

Let the **golden section** be x/y = r then, dividing through by y^2 gives the equation

$$r^2 - r - 1 = 0$$

Solving for r gives

$$r = \frac{1 \pm \sqrt{5}}{2}$$

Hence we have

$$r = \frac{1 + \sqrt{5}}{2}$$

Thus the golden section ratio is

r = 1.618 (3 decimal places)

==============================

The Pentagram

Around 500 BC, Pythagoras founded his school of philosophy at Croton in Southern Italy. The logo for his school was the **mystic pentagram.** It is formed by the diagonals of a regular pentagon. Why did he choose the pentagram for his logo?

The pentagram

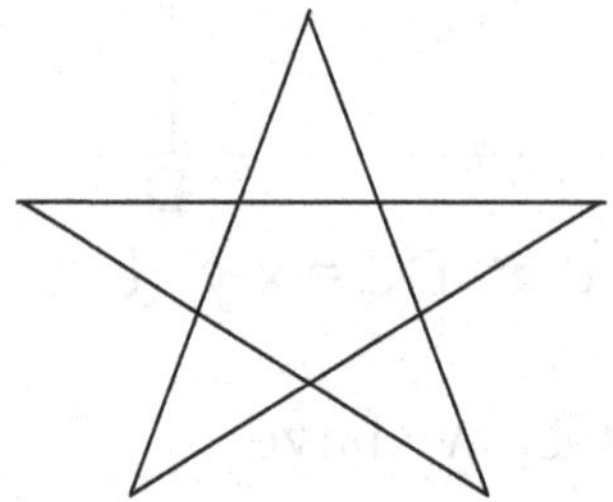

If we draw the outside pentagon then we have the following figure:

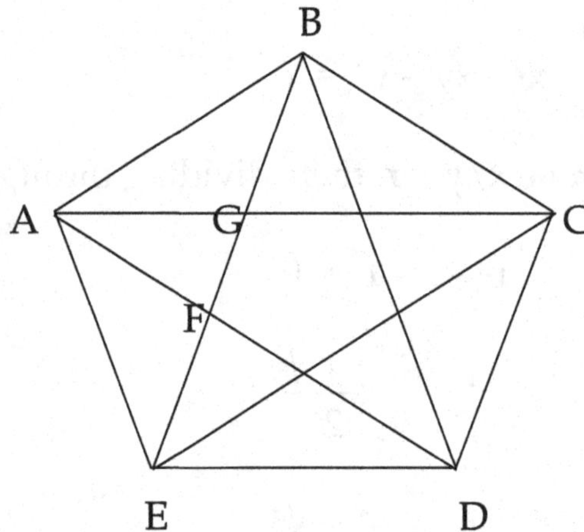

The acute angles in this figure are all, either 36 or 72 degrees.

Therefore triangles ABD / AFB are similar and $\frac{AB}{AF} = \frac{AD}{AB} = \frac{BD}{FB}$

These two triangles are also isosceles with AB = AF and BE = BD

Let AB = BF = x and FE = y then we have $\frac{x}{y} = \frac{x+y}{x}$

or (see above) $x^2 - xy - y^2 = 0$ which is the equation for the Golden Section

Therefore BF:FE is equal to the golden ratio

Further if we subtract the tops and the bottoms of the fractions we get

$$\frac{x+y}{x} = \frac{x}{y} = \frac{(x+y)-x}{x-y} = \frac{y}{x-y} = \frac{BG}{GF}$$

Therefore in the mystic pentagram we have

BG:GF = BF:FE = the golden ratio

No surprise then that Pythagoras chose the pentagram for his logo.

==============================

Returning to the Fibonacci rabbit sequence:

0, 1, 1, 2, 3, 5, 8, 13, 21, 34, 55, 89, 144, and calculate the ratios of consecutive terms (to 3 dp) .

The following sequence shows the ratios of the terms Un+1/Un

∞, 1, 2, 1.5, 1.67, 1.6, 1.625, 1.615, 1.619, 1.618, 1.618, 1.618, ...

The ratios of the terms seem to converge to the golden ratio r=1.618
Once again we need a more rigorous mathematical proof:

Theorem 2
The ratios of consecutive terms of the Fibonacci sequence converge to the golden section.

Proof
Suppose that there is a limiting value for Un+1/Un that we can call k so that we can write Un+1/Un → k as the value of n is made larger.
Now the recurrence relation for the Fibonacci sequence is

$$Un{+}2 = Un{+}1 + Un$$

and if we divide through by Un we get

$$\frac{Un{+}2}{Un} = \frac{Un{+}1}{Un} + 1$$

This equation can be rearranged to give

$$\frac{Un+2}{Un+1} \times \frac{Un+1}{Un} - \frac{Un+1}{Un} - 1 = 0$$

and if each of the fractions $\frac{Un+2}{Un+1}$ and $\frac{Un+1}{Un}$ approach the value k

then this equation gets closer and closer to

$$k^2 - k - 1 = 0$$

Now this is the same equation that we derived for the **golden section.** Solving this equation for k gives

$$k = \frac{1 \pm \sqrt{5}}{2}$$

the positive root is the golden section.
Therefore, the ratios of consecutive terms of the Fibonacci sequence converge to a limit which is equal to the **golden section.** ■

A Formula for Un

Recurrence relations such as $Un+2 = Un+1 + Un$ can sometimes be solved by trying

$$Un = x^n$$

To find possible values for x, we substitute

$$Un+2 = x^{n+2},\ Un+1 = x^{n+1} \text{ and } Un = x^n$$

which gives $$x^{n+2} = x^{n+1} + x^n$$

and dividing by x^n and rearranging gives the equation

$$x^2 - x - 1 = 0$$

again, we have the equation for the golden section, which has the two solutions

$$x = \frac{1+\sqrt{5}}{2} \qquad x = \frac{1-\sqrt{5}}{2}$$

This suggests a general term for the Fibonacci sequence of the form

$$A\left(\frac{1+\sqrt{5}}{2}\right)^n + B\left(\frac{1-\sqrt{5}}{2}\right)^n$$

We can use the values of U1 = 0 and U2 = 1 to find the values of A and B: n=1 gives

$$A\left(\frac{1+\sqrt{5}}{2}\right) + B\left(\frac{1-\sqrt{5}}{2}\right) = 0$$

i.e. $$A + B + \sqrt{5}(A - B) = 0 \qquad \ldots\ldots \; 1$$

n=2 gives

$$A\left(\frac{1+\sqrt{5}}{2}\right)^2 + B\left(\frac{1-\sqrt{5}}{2}\right)^2 = 1$$

i.e. $$A(6 + 2\sqrt{5}) + B(6-2\sqrt{5}) = 4$$

or $$3(A+B) + \sqrt{5}(A-B) = 2 \qquad \ldots\ldots\ldots \; 2$$

From equations 1 and 2 we find

$$2(A+B) = 2 \qquad A+B = 1$$

and $$2\sqrt{5}(A-B) = -2 \qquad A-B = -1/\sqrt{5}$$

adding $$2A = 1 - 1/\sqrt{5} \qquad A = \frac{\sqrt{5}-1}{2\sqrt{5}}$$

subtract $$2B = 1 + 1/\sqrt{5} \qquad B = \frac{\sqrt{5}+1}{2\sqrt{5}}$$

Thus we have

$$Un = \frac{\sqrt{5}-1}{2\sqrt{5}}\left(\frac{1+\sqrt{5}}{2}\right)^{n} + \frac{\sqrt{5}+1}{2\sqrt{5}}\left(\frac{1-\sqrt{5}}{2}\right)^{n}$$

$$= \frac{5-1}{4\sqrt{5}}\left(\frac{1+\sqrt{5}}{2}\right)^{n-1} + \frac{1-5}{4\sqrt{5}}\left(\frac{1-\sqrt{5}}{2}\right)^{n-1}$$

$$= \frac{1}{\sqrt{5}}\left(\frac{1+\sqrt{5}}{2}\right)^{n-1} - \frac{1}{\sqrt{5}}\left(\frac{1-\sqrt{5}}{2}\right)^{n-1}$$

$$= \frac{1}{\sqrt{5}}\left\{\left(\frac{1+\sqrt{5}}{2}\right)^{n-1} - \left(\frac{1-\sqrt{5}}{2}\right)^{n-1}\right\}$$

To prove that this is the formula for Un for an all values of n, we nail the proof down by an appeal to induction:

Suppose that Un is given by

$$Un = \frac{1}{\sqrt{5}}\left\{\left(\frac{1+\sqrt{5}}{2}\right)^{n-1} - \left(\frac{1-\sqrt{5}}{2}\right)^{n-1}\right\}$$

for all values of n from 1 up to k
then

$$Uk+1 = Uk + U\,k-1$$

$$= \frac{1}{\sqrt{5}}\left\{\left(\frac{1+\sqrt{5}}{2}\right)^{k-1} - \left(\frac{1-\sqrt{5}}{2}\right)^{k-1}\right\}$$

$$+ \frac{1}{\sqrt{5}}\left\{\left(\frac{1+\sqrt{5}}{2}\right)^{k-2} - \left(\frac{1-\sqrt{5}}{2}\right)^{k-2}\right\}$$

$$= \frac{1}{\sqrt{5}}\left\{\left(\frac{1+\sqrt{5}}{2}\right)^{k-2}\left(1 + \frac{1+\sqrt{5}}{2}\right) + \left(\frac{1-\sqrt{5}}{2}\right)^{k-2}\left(1 + \frac{1-\sqrt{5}}{2}\right)\right\}$$

$$= \frac{1}{\sqrt{5}}\left\{\left(\frac{1+\sqrt{5}}{2}\right)^{k-2}\left(\frac{3+\sqrt{5}}{2}\right) - \left(\frac{1-\sqrt{5}}{2}\right)^{k-2}\left(\frac{3-\sqrt{5}}{2}\right)\right\}$$

But $\left(\frac{1+\sqrt{5}}{2}\right)^2 = \left(\frac{6+2\sqrt{5}}{4}\right) = \frac{3+\sqrt{5}}{2}$ and $\left(\frac{1-\sqrt{5}}{2}\right)^2 = \frac{6-2\sqrt{5}}{4} = \frac{3-\sqrt{5}}{2}$

Therefore

$$U_{k+1} = \frac{1}{\sqrt{5}}\left\{\left(\frac{1+\sqrt{5}}{2}\right)^{k} - \left(\frac{1-\sqrt{5}}{2}\right)^{k}\right\}$$

Thus, if the formula for Un is correct for n= 1 up to k, then it is also correct for n=k+1.

Now putting **n=1** in the formula gives

$$U1 = \frac{1}{\sqrt{5}}\left\{\left(\frac{1+\sqrt{5}}{2}\right)^{0} - \left(\frac{1-\sqrt{5}}{2}\right)^{0}\right\} = 0$$

and putting **n=2** in the formula gives

$$U2 = \frac{1}{\sqrt{5}}\left\{\left(\frac{1+\sqrt{5}}{2}\right)^{1} - \left(\frac{1-\sqrt{5}}{2}\right)^{1}\right\} = 1$$

Therefore the formula is correct for n=1 and n=2 and so, by the principle of mathematical induction, the formula for all positive integral n is given by

$$Un = \frac{1}{\sqrt{5}}\left\{\left(\frac{1+\sqrt{5}}{2}\right)^{n-1} - \left(\frac{1-\sqrt{5}}{2}\right)^{n-1}\right\}$$

=======================================

The Sum of a Fibonacci Sequence

Since we now know that Un is given by

$$Un = \frac{1}{\sqrt{5}}\left\{\left(\frac{1+\sqrt{5}}{2}\right)^{n-1} - \left(\frac{1-\sqrt{5}}{2}\right)^{n-1}\right\}$$

we could simply sum the series U1 + U2 + U3 + ... + Un using the formula for

the sum of n terms of a Geometric progression: $Sn = \frac{a(r^n - 1)}{(r - 1)}$

Thus

$$\left(\frac{1+\sqrt{5}}{2}\right)^{0} + \left(\frac{1+\sqrt{5}}{2}\right)^{1} + \left(\frac{1+\sqrt{5}}{2}\right)^{2} + \dots + \left(\frac{1+\sqrt{5}}{2}\right)^{n-1} = \frac{\left(\frac{1+\sqrt{5}}{2}\right)^{n} - 1}{\frac{1+\sqrt{5}}{2} - 1}$$

and

$$\left(\frac{1-\sqrt{5}}{2}\right)^{0} + \left(\frac{1-\sqrt{5}}{2}\right)^{1} + \left(\frac{1-\sqrt{5}}{2}\right)^{2} + \dots + \left(\frac{1-\sqrt{5}}{2}\right)^{n-1} = \frac{\left(\frac{1-\sqrt{5}}{2}\right)^{n} - 1}{\frac{1-\sqrt{5}}{2} - 1}$$

However, a shorter solution can be found by going back to the recurrence relation definition for Un+2, from which we find:

$$\begin{aligned} Un{+}2 &= Un{+}1 + Un \\ Un{+}1 &= Un + Un{-}1 \\ &\;\;. \\ &\;\;. \\ U3 &= U2 + U1 \\ U2 &= 1 \\ U1 &= 0 \end{aligned}$$

adding and noting that many terms cancel on both sides, we have

$$Un{+}2 + U1 = Un + Un{-}1 + Un{-}2 + \dots\dots U1 + 1$$

i.e.

$$\sum_{1}^{n} Un = Un{+}2 - 1$$

Theorem 3 : The sum of n terms of a Fibonacci Sequence is always one less than term Un+2

Proved above ■

Gaps in the Fibonacci Sequence

Since $U_{n+2} - U_{n+1} = U_n$, the gaps between the Fibonacci numbers are themselves Fibonacci numbers:

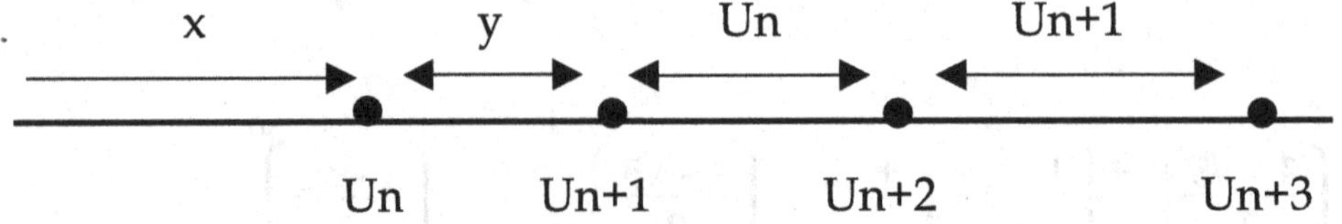

Suppose that the nth Fibonacci number is $U_n = x$ and that the gap between U_n and U_{n+1} is y. Then $U_{n-1} = U_{n+1} - U_n = (x+y) - x = y$

We can then write down

$$U_{n-1} = 0x + 1y$$

$$U_n = 1x + 0y$$

$$U_{n+1} = 1x + 1y$$

Adding the two previous terms: $U_{n+2} = 2x + 1y$

$$U_{n+3} = 3x + 2y$$

$$U_{n+4} = 5x + 3y$$

$$U_{n+5} = 8x + 5y$$

and we see that the coefficients of x and y are themselves Fibonacci numbers.

Remembering

U1	U2	U3	U4	U5	U6	U7 ......
0	1	1	2	3	5	8 ......

we see that $U_{n+4} = U_6.x + U_5.y$

$U_{n+5} = U_7.x + U_6.y$ etc.

or $U_{n+5} = U_7.U_n + U_6.U_{n-1}$

In general, we have

$$U_{n+p} = U_{p+2}.Un + U_{p+1}.Un\text{-}1$$

$$= (U_{p+1} + Up).Un + U_{p+1}(Un+1 - Un)$$

$$= Un.Up + Un+1.Up+1$$

Giving

Theorem 4

Un+p = Un.Up + Un+1.Up+1 for all p>=1

(Just proved) ■

Exercise 1

1. Note the values of the expression $Un+1.Un\text{-}1 - Un^2$.

For example, $U7.U5 - U6^2$

0, 1, 1, 2, 3, 5, 8, 13, 21,.....

$U7.U5 - U6^2 = 8x3 - 25 = -1$

Prove that **$Un+1.Un\text{-}1 - (Un)^2 = (-1)^{n+1}$**

2. Note the values of Un+2.Un-1 – Un+1.Un
For example

U7.U4 – U6.U5 = 8x2 – 5x3 = +1

Prove that **$Un+2.Un\text{-}1 - Un+1.Un = (-1)^{n+1}$**

3. Prove that

$Un+2.Un\text{-}1 - Un+1.Un = Un+1.Un\text{-}1 - (Un)^2$

4. Prove that **$U2n = (Un)^2 + (Un+1)^2$**

5. Prove that **$(Un+2)^2 = U2n + 2Un.Un+1$**

6. Prove that **$(Un+2)^3 = U3n + 3(Un)^2.Un+1$**

========================

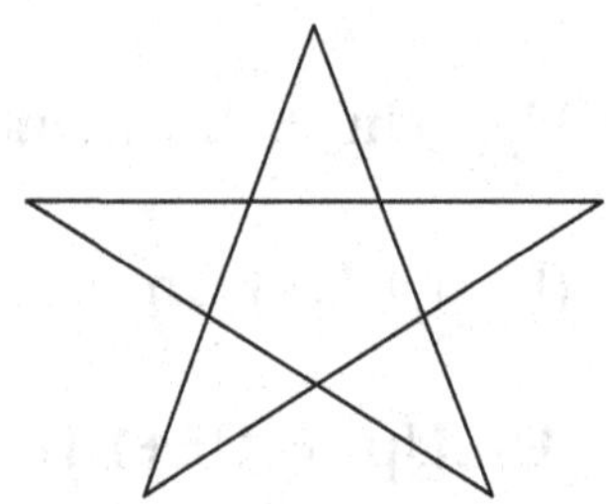

Generalised Fibonacci

Exercise 2

[1]

A sequence is defined by the recurrence relation

$$U_{n+2} = a.U_{n+1} + U_n$$

with $U_1 = 0$ and $U_2 = 1$

Prove that

(i) $U_n.U_{n+3} - U_{n+1}.U_{n+2} = a(-1)^n$

(ii) $U_n.U_{n+2} - (U_{n+1})^2 = (-1)^n$

[2]

If $U_{n+1} = \sum_{1}^{n} U_n$ with $U_1 = 0$ and $U_2 = 1$ then

Prove that $U_n = 2^{n-3}$ $(n \geq 3)$

[3]

If g is the golden section then the Fibonacci term Un is given by

$$U_n = (g^{n-1} - (1-g)^{n-1}) / \sqrt{5}$$

Fibonacci, Leonardo of Pisa, died at the age of about seventy around the year 1240.

=======================================

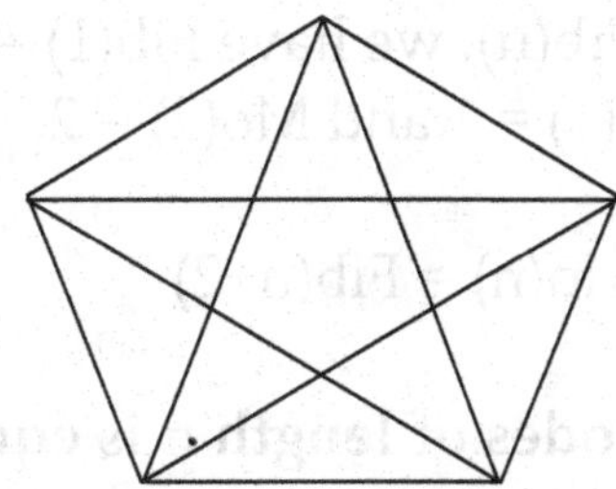

Samuel Finley Breeze Morse was born on the 27th April 1791 in Charlestown, Massachusetts. Morse studied at Yale University and graduated in 1810. He started his adult life as a painter and on a visit to England between 1811 and 1815 he exhibited at the Royal Academy. However, Samuel Morse is not known today for his paintings. In 1832 he became interested in the new electric telegraph and developed the single wire telegraph. Morse code became the European standard for telegraphy in 1851.

The code is made up of dots and dashes. A dash takes up the space of two dots.
We ask how many codes there are of different lengths, a unit length being the length of one dot.

		Number of codes
Codes of length 1	•	1 code
Codes of length 2	— ••	2 codes
Codes of length 3	••• •— —•	3 codes
Codes of length 4	•••• ••— •—• —•• ——	5 codes

Let us define the function Mo(n) as the number of Morse codes of length n.

A code of length n can

[1] start with a dot and be followed by a code of length n-1

or [2] start with a dash and be followed by a code of length n-2

Therefore $\quad Mo(n) = Mo(n-1) + Mo(n-2)$

But this is the recurrence relation for the Fibonacci sequence. If we call the Fibonacci numbers Fib(n), we have Fib(1) = 0, Fib(2) = 1, Fib(3) = 1, Fib(4) = 2.. whereas Mo(1) = 1 and Mo(2) = 2.

Therefore $$Mo(n) = Fib(n+2)$$

The number of Morse codes of length n is equal to Fibonacci number n+2.

==================================

The list of codes according to code length:

1	2	3	4	5	6	>6
• E	•• T	••• S	•••• H	••••• 5	••••••	•——— J
	— I	•— A	••— U	•••— V	••••— 4	——•— Q
		—• N	•—• R	••—• F	•••—•	•———— 1
			—•• D	•—•• L	••—••	••——— 2
			—— M	•—— W	••——	•••—— 3
				—••• B	•—•••	——••• 7
				—•— K	•—•— Y	———•• 8
				——• G	•——• P	————• 9
					—••••	————— 0
					—••— X	
					—•—• C	
					——•• Z	
					——— O	

Not all codes are used, but we can see that

Mo(3) = Mo(2) + Mo(1)

Mo(4) = Mo(3) + Mo(2)

Mo(5) = Mo(4) + Mo(3)

and Mo(6) = Mo(5) + Mo(4)

Some hints and solutions:

Exercise 1

[1] Let Un-1 = x and Un = y then we have:

Un-1	Un	Un+1	Un+2	Un+3 ……
x	y	x+y	x+2y	2x+3y ……

Evaluate Un+1.Un-1 –$(Un)^2$ to give x^2+xy-y^2

Skip along one place and evaluate Un+2.Un – $(Un+1)^2$ and we find the same value but opposite in sign. Therefore, the value of the expression is constant but alternating in sign.

[2] Use the same method.

[3] Both are $(-1)^{n+1}$ by [1] and [2].

[4] This is immediate from Theorem 4, putting p=n.

[5] U2n = $(Un)^2 + (Un+1)^2$ from [4]

= $(Un + Un+1)^2$ - 2Un.Un+1

= $(Un+2)^2$ – 2Un.Un+1 and now transpose

[6] U3n = U2n+n = U2n.Un + U2n+1.Un+1 using theorm 4
= $(Un^2 + Un+1^2)$.Un + (Un.Un+1 + Un+1.Un+2).Un+1

by [4] and using theorem 4 with p = n+1

= $Un^3 + 3Un+1^2.Un + Un+1^3$ using Un+2 = Un + U n+1

= $(Un + Un+1)^3 - 3Un^2.Un+1$ which gives the required result

Exercise 2

[1] Let Un = x and Un+1 = y then we have:

Part (i)

Un	Un+1	Un+2	Un+3	Un+4 ......
x	y	ay+x	$a^2y+ax+y$	$a^3y+a^2x+2ay+x$

Evaluate Un.Un+3 – Un+1.Un+2 to give $ay^2 - a^2xy - ax^2$

Skip one along the sequence and evaluate Un+1.Un+4 – Un+2.Un+3

The value changes sign showing that the values of the expression alternate in sign but keep the same magnitude.

Part (ii)

Show that Un.Un+3 – Un+1.Un+2 = $a(Un.Un+1 - Un+1^2)$

[2] Show that Un = 2Un-1

[3] Use the formula for Un with $g = \frac{1+\sqrt{5}}{2}$

==================================

One of the problems in **Liber Abaci,** Fibonacci's book on calculations, reads as follows:

Seven old women went to Rome.
Each woman had seven mules.
Each mule carried seven sacks;
Each sack contained seven loaves;
And with each loaf were seven knives;
Each knife was put into seven sheaths.

Presumably you were meant to calculate how many there were of each item.

In 1650 BC, an Egyptian scribe called Ahmes filled a roll of papyrus with solutions of mathematical problems that were known to the Egyptian mathematians of that time. The scroll is about 1 foot wide and 18 feet long and is stored in the British Museum. It is also called the Rhind papyrus after the Scottish archaeologist Henry Rhind who brought it to England. Problem 79 goes something like this:

> Seven houses:
> Each has seven cats,
> Each cat eats seven mice,
> Each mouse would have eaten seven ears of corn,
> Each ear of corn would have given seven measures of corn. (flour?)
>
> How much corn (flour?) was saved?

I wonder if Fibonacci had heard of this one?
(Ref: A History of Mathematics: Carl Boyer)

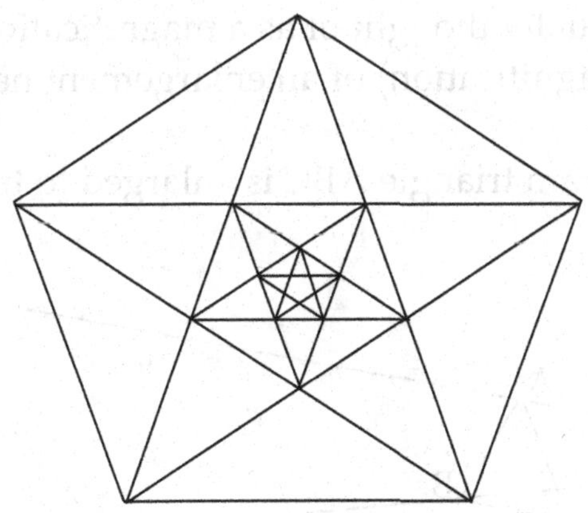

CHAPTER 19

Sin, Cos and Tan

Trigonon + metria (Greek) = triangle + measure = trigonometry

Trigonometry is about measuring triangles, or more accurately, calculating their sides and angles. Trig ratios depend fundamentally upon the idea of an enlargement or magnification. The usual introduction to trig ratios often starts with a right angled triangle and an angle in the triangle marked θ. When the triangle is magnified, we find that the ratios of various sides remain the same. There are six possible ratios for the sides of the triangle, so there are six trig ratios:

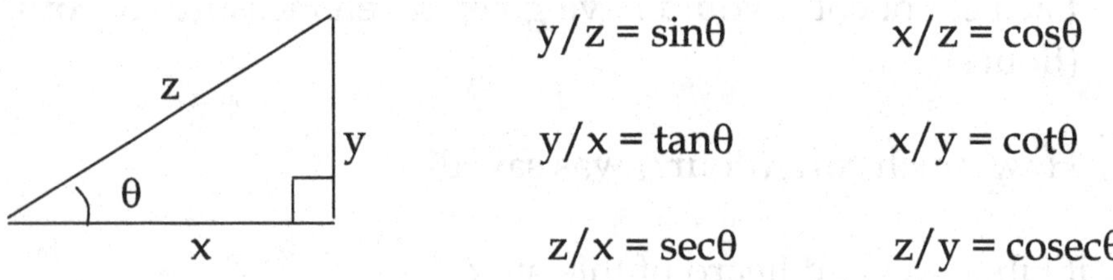

$y/z = \sin\theta$ $x/z = \cos\theta$

$y/x = \tan\theta$ $x/y = \cot\theta$

$z/x = \sec\theta$ $z/y = \text{cosec}\theta$

The reason for these strange names will be discused later, but first:

Enlargements

An enlargement is usually thought of as a magnification but in mathematics, the scale factor (or magnification) of an enlargement need not necessarily be greater than one.

In the following diagram triangle ABC is enlarged to triangle A′B′C′ with a scale factor of 3.

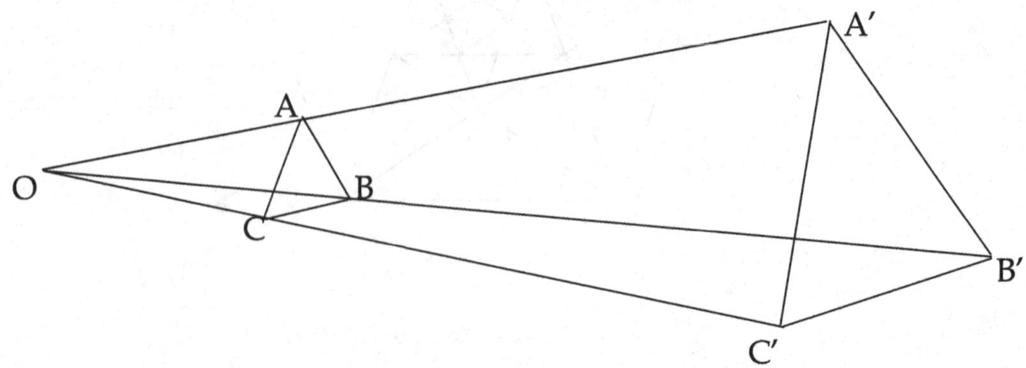

ΔABC is enlarged from centre O with scale factor 3 to ΔA′B′C′
(we can also say that the triangles are in perspective from center O)

Since the scale factor of the enlargement is 3 we have

A′B′=3AB **B′C′=3BC** **C′A′=3CA**

(but note that ΔA′B′C′ = 9ΔABC in area)

In the next figure, triangle ABC is "enlarged" from centre O by a scale factor ½

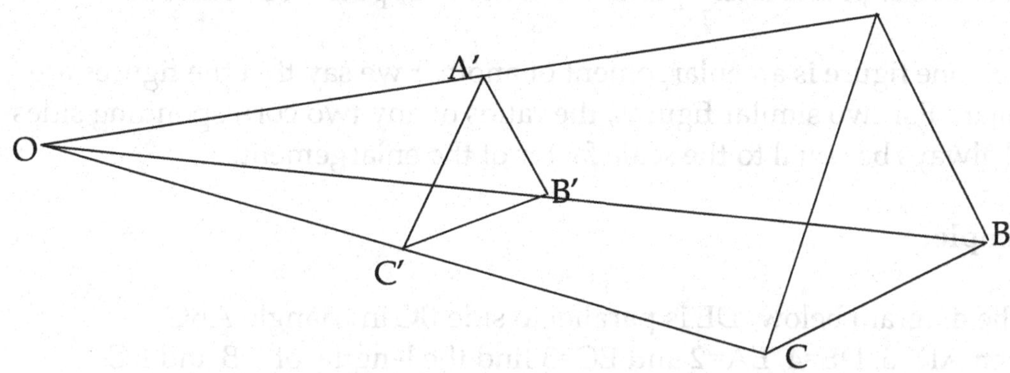

ΔABC is enlarged from centre O by a scale factor ½ to give ΔA′B′C′

Each length in ΔABC is multiplied by the scale factor so that

A′B′=½AB B′C′=½BC C′A′=½CA

Negative scale factors reverse the orientation of the object. In the above figures, with positive scale factors, A→B→ C and A′→B′→ C′ are both clockwise directions.

The next figure shows a "negative enlargement" with scale factor -2. In this figure, A→B→ C is clockwise but A′→B′→ C′ is anticlockwise.

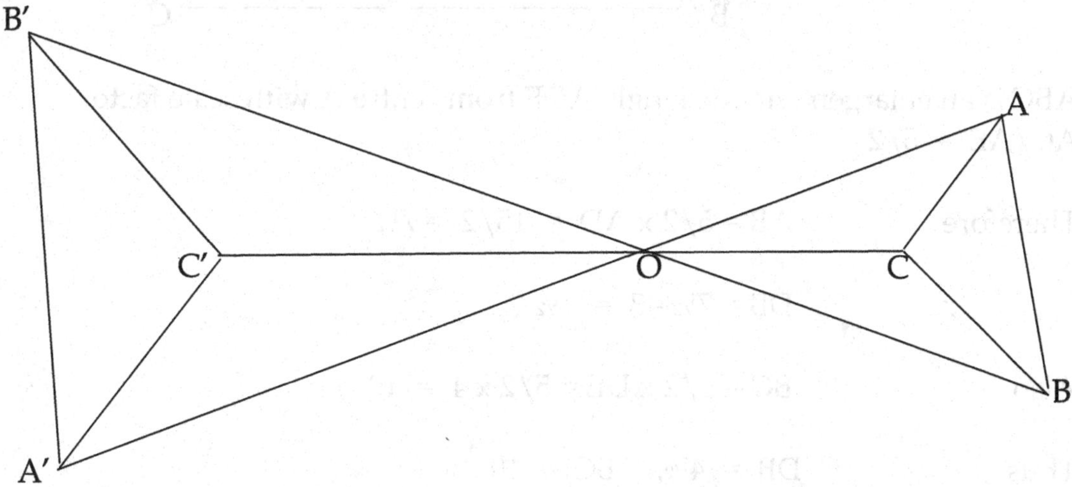

Each length in the object figure is multiplied by the scale factor –2.

A′B′= –2AB
B′C′= –2BC
C′A′= –2CA

The minus sign indicates "parallel but in the opposite direction".

When one figure is an enlargement of another we say that the figures are similar. For two similar figures, the ratios of any two corresponding sides will always be equal to the scale factor of the enlargement.

Example

In the diagram below, DE is parallel to side BC in triangle ABC.
Given AD=3, DE=4, EA=2 and EC=3 find the lengths of DB and BC.

Solution

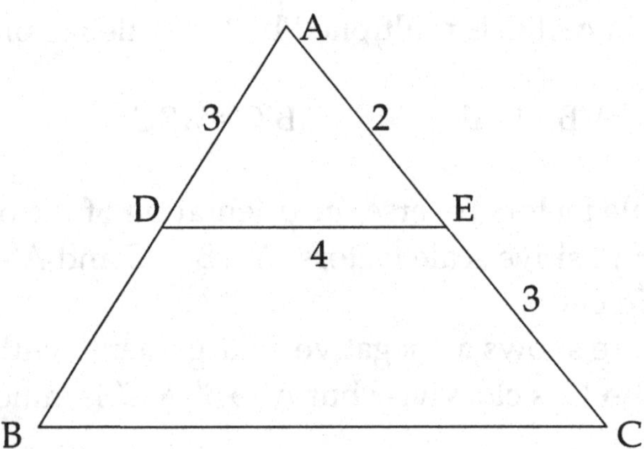

ABC is an enlargement of triangle ADE from centre A with scale factor AC/AE = 5/2

Therefore AB = 5/2 x AD = 15/2 = 7½

∴ DB = 7½ - 3 = 4½

Also BC = 5/2 x DE = 5/2 x 4 = 10

Thus DB = 4½, BC = 10

==========================

Alternatively, we could say that ADE is an "enlargement" of ABC with scale factor 2/5, giving

$$DE = 2/5 \times BC \qquad BC = 5/2 \times DE \quad \text{etc.}$$

Exercise 1

1.
In the diagram below, DE is parallel to side BC in triangle ABC.
Given AD=3, DE=3, EA=2 and EC=4 find the lengths of DB and BC.

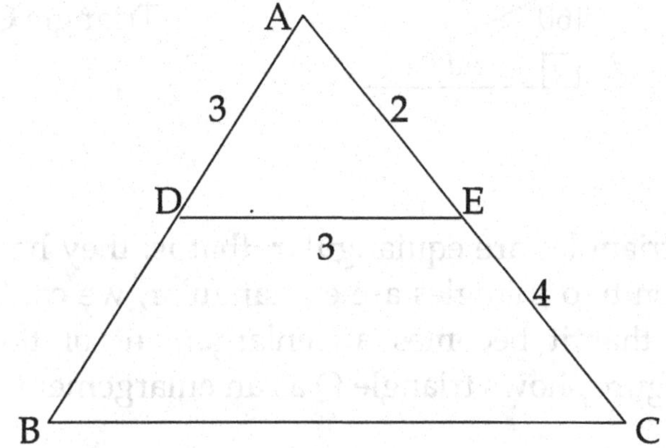

2.
In the diagram below, PQ is parallel to the side BC of triangle ABC.
Given AB=8, BC=10, CA=12 and CQ=18 find the lengths of PB and PQ.

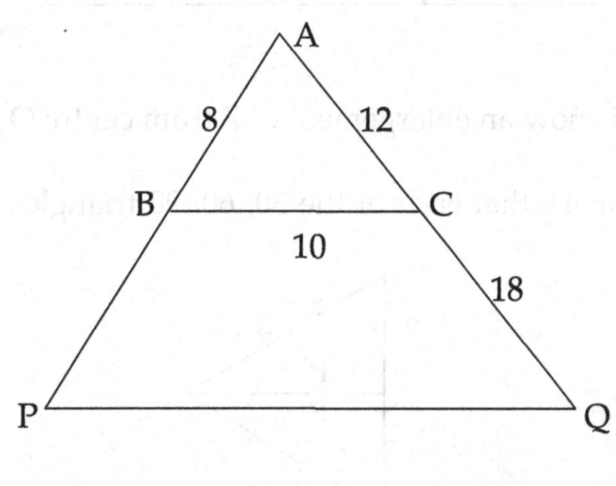

===================================

The 30, 60 90 triangle

In the next diagrams we have two 30°, 60°, 90° triangles labelled triangle P and triangle Q.

The shortest side of triangle P is 2 and the shortest side of triangle Q is 4.

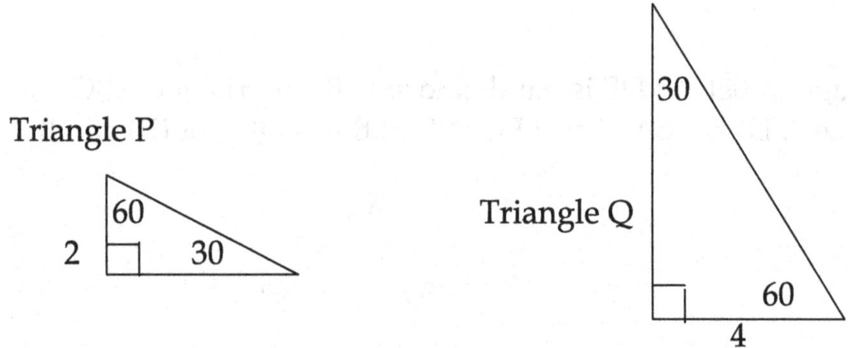

These two triangles are equiangular, that is, they have exactly the same size angles. When two triangles are equiangular, we can always move the larger triangle so that it becomes an enlargement of the smaller triangle. The following figure shows triangle Q as an enlargement of triangle P.

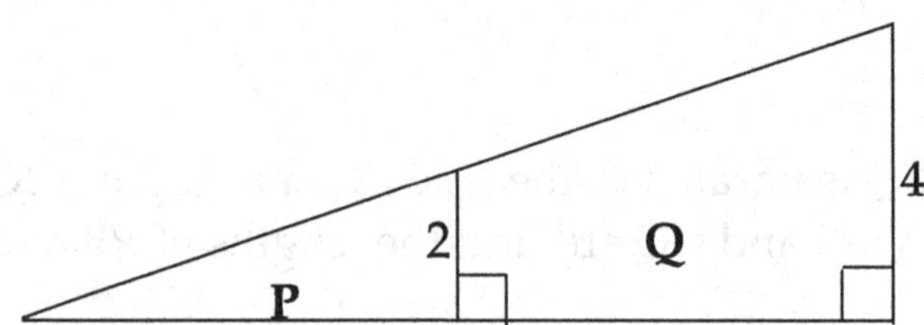

Triangle Q is now an enlargement of P from centre O with scale factor 2.

We now remark that each of the 30, 60, 90 triangles is half of an equilateral triangle:

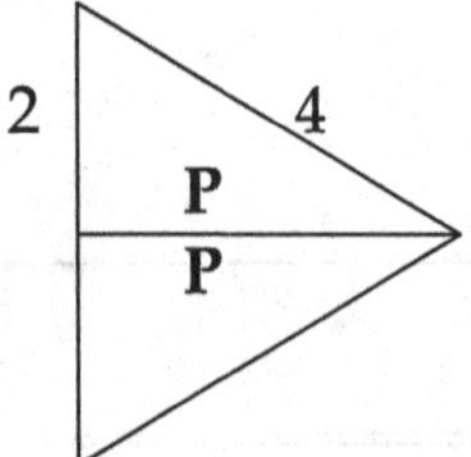

This means that we know that the longest side of triangle P must be 4 and using Pythagoras, the third side is $\sqrt{(16-4)} = \sqrt{12} = 2\sqrt{3}$.

Similarly Q is half of an equilateral triangle of side 8.

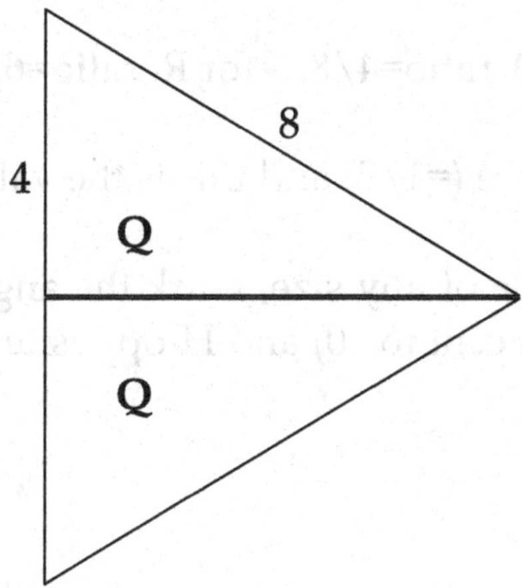

The sides of triangle Q will be 4, 8 and, by Pythagoras, $\sqrt{(64\text{-}16)} = \sqrt{48} = \sqrt{(16X3)} = 4\sqrt{3}$.

In general, if the shortest side of a 30, 60, 90 triangle is **k** then the longest side will be **2k** and the other side will be **k**$\sqrt{3}$
In the next figure we have drawn a diagram of four 30, 60, 90 triangles labeled **P, Q, R** and **S.**

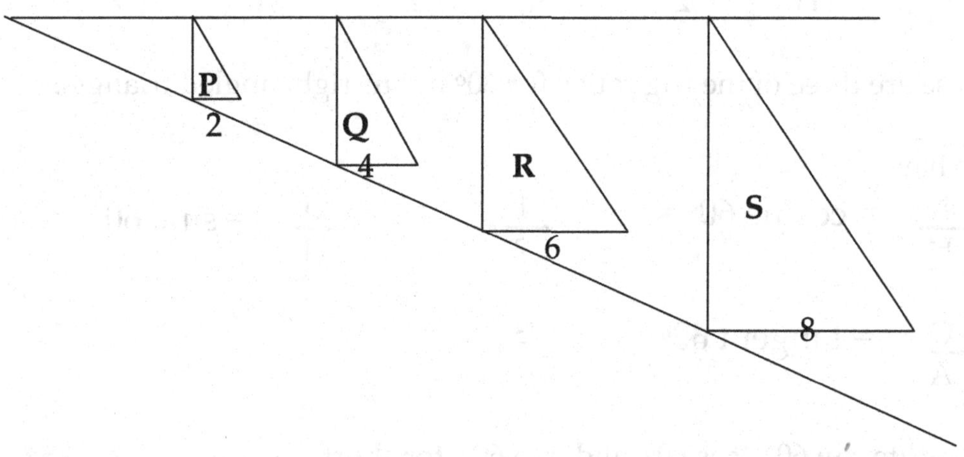

Suppose that the shortest sides of the triangles are 2, 4, 6 and 8 as shown, then since each one of the triangles is half of the corresponding equilateral triangle, we know that the longest sides of each triangle, the hypotenuse, will be 4, 8, 12 and 16. The third sides will be $2\sqrt{3}$, $3\sqrt{3}$, $6\sqrt{3}$ and $8\sqrt{3}$.

Now consider the ratio of the shortest side to the longest side for each triangle:

For P ratio = 2/4, for Q ratio=4/8, for R ratio=6/12, for S ratio=8/16

Each of the ratios the same (=1/2) and this is the value of cos 60.

Draw a 30, 60, 90 triangle of any size, mark the angle 60° and label the sides O(opposite to 60), A(adjacent to 60) and H(opposite 90=hypotenuse)

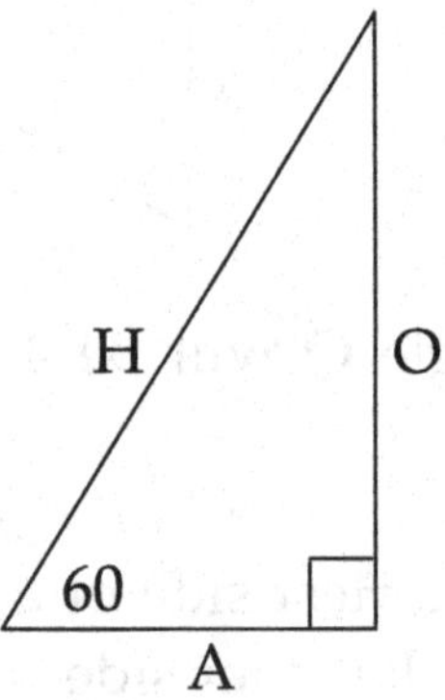

Then we always find the same ratios for

$$\frac{A}{H} = \frac{1}{2} \qquad \frac{O}{H} = \frac{\sqrt{3}}{2} \qquad \frac{O}{A} = \frac{\sqrt{3}}{2}$$

These are three of the trig ratios for 60° in our right angled triangle.

We have

$$\frac{A}{H} = \text{cosine } 60^\circ = \frac{1}{2} \qquad \frac{O}{H} = \text{sine } 60^\circ = \frac{\sqrt{3}}{2}$$

$$\frac{O}{A} = \text{tangent } 60^\circ = \sqrt{3}$$

We write sin 60°, cos 60° and tan 60° for short.

==============================

Exercise 2

By considering appropriate triangles, prove the following results

$\cos 60 = \frac{1}{2}$	$\sin 60 = \frac{\sqrt{3}}{2}$	$\tan 60 \quad \sqrt{3}$
$\cos 30 = \frac{\sqrt{3}}{2}$	$\sin 30 = \frac{1}{2}$	$\tan 30 = \frac{1}{\sqrt{3}}$
$\cos 45 = \frac{1}{\sqrt{2}}$	$\sin 45 = \frac{1}{\sqrt{2}}$	$\tan 45 \quad 1$

Following tradition, for the triangle drawn in the next figure, the side opposite the angle θ is labeled **O,** the side next to (or adjacent to) the angle is labeled **A**. The hypotenuse is labeled **H** (Greek/Latin combination of hypo and teino=stretch). **H** is always opposite the 90° and is always the longest side in the right-angled triangle.

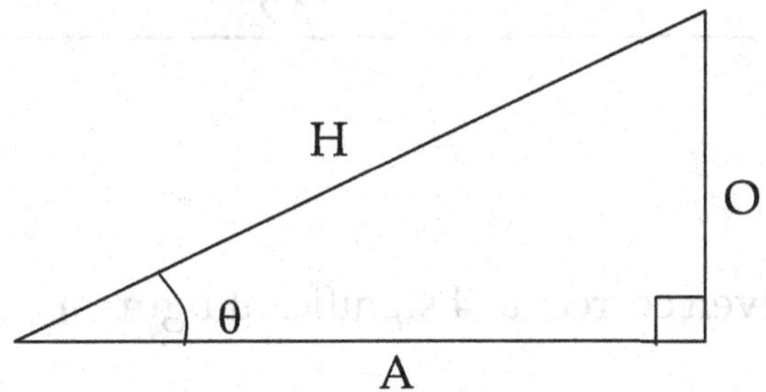

SOHCAHTOA

The exploding island of Krakatoa (1883) has helped many a student of trigonometry:

$$\sin = \frac{O}{H} \qquad \cos = \frac{A}{H} \qquad \tan = \frac{O}{A}$$

The reason that the trigonometry of sines, cosines and tangents works is because all right angled triangles with a given angle θ will be similar so that the ratios O/H, A/H and O/A will be the same for that angle θ whichever triangle we choose. In the following diagram, whichever right angle triangle we choose, the trig ratios for the angle θ will be the same:

$$\frac{O1}{H1} = \frac{O2}{H2} = \frac{O3}{H3} = \sin\theta$$

$$\frac{A1}{H1} = \frac{A2}{H2} = \frac{A3}{H3} = \cos\theta$$

$$\frac{O1}{A1} = \frac{O2}{A2} = \frac{O3}{A3} = \tan\theta$$

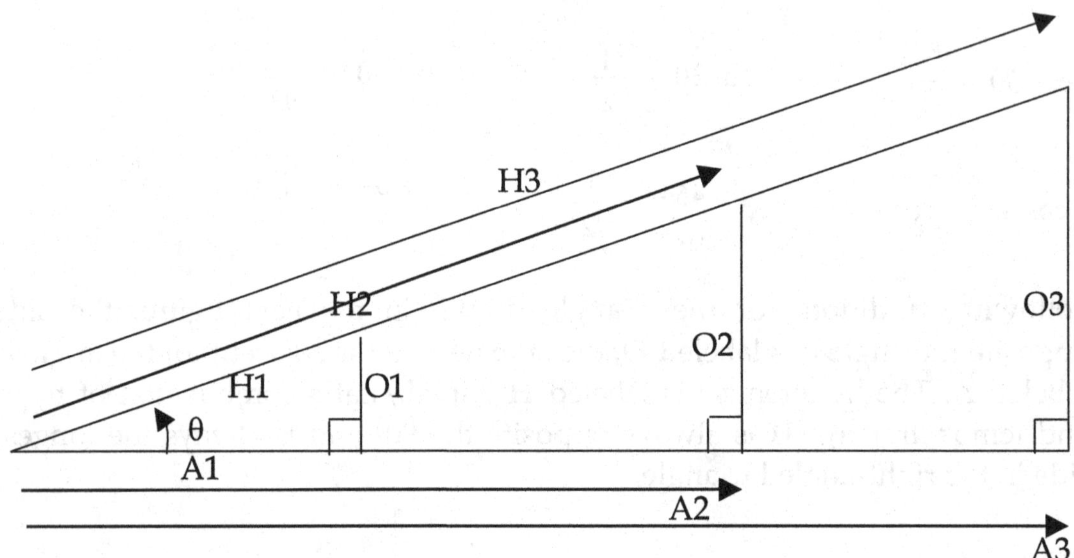

Using the Calculator

Examples: (answers are given correct to 4 significant figures)

1. Calculate x

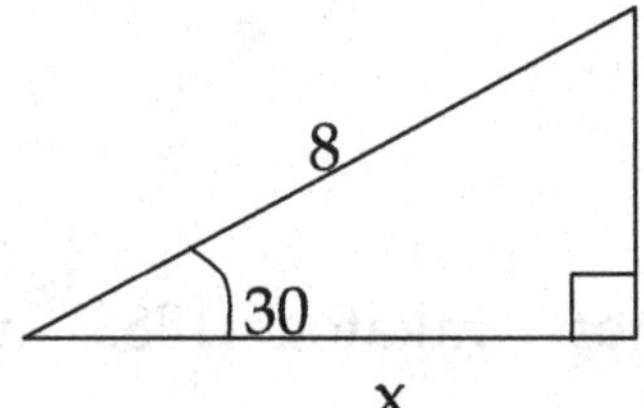

Solution $\frac{x}{8} = \frac{A}{H}$

use cosine $\frac{x}{8} = \cos 30$

$x = 8\cos 30$

using a calculator:

[8] [x] [cos] [3] [0] [=] $x = 6.928$ (4 s.f.)

2.

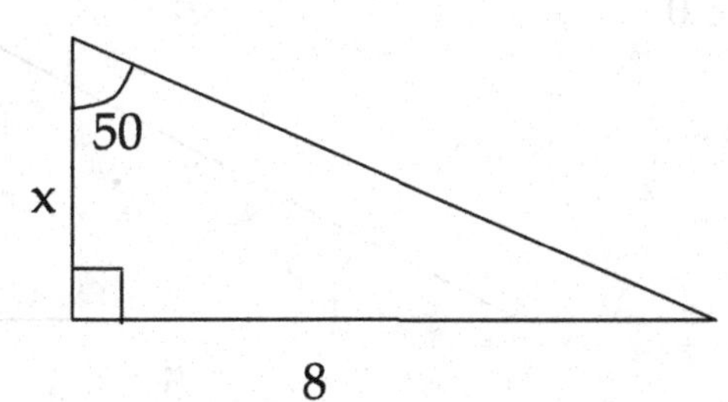

Calculate x

8

Solution $\frac{8}{x} = \frac{O}{A}$

use tangent $\frac{8}{x} = \tan 50$ $x = \frac{8}{\tan 50}$

using your calculator:

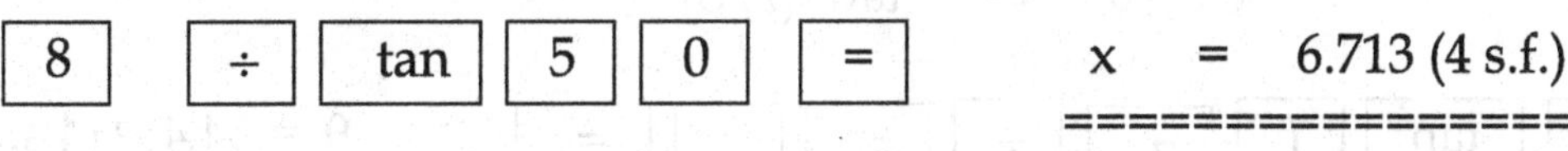

x = 6.713 (4 s.f.)

3. Calculate angle θ

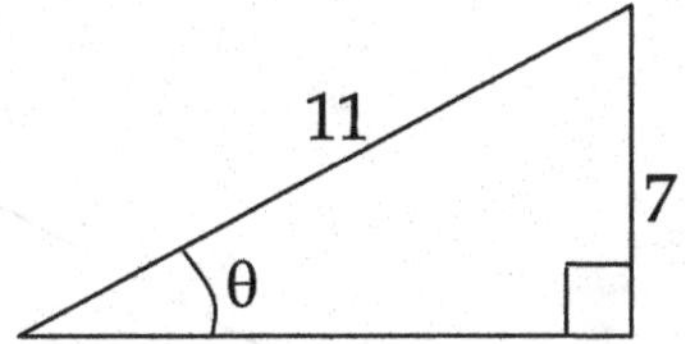

Solution

$\frac{7}{11} = \frac{O}{H}$

use sine $\sin\theta = \frac{7}{11}$

$\theta = \sin^{-1}(7/11)$

using a calculator:

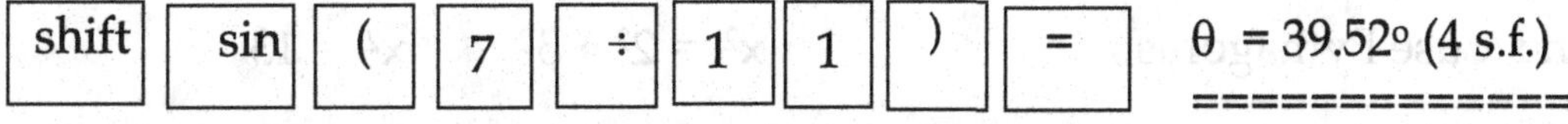

θ = 39.52° (4 s.f.)

4. Calculate angle θ

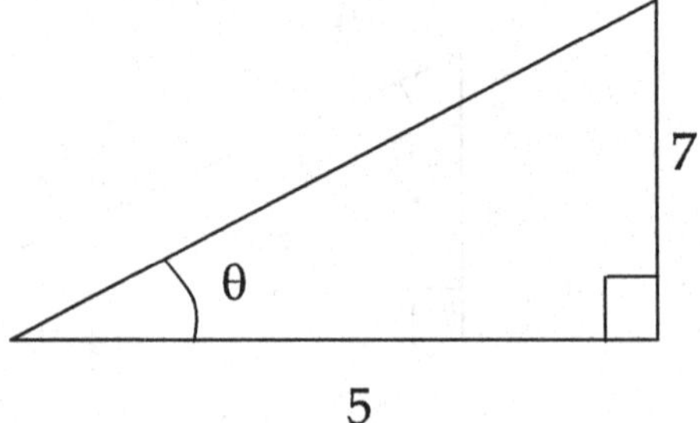

Solution

$$\frac{7}{5} = \frac{O}{A}$$

use tangent $\tan\theta = \frac{7}{5}$

$$\theta = \tan^{-1}(7/5)$$

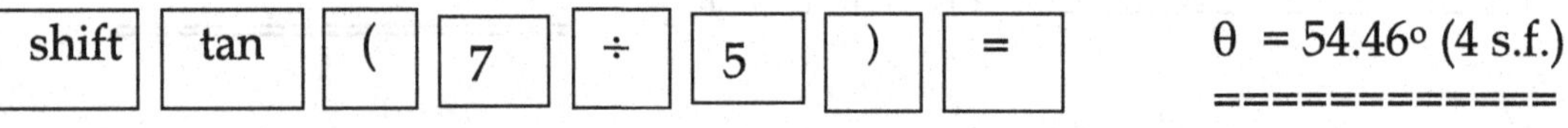

$\theta = 54.46^\circ$ (4 s.f.)

5. Calculate angle θ and side x

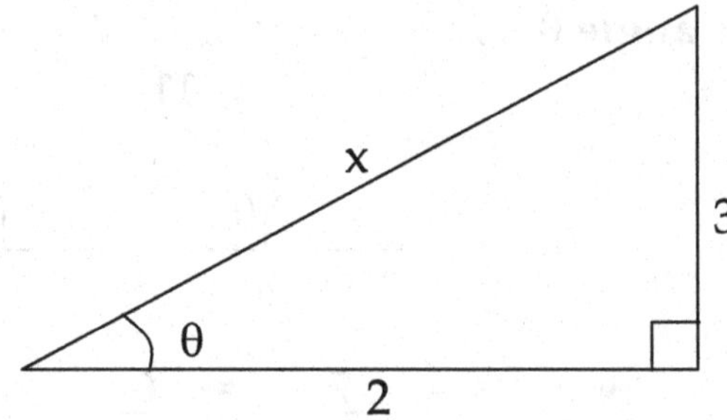

Solution

$$\tan\theta = \frac{3}{2}$$

$$\theta = \tan^{-1}(3/2)$$

shift | tan | (| 3 | ÷ | 2 |) | =

$\theta = 56.31^\circ$ (4 s.f.)

To find x use Pythagoras: $x^2 = 2^2 + 3^2$ $x^2 = 13$

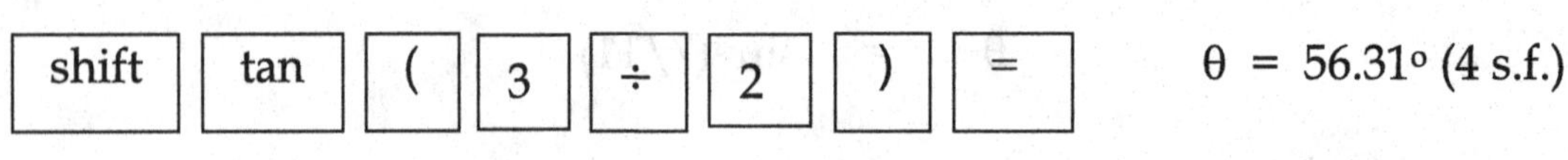

$x = 3.606$ (4 s.f.)

Note that we could calculate x as follows:

$$\cos\theta = 2/x$$

$$x = 2/\cos 56.31$$

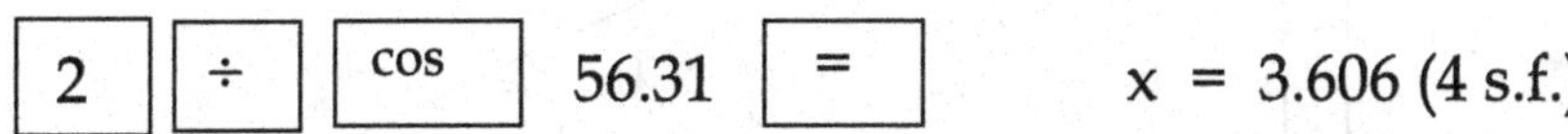

$x = 3.606$ (4 s.f.)

BUT it is better to avoid using a calculated value (i.e. $\theta = 56.31$) if possible.

==

Exercise 3

Find the unknown quantity in each of the following diagrams, correct to 4 significant figures.

[1] 50° 2.3 x

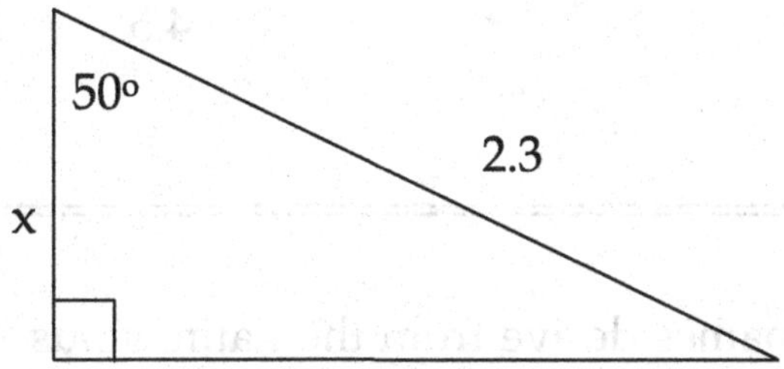

[2] 35° 5.5 x

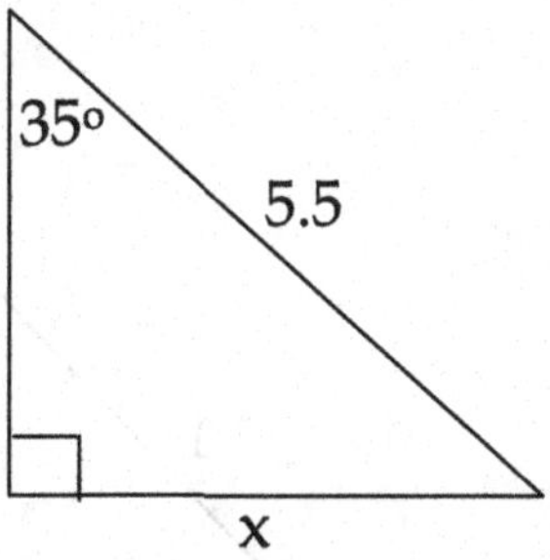

[3] 3.3 70 x

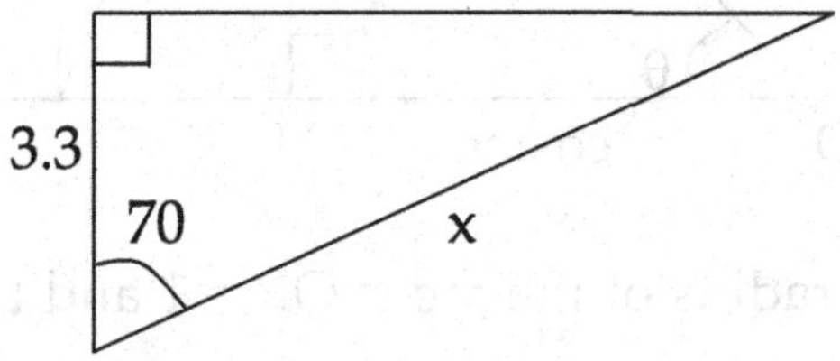

[4] θ 4.5 3.3

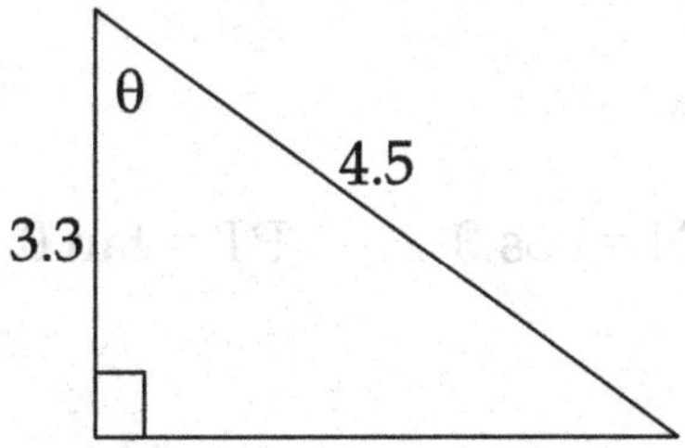

[5]

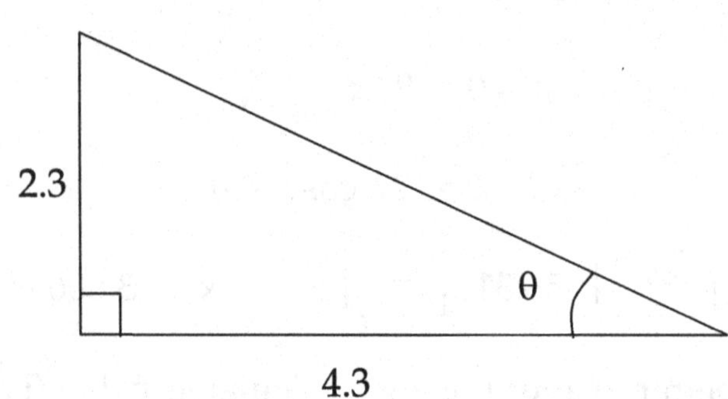

[6]

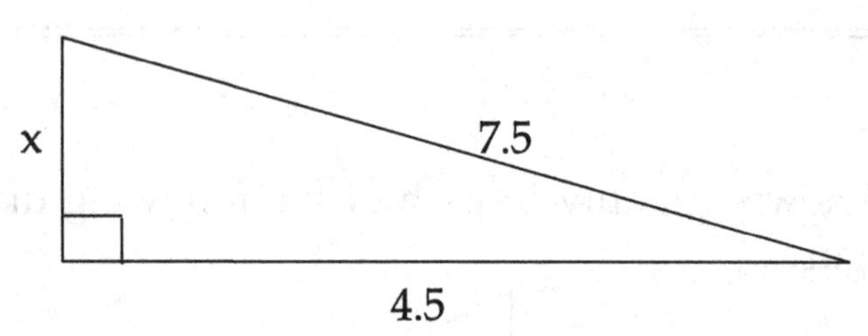

==

The strange names derive from the Latin: sinus = curve or arc, tango means to touch and seco means I cut.

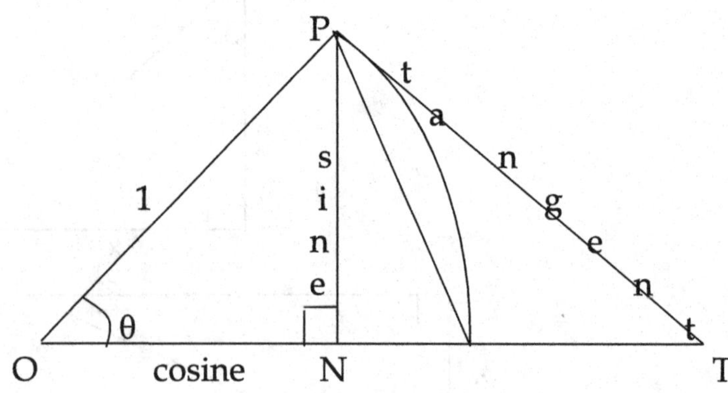

In the above figure, the radius of the arc is OP = 1 and the angle of the arc is θ.

We have:

$$PN = \sin\theta \qquad ON = \cos\theta \qquad PT = \tan\theta$$

Secant, Cosecant and Cotangent

At this point we introduce the three remaining trig ratios to give the complete the set of six.

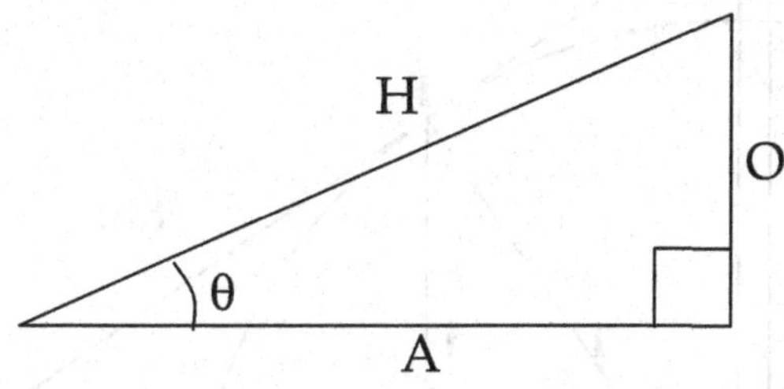

The six possible trig ratios for this right-angled triangle are given by

$$\frac{O}{H} \quad \frac{A}{H} \quad \frac{O}{A} \quad \frac{H}{O} \quad \frac{H}{A} \quad \frac{A}{O}$$

The first three of these ratios are the familiar sin = O/H, cos = A/H and tan = O/A.
The three new ratios are the inverses of these three:

secant = 1/cos, cosecant = 1/sin and cotangent = 1/tan

The word secant derives from the Latin (secare = to cut), hence the English words section, sector and secateurs.
The following diagram shows the secant for our unit circle:

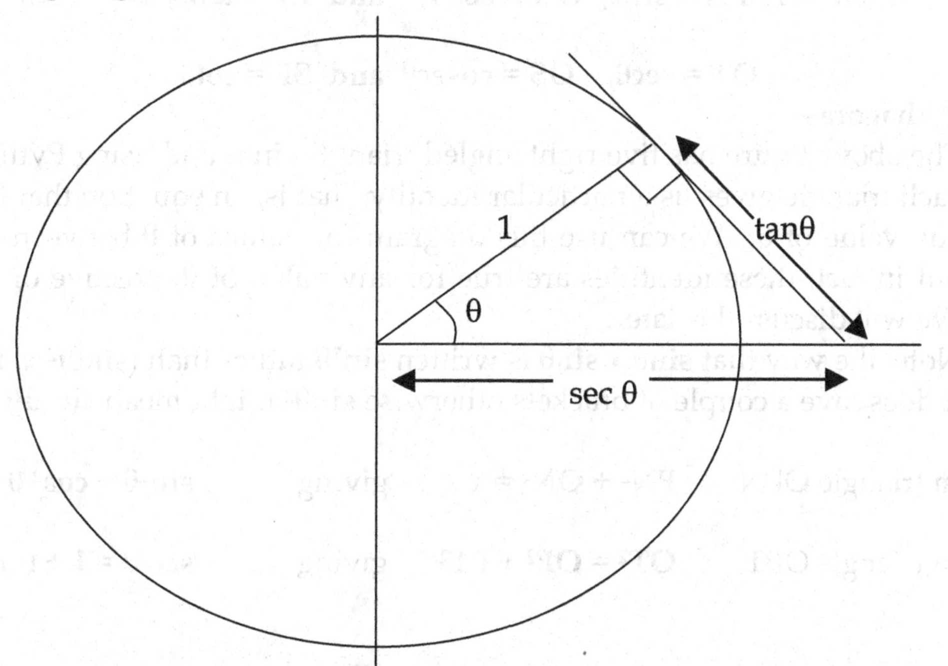

and the following diagram lumps all six ratios on the same figure:

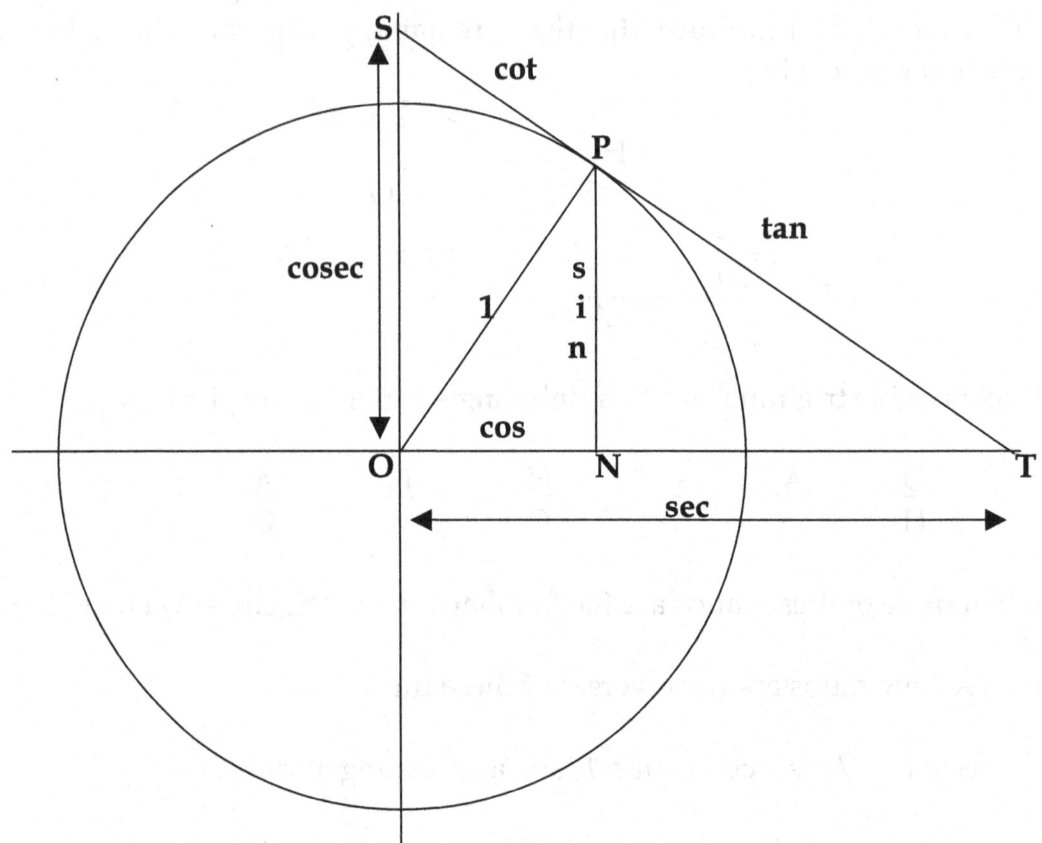

In this figure, we have

$$OP = 1, \; PN = \sin\theta, \; ON = \cos\theta \quad \text{and} \quad PT = \tan\theta,$$

$$OT = \sec\theta, \quad OS = \text{cosec}\theta \; \text{and} \; SP = \cot\theta$$

Pythagoras

The above figure has five right angled triangles in it and using Pythagoras in each triangle gives us a particular identity, that is, an equation that is true for any value of θ. We can use this diagram for values of **θ** between 0 and 90, but in fact, these identities are true for any value of **θ,** positive or negative. We will discuss this later.

Note: the way that **sinθ x sinθ** is written **sin²θ** rather than **(sinθ)²** .. I suppose it does save a couple of brackets otherwise **sinθ²** might mean **sin(θ²)** !

In triangle OPN	$PN^2 + ON^2 = 1$	giving	$\sin^2\theta + \cos^2\theta = 1$
In triangle OPT	$OT^2 = OP^2 + PT^2$	giving	$\sec^2\theta = 1 + \tan^2\theta$

In triangle OPS $OS^2 = OP^2 + SP^2$ giving $\mathbf{cosec^2\theta = 1 + cot^2\theta}$

Exercise 4
Using this figure and by multiplying out, show that

[1] $(\sec\theta - \cos\theta)^2 + \sin^2\theta = \tan^2\theta$

[2] $(\cot\theta + \tan\theta)^2 = \sec^2\theta + \mathrm{cosec}^2\theta$

Trigonometry as we have described it depends on 90 degree triangles and later, in a more general sense, on rectangular Cartesian axes.

The x coordinate of P is **cosθ**

The y coordinate of P is **sinθ**

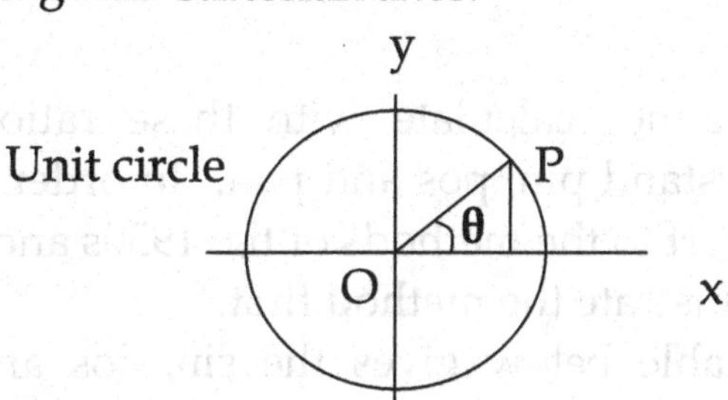

However, it is possible to develop a trigonometry based on a triangle that is not a 90 degree triangle and with a corresponding oblique set of axes.

Pin, Pos and Pan (an unorthodox treatment of three new trig ratios)

The ideas illustrated above do not apply only to right angled triangles. We could, for example, form a set of similar triangles as we have done in the diagrams above, but using 60° instead of 90°.
In the following figure, we have an angle θ in three 60° triangles.

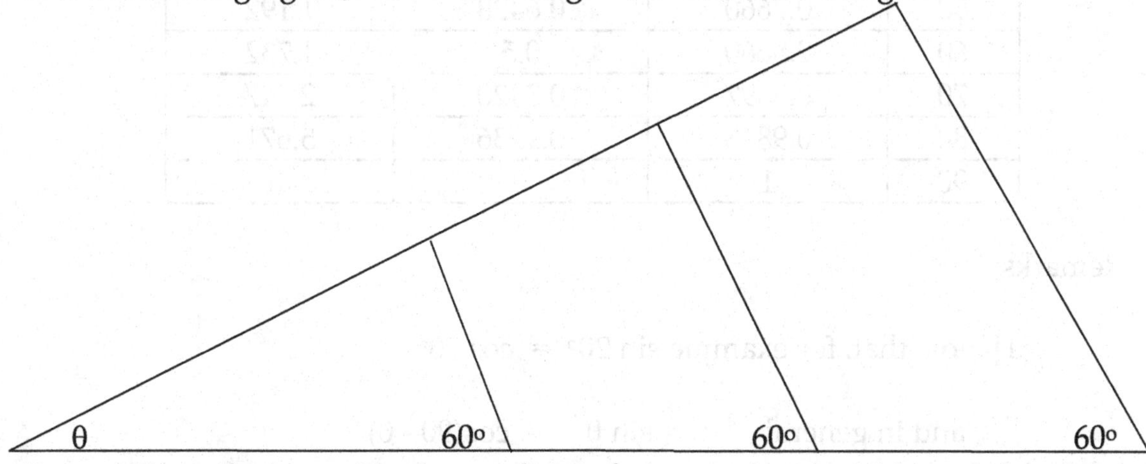

To avoid confusion with the traditional names for the sides of the right angled triangle, O, A and H we call the opposite side for the 60° triangle Oh, the adjacent side Ah and the "hypotenuse" which is now opposite a 60° angle, Hh.

The ratios are no longer the traditional sin, cos and tan and so we invent new names. We shall call the ratios pin, pos and pan.
Thus for the 60° triangle we have:

pin θ = $\frac{Oh}{Hh}$ **pos θ** = $\frac{Ah}{Hh}$ **pan θ** = $\frac{Oh}{Ah}$

We cannot calculate with these ratios because our calculator does not understand pin, pos and pan. In order to use these new trig ratios we have to revert to the methods of the 1950s and use four figure tables.
We illustrate the method first.
The table below gives the sin, cos and tan trig ratios for our usual 90° triangle.

Four Figure tables for 90° triangles

	sin θ	cos θ	tan θ
10	0.1736	0.9848	0.1763
20	0.3240	0.9397	0.3640
30	0.5	0.8660	0.5774
40	0.6428	0.7660	0.8391
50	0.7660	0.6428	1.192
60	0.8660	0.5	1.732
70	0.9397	0.3420	2.747
80	0.9848	0.1736	5.671
90	1	0	∞

Remarks:

[1] Note that, for example sin 20° = cos 70°

and in general sin θ = cos(90 - θ)

Draw a suitable triangle to illustrate this relation.

[2] Check numerically that $\tan\theta = \dfrac{\sin\theta}{\cos\theta}$

[3] Draw diagrams to show that $\cos 60 = \sin 30 = \frac{1}{2}$ and $\tan 60 = \sqrt{3}$

Note how, when θ gets very small (i.e. $\theta \to 0$) then the values of sin θ and tan θ become the same:

$\tan 2^\circ = 0.03492$ $\tan 1^\circ = 0.017455$ $\tan 0.1^\circ = 0.001745331$

$\sin 2^\circ = 0.03489$ $\sin 1^\circ = 0.017452$ $\sin 0.1^\circ = 0.001745328$

Investigation:
For what small angle θ does your calculator think that $\sin\theta = \tan\theta$?

(mine gives sin 0.001 = 0.00001745329252 and tan 0.001 the same)

Using the four figure tables:

Example

Solve the triangle (i.e. find all the sides and angles)

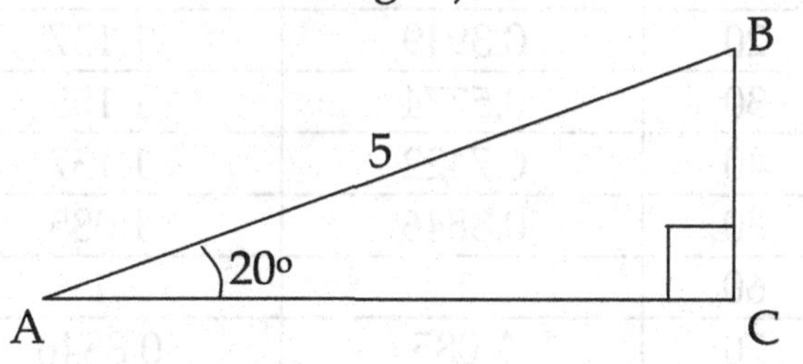

Solution

(i) to find BC $\quad BC/5 = \sin 20$

$BC = 5\sin 20$

$= 5 \times 0.3420$ (from tables)

$BC = 1.71$ (3 s.f.)

(ii) to find AC $\quad AC/5 = \cos 20$

$$AC = 5 \cos 20$$

$$= 5 \times 0.9397 \qquad \text{(from tables)}$$

$$AC = 4.70 \text{ (3 s.f.)}$$

Solution

AB=5, BC=1.71, CA=4.70, ∠A=20°, ∠B=70°, ∠C=90°

Let us now see how we could perform similar calculations using pin, pos and pan.

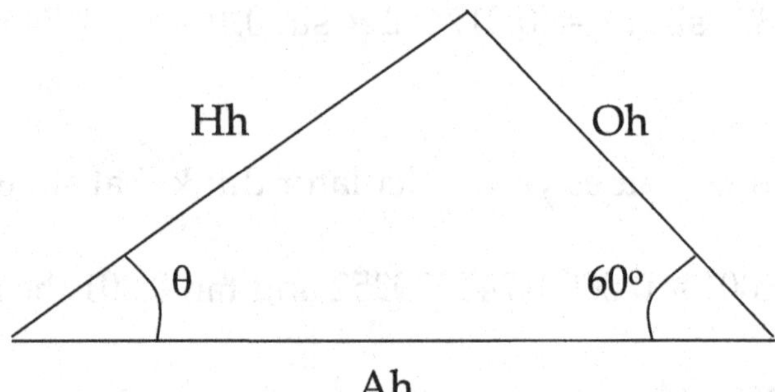

Four Figure tables for 60° triangles

	pin θ	pos θ	pan θ
10	0.2005	1.085	0.1848
20	0.3949	1.137	0.3473
30	0.5774	1.155	0.5
40	0.7422	1.137	0.6528
50	0.8846	1.085	0.8153
60	1	1	1
70	1.085	0.8846	1.2265
80	1.137	0.7422	1.5319
90	1.155	0.5774	2

Remarks

[1] Note, for example, that pos 10 = pin 70 = pos 50

In general, for $\theta<60$,

$$\text{pos}\,\theta = \text{pin}(60+\theta) = \text{pos}(60-\theta)$$

for $\theta>60$

$$\text{pos}\,\theta = \text{pin}(120-\theta) = \text{pos}(\theta-60)$$

[2] Check numerically that $\text{pan}\,\theta = \dfrac{\text{pin}\,\theta}{\text{pos}\,\theta}$

[3] Explain why $\text{pos}\,60 = \text{pin}\,60 = \text{pan}\,60 = 1$

[4] Draw a diagram to show that

$$\text{pos}\,30 = \frac{\sqrt{3}}{2} \qquad \text{pin}\,30 = \frac{1}{2} \qquad \text{pan}\,30 = \frac{1}{\sqrt{3}}$$

[5] Explain why $\text{pos}\,\theta \to 1$ and $\text{pin}\,\theta \to 0$ as $\theta \to 0$

==

Example:

Solve the triangle

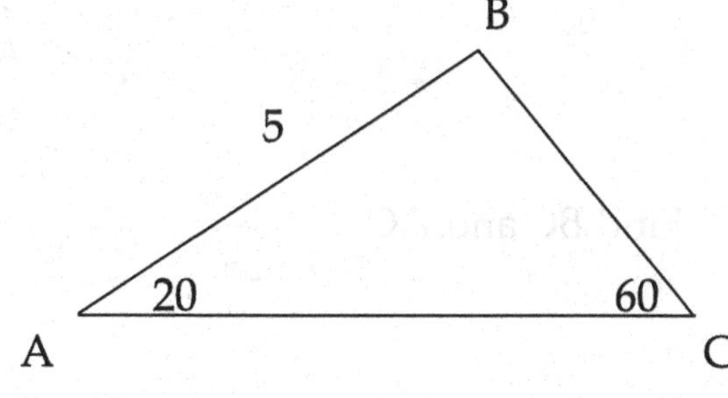

Solution

To find BC: $BC/5 = \text{pin}\,20$

$BC = 5\,\text{pin}\,20$

$= 5 \times 0.3949$ (from tables)

$= 1.9745$

$BC = 1.97$ (3s.f.)

To find AC:

$$AC/5 = \text{pos } 20$$

$$AC = 5 \text{ pos } 20$$

$$= 5 \times 1.137 \qquad \text{(from tables)}$$

$$= 5.685$$

$$AC = 5.69 \ \ (3\text{s.f.})$$

Solution

AB=5, BC=1.97, CA=5.69, $\angle A=20^{\circ}$, $\angle B=100^{\circ}$, $\angle C=60^{\circ}$

Exercise 5

Use the pin, pos, pan tables to solve the following triangles

[1]

Find AC and BC

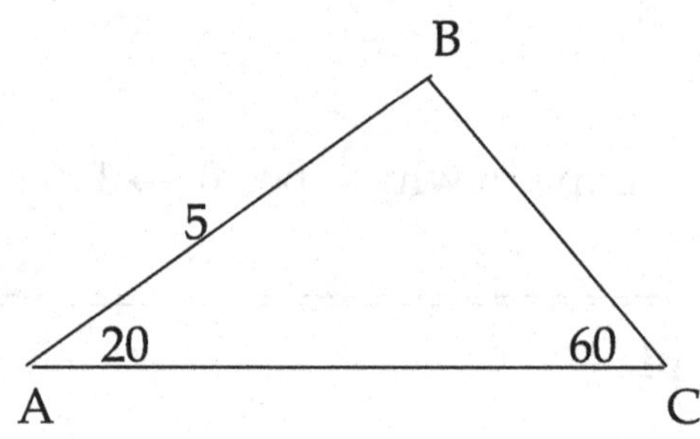

[2]

Find BC and AC

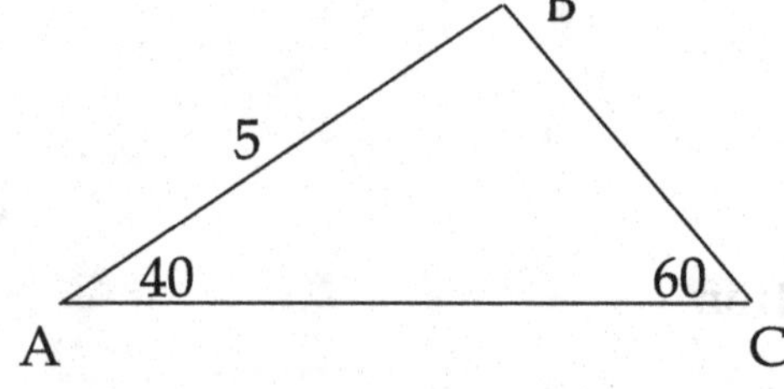

[3] Prove that

$$\text{pin } \theta = \frac{\sin \theta}{\sin 60}$$

[4] Prove that

$$\text{pos } \theta = \cos \theta + \sin \theta \,.\, \tan 30$$

$$= \cos \theta + \text{pin } \theta \,.\, \sin 30$$

[5] Prove that

$$\text{pin } \theta = \frac{\sin \theta}{\cos 30}$$

[6] Prove that

$$\text{pan}\ \theta = \frac{2\tan\theta}{\sqrt{3} + \tan\theta}$$

End of the digression into pin, pos and pan.

==============================

Angles Greater than 90

We cannot have an angle greater than 90° in a right angled triangle, but if you ask your calculator to display sin 150° it will tell you that

$$\sin 150° = 0.5$$

it will also tell you for example that cos 3000 = -0.5, tan –400 = -0.839099631 and that sin 600030 = -1

The definitions of the trig ratios can be extended so that they can deal with angles of any size, positive or negative and yet include the SOHCAHTOA definitions we already have for the 90° triangle.

The trick is to use coordinates on a unit circle:

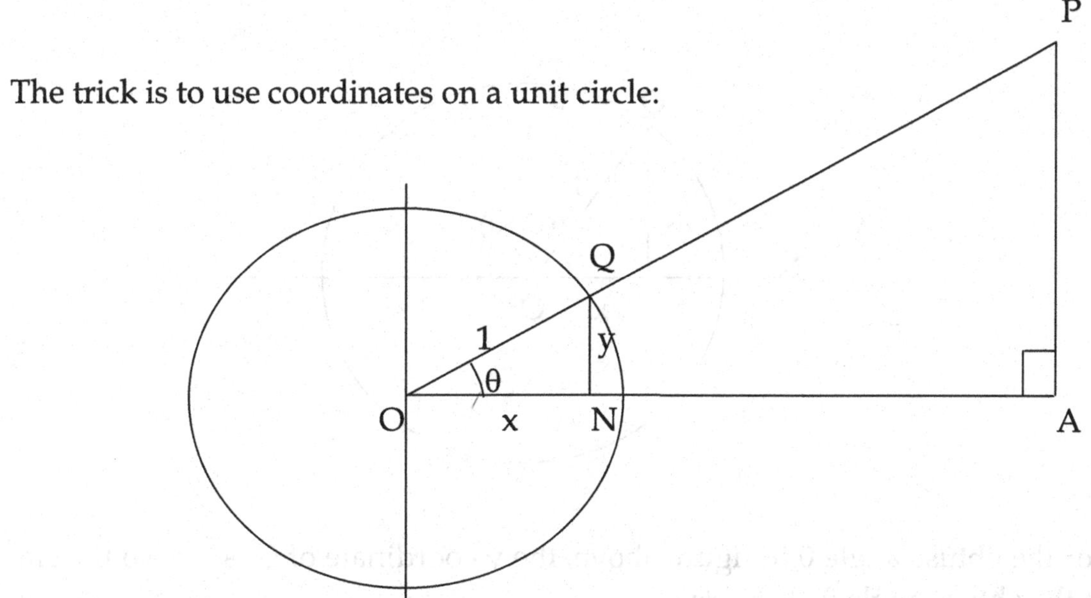

The circle, centre O has radius OQ=1.
OA is some initial line from where we start measuring the angle θ=∠AOQ.

General Definitions:

We define: $\sin\theta$ = the y coordinate of Q

$\cos\theta$ = the x coordinate of Q

$\tan\theta$ = the gradient of OQ

so that $\sin\theta = y$, $\cos\theta = x$ and $\tan\theta = y/x$.

Since triangle POA is an enlargement of triangle QON, we have:

PA = scale factor x QN and OP = scale factor x OQ so PA/OP=QN/OQ

$$\sin\theta = O/H = PA/OP = QN/OQ = y/1 = y$$

$$\cos\theta = A/H = OA/OP = ON/OQ = x/1 = x$$

$$\tan\theta = O/A = PA/OA = QN/ON = y/x$$

Our right-angled triangles can be "solved" as before but we now have trig ratios of angles of any size to play with.

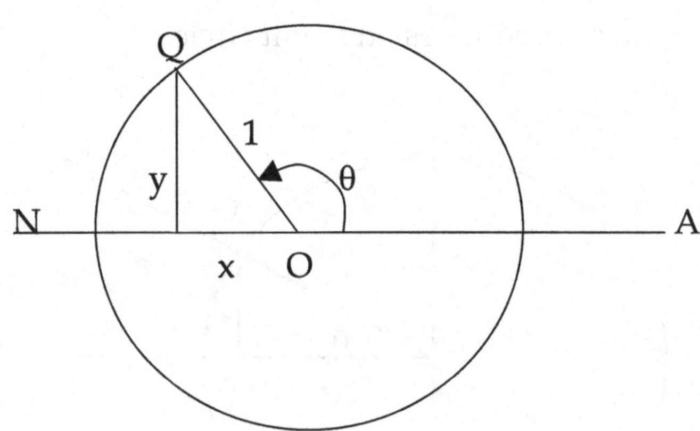

For the obtuse angle θ in figure above, the y coordinate of Q is +ve so the sin of the obtuse angle θ is also +ve.

The x coordinate of Q is negative so the cosine of the obtuse angle θ is negative.

Notes:

[1] By Pythagoras $x^2 + y^2 = 1$ for any angle θ

hence $\mathbf{\cos^2\theta + \sin^2\theta = 1}$ for any angle θ

[2] $\tan\theta = y/x$ (definition)

hence $\mathbf{\tan\theta = \sin\theta / \cos\theta}$ for any angle θ

[3] $$\mathbf{\sec\theta = 1/\cos\theta}$$

$$\mathbf{\operatorname{cosec}\theta = 1/\sin\theta}$$

$$\mathbf{\cot\theta = 1/\tan\theta}$$

[4] $$\sin^2\theta + \cos^2\theta = 1$$

gives

$$\frac{\sin^2\theta}{\cos^2\theta} + \frac{\cos^2\theta}{\cos^2\theta} = \frac{1}{\cos^2\theta}$$

leading to

$$\mathbf{\sec^2\theta = 1 + \tan^2\theta}$$

also

$$\frac{\sin^2\theta}{\sin^2\theta} + \frac{\cos^2\theta}{\sin^2\theta} = \frac{1}{\sin^2\theta}$$

leading to

$$\mathbf{\operatorname{cosec}^2\theta = 1 + \cot^2\theta}$$

Special Angles

Useful relations can be found between the trig ratios for angles outside the range $0<\theta<90$ by considering various transformations (reflections and rotations). Many text books "prove" these relationships using particular diagrams but it must be appreciated that the relations are true for angles of any size for example millions of degrees, positive or negative. Using transformations allows us to prove the relations in a more general way.

The first of these is the relationship between the trig ratios of the angles θ and -θ (or 360-θ).

Transformation 1: A reflection in the x-axis.

If Q(x,y) is any point on the unit circle with angle θ, then, by definition,

$$\sin\theta = y,\quad \cos\theta = x, \text{ and } \tan\theta = y/x.$$

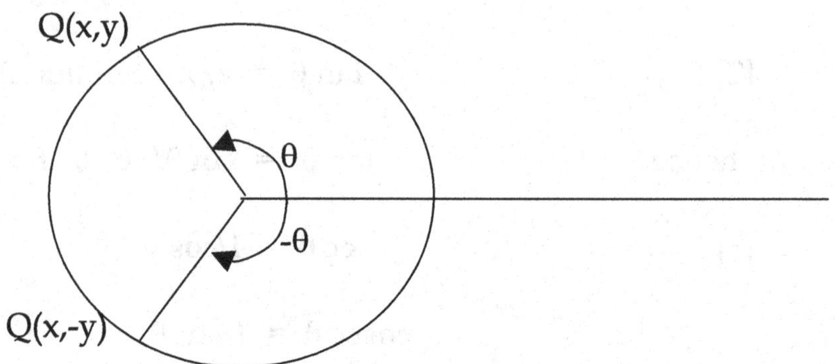

If Q(x,y) on the unit circle has angle θ, then its refection in the x axis will be Q′(x,-y) with angle -θ (or 360-θ).
We therefore have

sin(-θ) = y coordinate of Q′

= -y

= -(y coordinate of Q)

= -sin θ

cos(-θ) = x coordinate of Q′

= x

= cos θ

tan(-θ) = (y coordinate ofQ′)/(x coordinate of Q′)

= (-y)/(x) = -(y/x) = -tan θ

alternatively, we could write

$$\tan(-\theta) = \sin(-\theta)/\cos(-\theta) = (-\sin\theta)/(\cos\theta) = -\tan\theta$$

Summary
Thus we have the following relations, for any angle θ,

$$\sin(-\theta) = -\sin\theta$$

$$\cos(-\theta) = \cos\theta$$

$$\tan(-\theta) = -\tan\theta$$

Transformation 2: Reflection in the y-axis

The next transformation to consider is a reflection in the Oy axis. This transformation relates the angles θ and 180-θ.

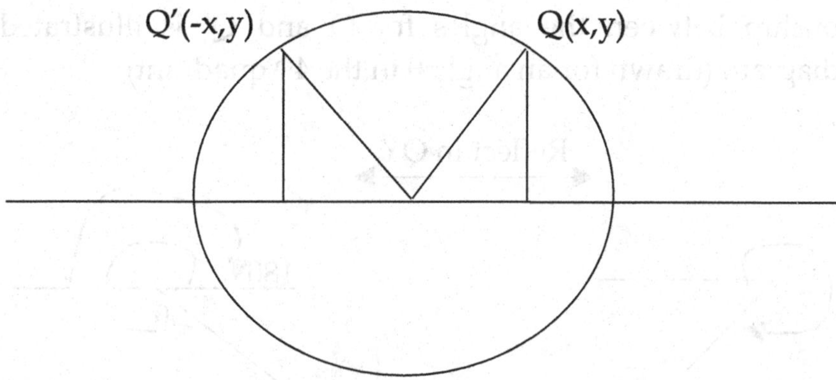

The following diagrams are simply to confirm that the transformation does give two points with angles θ and 180-θ

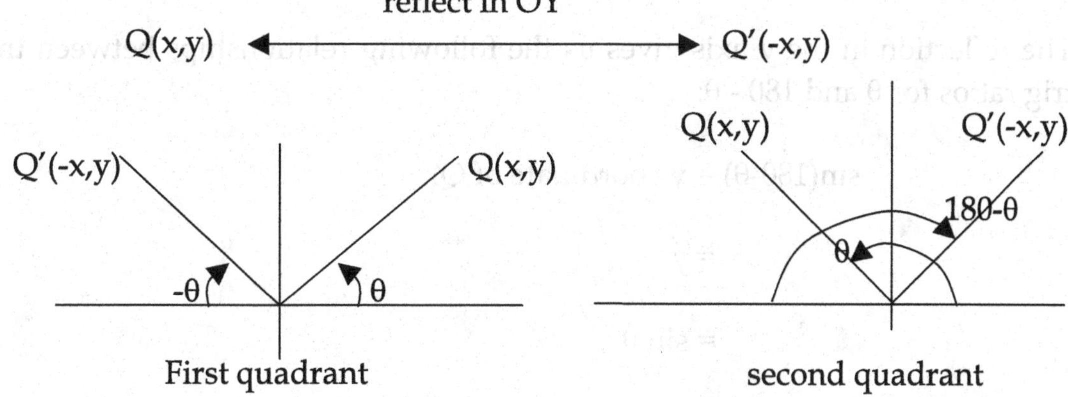

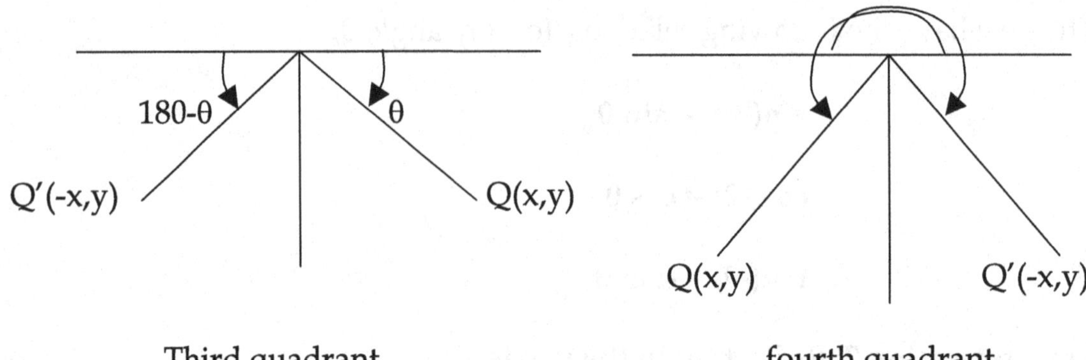

Third quadrant fourth quadrant

Whichever quadrant Q(x,y) lies in, its refection in the y axis will be Q′(-x,y). Also, if the angle for Q is θ, of any size whatever, then the angle for Q′ will be (180- θ).
The relationship between the angles for Q and Q′ is illustrated in the following diagram (drawn for an angle θ in the 4th quadrant).

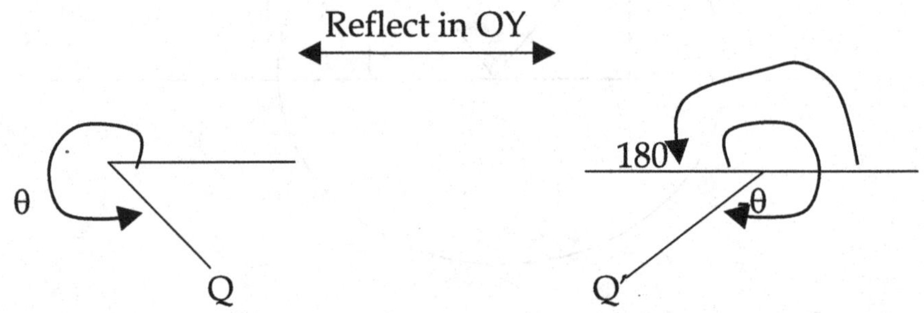

The angle for Q is θ and the angle for Q′ is 180 - θ

The reflection in the y axis gives us the following relationships between the trig ratios for θ and 180 - θ:

sin(180-θ) = y coordinate of Q′

= y

= sin θ

cos(180-θ) = x coordinate of Q′

$= -x$

$= -\cos\theta$

$\tan(180-\theta) = (\text{y coordinate of q}')/(\text{x coordinate of Q}')$

$= (y)/(-x)$

$= -\tan\theta$

Summary:

$$\sin(180-\theta) = \sin\theta$$
$$\cos(180-\theta) = -\cos\theta$$
$$\tan(180-\theta) = -\tan\theta$$

==================================

Transformation 3: Rotate 90°

A rotation about O through an angle θ gives relationships between the trig ratios for angles θ and (90 + θ).

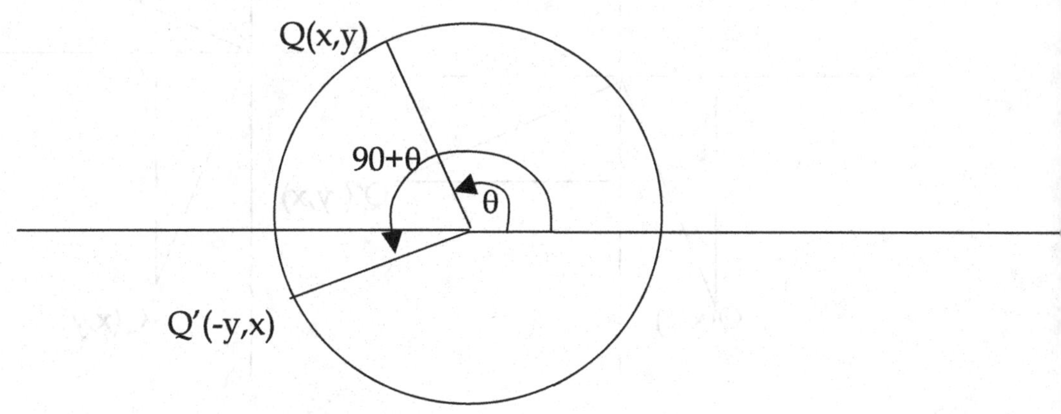

Rotating through 90° interchanges the two coordinates. The old x coordinate becomes the new y coordinate and the old y coordinate becomes the new x coordinate but points in the wrong direction.

The mapping $Q(x,y) \xrightarrow{\text{Rotate } 90^\circ} Q'(-y,x)$

is illustrated in the following diagrams in which the values of x and y are positive or negative according to which quadrant Q lies in:

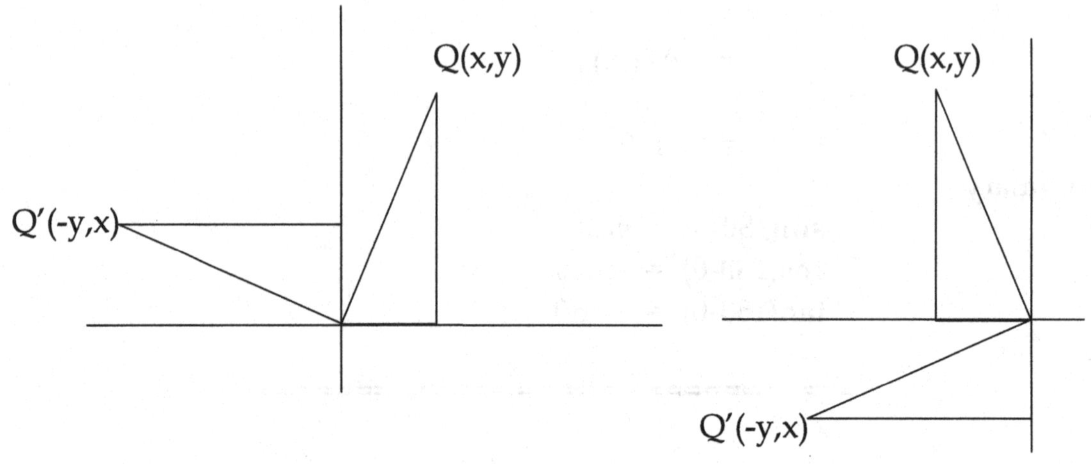

First Quadrant   Second Quadrant

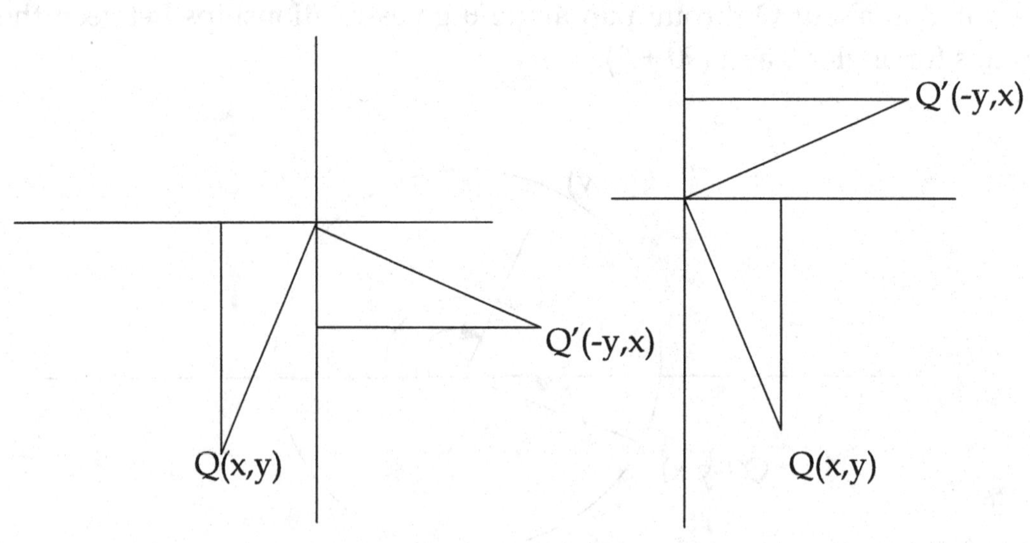

Third Quadrant   Fourth Quadrant

If the angle for Q is θ then the angle for Q′ is (90+θ) hence we have;

$$\sin(90+\theta) \;=\; y \text{ coordinate of } Q'$$

$= x \quad = x \text{ coordinate of } Q \quad = \cos\theta$

$\cos(90+\theta) = x \text{ coordinate of } Q'$

$= -y = -(y \text{ coordinate of } Q) = \sin\theta$

$\tan(90+\theta) = (y \text{ coordinate of } Q')/(x \text{ coordinate of } Q')$

$= (x)/(-y)$

$= -(x/y) = \cot\theta$

Summary:

$$\sin(90+\theta) = \cos\theta$$
$$\cos(90+\theta) = -\sin\theta$$
$$\tan(90+\theta) = -\cot\theta$$

==================================

Transformation 4: Rotate 180°

A rotation of 180° gives the relationships between the trig ratios for angle θ and (180-θ).
The diagrams for each quadrant are as follows:

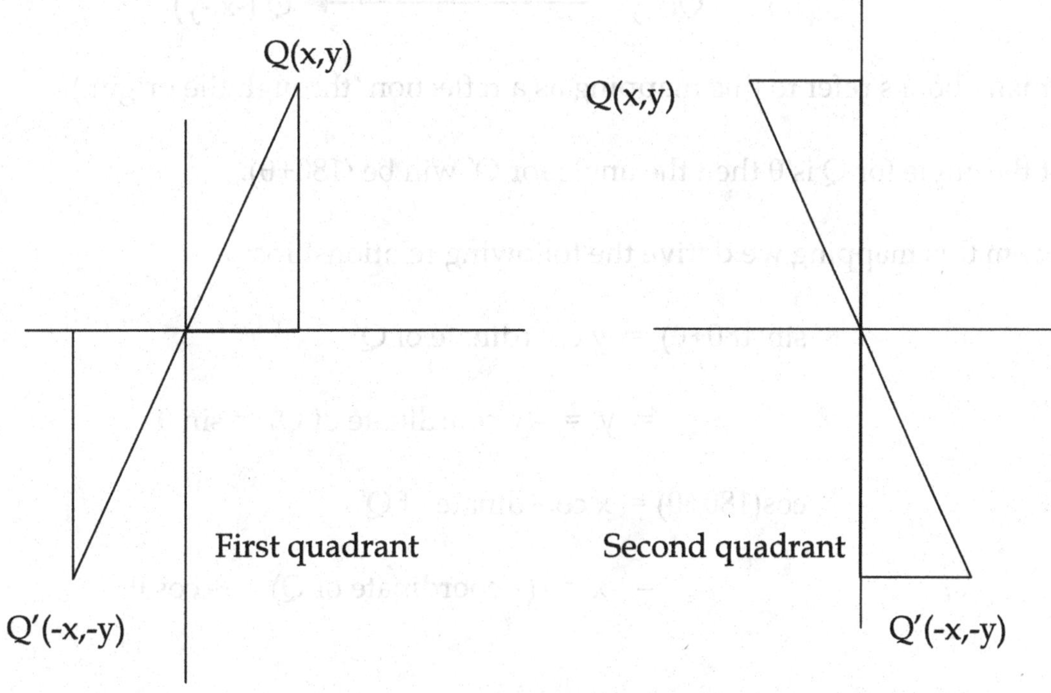

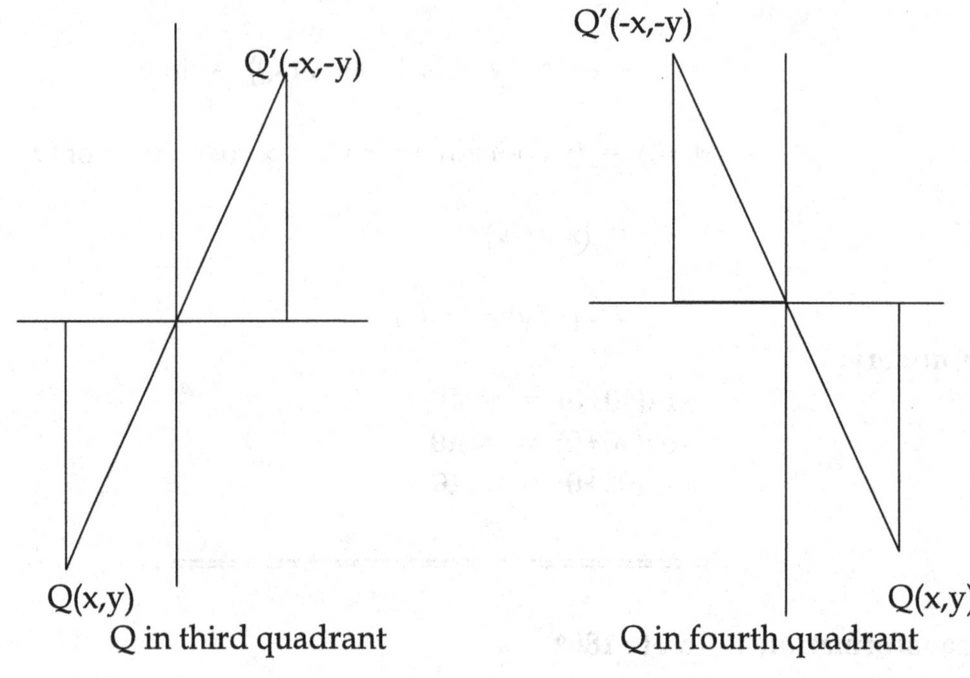

Q in third quadrant

Q in fourth quadrant

We can express these results in the form of the mapping

$$Q(x,y) \xrightarrow{\text{Rotate } 180^{\circ}} Q'(-x,-y)$$

(many books refer to this mapping as a reflection 'through the origin').

If the angle for Q is θ then the angle for Q′ will be (180+θ).

From this mapping we derive the following relationships:

$$\sin(180+\theta) = \text{y coordinate of } Q'$$

$$= -y = -(\text{y coordinate of Q}) = \sin\theta$$

$$\cos(180+\theta) = \text{x coordinate of } Q'$$

$$= -x = -(\text{x coordinate of Q}) = -\cos\theta$$

$\tan(180+\theta)$ = (y coordinate of Q′)/(x coordinate of Q′)

= (-y)/(-x)

= y/x = (y coordinate of Q)/(x coordinate of Q)

= $\tan\theta$

Summary:

$$\sin(180+\theta) = -\sin\theta$$
$$\cos(180+\theta) = -\cos\theta$$
$$\tan(180+\theta) = \tan\theta$$

Transformation 5: Rotate 270°

This transformation gives the relationship between the trig ratios for the angles θ and 270+θ.

Exercise 6

Draw the diagrams for the four quadrants showing Q(x,y) with angle θ and the rotation of OQ by 270° to OQ′ with angle (270+θ).

Verify that the mapping for the rotation through 270° can be expressed as

$$Q(x,y) \xrightarrow{\text{Rotate } 270^\circ} Q'(y,-x)$$

Deduce the following relationships:

$$\sin(270+\theta) = -\cos\theta$$
$$\cos(270+\theta) = \sin\theta$$
$$\tan(270+\theta) = -\cot\theta$$

The final set of relationships between the trig ratios of these special angles are produce by

(i) reflection in the line y=x

and (ii) reflection in the line y= -x

Transformation 6: Reflect in the line y=x

Reflection in the line y=x gives relationships between the trig ratios for angle θ and angle (90-θ)

Exercise 7
Draw diagrams for each of the four quadrants to show that the mapping for a reflection in the line y=x is given by

$$Q(x,y) \xrightarrow{\text{Reflect in } y=x} Q'(y,x)$$

The relationship between the angles for Q and Q′ is illustrated in the following diagram draw for an angle θ in the fourth quadrant:

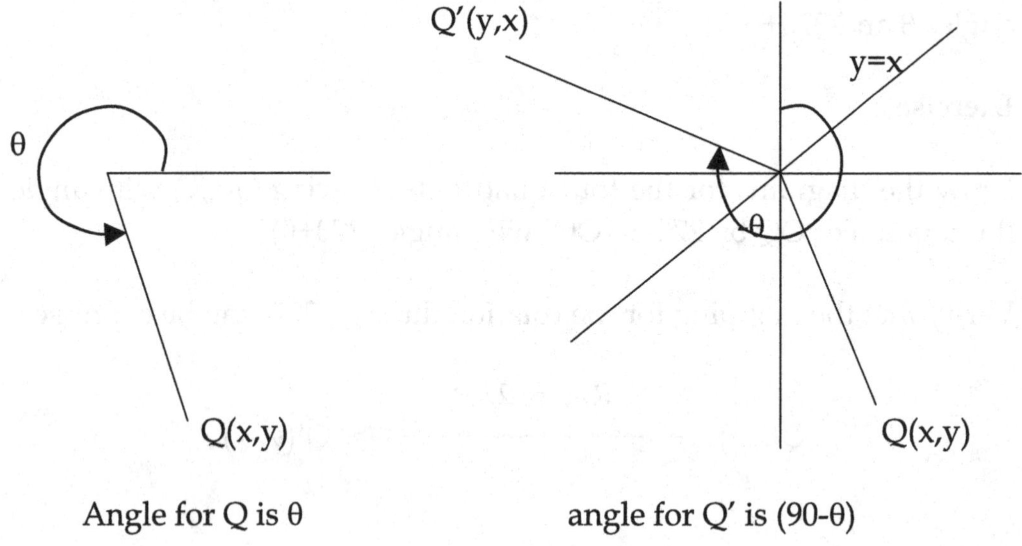

Angle for Q is θ

angle for Q′ is (90-θ)

Thus

sin(90-θ) = y coordinate of Q′

= x = x coordinate of Q = cos θ

cos(90-θ) = x coordinate of Q′ = y = y coordinate of Q = sin θ

$\tan(90\text{-}\theta) = \sin(90\text{-}\theta)/\cos(90\text{-}\theta) = \cos\theta/\sin\theta = \cot\theta$

Summary

$$\sin(90\text{-}\theta) = \cos\theta$$
$$\cos(90\text{-}\theta) = \sin\theta$$
$$\tan(90\text{-}\theta) = \cot\theta$$

Transformation 7: reflect in the line y=-x

Reflection in the line y=-x gives relationships between the trig ratios for angle θ and angle (270-θ)

Exercise 8

Draw diagrams for each of the four quadrants to show that the mapping for a reflection in the line y = -x is given by

$$Q(x,y) \xrightarrow{\text{Reflect in } y = -x} Q'(-y,-x)$$

The relationship between the angles for Q and Q′ is illustrated in the following diagram which is again drawn for an angle θ in the fourth quadrant.

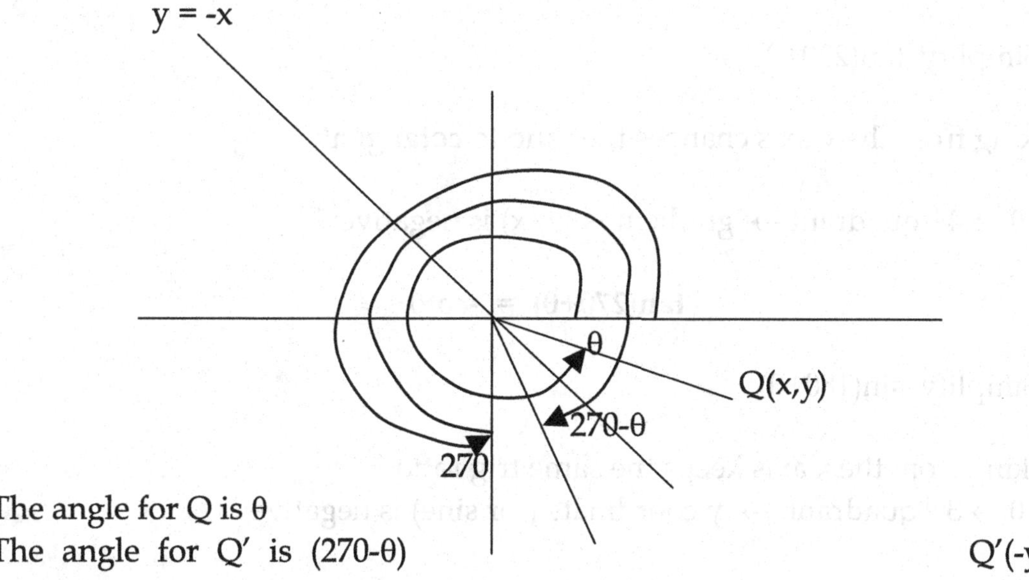

The angle for Q is θ
The angle for Q′ is (270-θ)

Thus $\sin(270\text{-}\theta)$ = y coordinate of Q′ = -x = $-\cos\theta$

$\cos(270-\theta) = x$ coordinate of $Q' = -y = -\sin\theta$

$\tan(270-\theta) = \sin(270-\theta)/\cos(270-\theta) = (-\cos\theta)/(-\sin\theta) = \cot\theta$

Summary

$$\sin(270-\theta) = -\cos\theta$$
$$\cos(270-\theta) = -\sin\theta$$
$$\tan(270-\theta) = \cot\theta$$

These relationships between can be combined into the following rules:

Working from the x axis:

The trig ratio remains the same
Get the sign from the coordinates in the acute angle case

Working from the y axis:

The ratio changes to the co-ratio
Get the sign from the coordinates in the acute angle case

Examples:

[1] Simplify $\tan(270+\theta)$

working from the y axis changes the ratio to **cotangent**

$270+\theta \rightarrow 4^{th}$ quadrant $\rightarrow$ gradient is (y/x) is negative:

$$\tan(270+\theta) = -\cot\theta$$

[2] Simplify $\sin(180+\theta)$

Working from the x axis keeps the same trig ratio
$180+\theta \rightarrow 3^{rd}$ quadrant $\rightarrow$ y coordinate (for sine) is negative:

$$\sin(180+\theta) = -\sin\theta$$

Exercise 9 Simplify

1. $\sin(90+\theta)$ 2. $\cos(90+\theta)$ 3. $\tan(90-\theta)$ 4. $\sin(180+\theta)$

5. $\cos(180-\theta)$ 6. $\tan(180+\theta)$ 7. $\sin(270-\theta)$ 8. $\cos(270+\theta)$

9. $\tan(270-\theta)$ 10. $\sin(270+\theta)$

Standard Trig Notation

Let ABC denote any triangle, then the standard labels for the various side and angles are:

$\angle ABC = B$, $\angle BCA = C$, $\angle CAB = A$ $BC = a$, $CA = b$, $AB = c$

The area of the triangle is denoted by Δ.
The radius of the circle that can be drawn through A, B and C is R.
The radius of the circle that can be drawn to touch the three sides is r.

The Area of Triangle ABC

Let the height of the triangle from B to AC be h, then we have the following possible diagrams:

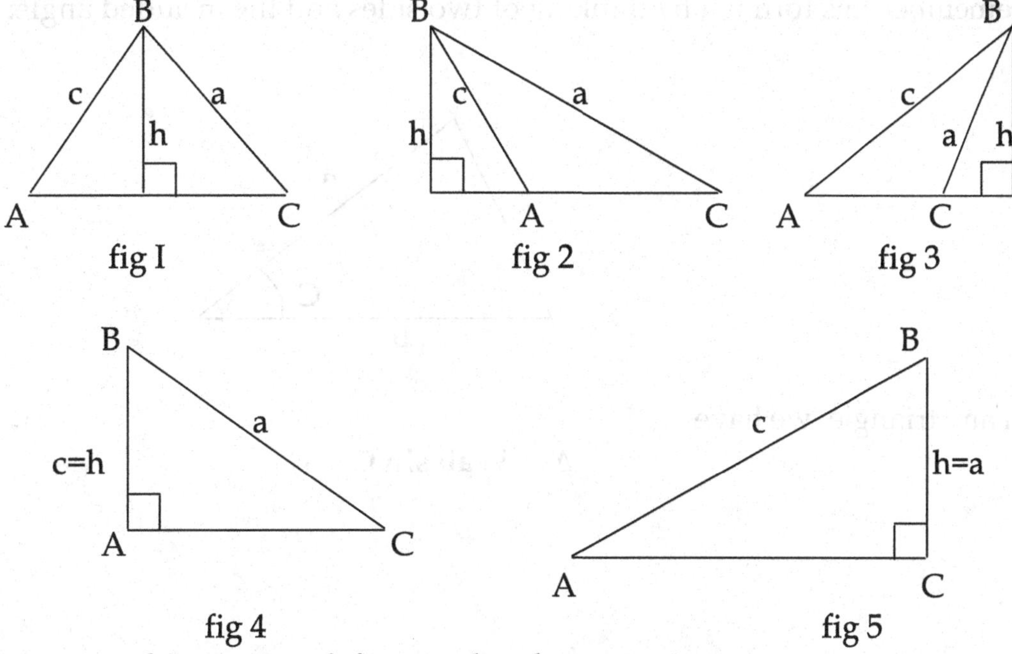

fig I fig 2 fig 3

fig 4 fig 5

In figures 1 and 2, $\sin C = h/a$ so that $h = a \sin C$

In figure 3 $\sin(180-C) = h/a$ but we know that $\sin(180-C)=\sin C$

Therefore, in figure 3, we have again, $h = a \sin C$

In figure 4 $\sin C = h/a$ hence $h = a \sin C$

In figure 5 $\sin C = 1$ and $h/a = 1$ because $h=a$

therefore, in figure 5, we have $h= a \sin C$

In all cases, the height of the triangle drawn from B to AC is $h=a \sin C$

The area of the triangle is half base times height so, in each of these diagrams, the area of the triangle is given by

$$\Delta = \tfrac{1}{2} AC. h$$

$$= \tfrac{1}{2} b.a \sin C$$

$$= \tfrac{1}{2}ab \sin C$$

Remember this formula by thinking of two sides and the included angle:

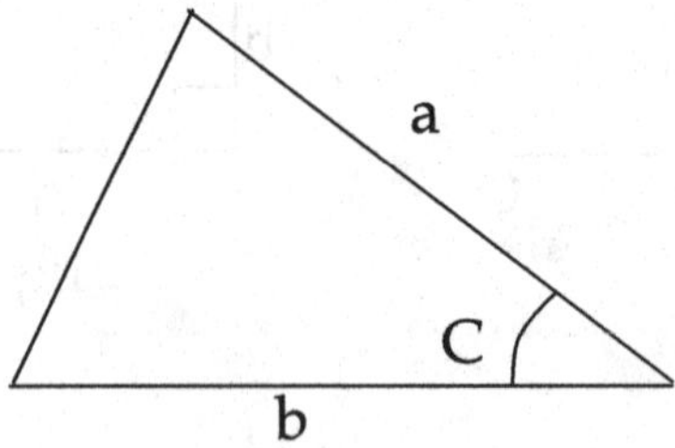

In any triangle, we have

$$\mathbf{\Delta = \tfrac{1}{2} ab \sin C}$$

The Sine Rule

By considering the perpendicular from C to BA we would get the formula
$\Delta = \frac{1}{2}bc \sin A$
and by considering the perpendicular from A to CB we would get the formula $\Delta = \frac{1}{2}ca \sin B$
Notice that we can derive these other formulae for Δ just by cycling the letters round: A→ B→ C and a→b→c. This is frequently a useful dodge!
Thus we have:

$$\Delta = \tfrac{1}{2}ab \sin C = \tfrac{1}{2}bc \sin A = \tfrac{1}{2}ca \sin B$$

$$ab \sin C = bc \sin A = ca \sin B$$

divide these equations by abc to get

$$\frac{\sin C}{a} = \frac{\sin A}{b} = \frac{\sin B}{c}$$

Turn the whole thing upside down
(this is OK because if p/q = r/s then q/p = s/r) and we get

$$\frac{a}{\sin A} = \frac{b}{\sin B} = \frac{c}{\sin C}$$

This is the **sine rule**, and is true for any triangle ABC.

================================

An alternative proof of the sine rule uses the circumcircle of the triangle ABC

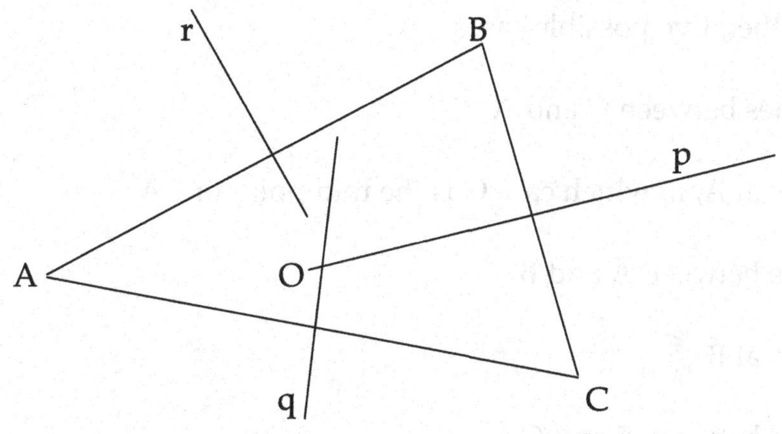

The perpendicular bisector of BC is labeled p
The perpendicular bisector of CA is labeled q
The perpendicular bisector of AB is labeled r
Each point on p is equidistant from B and C
Each point on q is equidistant from C and A
Let p and q meet at O, then OB = OC and OC = OA

Therefore OB = OA so that O is equidistant from A and B. Therefore O lies on the perpendicular bisector of AB which is the line r. The three lines p, q and r are therefore concurrent, meeting at the point O.
Also, OA = OB = OC so that O is the centre of a circle that may be drawn through the three points A, B and C.

O is called the circumcentre of triangle ABC.
The radius of this circle is OA = OB = OC = R.

================================

A Proof of the Sine Rule using R

Let ABC be any triangle.

The order A→B→C is taken as clockwise in this proof. If you drew the diagram with A→B→C anticlockwise, then turn the paper over and anticlockwise becomes clockwise!

Draw the circumcircle of ΔABC let the centre be O
Draw the diameter CON and join BN.

There are then five possible cases:

[1] N lies between C and A

[2] N is at A, in which case O is the mid point of CA

[3] N is between A and B

[4] N is at B

[5] N is between B and C

The five possible diagrams are shown in figures 1,2,3,4 and 5 below:

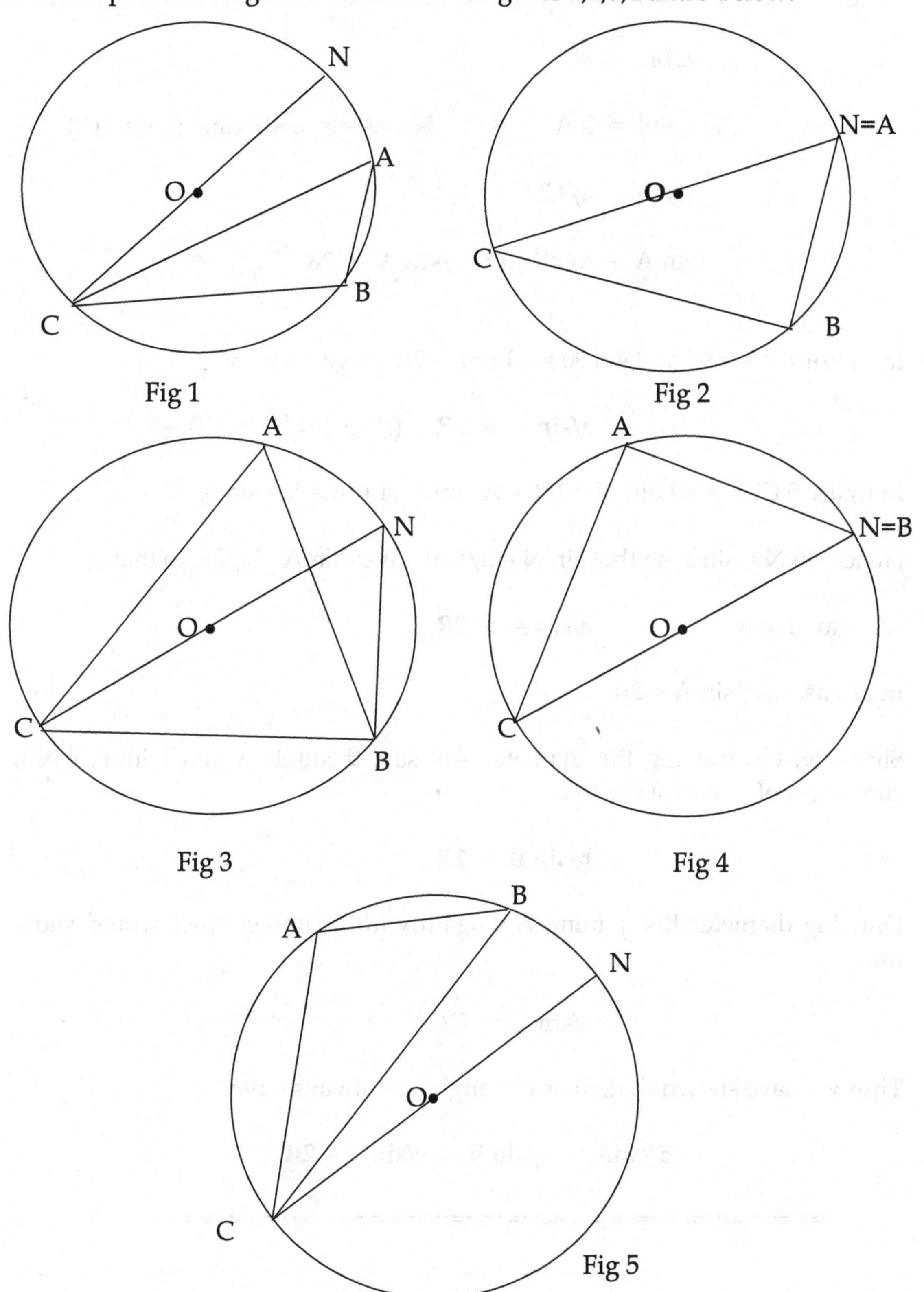

Fig 1

Fig 2

Fig 3

Fig 4

Fig 5

In figures 1,2 and 3 we have

$$\angle CBN = 90$$

$$\angle N = \angle A \qquad \text{(same segment..same point in 2)}$$

$$\sin N = a/CN$$

$$\therefore \sin A = a/2R \quad \text{or} \quad \mathbf{a/sin\ A = 2R}$$

In figure 4 $\angle A = 90$ and sin 90 = 1 but a = 2R so we do have

a/sin A = 2R (since 2R/1 = 2R)

In figure 5 CBN = 90 but N = 180 – A and sin 180-A = sin A

Hence sin N = sin A so that sin N = a/CN gives sin A = a/2R so that

we again have **a/sin A = 2R**

In all cases, a/sin A = 2R

Similarly, by drawing the diameter AN say, through A and joining CN a similar proof would show that

b/sin B = 2R

Drawing diameter BN, joining AN again with a similar proof would show that

c/sin C = 2R

Thus we have shown that, in any triangle, we always have

a/sinA = b/sin B = c/sin C = 2R

==

Examples

[1] Find x

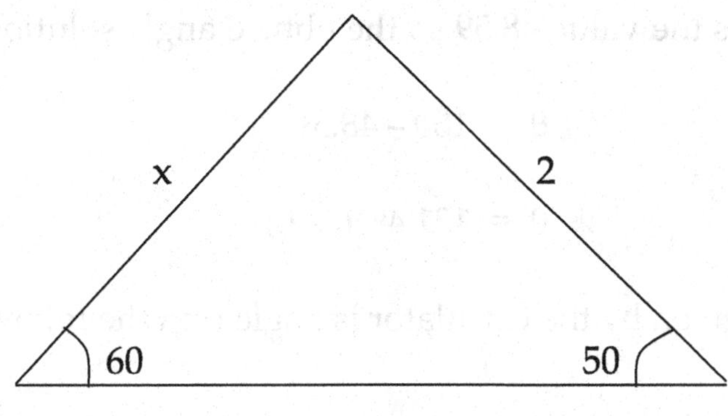

Solution

$$\frac{x}{\sin 50} = \frac{2}{\sin 60}$$

$$x = 2 \sin 50 \,/\, (\sin 60)$$

2	sin	5	0	÷			sin	=

$$x = 1.769 \quad (4 \text{ s.f.})$$

================================

[2] Find θ

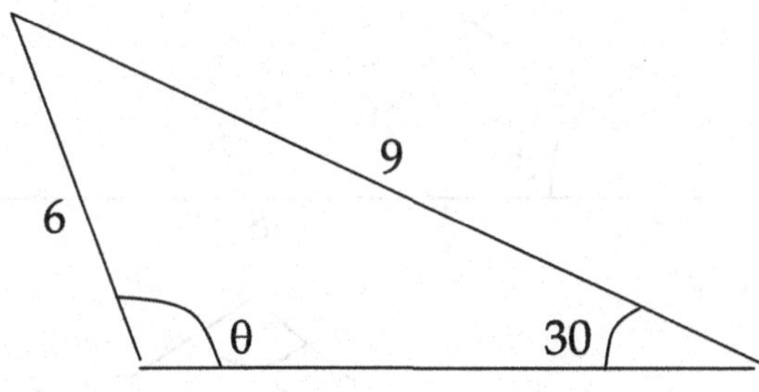

Solution

$$\frac{9}{\sin \theta} = \frac{6}{\sin 30}$$

$$\sin \theta = 9 \sin 30 / 6$$

$$\theta = \sin^{-1}(9 \sin 30 \,/\, 6)$$

shift	sin	(	9	X	sin	3		÷	6	)	=

the calculator gives the value 48.59 so the obtuse angle solution is 180 - 48.59

$$\theta = 180 - 48.59$$

$$\theta = 131.4^\circ \text{ (4s.f.)}$$

Note: The value given by the calculator is angle α in the following figure:

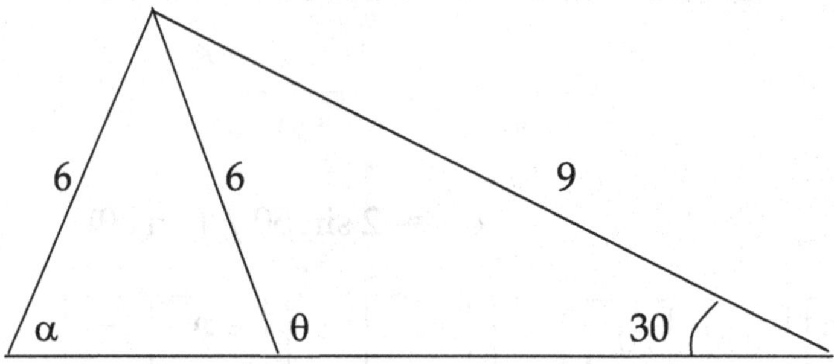

Exercise 10 (give answers to 4 significant figures)

1. Find x

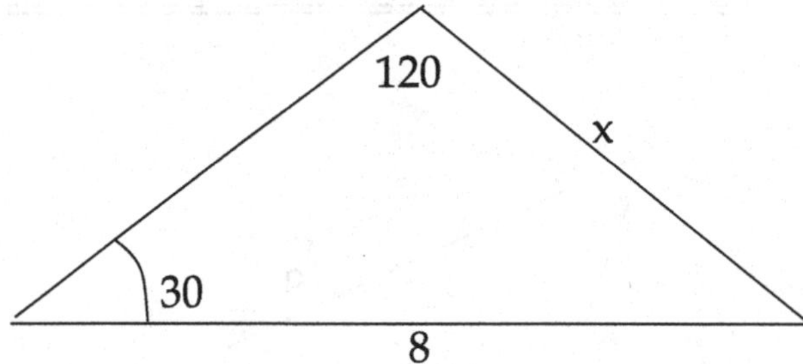

2. Find θ

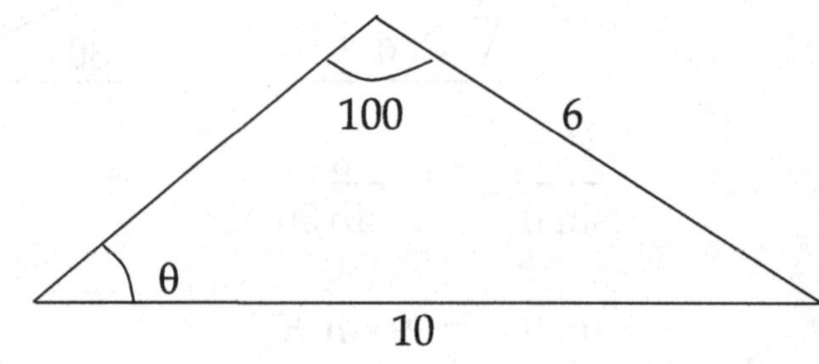

3. Find θ

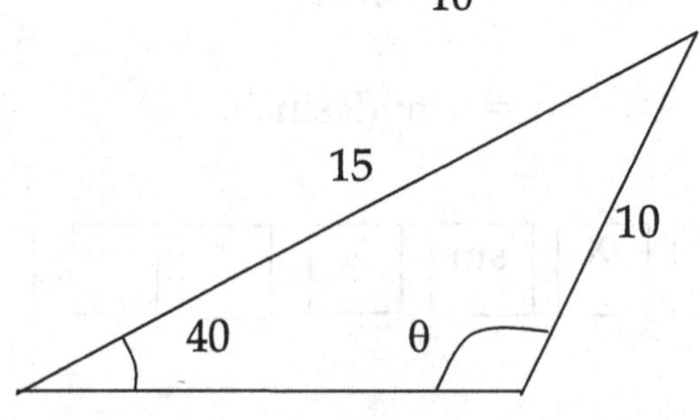

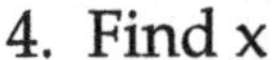

4. Find x

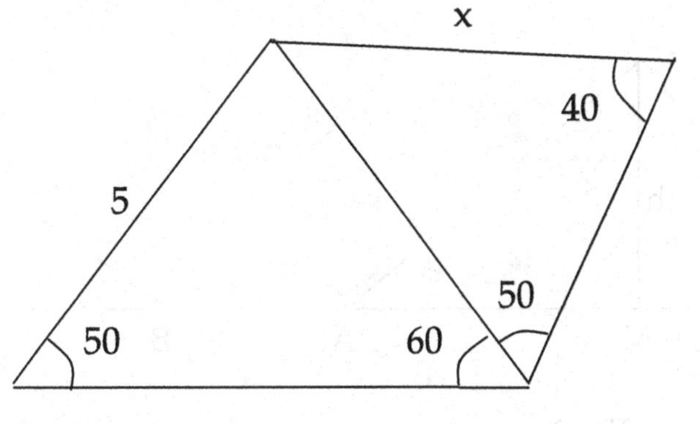

==

The Cosine Rule

In triangle ABC, let h be the length of the perpendicular CN, from C to AB. Let AN = x

(we assume that the triangle ABC is labelled in a clockwise direction: if you label the triangle ABC in the anti clockwise direction, turning over the paper will change anticlockwise to clockwise).

We then have the following possible cases:

[1] N is to the left of B
[2] N is at B
[3] N is between A and B
[4] N is at A
[5] N is to the right of A

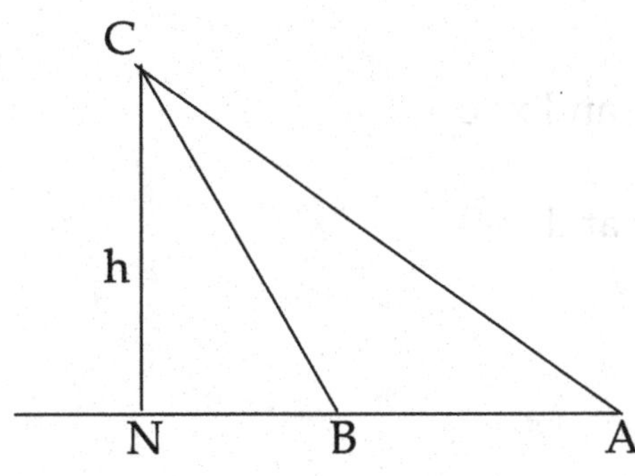

Fig 1.

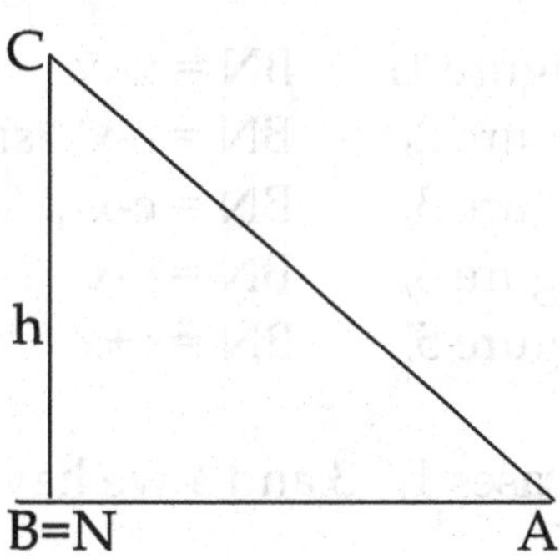

Fig 2.

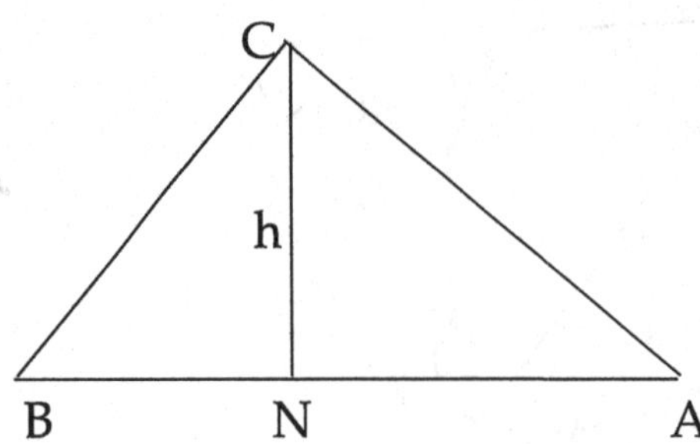

Fig 3.

Fig 4.

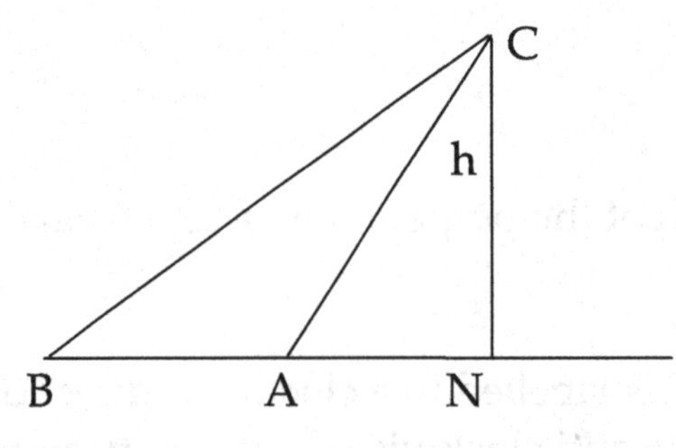

Fig 5.

In each diagram, Let AN=x.
In figure 2 B=N so that AN=AB.
In figure 4 A=N so that AN=0.

By Pythagoras, $AC^2 = h^2 + AN^2$ $\therefore$ $b^2 = h^2 + x^2$

Also we have $a^2 = h^2 + BN^2$ (in fig 2 b=h and BN=0)

Now in figure 1, BN = x-c,
in figure 2, BN = c-x since BN=0 and x=c
in figure 3, BN = c-x
in figure 4, BN = c-x since BN=c and x=0
in figure 5, BN = c+x

Thus in cases 1,2,3 and 4 we have

$$a^2 = h^2 + (c-x)^2$$

$$= b^2 - x^2 + c^2 - 2cx + x^2$$

$$= b^2 + c^2 - 2cx$$

Now in each of figures 1,2,3 we have cos A = x/b,

In fig. 4 where cos A = cos 90 and x=0 we have cos A = x/b = 0

so that x = b cos A in cases 1,2,3 and 4

Therefore, in figures 1,2,3 and 4 we have

$$a^2 = b^2 + c^2 - 2bc \cos A$$

In figure 5, BN = c+x and cos(180-A) = x/b which gives us

$$a^2 = b^2 + c^2 + 2cx$$

$$= b^2 + c^2 + 2bc \cos(180-A)$$

But cos(180-A) = -cos A hence in figure 5 we again have,

$$a^2 = b^2 + c^2 - 2bc \cos A$$

Thus we have proved that, in any triangle ABC

$$\mathbf{a^2 = b^2 + c^2 - 2bc \cos A}$$

By considering perpendiculars from A to BC and from B to CA, we also have

$$\mathbf{b^2 = c^2 + a^2 - 2ca \cos B}$$

and $$\mathbf{c^2 = a^2 + b^2 - 2ab \cos C}$$

==

Exercise 11

Prove that

$$\mathbf{a^2 + b^2 + c^2 = ab \cos C + bc \cos A + ca \cos B}$$

This alternative proof avoids Pythagoras:

Draw the perpendicular from A to BC

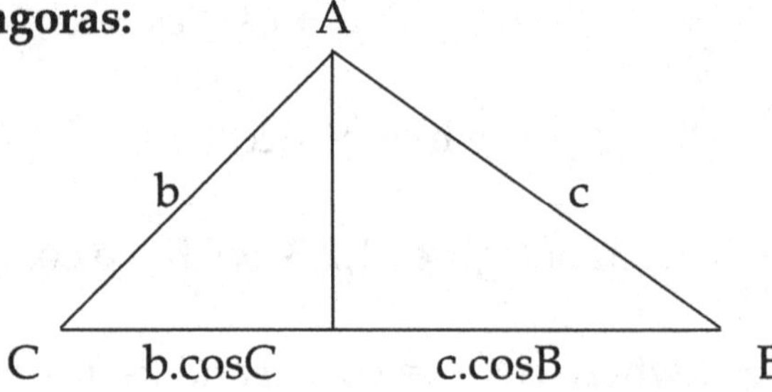

We have $a = b.\cos C + c.\cos B$

and similarly,

$$b = c.\cos A + a.\cos C$$

And

$$c = a.\cos B + b.\cos A$$

Substitute for cosC and cosB from the last two equations into the first and we have;

$$a = b.\frac{(b - c.\cos A)}{a} + c.\frac{(c - b.\cos A)}{a}$$

giving $a^2 = b^2 - bc.\cos A + c^2 - bc.\cos A$

$$\mathbf{a^2 = b^2 + c^2 - 2bc.\cos A}$$

==================================

Exercise 12

Verify the above proof for these other possible figures:

Fig 1

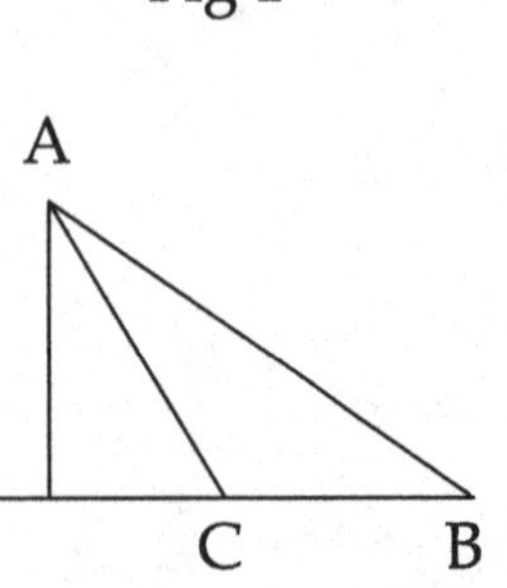

fig 2

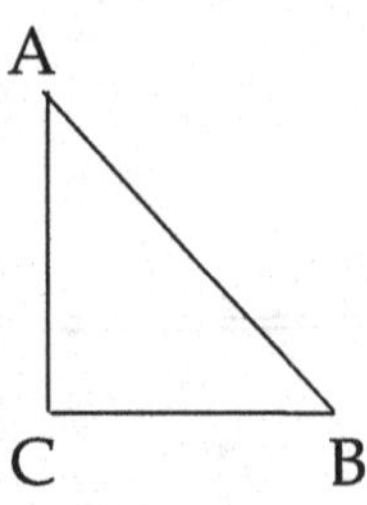

fig 3

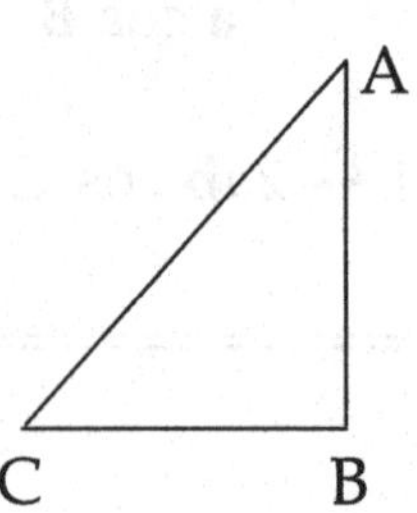

fig 4

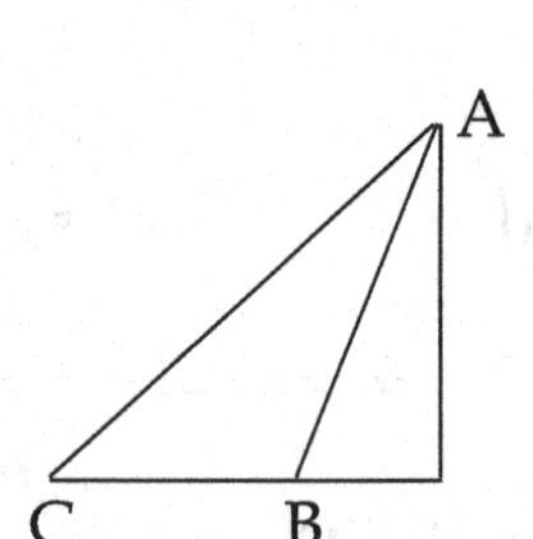

Example

Find x

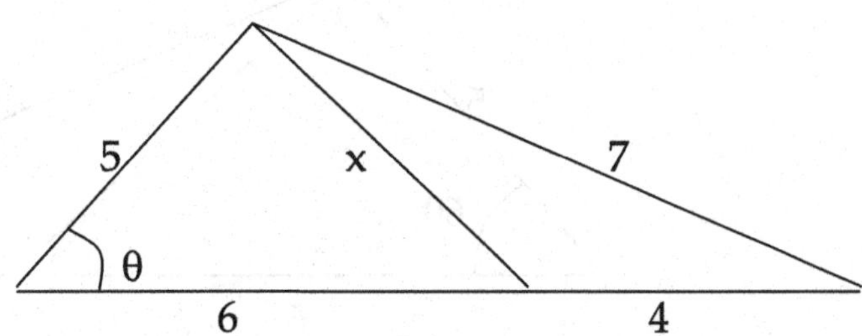

Solution

One application of the cosine rule allows us to find cos θ:

$$7^2 = 5^2 + 10^2 - 2x5x10 \cos \theta$$

$$100 \cos \theta = 25 + 100 - 49$$
$$\cos \theta = 76/100$$

A second application of the cosine rule now allows us to find x:

$$x^2 = 5^2 + 6^2 - 2x5x6 \cos \theta$$

$$= 25 + 36 - 60x76/100$$

$$= 15.4$$

x = 3.924 (4 s.f.)

====================================

Exercise 13

Find x and θ in each of the following diagrams, correct to 4 significant figures.

[1]

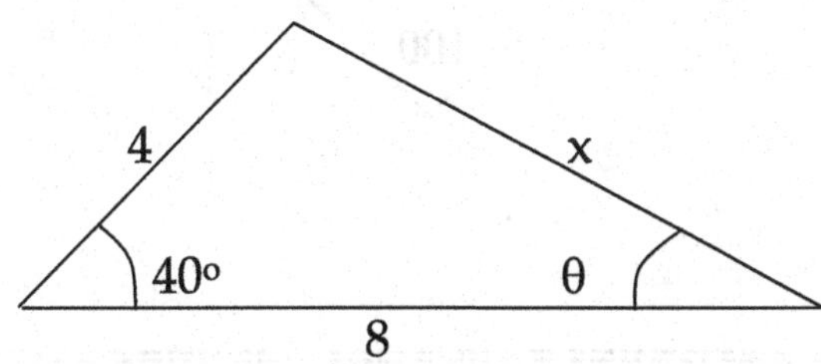

[2]

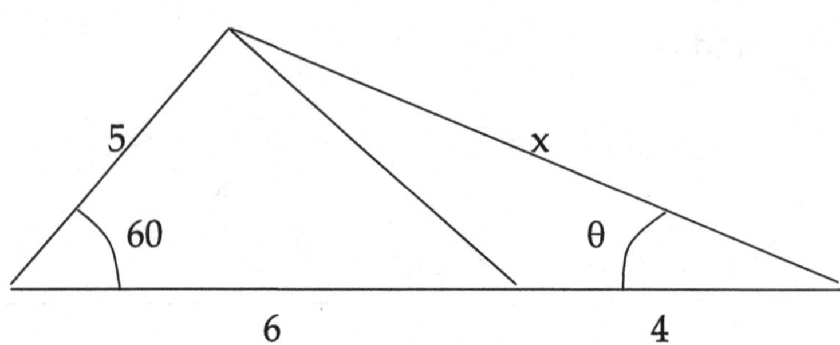

[3]

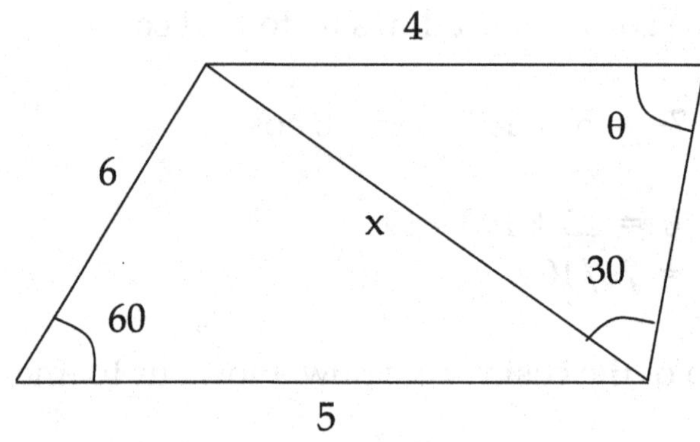

[4]

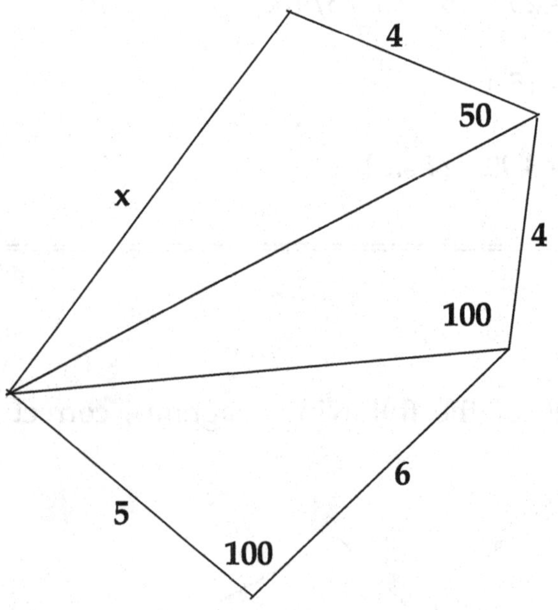

====================================

Answers:

Exercise 1 [1] DB=6, BC=9 [2] PB=12, PQ=25

Exercise 2

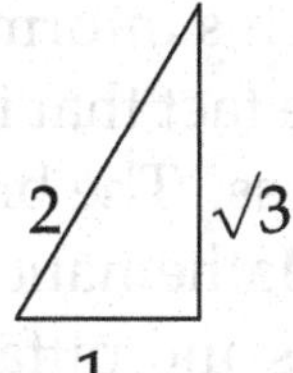

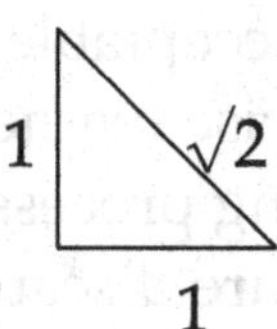

Exercise 3 [1] x=1.478 [2] x=3.155 [3] x=9.649

[4] θ = 42.83 [5] θ = 28.14 [6] x=6

Exercise 5 [1] AC = 5.685, BC = 1.9745 [2] BC = 3.711, AC = 5.685

Exercise 9 [1] cosθ [2] -sinθ [3] cotθ [4] -sinθ

[5] -cosθ [6] tanθ [7] -cosθ [8] sinθ

[9] cotθ [10] -cosθ

Exercise 10 (on the sine rule)

[1] x = 4.619 [2] θ = 36.22° [3] θ = 105.4° [4] x = 5.271

Exercise 11

Add the three cosine rule equations.

Exercise 13 (on the cosine rule)

[1] x = 5.565, θ = 27.52° [2] x = 5√3 = 8.66, θ = 30°

[3] x = 5.568, θ = 44.10° [4] x = 7.997

==================================

CHAPTER 20

Areas, Volumes and Pi

Students of mathematics often use such phrases as "in the limit", "when n goes to zero" or "when n goes to infinity". This informal jargon for "going to a limit" is generally acceptable in spite of the fact that it is really a camouflage for a precise mathematical process. The branch of mathematics that deals with limiting processes is called Mathematical Analysis. If precision and rigour are discarded then traps and pitfalls can easily be found, however, in spite of this we will often use this kind of jargon. Ignorance of the precise definitions of limiting processes did not hinder Newton or Leibnitz in the invention of the calculus (see chapter 26). They showed that progress can be made in mathematical understanding using only an intuitive grasp of limiting processes.

Examples:

[1]

(a) As $n \to 0$, $\frac{n+1}{n+4} \to \frac{1}{4}$

(b) "As n becomes infinitely small, the value of $\frac{n+1}{n+4}$ becomes closer to ¼."

(c) $\lim_{n \to 0} \frac{n+1}{n+4} = \frac{1}{4}$

Formally, (a), (b) and (c) all mean that, given any small number $\varepsilon > 0$, we can find a number δ such that $|n| < \delta \Rightarrow \left| \frac{n+1}{n+4} - \frac{1}{4} \right| < \varepsilon$

In this example, if we would like to make $(n+1)/(n+4) - \frac{1}{4} < \varepsilon$ then we need to choose $n < 16\varepsilon/(3-4\varepsilon)$

[2]

(a) As $n \to \infty$ $\frac{n+1}{2n-4} \to \frac{1}{2}$

(b) As n goes to infinity, $\frac{n+1}{2n-4}$ goes to ½

(c) $\lim_{n \to \infty} \frac{n+1}{2n-4} = ½$

Formally, (a), (b) and (c) would be expressed as

Given any small number $\varepsilon > 0$, we can find a number N such that $n > N$ implies that

$$\left| \frac{n+1}{2n-4} - \frac{1}{2} \right| < \varepsilon$$

In this case, the value of N can be taken as $2 + 1/\varepsilon$

[3] If $n \to 1$ then $\frac{n^3 - n^2}{n^2-n} \to 1$

Although we have a limit of the form 0/0, we can write

$$\lim_{n \to 1} \frac{n^3-n^2}{n^2-n} = \lim_{n \to 1} \frac{n^2(n-1)}{n(n-1)} = \lim_{n \to 1} n = 1$$

We do not need complicated formal proofs to show that

[4] $\lim_{n \to 0} \frac{n^2+n+1}{2n^2+3n+4} = \frac{0+0+1}{0+0+4} = ¼$

[5] $\lim_{n \to \infty} \frac{n^2+n+1}{2n^2+3n+4} = \lim_{n \to \infty} \frac{1+1/n+1/n^2}{2+3/n+4/n^2} = \frac{1+0+0}{2+0+0} = 1/2$

Geometrical examples

Inverted triangles

We start with a triangle, which, for convenience, we assume has a unit area. We then use the mid points to create an inverted triangle that will have an area of ¼. We continue in this way, creating inverted triangles with areas 1/4, 1/16, 1/64 etc... and try to sum the areas of the inverted triangles:

The Area of the inverted triangles

The sums of the areas of the shaded triangles are given below each figure

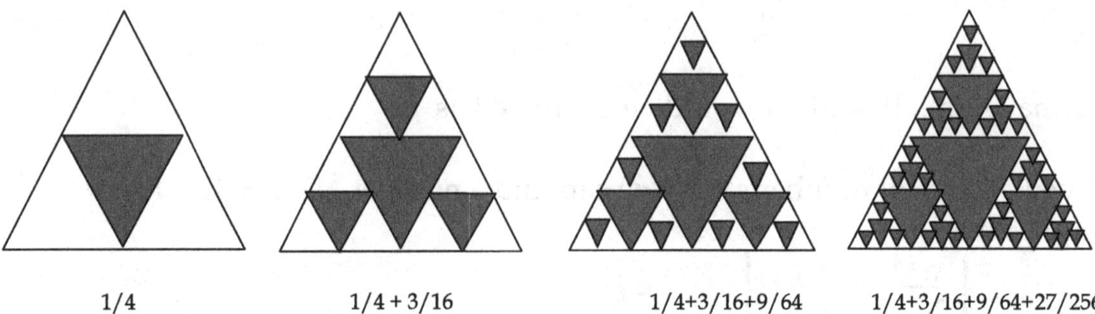

1/4 1/4 + 3/16 1/4+3/16+9/64 1/4+3/16+9/64+27/256

Each time we create a new set of inverted triangles, the number of triangles is 3 times the previous set but the area of each is one quarter the previous area.

If we continue to create more and more blue triangles ad infinitum, then the sum of the areas will be the sum of the geometric series:

1/4+3/16+9/64+27/256+81/1024+...

Using $\frac{a}{1-r}$ for the sum to infinity, the sum of the blue areas is $\frac{1/4}{1-3/4} = 1$

If we could continue to add more inverted triangles for the rest of time we would get closer and closer to colouring the whole triangle blue.

==================================

The Perimeter of the inverted triangles

In this example, instead of calculating the areas of the inverted triangles, we calculate their total perimeter.

For convenience, in this example we assume that then length of the side of the first triangle is one. The perimeters are dotted and the values written below each figure:

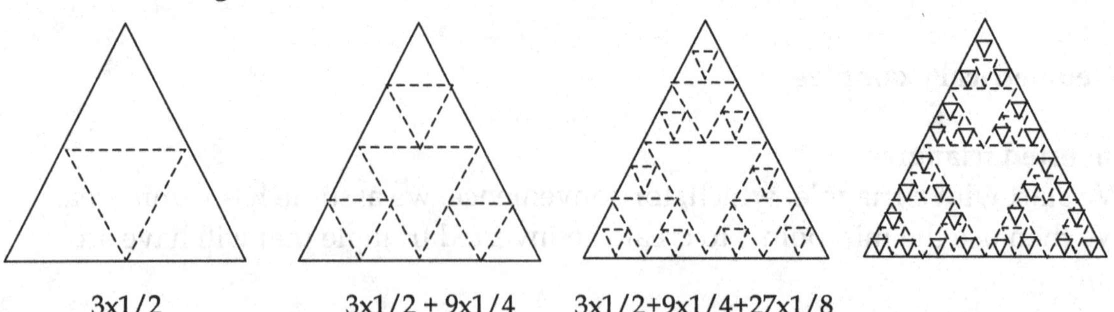

3x1/2 3x1/2 + 9x1/4 3x1/2+9x1/4+27x1/8

The length of the side of a triangle is halved each time but the number of triangles is multiplied by three. Therefore the sum of the lengths is the sum of a Geometric Series with common ratio 3/2

3x1/2+9x1/4+27/8+81/16 ....

and the sum of this G.P. goes to infinity.

Thus, the inverted triangles form an region with a finite area but an infinite boundary.

==============================

The Snowflake

We start with an equilateral triangle of side 1

Length of curve = 3

Now trisect each side and add another point to the snowflake:

Length of curve is now 12x1/3 = 4

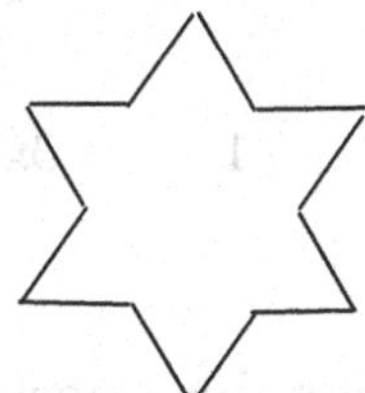

Continue trisecting the sides and adding another point to the snowflake:

The length of this snowflake is 48x1/9 = 5.333...

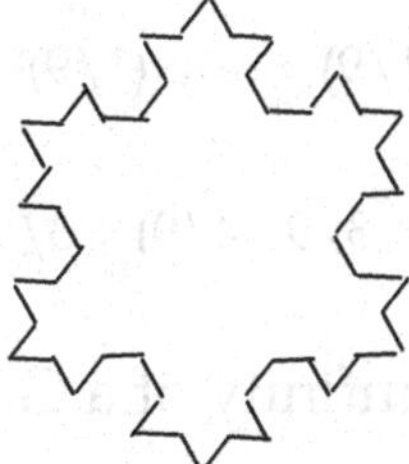

Each time we add new points, the length of each line is divided by 3, but the number of lines is multiplied by 4.

Therefore, the sequence of values for the perimeter of the snowflake runs like this:

3x1, 3x4x1/3, 3x4x4x1/9, 3x4x4x4x1/27, 3x4x4x4x4x1/81, ...

The length of the perimeter of the of the snowflake is of the form $3x\left(\frac{4}{3}\right)^n$
and this becomes infinite.
If we let n→ infinity we have what we will call the **snowflake curve.**
If an ant were to start walking round the perimeter of the snowflake curve then it could walk for the rest of time but still remain in the same place for every part of the snowflake curve is infinite in length. Of course, in order to be able to do this, the ant would have to be infinitely small.

The area of the snowflake

Let us take the area of the initial triangle to be 1.

Each time we add another set of points to the snowflake, we add a number of small triangles to each side.
The area of any triangle that we add is 1/9 th the area of the previous triangles that were added and the number of triangles we add is equal to the number of sides.
We have:

Number of sides	3	3x4	3x4x4	3x4x4x4	3x4x4x4x4
Area of new triangles added	1	3x1/9	$3x4x(1/9)^2$	$3x4x4x(1/9)^3$	$3x4x4x4x(1/9)^4$

None of these triangles overlap so we can get the area of the snowflake by adding up this series:

$$\text{Area of snowflake} = 1 + 3x(1/9) + 3x4x(1/9)^2 + 3x4x4x(1/9)^3 + 3x4x4x4x(1/9)^4 + ..$$

$$= 1 + 3/9 + 3/9x(4/9) + 3/9x(4/9)^2 + 3/9x(4/9)^3 + 3/9x(4/9)^4 + ..$$

which is 1 plus the sum to infinity of a G.P. Using a/(1-r) for the sum to infinity,

$$\text{area of snowflake curve} = 1 + \frac{3/9}{1 - 4/9} = 1 + 3/5 = 1.6 \text{ (exactly)}$$

The snowflake curve therefore has an area exactly 1.6 of the original triangle but has an infinite perimeter.

The Party Hat

The snowflake is interesting because it continues to grow but never reaches the limiting area 1.6.
The party hat is a much simpler example of a shape that has a finite area but an infinite perimeter.
We start with a unit square and divide one side into 2. Then we divide the same side into 4, then 8 then 16 and so on, each time completing a jagged edge to give a shape like the paper hats we sometimes find in a party cracker but keeping the same area:

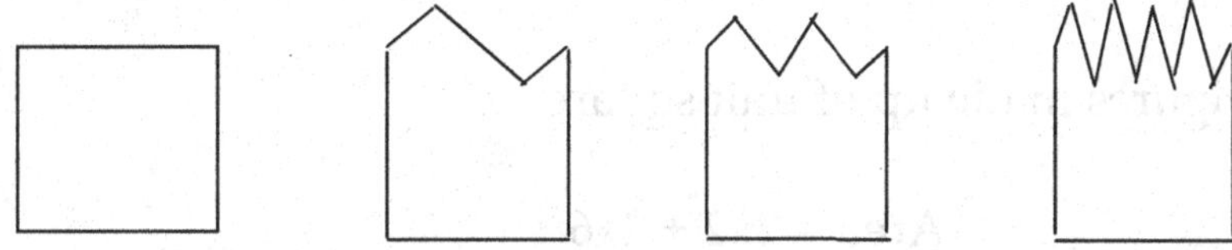

Let each line of the jagged edge be of length x, then the perimeters are:

4	3+4x	3+8x	3+16x

The formula for the perimeter is $3+2^n x$ which goes to infinity.

The area of the party hat stays at 1.

Exercise 1 **The Squareflake**

Start with a unit square.
Trisect each side and grow a smaller square on the middle third:

Prove that the perimeter of the squareflake becomes infinite but that the area becomes 2.
(The perimeter involves a GP with common ratio 5/3 and the area involves a GP with common ratio 5/9)

==================================

Measuring Area

To find the area of a flat shape, we count the number of unit squares that we can find within the boundary of the shape.

The area of this rectangle is 15 unit squares because we find 3 rows of 5 squares each, making 3x5 = 15 unit squares.

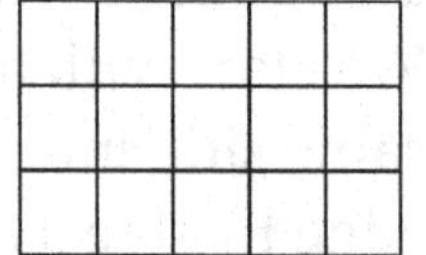

The areas of more complicated shapes can sometimes be found by adding or subtracting the areas of different parts.

Examples: Figures made up of unit squares

[1]

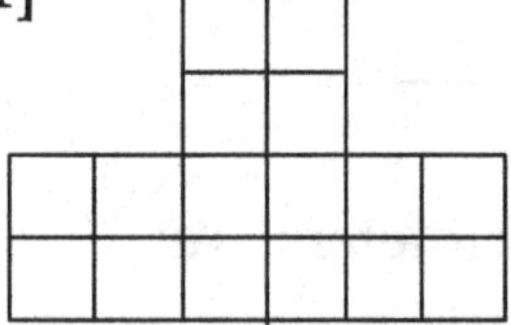

Area = 2x2 + 2x6

= 4 + 12 = 16 unit squares

[2]

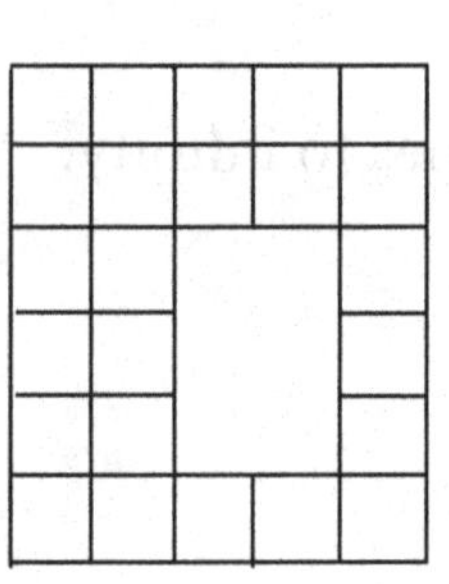

Area = 5x6 – 2x3

= 30 – 6 = 24 unit squares

Exercise 2

Find the areas of the following shapes:

[1]

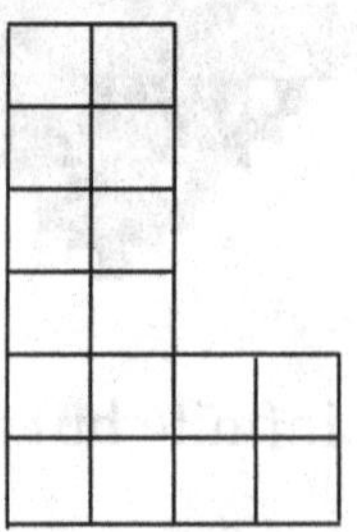

[2]

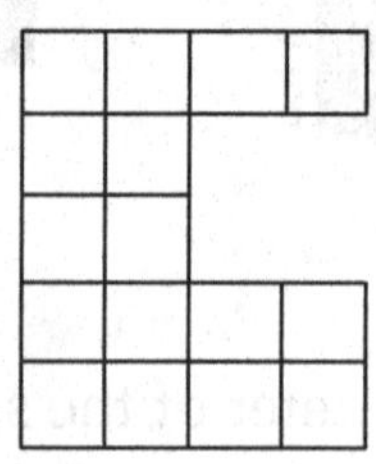

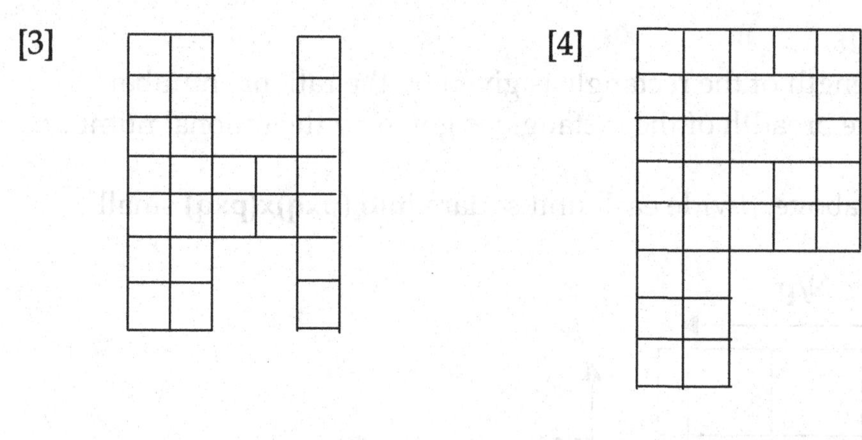

====================================

Area of a Rectangle

Example

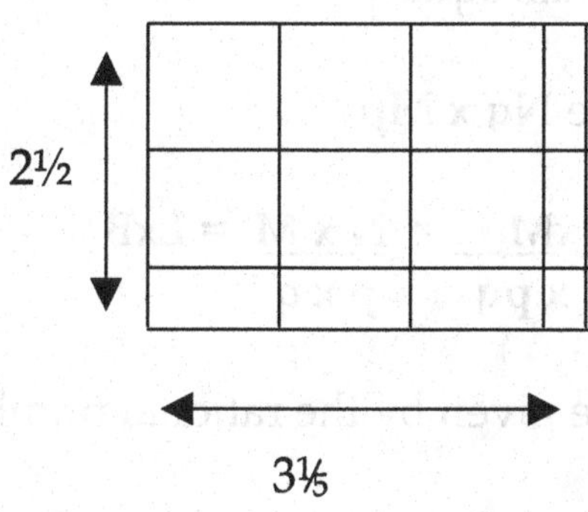

This rectangle measures 2½ by 3⅕

In order to "count the squares" in this figure, we need to divide each unit square into 10x10 = 100 small squares. (10 is the L.C.M of 2 and 5).

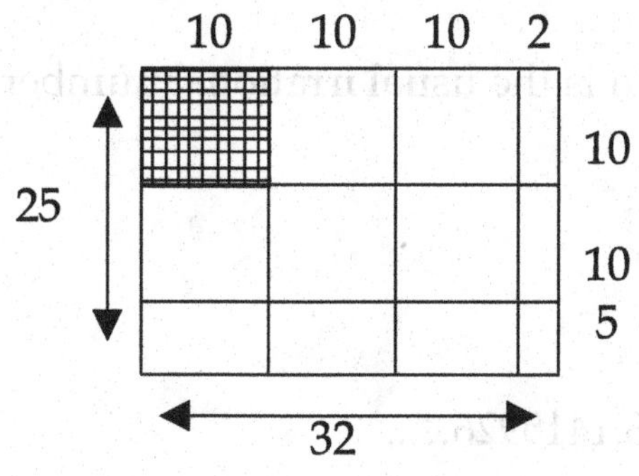

Using the smaller squares:
Length = 30 + 2
Width = 20 + 5

Number of small squares = 32x25

Number of unit squares = $\frac{32 \times 25}{100}$

= 8

The General Result

Suppose that the length of the rectangle is given by the rational number **L= N/p** and that the breadth of the rectangle is given by the rational number **B=M/q**

As in the example above, divide each unit square into **(pxq)x(pxq)** small squares.

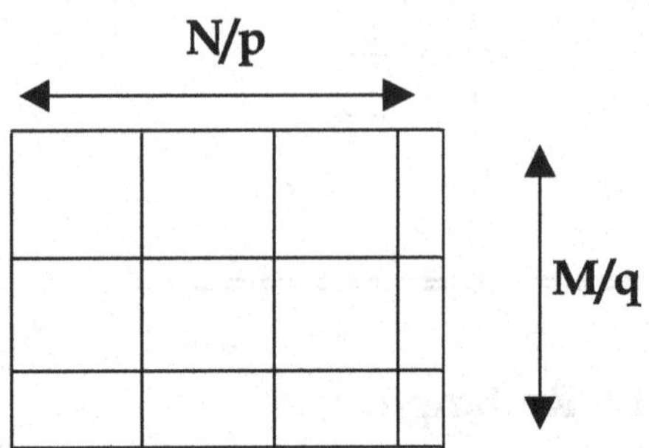

Now **N/p = Nq/pq** giving a length of **Nq** small squares
and **M/q = Mp/qp** giving a width of **Mp** small squares

The total number of small squares is therefore **Nq x Mp**

The total area in unit squares is therefore $\frac{NqxMp}{pq \times pq} = \frac{N \times M}{p \times q} = LxB$

Therefore, if the dimensions of a rectangle are given by the rational numbers length=**L** and breadth =**B**, then

Area = LxB

==============================

Irrational dimensions

Suppose we have a π by π square, where π is the usual irrational number 3.1415926…….

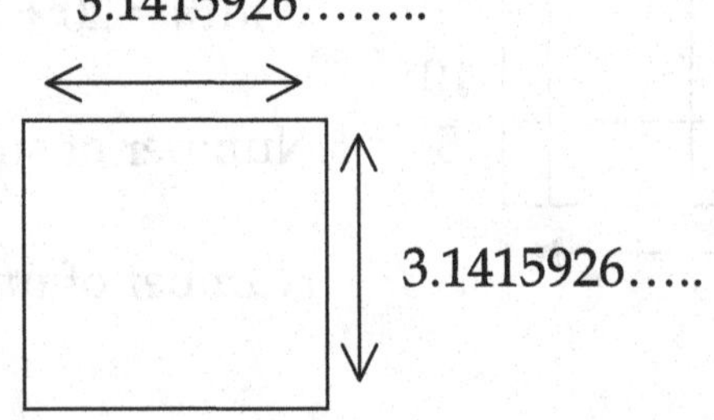

If the area of this π by π square is A, then we can say,

$$3.1^2 < A < 3.2^2$$
$$3.14^2 < A < 3.15^2$$
$$3.141^2 < A < 3.142^2$$
$$3.1415^2 < A < 3.1416^2 \text{ etc...}$$

The values on the left all represent the areas of squares with rational sides that are less in area than A and the values on th right all represent the areas of squares with rational sides that are each greater in area than A. Both of these sequences of numbers tend to the value $\pi x \pi$.
Thus A is squashed between two sequences of numbers that each tend to the value π x π and therefore,

$$\mathbf{A = \pi x \pi}$$

This kind of argument shows that the formula for the area of the rectangle still applies when one, or both, of the sides is an irrational number.

The above results show that for any rational or irrational dimensions:

Theorem 1 The area of a Rectangle is LengthxBreadth

==============================

The Area of a Curved Shape

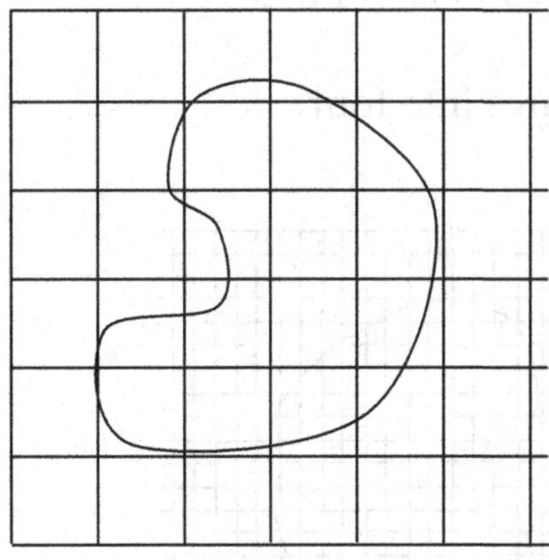

This curved shape is drawn on a 6x6 grid of unit squares. The curved shape overlaps 18 of these squares. Only 2 of the unit squares are completely inside the curved shape.

If A is the area of the curved shape, then, by counting unit squares, we see that

$$2 < A < 18$$

Now divide the unit squares into four quarters:

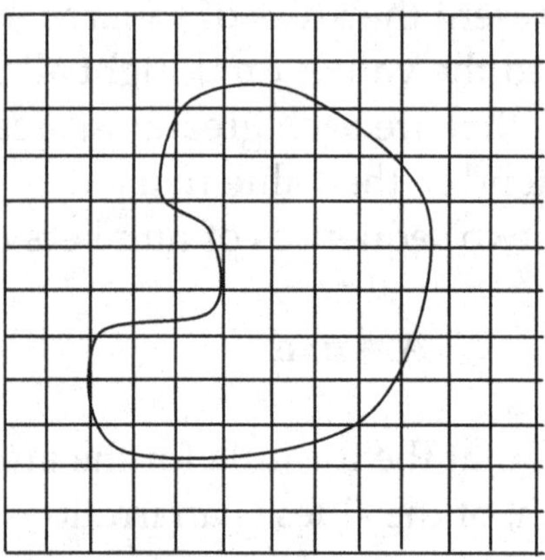

The curved shape now overlaps 59 of these squares and there are 27 squares inside but each of these small squares has an area ¼.

$$59 \times \tfrac{1}{4} = 14.75 \text{ and } 27 \times \tfrac{1}{4} = 6.75$$

Therefore we can say:

$$6.75 < A < 14.75$$

Now divide these small squares into four:

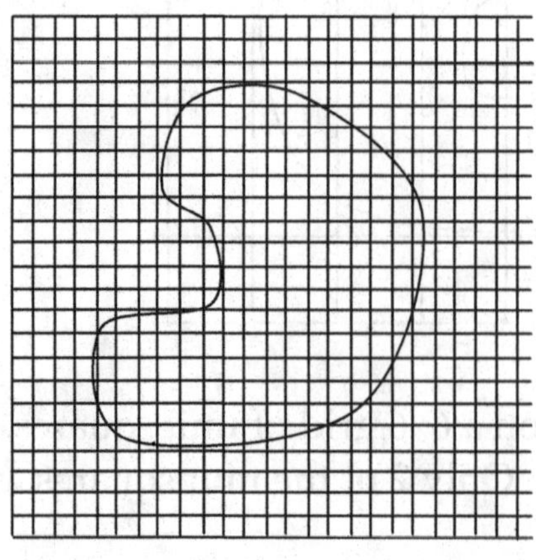

The curved shape now overlaps 194 squares and 135 squares are inside. Each one of these squares has an area 1/16.

$$194/16 = 12.125 \text{ and } 135/16 = 8.4375$$

Therefore we can say that

$$9.1875 < A < 13.3125$$

We continue in this way, counting squares and then dividing the squares into four smaller squares. The area A becomes squashed between two sequences of numbers, in our exercise, 2, 7.25, 9.1875….is an increasing sequence while 18, 15.75, 13.3125…. is a decreasing sequence. Hopefully the two sequences will approach the same limit which will be the area of the curved shape.

Isometries and Area

Isometries are transformations that preserve the size and shape of a figure. The basic isometries are:

Translation:

Rotation:

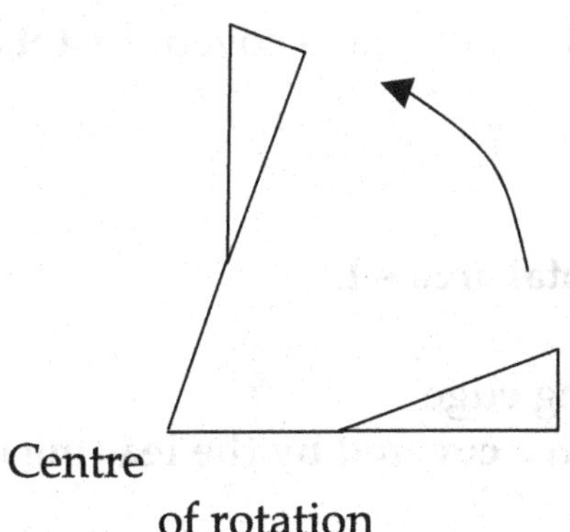

Reflection in a mirror line:

Mirror
Line

The number of unit squares in a figure is not changed by an isometry so the area of the figure will be preserved.

Theorem 2

In a translation, the area covered over by the leading edge is equal to the area uncovered by the trailing edge.

Proof

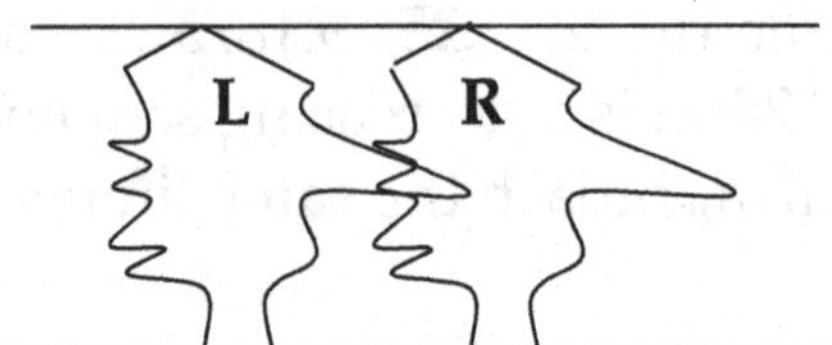

Figure **L** is translated to give figure **R**

Let the area between the trailing edge of **L** and the leading edge of **R** be called the **total area** then:

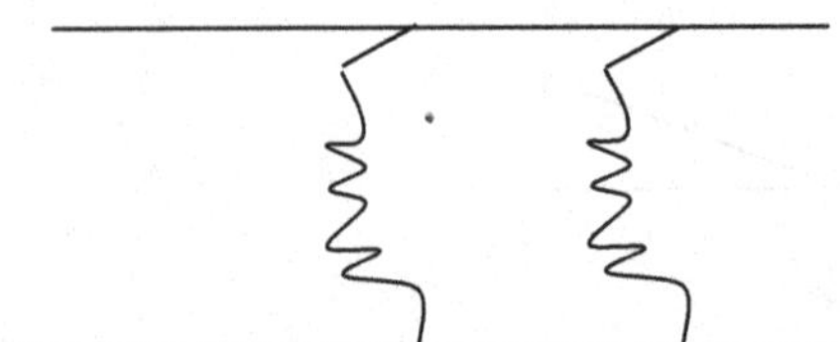

Area uncovered = **total area – R**

area covered = **total area – L**

Since this is a translation **L=R**

Therefore **total area – R = total area – L**

∴ **area uncovered by the trailing edge**
= area covered by the leading edge

==================================

The Parallelogram

A parallelogram is a four sided figure with its opposite sides parallel.

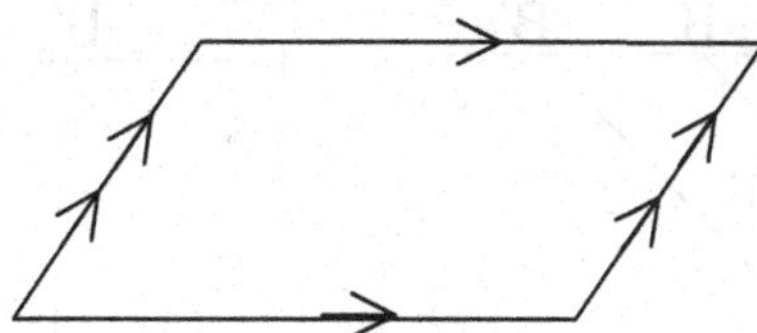

Theorem 3 (i) The Opposite sides of a parallelogram are equal.
(ii) The Opposite angles of a parallelogram are equal.

Proof

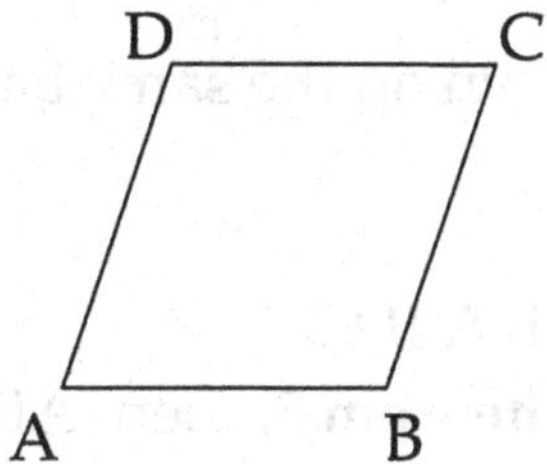

ABCD is any parallelogram

Let O be the mid point of DB

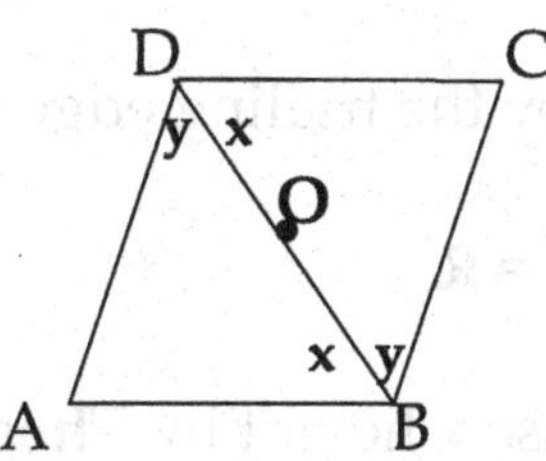

Consider a rotation of ΔABD by 180 degrees about O

Then B ⟶ D and since AB//DC, AB ⟶ CD,
and since BC//AD, BC ⟶ DA

Therefore, ΔABD rotates onto ΔCDB.

The rotation preserves distances and angles AB = DC , AD = BC and ∠A = ∠C

We may also observe that the angles marked x and y are equal (alternate angles) proving that ∠ADC = ∠ABC since both are x+y.

==================================

Theorem 4 Area of a Parallelogram = base x height

Proof

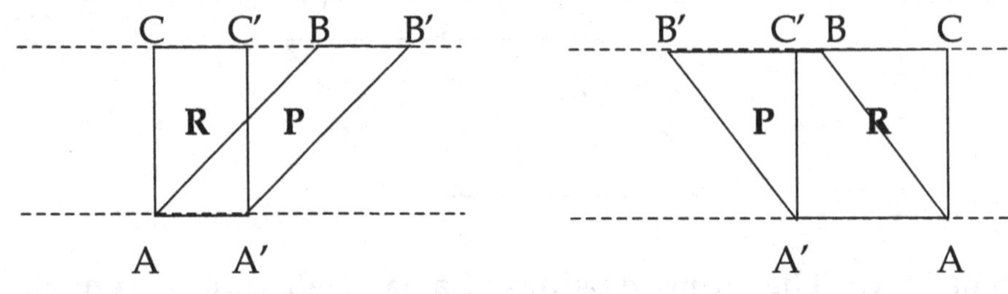

AA′B′B is parallelogram **P**

AA′C′C is rectangle **R** drawn on the same base AA′ and with the same height AC

Consider triangles ABC and A′B′C′
Since AA′ = BB′ = CC′ by **Theorem 3,** then ABC ⟶ A′B′C′ is a translation, with A ⟶ A′ , B ⟶ B′ and C ⟶ C′

The area covered by the leading edge is parallelogram **P**

The area uncovered by the trailing edge is rectangle **R**

Therefore, by **theorem 2, P = R**

The area of rectangle **R** is base x height by **Theorem 1**

$\therefore$ **area of parallelogram P = base x height**

==

Theorem 5 **Area of a triangle = ½ base x height**

Proof

Let ABC be any triangle

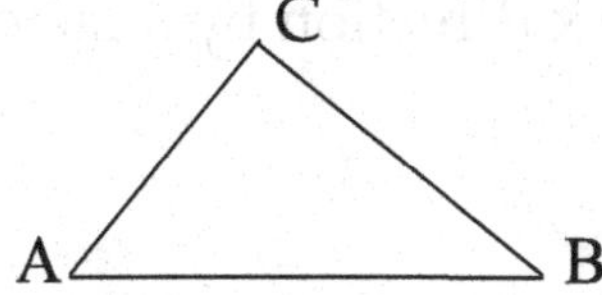

Complete parallelogram ABCD

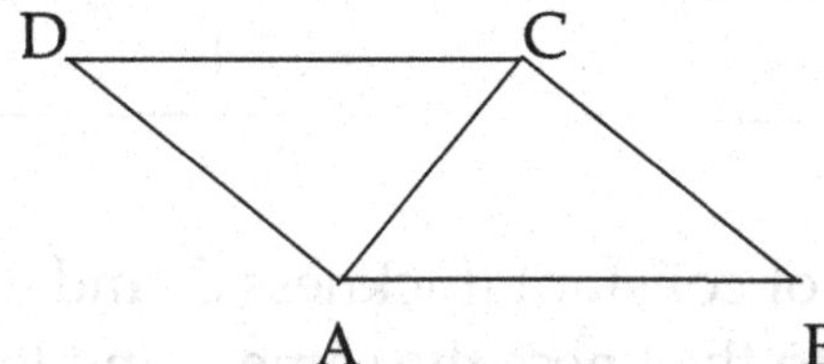

Triangles ABC and ACD are congruent (three sides, using **theorem 3**) therefore their areas are equal.

The area of parallelogram ABCD is **base x height (theorem 4)**

Therefore the **area of triangle ABC = ½ base x height**

==========================

Theorem 6. **The area of a trapezium= ½(a+b).h** where a and b are the lengths of the parallel sides and h is the distance between them.

Proof

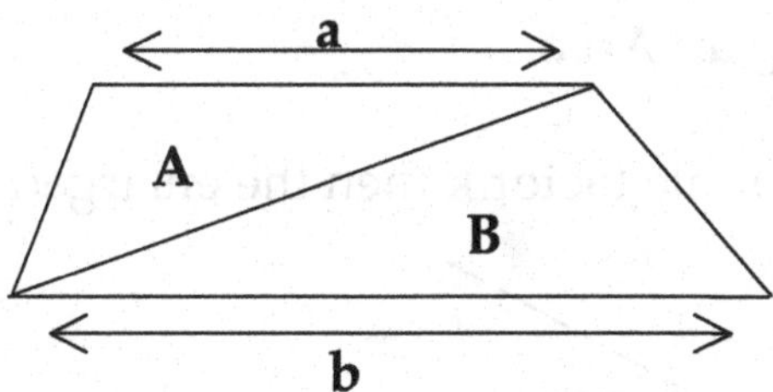

The figure shows a trapezium with parallel sides of lengths **a** and **b.**

A diagonal has been drawn, dividing the trapezium into two triangles **A** and **B.**

Area of trapezium = A + B = ½ ah + ½ bh (theorem 5)

= ½(a+b).h

=======================================

Theorem 7 **Stretching an Area**

If area A is stretched in the x direction by a factor k, then the area becomes kA.

Proof

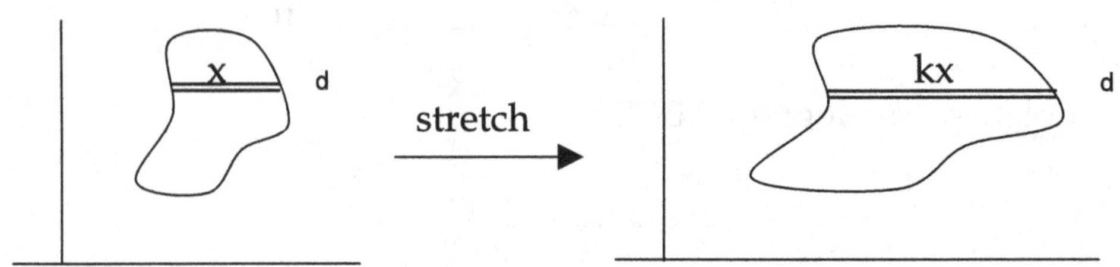

Let A be divided into strips of constant thickness d, and length x.
Each strip will be stretched to thickness the same d and length kx.
Therefore, the area of the stretched area will approximate to the sum of the areas of these strips which is

$$\Sigma\, d.kx$$

where d is the constant thickness, the same for each strip, but x varies.
The stretched area is therefore

$$\lim_{d \to 0} \Sigma\, d.kx = k. \lim_{d \to 0} \Sigma\, d.x = k.A$$

==============================

Theorem 8 **Enlarging an Area**

If an area A is magnified by a scale factor k then the enlarged area is k^2A

Proof

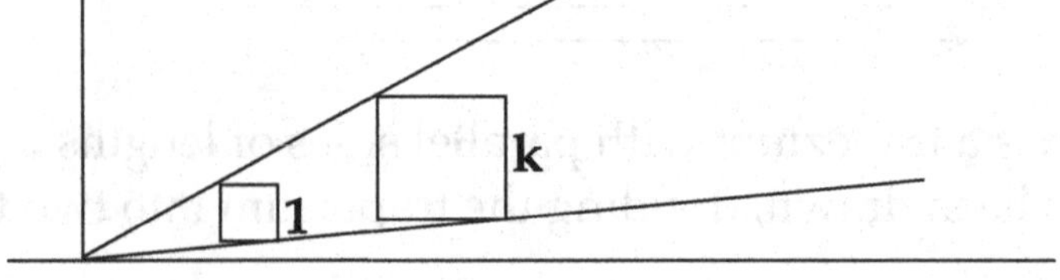

Each unit square involved in the calculation of area A becomes a k by k square in the enlarged square.
therefore, if A = Σ (areas of squares and part squares that make up A)
then
enlarged area A = Σ k^2(areas of squares and part squares that make up A)
= k^2 Σ (areas of squares and part squares that make up A)
= k^2A

Shear

Consider 4 congruent books piled on top of one another:

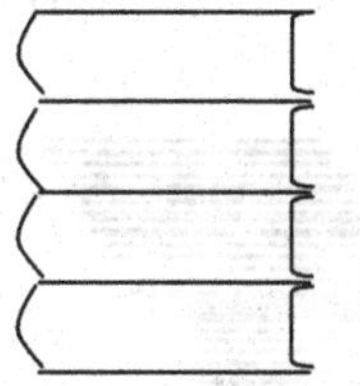

Suppose that the thickness of each book is **t** and its width **L**.
Then the area of each book is **tL**.
The area of the front of the pile is **4tL**.

Now push the books over a little:

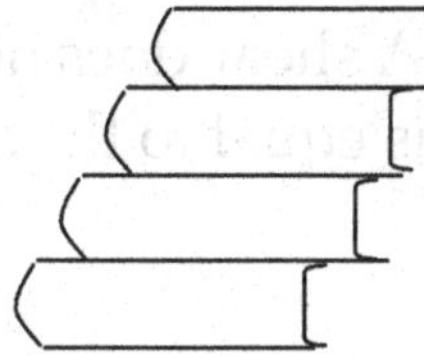

We say that the pile has been "sheared over".
The area of the front of the pile has not changed, it is still A = 4tL.

Suppose that now, we pile up 8 books that are only half as thick.
The area of this pile will be 8x ½tL = 4tL.
The area of the pile has not changed:

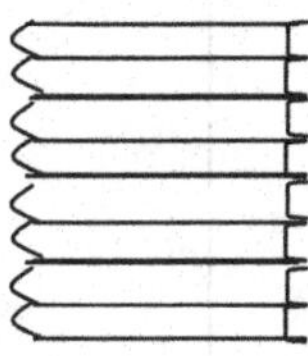

Shear this pile over:

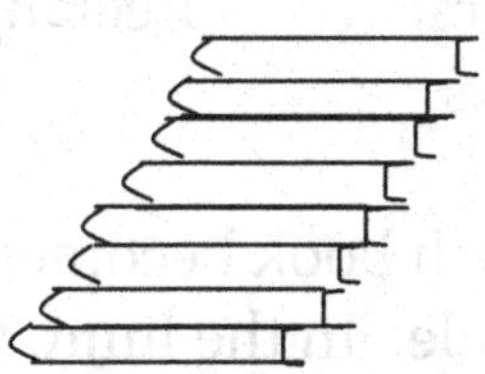

The area of both piles of 8 books is still 4tL.

If we generalize this, we may take N times as many books, i.e. 4N books, but make each book "one Nth" the thickness, that is thickness t/N.
The height of the pile is 4N x t/N = 4t, as before.
The area of the pile is 4N x t/N x L = 4tL as before.
To take this to the extreme, we let N→∞. However many books we have, the area of the pile will always be 4tL.

"In the limit" the front of the pile of books looks like a rectangle and when sheared over, the pile looks like a parallelogram.
Shearing over the pile does not change the area. This indicates that the area of the parallelogram is equal to the area of the rectangle.

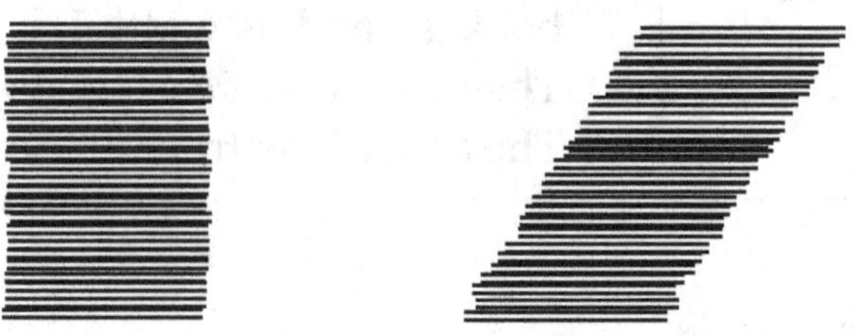

The parallelogram is the sheared rectangle. A shear does not affect the area and therefore the area of the parallelogram is equal to the area of the rectangle.
Therefore,

Area of parallelogram = base x height

The deformed pile of books does not, of course, need to be straight edged.

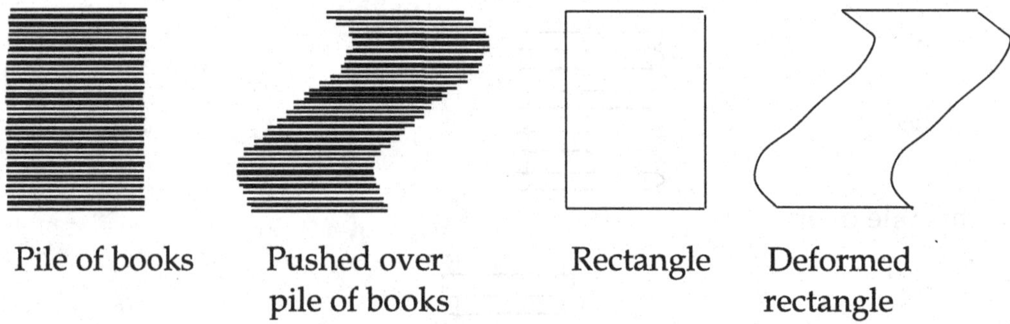

Pile of books | Pushed over pile of books | Rectangle | Deformed rectangle

In the limit, as the thickness of each book becomes smaller, the pile of books approaches the shape of a rectangle. In the limit, the pushed over pile of books approaches the shape of the deformed rectangle with smooth sides. These are special cases of **Cavalieri's principle.**
Bonaventura Cavalieri (1598-1647) was a student of Galileo and he was appointed to the post of professor of Mathematics at the University of Bologna in 1629.

==============================

Theorem 9 **Cavalieri's Principle for Areas.**

If two shapes, standing on a horizontal base, have the same width at any height above the base then the two shapes are equal in area.

Proof

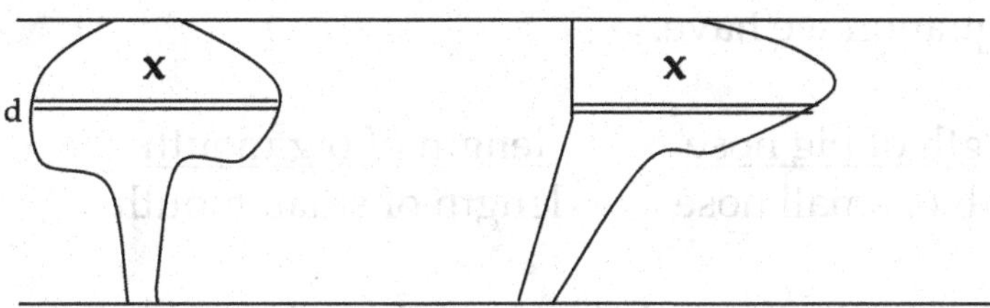

The diagram shows two shapes which are such that the width x, at any height above the base is the same in both figures.
Divide each shape into narrow strips of width x, which varies at different heights, but with the same width d.
The area of each shape is then approximately Σ **d.x**
Therefore, the area of both shapes is $\lim_{d \to 0} \Sigma\, d.x$

==============================

Similar shapes
When two shapes are similar, but not the same size, they can always be placed so that the larger shape is an enlargement of the smaller.

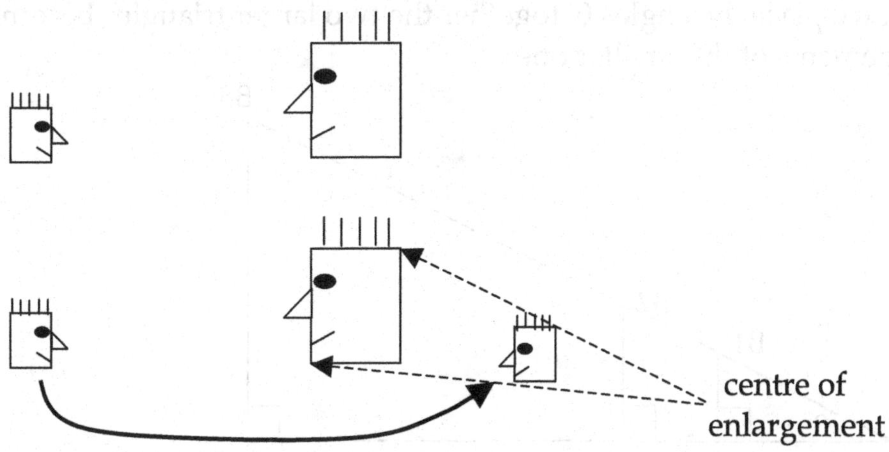

When two shapes are similar, then the ratio of two corresponding lengths in each figure is the same, for example, here:

$$\left(\frac{\text{length of nose}}{\text{length of mouth}}\right)_{\substack{\text{large}\\\text{figure}}} = \left(\frac{\text{length of nose}}{\text{length of mouth}}\right)_{\substack{\text{small}\\\text{figure}}}$$

Rewriting this equation we have:

$$\frac{\text{length of big nose}}{\text{length of small nose}} = \frac{\text{length of big mouth}}{\text{length of small mouth}}$$

and each of these ratios is the scale factor of the enlargement.

======================================

Elementary trigonometry relies of the fact that all right angle triangles that have a given angle, say θ, will be similar.

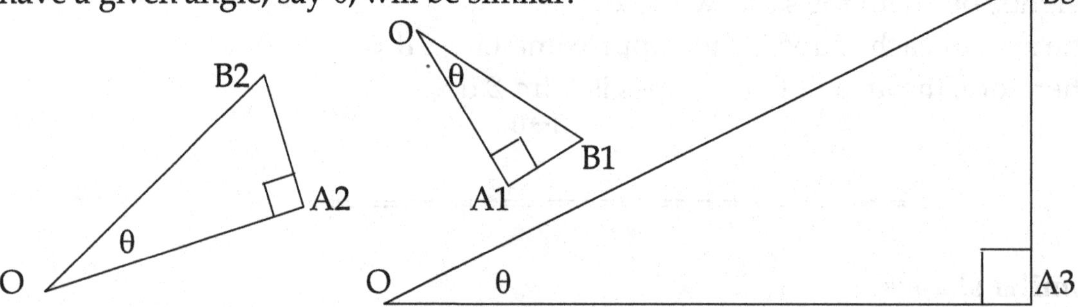

These three triangles have a right angle at angle A and angle θ at O. When we place the angles θ together the two larger triangles become enlargements of the smaller one:

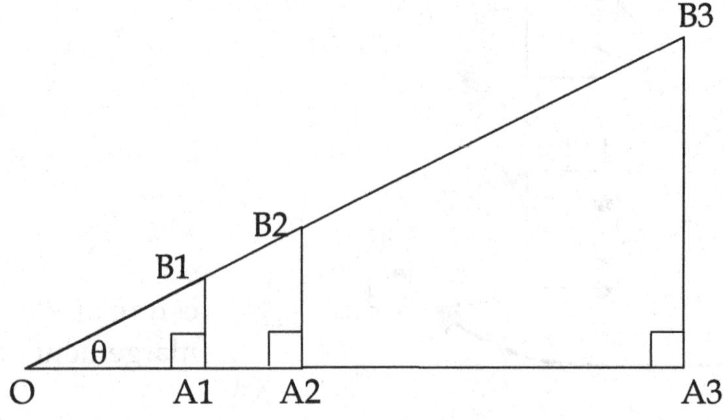

The triangles are similar and the ratios of corresponding sides are equal, for example

$$\frac{A1B1}{OB1} = \frac{A2B2}{OB2} = \frac{A3B3}{OB3}$$

The value of this ratio is **sinθ**

==============================

Perimeter and Diameter

π(4) **The valueof pi for a square**

All squares are similar to each other and therefore, the ratio of any two corresponding lengths in any square will always be the same.
We choose the two lengths to be
[i] the perimeter of the square
and [ii] the diameter of the square
where perimeter has its usual meaning, the sum of the lengths of the sides, but diameter means the diameter of the circle that can be drawn through the four corners of the square.

Assume that the side of the square is of unit length so that

perimeter = 4

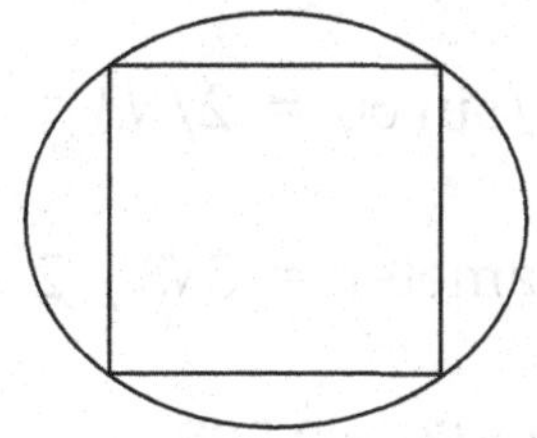

By Pythagoras, the diameter of the square = √(1+1) = √2

If we calculate the ratio $\frac{\text{perimeter}}{\text{diameter}}$

Then for any square we have $\frac{\text{perimeter}}{\text{diameter}} = \frac{4}{\sqrt{2}} = 2.828$

We shall call this value **π(square) = π(4) = 2.828**

==================================

π(3) **The value of pi for a triangle**

All equilateral triangles
are similar.
For convenience,
take the length of the side
to be one, then

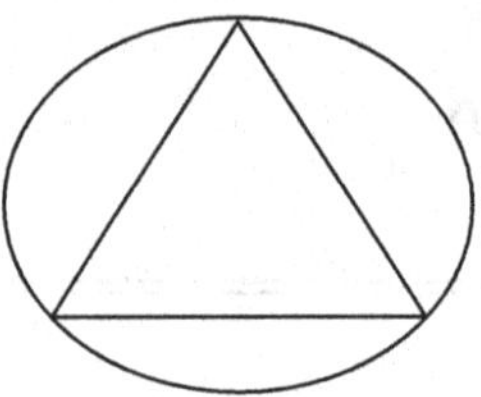

perimeter = 3

The angle marked **a** at the centre
Is 360/3 = 120
Therefore, the angle marked
b is 120/2 = 60

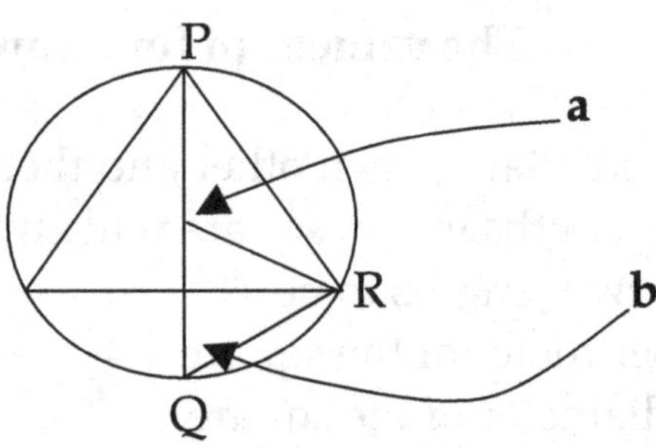

PQ is the diameter and so there is a right angle at R
Therefore,

$$\sin b = \sin 60 = PR/PQ = 1/\text{diameter}$$

$$\therefore \text{ diameter} = 1/\sin 60 = 2/\sqrt{3}$$

$$\therefore \text{ perimeter/diameter} = 3\sqrt{3}/2 = 2.598$$

and we write **π(triangle) = π(3) = 2.598**

and since all equilateral triangles are similar, this is a constant for any equilateral triangle

=====================================

π(6) **The value of pi for a hexagon**

The diameter of a regular hexagon
with side=1, will be 2.

Therefore,

π(hexagon) = π(6) = 6/2 = 3

=================================

Theorem 10 $\pi(n) = n.\sin(180/n)$

Proof

We calculate $\frac{\textbf{perimeter}}{\textbf{diameter}}$ for a regular n-agon.

Consider a regular **n** sided polygon **(n-agon)** inscribed in a circle: Suppose that the side of the **n-agon** is of Unit length and that the radius of the circle is **r**.

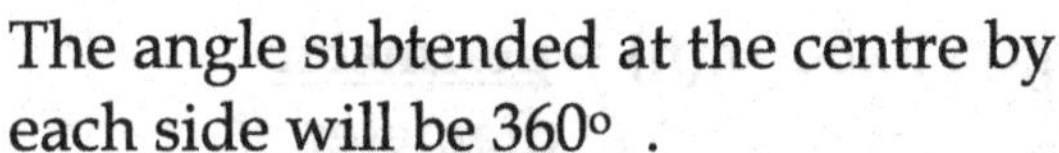

The angle subtended at the centre by each side will be $\frac{360^{\circ}}{n}$.

Hence this angle is $\frac{180}{n}$

therefore $\sin\left(\frac{180}{n}\right) = \frac{½}{r} = \frac{1}{2r} = \frac{1}{\text{diameter}}$

The diameter of the regular n-agon is therefore given by

$$\textbf{diameter} = \frac{1}{\sin(180/n)}$$

Thus the general formula for **π(n-agon)** is

$$\pi(\text{n-agon}) = \pi(n) = n.\sin(180/n)$$

=======================================

This formula gives us the following results:

$\pi(3) = 2.598076$

$\pi(4) = 2.828427$

$\pi(5) = 2.938926$

$\pi(6) = 3$

$$\pi(100) = 3.141075$$

.

$$\pi(1000) = 3.141587486$$

.

$$\pi(10000) = 3.141592602$$

.

$$\pi(1000000) = 3.141592654$$

================================

Theorem 11 **The Area of the n-agon = $\pi(n).r^2 \cos(180/n)$**

Proof

We have defined **$\pi(n)$** for the regular n-agon to be **$\pi(n) = \frac{\text{perimeter}}{\text{diameter}}$**

Let AB be one side of the n-agon
Let x be the length of the side of the n-agon.
Let r be the radius of the circle centre O.

The angle at the centre is 360/n
This angle is 180/n

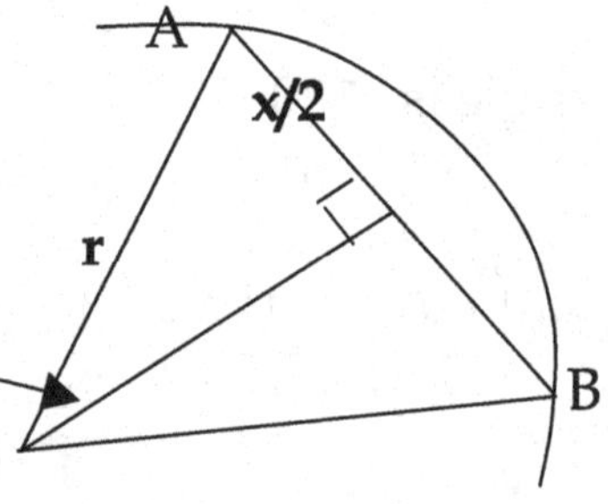

$$\textbf{Area of n-agon} = n.\ \Delta OAB = n.\ \frac{x}{2}\, r \cos(180/n)$$

$$= n.\ r \sin(180/n).r \cos(180/n)$$

but we have shown that

$$\pi(n) = \frac{\text{perimeter}}{\text{diameter}} = n.\sin(180/n)$$

The area of the n-agon is therefore **$\pi(n).r^2 \cos(180/n)$**

================

Exercise 3 Prove the following

[1] Area of an equilateral triangle = $\frac{1}{2}\,\pi(3)\, r^2$

[2] Area of a square = $\frac{1}{2}\sqrt{2}\ \pi(4)\, r^2$

[3] Area of a pentagon = $\frac{1}{4}(\sqrt{5} + 1)\ \pi(5)\, r^2$

[4] Area of a hexagon = $\frac{1}{2}\sqrt{3}\ \pi(6)\, r^2$

=========================

Rectifiable curves

Straight lines can be measured with a ruler but how should we measure a curve? Practically, we could lay a thread along the curve and then measure the thread when it is straightened out and laid alongside the ruler, or, we could roll the curve along the ruler and see where it starts and where it finishes.

Mathematically, we need to define what we mean by the length of a bent curve.

Given the bent curve shown in the next figure, we have inscribed an open polygon P1P2P3P4 in the curve:

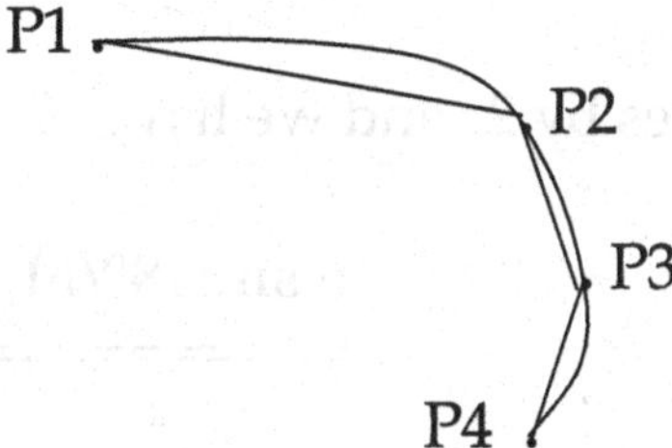

As we add more and more points, the length of the open polygon increases:

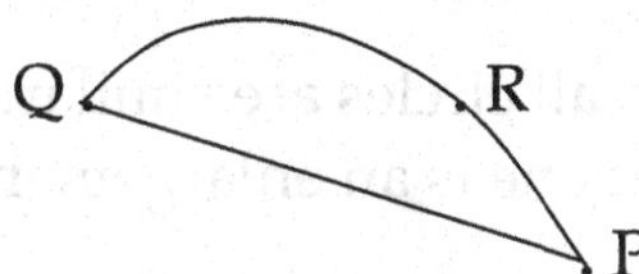

In this figure, if R is added between P and Q then PR+RQ > PQ

If there is an upper bound to the length of the open polygon P1P2P3....Pn as n goes to infinity, then this upper bound is defined to be the length of the curve.

Using this definition for the length of a curve, the circumference of a circle will be the limit of the perimeter of inscribed polygons, as the number of sides goes to infinity.

 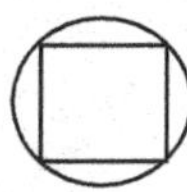  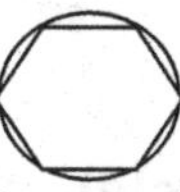 

Perimeter of n-agon ⟶ circumference of circle

Theorem 12 $n \sin(180/n) \rightarrow \pi$

Proof

The perimeter of the n-agon is

$$n.AB = n.2r\sin(180.n)$$

Assuming that the circle is a rectifiable curve, the perimeter of the n-agon must approach the length of the circle as $n \rightarrow \infty$

Therefore $n.2r \sin(180/n) \rightarrow 2\pi r$

Divide both sides by 2r and we have

$$n \sin(180/n) \rightarrow \pi$$

==============

Values of Pi

Just as all squares are similar, all circles are similar. Given any two circles, we can always place them so that one is an enlargement of the other:

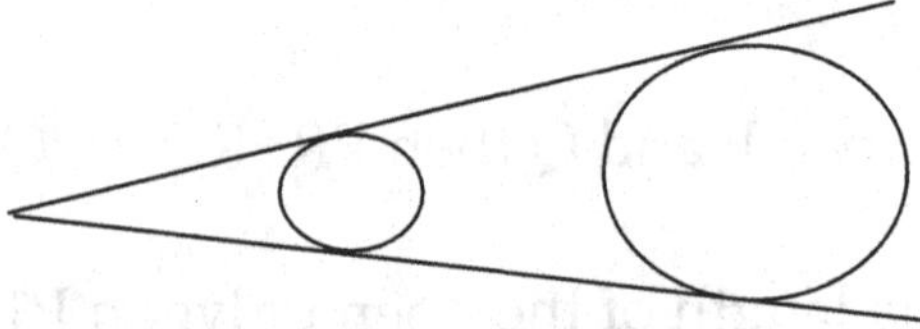

Further, just as, for all squares

$$\frac{\text{perimeter}}{\text{diameter}} = \pi(\text{square}) = \pi(4) = 2.828427$$

we have, for all circles

$$\frac{\text{perimeter}}{\text{diameter}} = \pi(\text{circle}) = \pi(\infty) = \pi$$

The value of π has been refined over the centuries and ancient documents have revealed:

Discoverer	found what	when
Babylonians	$3\frac{1}{8} = 3.125$ $\sqrt{10} = 3.16$	2000 B.C.
Egyptians	$\left(\frac{16}{9}\right)^2 = 3.1605$	1650 B.C. (Rhind papyrus)
Chinese	3	1200 B.C.
Old Testament	3	950 B.C.
Archimedes	$3\frac{10}{71} < \pi < 3\frac{1}{7}$ $(3.141 < \pi < 3.143)$	225 B.C.
Ptolemy	$\frac{377}{120} = 3.14166$	150 A.D.
Zu Chongzhi	$\frac{355}{113} = 3.1415929$	430-501 A.D.
Aryabhata (Hindu)	$\frac{62832}{20000} = 3.146$	c500 – c600 A.D.

==============================

The Greek letter pi was first used by the Welshman William Jones in 1706 and was adopted by the Swiss mathematician Euler (1707-1783)

The value of π is 3.14159 26535 89793 23846 26433 83279 50288 4197…

Pi is available to 200 million digits on the Web (March 2008)

Theorem 13 **The Area of a Circle = πr^2**

For any circle, we know that $\dfrac{\textbf{perimeter}}{\textbf{diameter}} = \pi$

for convenience, write P(x) for "the perimeter of x"

so that we have P(circle) = $2\pi r$

Let OAB be one triangle of the inscribed n-agon.

Then the area of the inscribed n-agon is given by

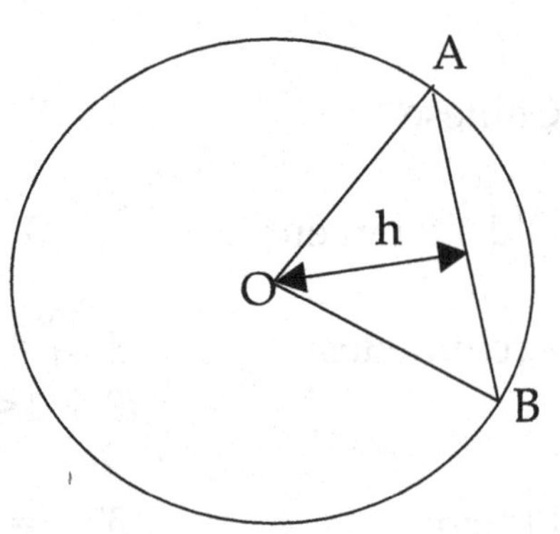

$$n.\Delta AOB = n.\ \tfrac{1}{2}\ AB.h$$

$$= \tfrac{1}{2}\ nAB.\ r \cos(180/n)$$

$$= \tfrac{1}{2}\ .P(\text{n-agon}).r \cos(180/n)$$

Now let n go to infinity, then P(n-agon) → P(circle) and cos(180/n) → 1

Therefore area of the inscribed n-agon → $\tfrac{1}{2}.2\ \pi\ r.r = \pi r^2$

Now consider a regular n-agon circumscribed about the circle:

Area of circumscribed n-agon

$$= n.\ \tfrac{1}{2}\ A'B'.r$$

$$= n.\ r \tan(180/n).r$$

$$= \frac{n.\sin(180/n).r^2}{\cos(180.n)}$$

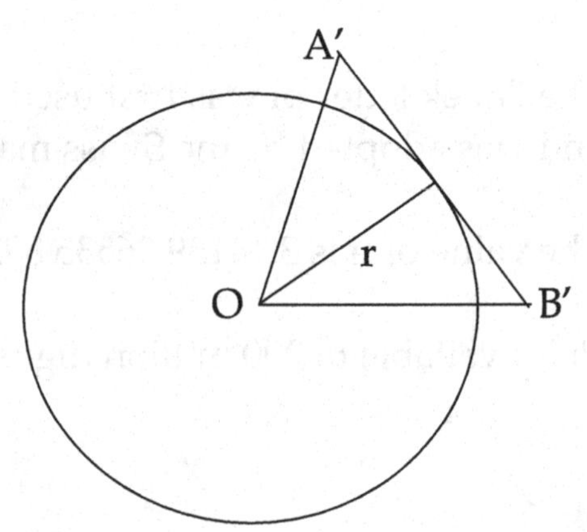

Now n.sin(180/n) → π and cos(108/n) → cos(0) = 1 as n → ∞

Therefore, the area of the circumscribed n-agon → πr^2

But: inscribed n-agon < circle < circumscribed n-agon

and both the inscribed the n-agon and the circumscribed n-agon → πr^2

therefore, **the area of the circle = πr^2**

======================================

Informal demonstrations

[1] Slice the circle up into equal sectors

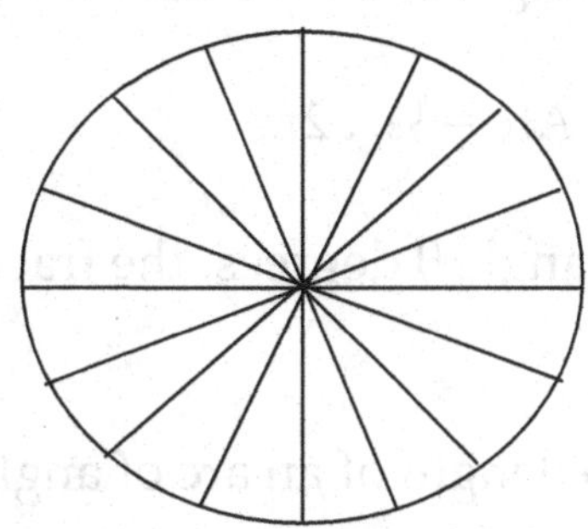

Now rearrange these sectors into a 'bumpy' rectangle shape:

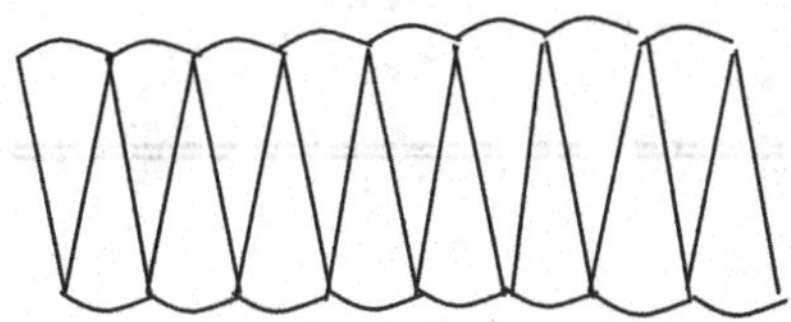

The width is equal to the radius and the length is roughly half of the circumference.

Therefore, the area is roughly r x πr = πr^2

================================

[2] Slice the circle up into equal sectors

Each sector is roughly a triangle of height r.

Add all these triangles to get

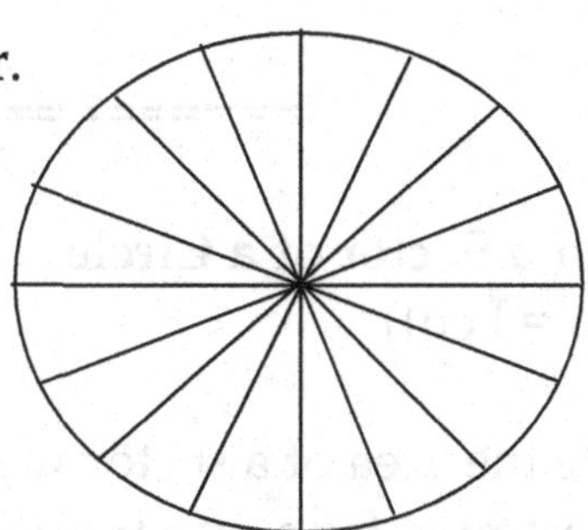

Area of circle = sum of (½ base x height)

= ½ (sum of bases) x r

= ½ (perimeter) x r

= ½ (2πr) x r = πr^2

The Length of an Arc of a Circle

To find the length of an arc we need to know what fraction of the circumference we want to measure, for example, for a quarter circle (a quadrant), with angle 90°

$$\text{Arc} = \tfrac{1}{4} \times 2\pi r$$

For an arc of angle 45°, the fraction is $\tfrac{1}{8}$

45°

$$\text{Arc} = \tfrac{1}{8} \times 2\pi r$$

For an arc of angle θ degrees, the fraction is $\dfrac{\theta}{360}$

Therefore, the length of an arc of angle θ is given by

$$\textbf{arc} = \frac{\theta}{360} \times 2\pi r$$

====================

Exercise 4

Find the perimeter of each of these figures;

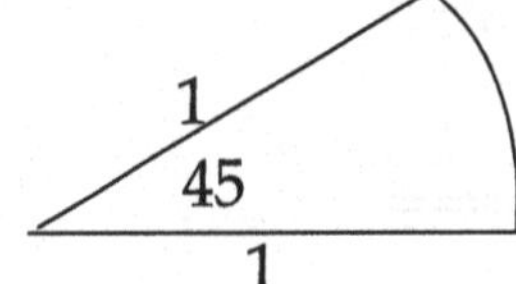

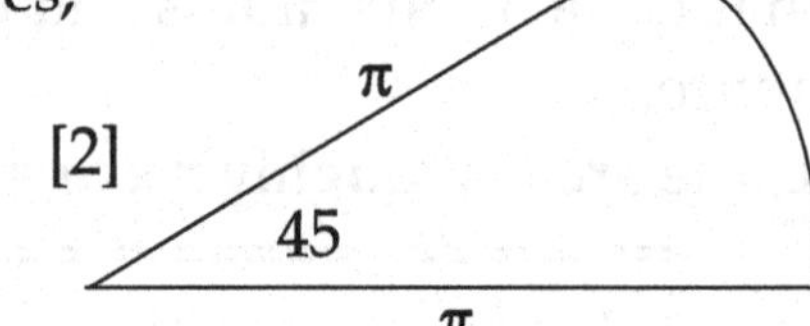

==========================

The Area of a Sector of a Circle
(Latin: seco = I cut)

To calculate the area of a sector we work out what fraction it is of the whole circle, for example for a quadrant (a quarter circle) of angle 90° the fraction will be $\tfrac{1}{4}$.

$$\text{sector} = \tfrac{1}{4} \times \pi r^2$$

For a sector of angle θ degrees, the fraction is $\frac{\theta}{360}$

Therefore, the area of a sector of angle θ is

$$\text{sector} = \frac{\theta}{360} \times \pi r^2$$

====================

Exercise 5

Find the perimeter of each of these figures;

[1]

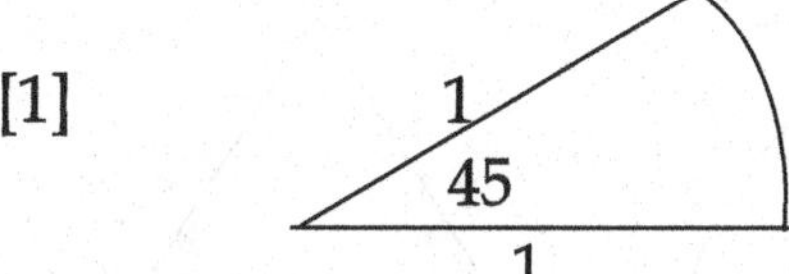

[2]

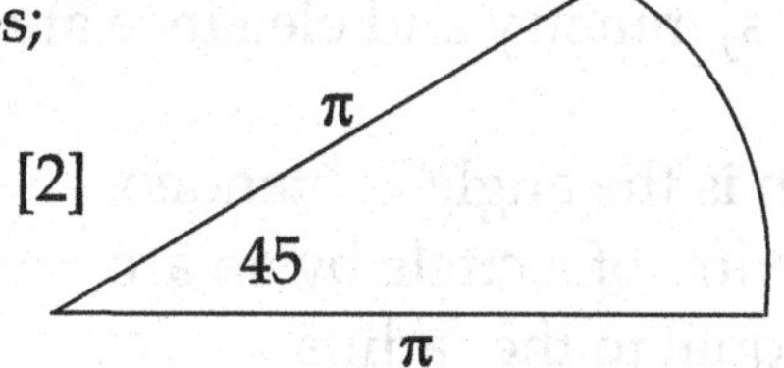

[3]

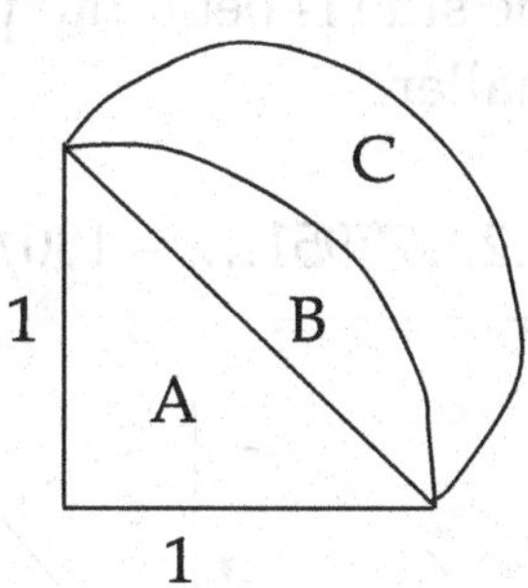

This figure is made up of:
a 1x1 right angle triangle,
a quadrant and a semi circle
Find the three area A,B and C

Radian measure

Most calculators have three different measures for angle. The most common is the degree.

360 degrees = 1 revolution

This comes down to us from the Babylonian mathematicians and astronomers of 2000 B.C. who divided their year into 360 days and used a base 60 number system. Thus the most popular measure of angle has 60 seconds of arc = 1 minute of arc, 60 minutes of arc = 1 degree and 360 degrees = 1 full turn.

You will also find the Grad. Whereas there are 90 degrees in one corner, there are 100 grads (or grades). This was an unnecessary attempt at decimalizing the units for measuring angles introduced by a Frenchman called Dupre in 1801. Thank goodness that nobody took much notice and we have been spared from using decimalized angles! Make sure that your calculator is NOT set to grad mode.

The third measure of angle is the **radian** and this **is** important. Radian measure enormously simplifies differential calculus and if radians were not used then many formulae would look clumsy an inelegant for we know that pattern, symmetry and elegance are vital if progress is to be made.

A radian is the angle subtended at the centre of a circle by an arc that is equal to the radius.

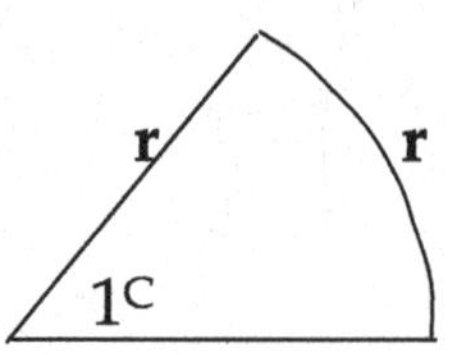

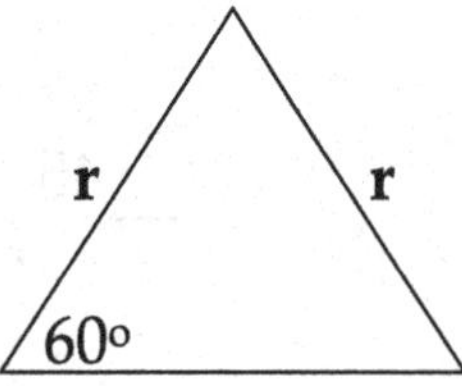

An equilateral triangle has angles of 60° so if one side is bent into part of a circle then the opposite angle will get a little smaller.

One radian is about 57° . The actual value is 57.29577951…. = 180/π

============================

Conversion

AB is an arc of length r that subtends one radian at the centre of the circle.
If this arc that is length **r** is multiplied in length by **2π** then it becomes an arc of length **2πr**, that is, the complete circumference.

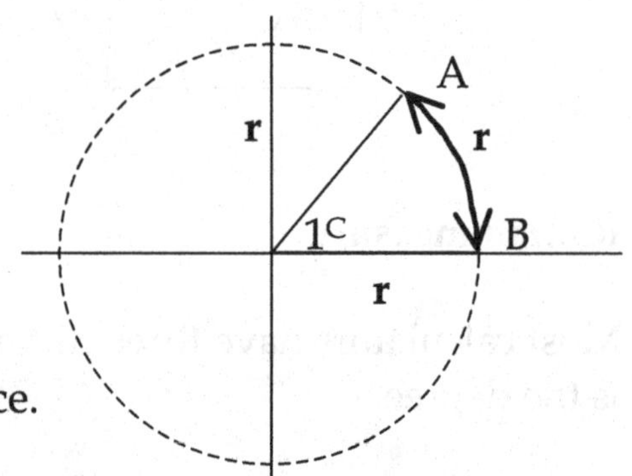

The angle at the centre is also multiplied by **2π** and therefore measures **2π** radians.
But the angle at the centre is also 360°

Therefore $\quad 2\pi \text{ radians} = 360^\circ$

$\pi \text{ radians} = 180^\circ$

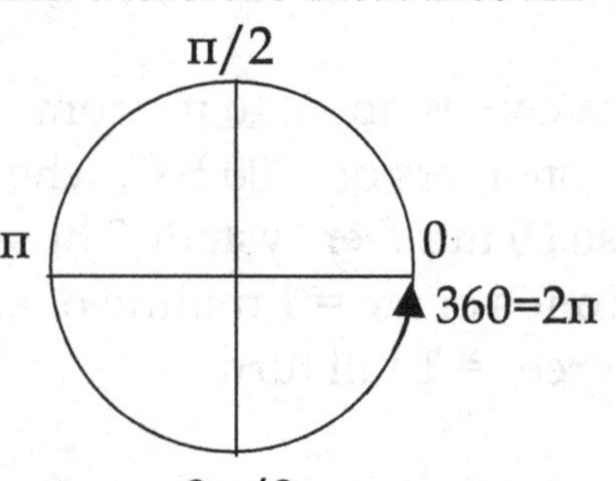

Length of Arc using Radian Measure

1 radian gives an arc of length r

$\therefore$ θ radians gives an arc of length rθ

therefore **arc = rθ**

====================

Area of a Sector using Radian Measure

2π radians gives an area πr^2 (the whole circle)

$\therefore$ θ radians gives an area $\pi r^2 \times \frac{\theta}{2\pi}$

therefore **sector = ½ $r^2\theta$**

====================

Example The shortest path

Dr Fred's house is a short distance from the college on the North side of a circular road which goes past the college gate. To get to college, Dr Fred can either walk round the road on the North side or he could cross the road and walk round the path on the South side before crossing back.

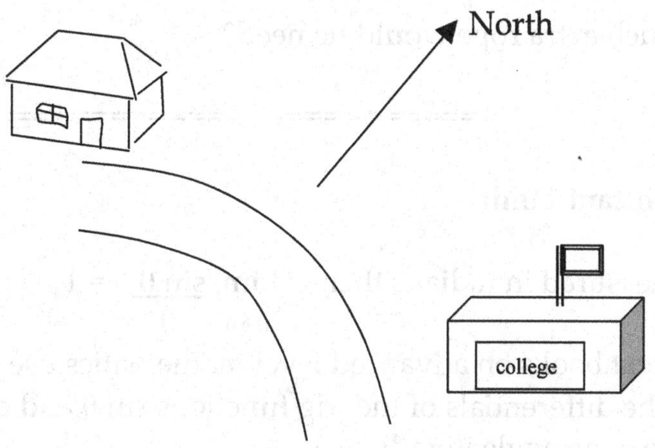

Dr Fred would like to know the shortest route.

Solution

let the angle of the arc be θ (radians)

Let the inside radius be r

Let the width of the road be d

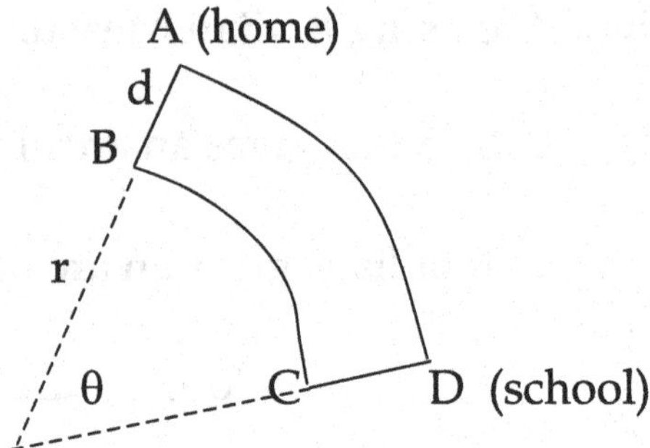

Length of outside path = arcAD = (r+d)θ

Length of inside path = AB + arcBC + CD = rθ + 2d

Outside path – inside path = (r+d)θ - (rθ+2d) = d(θ-2)

Therefore, the outside path is longer than the inside path if θ > 2 radians.

If the angle is greater than 115 degrees he should cross to the inside.
If the angle is less than 115 degrees he should go round the outside.

====================

Exercise 6

The radius of the Earth is about 3863 miles at the equator.
Dr Fred has a long rope that he says will go exactly round the equator.
To stop the rope getting wet and dirty, however, Dr Fred decides that the rope should be raised up one foot above the ground all the way round the Earth.
How much extra rope would he need?

============================

An important Limit

If θ is measured in radians then $\lim_{\theta \to 0} \frac{\sin \theta}{\theta} = 1$

(Many text books on advanced level mathematics use this limit in order to derive the differentials of the trig functions sin θ and cos θ where θ is the measure of an angle in radians.)

Theorem 14 If θ is in radians then $\lim_{\theta \to 0} \frac{\sin\theta}{\theta} = 1$

Proof

Let AB be one side of the inscribed polygon of a circle radius r, centre O.

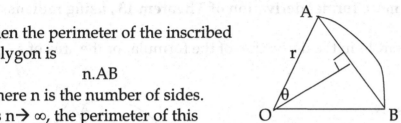

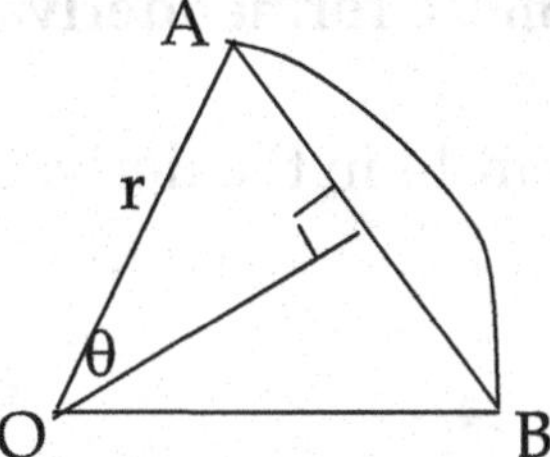

Then the perimeter of the inscribed polygon is

n.AB

where n is the number of sides.
As $n \to \infty$, the perimeter of this polygon goes to the circumference of the circle (see page 25: Rectifiable curves)
therefore

$$n.AB \to 2\pi r \qquad \text{(as n gets larger)}$$

Now $$n.AB = n.2r\sin\theta$$
Therefore

$$n.2r\sin\theta \to 2\pi r$$

so that $$\frac{\sin\theta}{\pi/n} \to 1 \qquad \text{(dividing by } 2\pi r\text{)}$$

But angle AOB = $2\pi/n$, therefore angle $\theta = \pi/n$

$$\therefore \quad \frac{\sin\theta}{\theta} \to 1$$

but as $n \to \infty$ then $\theta \to 0$

therefore $$\lim_{\theta \to 0} \frac{\sin\theta}{\theta} = 1$$

============================

Exercise 7 **[1]** **Prove that** $\pi(8) = \dfrac{8}{\sqrt{(4+2\sqrt{2})}} = 3.061$ **(4 s.f.)**

[2] **Prove that** $\pi(12) = \dfrac{6}{\sqrt{(2+\sqrt{3})}} = 3.106$ **(4 s.f.)**

====================================

Area of a circle (A more formal derivation of Theorem 13 , using radians)

Here, we use theorem 14 in the derivation of the formula for the area of a circle:

Working in radians

Let AB be one side of the inscribed n-agon

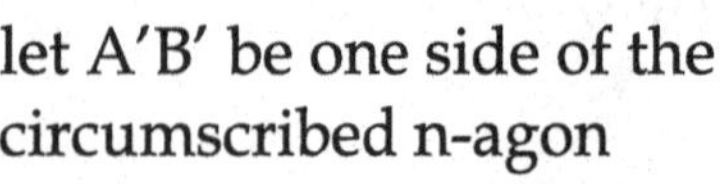

let A′B′ be one side of the circumscribed n-agon

then inscribed n-agon < circle < circumscribed n-agon

$$n.\Delta\, AOB \quad < \text{circle} < \; n.\Delta\, A'OB'$$

$$n.r \sin\theta . r \cos\theta \;<\text{circle}\;<\; n.r \tan\theta\, .\, r$$

$$\frac{\pi\, r^2 \sin\theta .\, \cos\theta}{\pi/n} < \text{circle} < \frac{\pi\, r^2 \tan\theta}{\pi/n}$$

but $\theta = \pi/n$

therefore $$\frac{\pi\, r^2 \sin\theta .\, \cos\theta}{\theta} < \text{circle} < \frac{\pi\, r^2 \sin\theta}{\theta . \cos\theta}$$

now let $n \to \infty$ so that $\theta \to 0$ then $\sin\theta/\theta \to 1$

therefore $$\pi\, r^2 \cos\theta \;<\text{circle}\;<\; \frac{\pi\, r^2}{\cos\theta}$$

thus the area of the circle is squashed between two limits, both of which become $\pi\, r^2$ therefore

area of the circle = $\pi\, r^2$

A Circular Argument

Consider this proof that $\frac{\sin\theta}{\theta} \to 1$

OAB is a sector of a circle of angle θ radians and radius r
AT is the tangent at T

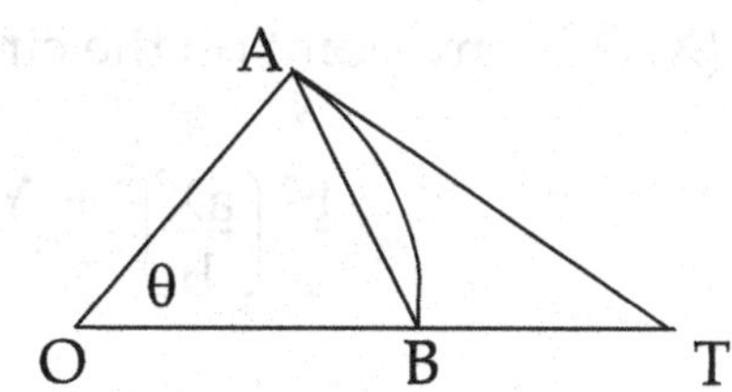

We have $\Delta OAB < \text{sector } OAB < \Delta OAT$

$$\tfrac{1}{2} r^2 \sin\theta < \tfrac{1}{2} r^2 \theta < \tfrac{1}{2} r^2 \tan\theta$$

$$\sin\theta < \theta < \tan\theta$$

divide by sin θ $$1 < \frac{\theta}{\sin\theta} < \frac{1}{\cos\theta}$$

since $\cos\theta \to 1$, $$\frac{\theta}{\sin\theta} \to 1$$

therefore $$\frac{\sin\theta}{\theta} \to 1$$

Unfortunately, the derivation for the area of a circle = πr^2 could use this limit and we need πr^2 in order to give the area of the sector.
So this "proof" could well be an example of a **circular argument.**
Many textbooks in advanced mathematics give this as a proof of this important limit. If so, it should also be shown how the formula πr^2 can be derived rigorously without using the limit.

================================

Stretching a Circle. The Area of an Ellipse

The following short proof verifies that if we stretch the circle $x^2+y^2=b^2$ by the right amount, we get the standard ellipse $\frac{x^2}{a^2} + \frac{y^2}{b^2} = 1$

Let the stretch factor be $\frac{a}{b}$ so that the point $P(x,y) \rightarrow P'(\frac{ax}{b}, y)$

then if P(X,Y) is any point on the circle, we have $X^2+Y^2=b^2$

so that $$\frac{b^2}{a^2}\left(\frac{aX}{b}\right)^2 + Y^2 = b^2$$

or, $$\frac{1}{a^2}\left(\frac{aX}{b}\right)^2 + \frac{Y^2}{b^2} = 1$$

showing that $\left(\frac{aX}{b}, Y\right)$ is a point on the ellipse $\frac{x^2}{a^2} + \frac{y^2}{b^2} = 1$

This shows that the stretch $P(x,y) \rightarrow P'\left(\frac{ax}{b}, y\right)$

carries every point on the circle to a point on the ellipse.

Reversing the argument, if P'(X,Y) is any point on the ellipse,

we find that $P\left(\frac{bX}{a}, Y\right)$ is a point on the circle and here, P "stretches" to P'.

Therefore, the ellipse consists of all of the stretched points of the circle, and only those points.

=======================================

The Area of an Ellipse.

The ellipse $\frac{x^2}{a^2} + \frac{y^2}{b^2} = 1$ is the set of points on the circle $x^2+y^2=b^2$ but

stretched by a factor a/b.

The area of the circle is πb^2. Therefore, using the stretch theorem for areas, the area of the ellipse is $\pi b^2 \times a/b$

hence **the area of the ellipse is πab**

Measuring Volume

Counting Cubes

To find the volume of a solid shape we count the number of unit cubes that we can find inside the surface of the shape.

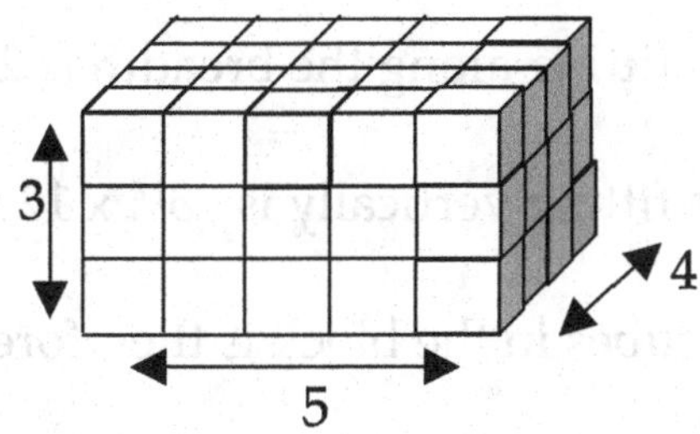

The volume of this 3 by 4 by 5 block is 3x4x5 = 60 because there are three layers each with 4 rows of 5 cubes.

A cuboid has length L, breadth B and height H where L,B and H are whole numbers.

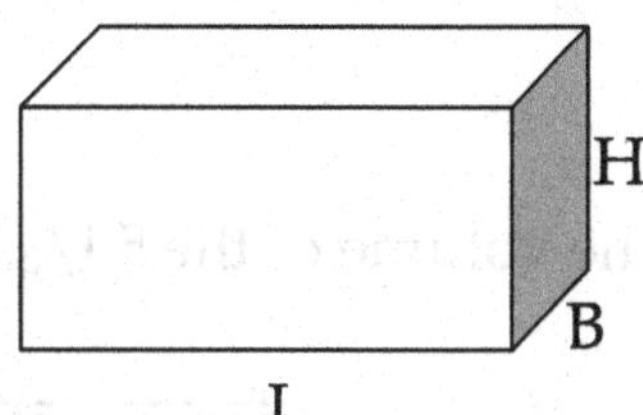

Theorem 15a
The volume of the cuboid is LxBxH

Proof

Each layer holds B rows of L cubes each giving LxB cubes per layer.
There are H layers, giving a total LxBxH

======================

Example
What is the volume of a rectangular block measuring $5\,{}^{1}/_{3}$ by $2\frac{1}{2}$ by $3\frac{1}{2}$

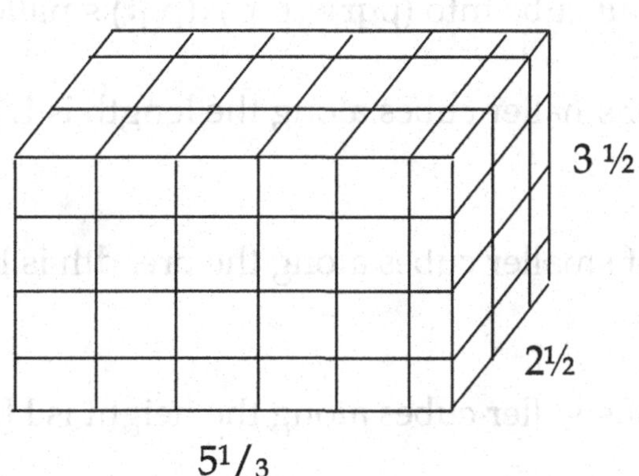

Solution

The product of the denominators of the fractions is 3x2x2 = 12

We therefore divide each unit cube into 12x12x12 smaller cubes.

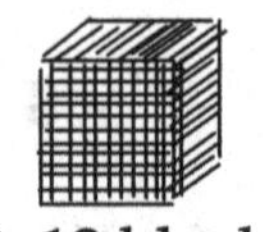

12x12x12 block

The number of smaller cubes fitting along the length is $5^1/_3 \times 12 = 60+4 = 64$

The number of smaller cubes fitting along the breadth is $2½ \times 12 = 24+6 = 30$

The number of smaller cubes fitting vertically is $3½ \times 12 = 36+6 = 42$

The total number of smaller cubes in the block is therefore 64x30x42 (by theorem 15a)

There are 12x12x12 smaller blocks in the unit cube, therefore the number of unit cubes in the block is

$$\frac{64 \times 30 \times 42}{12 \times 12 \times 12} = 46\ ^2/_3$$

The volume of the $5\ ^1/_3$ by 2½ by 3½ block is therefore $46\ ^2/_3$ unit cubes.

==================================

Theorem 15b

If the length, breadth and height of a rectangular block are given by the rational numbers L= P/p , B = Q/q and H=R/r , where P,Q,R and p,q,r are all positive integers, then the volume of the block is again given by LxBxH

Proof

Divide each unit cube into (pqr)x(pqr)x(pqr) smaller cubes, then

The number of smaller cubes along the length is $L.pqr = \frac{P.pqr}{p}$

The number of smaller cubes along the breadth is $B.pqr = \frac{Q.pqr}{q}$

The number of smaller cubes along the height is $H.pqr = \frac{R.pqr}{r}$

The total number of small cubes in the block is therefore $\frac{P.pqr}{p} \times \frac{Q.pqr}{q} \times \frac{R.pqr}{r}$

Now there are $(pqr)^3$ small cubes in each unit cube, therefore, the number of unit cubes in the block is

$$\frac{P(pqr)}{p} \times \frac{Q(pqr)}{q} \times \frac{R(pqr)}{r} \Big/ (pqr)^3 = \frac{P.Q.R}{p.q.r} = L \times B \times H$$

==============================

If the block has sides with irrational lengths we have to go through a limiting process in a way similar to finding the area of the π by π square.

The π^3 cube.

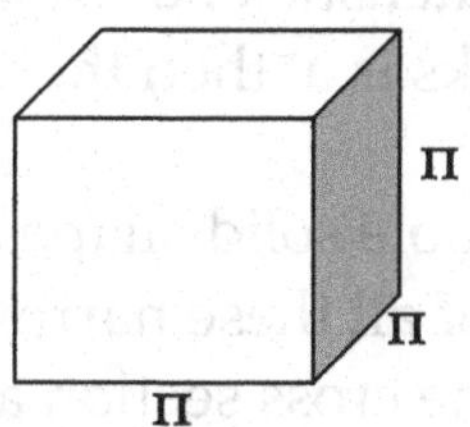

Let the volume of the cube be V.

Take π = 3.14159…..

Then

$$3.1^3 < V < 3.2^3$$
$$3.14^3 < V < 3.15^3$$
$$3.141^3 < V < 3.142^3$$
$$3.1415^3 < V < 3.1416^3$$
$$3.14159^3 < V < 3.14160^3$$

. .

and V is squeezed between two series of numbers, each of which tends to the limit π^3

Therefore $V = \pi^3$

==============================

Theorem 15

Thus we have demonstrated that the volume of any cuboid, of length L, breadth B and height H is given by the formula

V = L.B.H

==============================

Theorem 16 **Stretching a volume**

Suppose that we have a solid figure of volume V residing in x,y z coordinates:

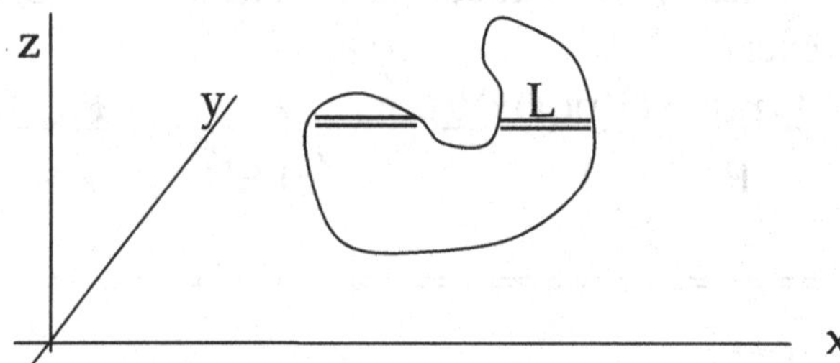

Divide the solid up into thin matchsticks parallel to the x axis and let the length of a typical matchstick be L. Suppose that the area of cross section of each of the matchsticks is **a**, then the volume of a typical matchstick will be **a.L**

The volume of the whole solid shape can be expressed approximately as the sum of the volumes of all these narrow matchsticks, getting closer to the actual volume **V** as the cross section area of the matchsticks is made smaller. Thus

$$V = \lim_{a \to 0} \left(\Sigma\, a.L \right)$$

Now stretch the solid by a stretch factor k, parallel to the x axis:

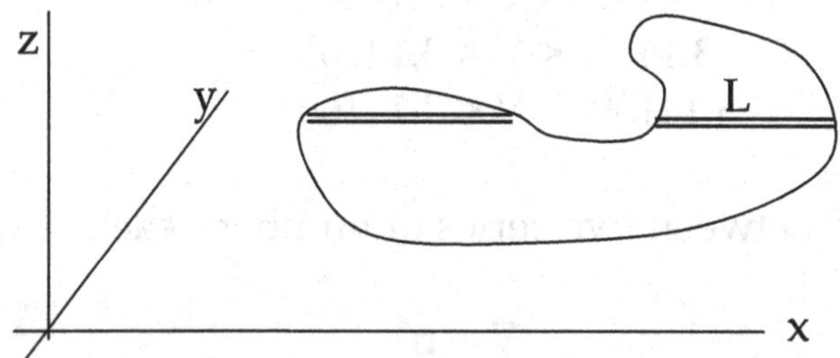

The matchstick is **k** times as long but still has the same thickness so the volume of a typical stretched matchstick is now **a.kL.**

The volume of the whole stretched shape is now

$$V(\text{stretched}) = \lim_{a \to 0} \left(\Sigma\, a.kL \right)$$

$$= k.\lim_{a \to 0} \left(\Sigma\, a.L \right) \qquad = \mathbf{k.V}$$

Thus, if we stretch a solid shape of volume V, by a stretch factor k, the volume will be increased to kV.

(alternatively, we could argue more intuitively, that each unit square (or part of), is stretched from a 1x1x1 cube to a kx1x1 block. In the stretched shape, instead of adding 1x1x1 cubes, we add the same number of kx1x1 blocks and get k times the volume.)

==============================

Theorem 17
An enlargement by scale factor k magnifies the volume by k^3

Proof
(Here we shall use the "intuitive" argument)

Suppose the solid shape of volume V is enlargded from centre O by a scale factor k.

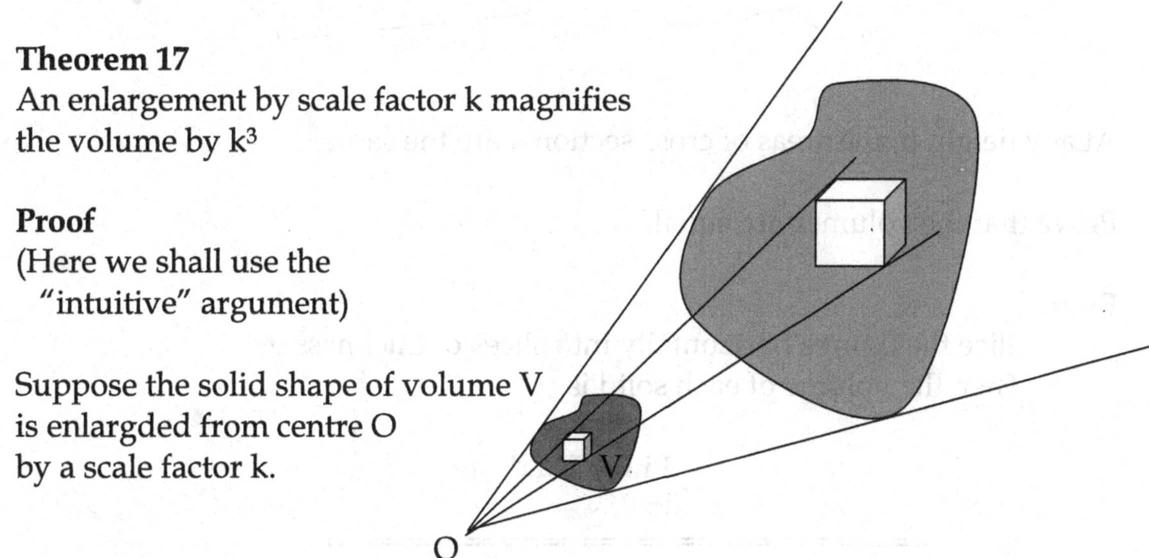

Consider one of the unit cubes that makes up the volume V:

This 1x1x1 unit cube will be enlarged into a kxkxk cube.

Thus each unit cube (or part of) that makes up the volume V will be enlarged by a factor k^3

The volume of the magnified solid shape is therefore k^3.V

======================================

Theorem 18 **Cavalieri's Principle for Volumes**

If two solids shapes have equal cross section areas at every height above the base, then they are equal in volume.

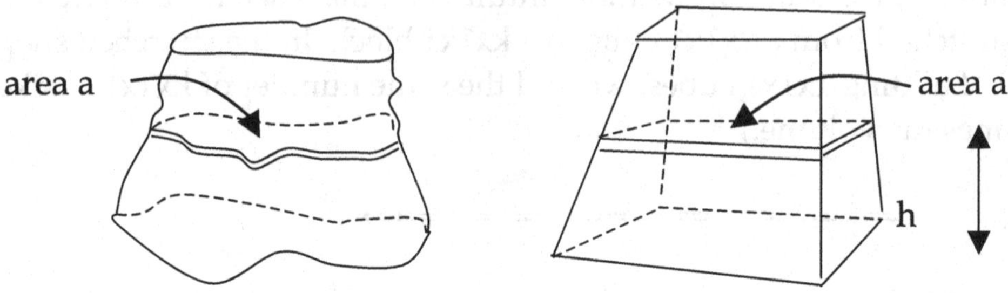

At any height **h**, the areas of cross section **a** are the same.

Prove that the volumes are equal.

Proof

Slice the figures horizontally into slices of thickness d.
Then the volume of each solid is

$$\lim_{d \to 0} \left(\Sigma \, a.d \right)$$

==================================

If a volume is sheared over parallel to its base then its volume does not change

This is simply a special case of Cavalieri's Principle.

In a shear transformation of a solid volume, the areas of cross section at the same height, will be congruent shapes.

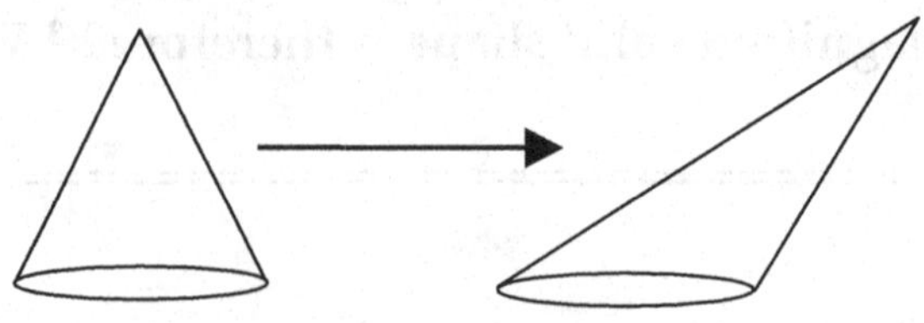

Shear preserves volume

==============================

Solid Prisms

A prism is usually regarded as a transparent solid with rectangular sides used for splitting up light into a spectrum of different colours.

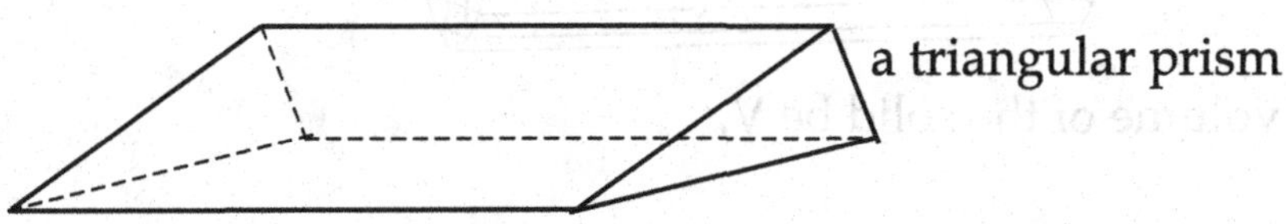

In general, a prism has two polygonal ends with parallelogram sides.

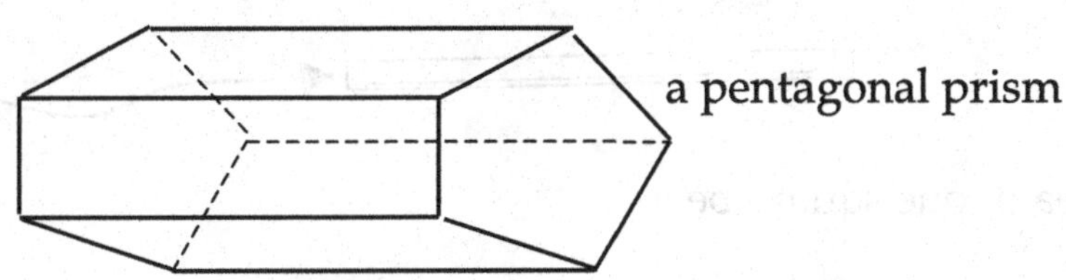

We will use the term "prism like" solid for any solid that has a constant area of cross section.

Prism like solids:

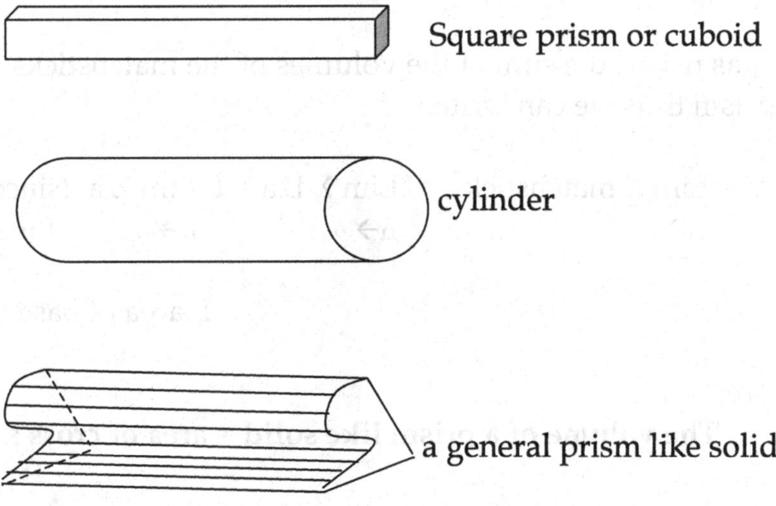

Theorem 19 The Volume of a Prism Like Solid = Length x Area

Proof

Let the volume of the solid be V.

Let the length of the solid be L .

Divide the base of the solid into many small squares (or part squares) and imagine the solid sliced into a large number of thin matchsticks:

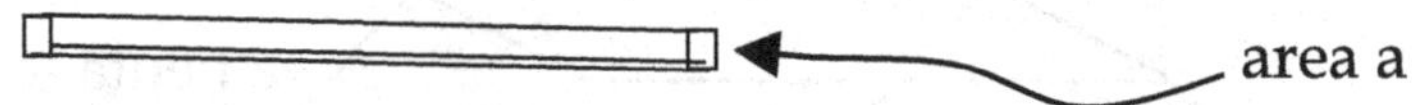

Let the area of one square be a

Again, we argue intuitively:
As the number of squares, $n \to \infty$ then the sum of the areas of all the squares $\to$ the area of the base of the prism.

We can write $\Sigma a \to$ area of base or $\lim_{n\to\infty} \Sigma a =$ area of base

Further, as $n\to\infty$, the sum of the volumes of the matchsticks $\to$ the volume of the prism thus we can write

$$V = \lim_{n\to\infty} \Sigma \text{ matchsticks} = \lim_{n\to\infty} \Sigma L.a = L.\lim_{n\to\infty} \Sigma a \quad \text{(since L is the same for all matchsticks)}$$

$$= L.\text{area of base}$$

Therefore:

The volume of a prism like solid = area of cross section x length

========================

Example: the volume of a cylinder = $\pi r^2.L$

Theorem 20 The Volume of a Square Pyramid = ⅓ base x height

Proof

A cube can be dissected into three congruent square pyramids.

In the following figures we see a cube of side x with diagonal OV and the three square pyramids that meet along OV:

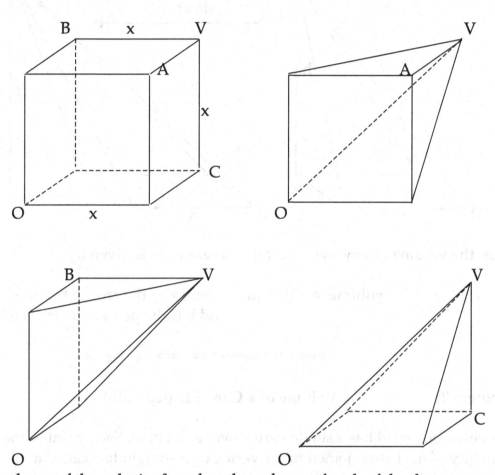

The volume of the cube is x^3 and so the volume of each of the three square pyramids is $^1/_3 x^3$.

Stretching the Pyramid

Take the last of these figures and stretch VC till we reach a height h.

The stretch factor is then h/x.

The volume will now be $^1/_3 x^3$ times the stretch factor

$= {}^1/_3 x^3 . h/x$

The volume of this square pyramid is $^1/_3 x^2 . h$

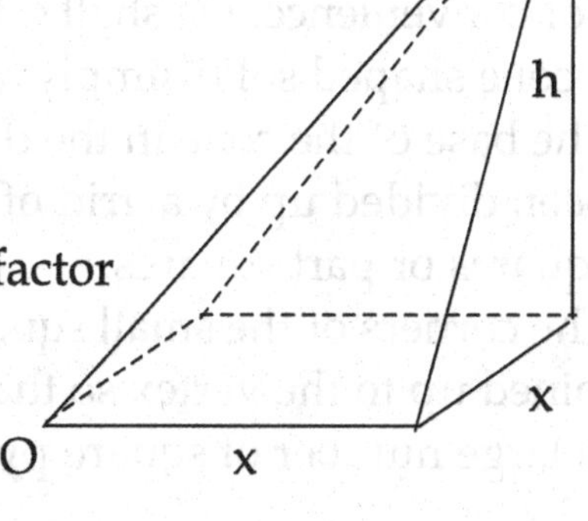

Shearing the Pyramid
By shearing the pyramid horizontally, we can put the vertex of the square pyramid at any point we like, at the same height above the base, without changing the volume:

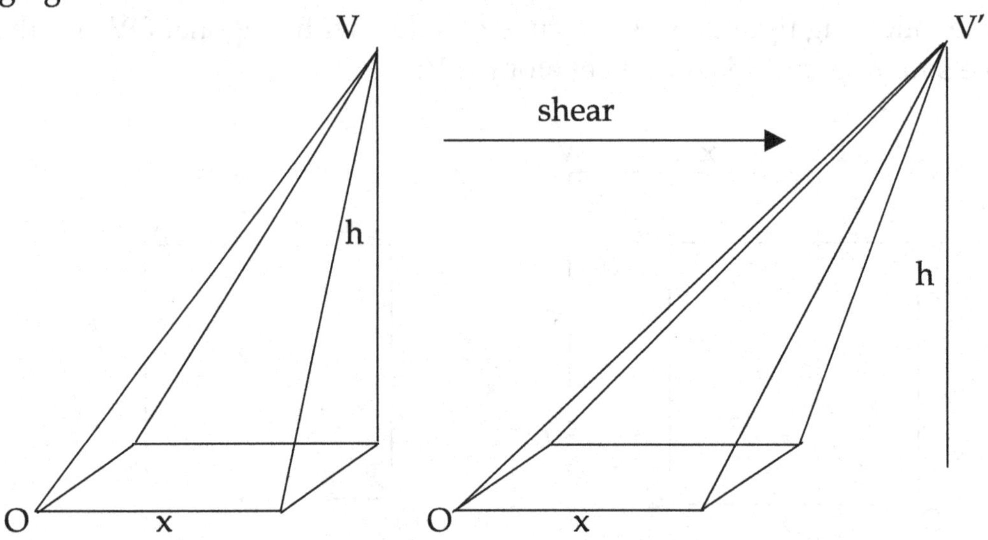

Thus, the volume of any pyramid on a square base is given by

volume = $\frac{1}{3}x^2.h$ where x^2 is the area of the base and h is the perpendicular height.

==========================

Theorem 21 Volume of a Cone Shaped Solid

A cone shaped solid has a single vertex over a flat base. Each point of the boundary of the base is joined to the vertex by a straight line called a generator of the cone.
A cone on a circular base is called a circular cone. A cone on a square base is called a square pyramid.
For convenience, we shall often refer to a cone shaped solid simply as a cone.
The base of the cone in the diagram has been divided up by a grid of small squares or part squares.
The corners of the small squares are joined up to the vertex so that we have a large number of square pyramids.

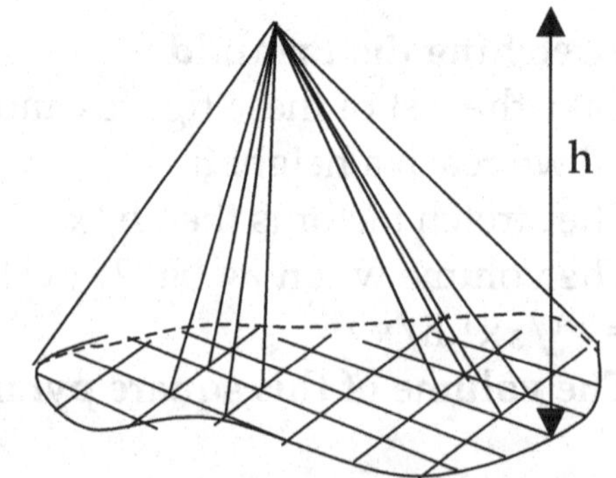

If h is the height of the cone, then the height of each of the square pyramids is also h and so, the volume of each of these square pyramids is ⅓ area of base x h

Thus the sum of the volumes of the square pyramids = ⅓ {sum of areas of bases} .h

Now if we examine the grid of squares slicing up the base of the pyramid we see that some of the squares are totally inside the base and some of the squares are overlapped by the base:

In this diagram 20 squares are inside and 46 squares are overlapped by the base.

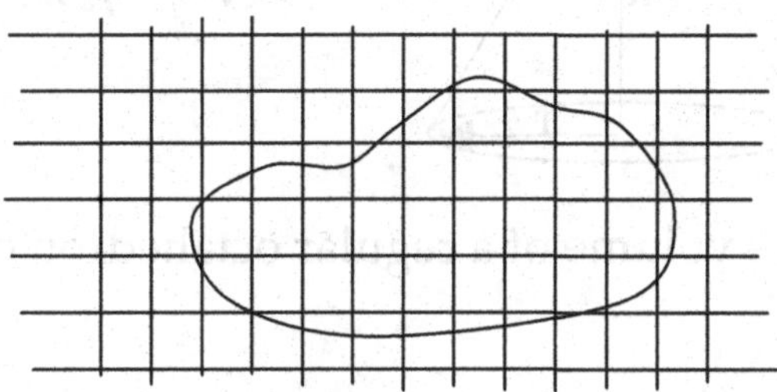

(The area of the base is therefore between 20 squares and 48 squares)

Let the sum of the pyramids on the overlapped squares be V(outside)

Let the sum of the pyramids on the inside squares be V(inside)

Let the volume of the cone be V.

Then $V(\text{inside}) < V < V(\text{outside})$

$\therefore$ ⅓ (inside base).h < V < ⅓ (outside base).h

now let the number of squares $\rightarrow \infty$

then (inside base) and (outside base) each → area of the base of the cone (see page 11)

Therefore V is squashed between two limit, both of which → ⅓ (base).h

Therefore **Volume of a cone = ⅓ (base).h**

Square pyramid

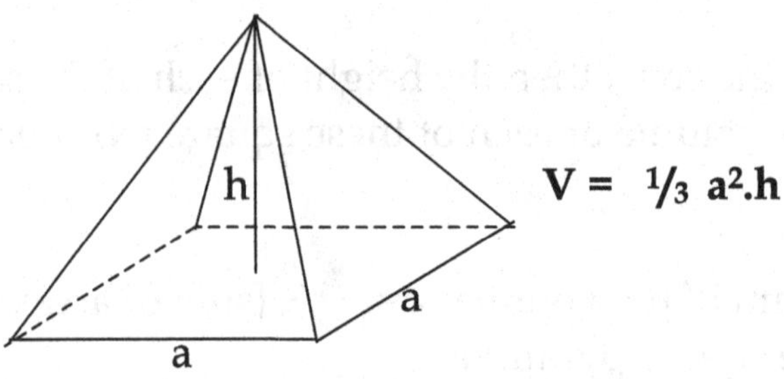

$V = \frac{1}{3} a^2.h$

Cone

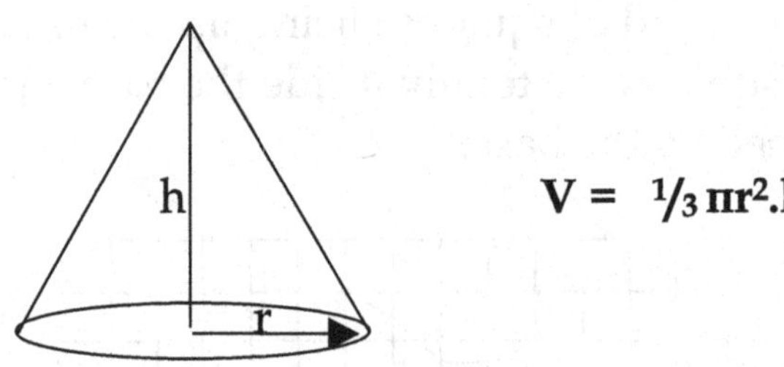

$V = \frac{1}{3} \pi r^2.h$

Exercise Prove that the volume of a regular octahedron of edge **a**, is $\frac{\sqrt{2}\, a^3}{3}$

The Sphere

(an intuitive argument)

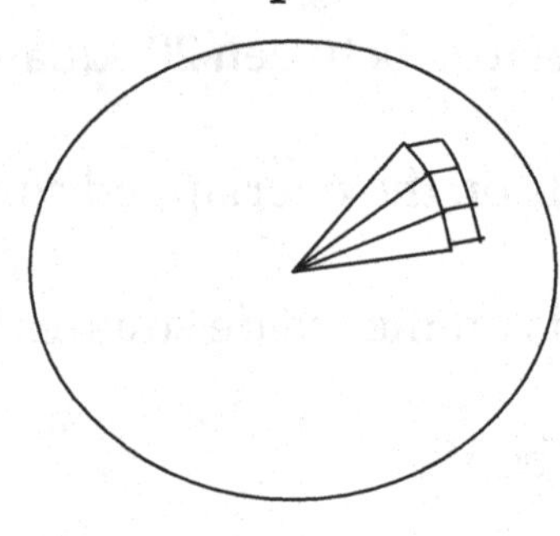

volume of sphere

= sum of pyramids

$= \Sigma\,(\frac{1}{3}$ base . height r)

$= \frac{1}{3}$ r. Σ (bases)

$= \frac{1}{3}$ r. surface area

We divide the surface of the sphere into small patches, e.g. using latitudes and longitudes. Join the corners of all the patches to the centre. Then the volume of the sphere approximates to the sum of the volumes of all these pyramids.
Since the height of each pyramid is the radius of the sphere, the sum of the volumes of the pyramids becomes $\frac{1}{3}$ r. sum of bases = $\frac{1}{3}$ r. surface area.

Thus **Volume of a sphere = $\frac{1}{3}$ radius x surface area**

Theorem 22 The Surface Area of a Sphere = $4\pi r^2$

Place the sphere of radius r inside a cylinder of radius r (see figure.)

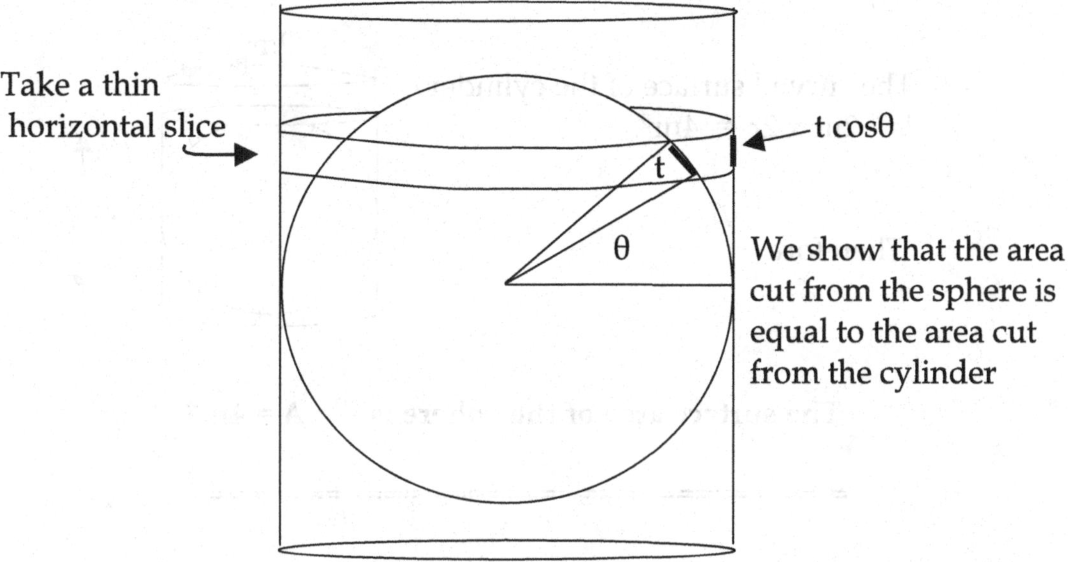
Take a thin
horizontal slice
t cosθ
t
θ
We show that the area
cut from the sphere is
equal to the area cut
from the cylinder

Proof

(We use an "intuitive" proof in the sense that a length t is defined and formulae deduced in terms of t, that can only become accurate when $t \to 0$)

Let the latitude of the slice be θ.

Let the width of the slice on the sphere be t.

The radius of the lower edge of band on the sphere is $r.\cos\theta$ so its length is $2\pi r.\cos\theta$.

The area of the band on the sphere can be taken as therefore $2\pi r.\cos\theta.t$ (length x width) becoming more accurate as $t \to 0$.

The width of the band on the cylinder is $t.\cos\theta$ (as $t \to 0$)

The radius of the band on the cylinder is r so its length is $2\pi r$.

The area of the band on the cylinder is therefore $2\pi r.t.\cos\theta$ (using lengthxwidth, becoming more accurate as $t \to 0$)

The areas of both bands are therefore $2\pi r.t.\cos\theta$ (as $t \to 0$)

From this we deduce that the surface area of the sphere is equal to the curved surface of the cylinder that it fits into.

The curved surface of the cylinder is $2\pi r \times 2r = 4\pi r^2$

$2\pi r$

$2r$

Therefore

The surface area of the sphere is $A = 4\pi r^2$

==================================

Intuitive arguments

(Intuitive = immediate insight without reasoning (Latin intueri = look into))

We have used a number of "intuitive arguments" in this chapter that, while not mathematically rigorous to the standard that a professional mathematician would require, nevertheless, clarify ideas for the student. My mathematics master at school used to say "nothing is obvious" and so we should not regard intuitive arguments as concrete proofs.

Here are two examples of "intuitive" arguments that we have mentioned:

Area of a circle

[1]
Slice up the circle into sectors and reform it into a "rectangle" of width r.
Since the circumference is $2\pi r$, the length of the rectangle will be πr.

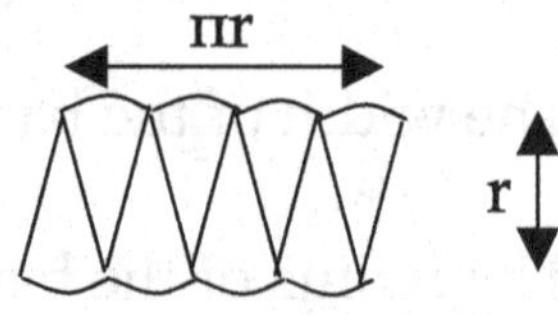

Rectangle = $\pi r \times r$, therefore area of circle = πr^2

==========================

[2] Divide the circle into sectors. Each sector is roughly a triangle of height r.

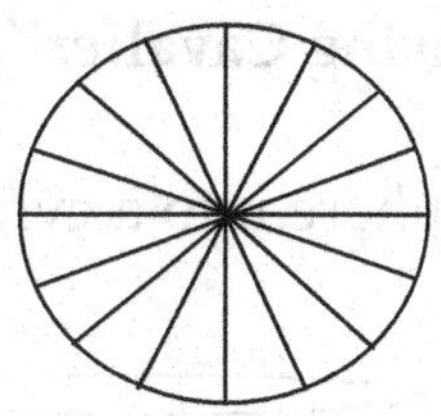

Add up the triangles and we get

area = sum of (½ base x height) = sum of (½ base x r) = ½ r (sum of bases)

$= \frac{1}{2} r \times \text{circumference} = \frac{1}{2} r \times 2\pi r = \pi r^2$

Therefore the area of circle = πr^2

=========================

We now present a similar "intuitive argument" for the volume of a sphere:

Volume of a Sphere

Theorem 23 **The Volume of a Sphere, radius r is** $\frac{4}{3}\pi r^3$

[1] "an intuitive proof"

Draw n equally spaced longitudes:

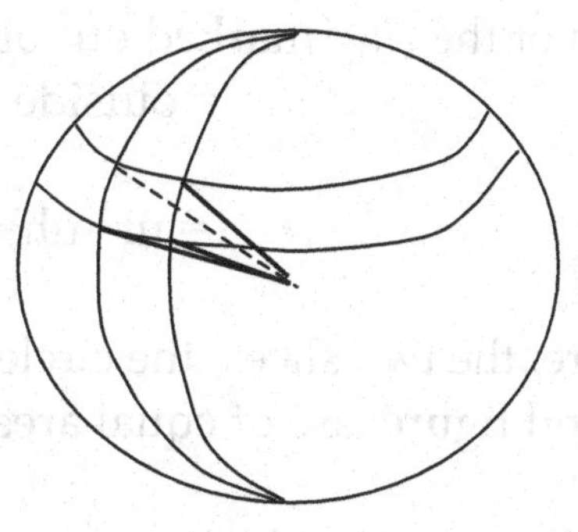

Draw (n-1) circles of latitude:

(Just 2 of each have been drawn here)

The n longitudes and (n-1) latitudes
Will divide the surface area into n^2 patches.

Join the corners of each patch to the centre to form a pyramid on a slightly curved base of area b say.
The sum of the volumes of these pyramids is $\Sigma\, {}^1/_3\, b \,.\, r = {}^1/_3\, r . \Sigma b$

but in the limit, Σ b will be the surface area of the sphere, therefore, as $n \rightarrow \infty$, the sum of the volumes of the pyramids becomes

$${}^1/_3\, r \times (\text{surface area})$$

$$= {}^1/_3\, r \times (4\pi r^2) = \frac{4}{3}\pi r^3$$

====================================

[2] **Proof using Cavalieri's Principle.**

We fit the sphere into a cylinder of length 2r and radius r:

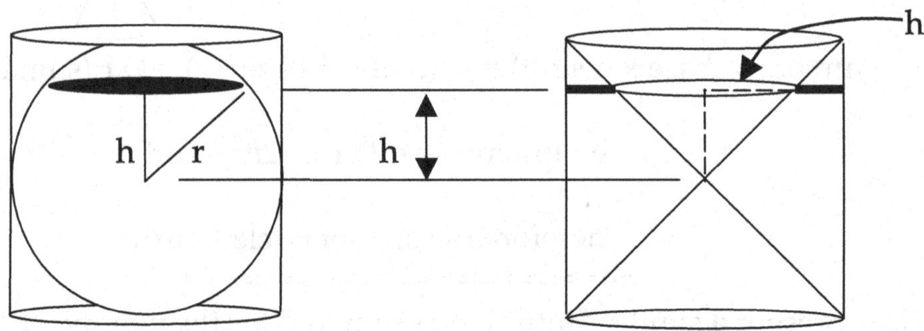

The circular slice of the sphere shown in the left hand figure a radius r^2-h^2 (by Pythagoras)
Therefore the area of the circle

$$= \pi(r^2-h^2)$$

The area of the ring marked out black in the right hand figure

= outside circle – inside circle

$= \pi r^2 - \pi h^2$ (since both dotted lines are length h)

Therefore, the two slices, the circle in the left hand figure and the ring in the right hand figure, are of equal area at all heights above the middle line.

By Cavalieri's principle, the volumes of the two solids are equal.
Doubling the two volumes, we see that

Sphere = cylinder – two cones

$$= \pi r^2(2r) - 2.(\tfrac{1}{3}.\pi r^2.r)$$

$$= 2\pi r^3 - \tfrac{2}{3}\pi r^3$$

Therefore, **volume of the sphere =** $\mathbf{\dfrac{4\pi r^3}{3}}$

======================================

Theorem 24 **The Volume of an Ellipsoid** $\frac{4\pi ab^2}{3}$

Proof

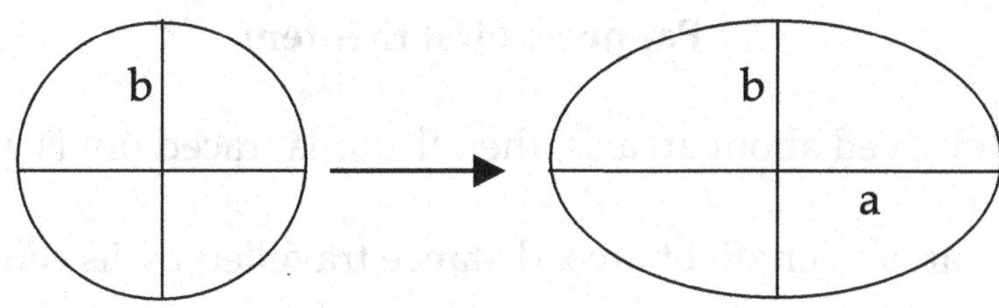

We know that the ellipse on the right is a stretch by factor a/b, of the circle on the left.

If we rotate both of these figures about the horizontal axis, we get a sphere on the left and an ellipsoid on the right.

The volume of the sphere is $\frac{4\pi b^3}{3}$

By Theorem 17, if we stretch by a factor a/b and we have the stretched volume

$$= \frac{a}{b} \times \frac{4\pi b^3}{3}$$

Volume of ellipsoid about the major axis $= \frac{4\pi ab^2}{3}$

======================

Example
If the ellipse is rotated about its minor (y) axis, what will its volume be?

Solution
Using a squash factor b/a on the sphere $\frac{4\pi a^3}{3}$ we get $\frac{4\pi a^2 b}{3}$

================================

Pappus of Alexandria was a remarkable Greek mathematician who lived from around 290 AD to 350 AD. We quote without proof here Pappus's theorems on volumes and areas of revolution:

Pappus's First theorem

If an arc is revolved about an axis, then the area traced out is given by

area = length of arc x distance travelled by its centroid

The Surface of a Cone

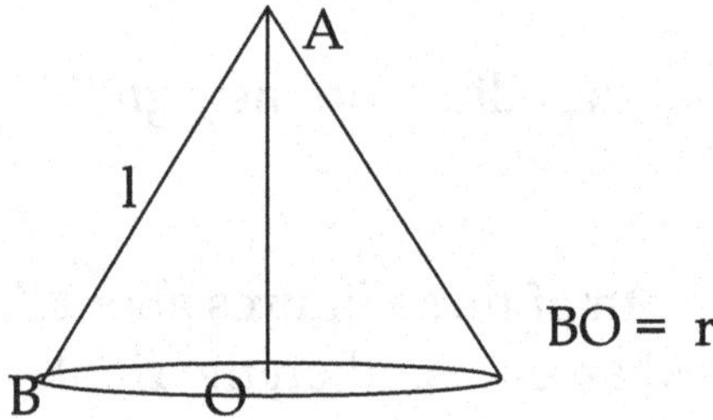

The line AB, of length l is rotated about the vertical line AO to generate a cone of base radius r

If the radius of the base is r, then the centroid of AB (its mid point) travels round a circle of length $2\pi(r/2) = \pi r$

By Pappus's first theorem, the **area of the curved surface of the cone is πrl**

==================================

The area of a doughnut

If a circle is rotated about an axis that does not intersect its circumference then it generates a doughnut shaped solid called a torus.
(Latin: torus=cushion).
A circle of radius r is rotated about such an axis so that its centre goes round a circle of radius R.
The centroid of the circle (its centre) travels round a circle of length $2\pi R$.
The length of the circle is $2\pi r$.
Therefore,
by Pappus's first theorem, the area of the doughnut is $2\pi r \times 2\pi R$

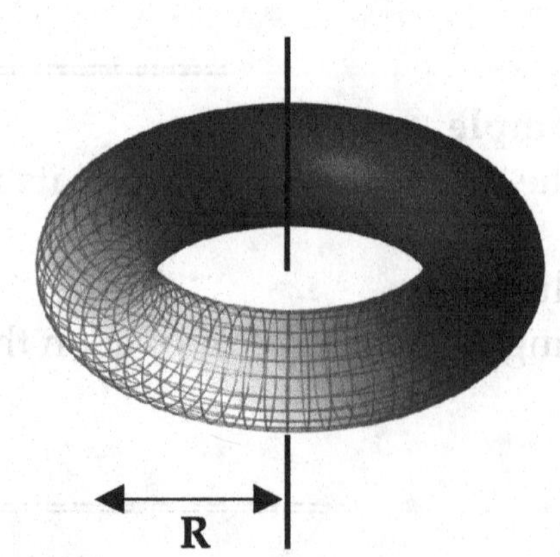

Surface of doughnut= $4\pi^2 rR$

Pappus's Second Theorem

If an area is rotated about an axis then the volume traced out is given by

Volume = area x distance travelled by the centroid

The Volume of a Doughnut

The circle of radius r is rotated round a circle of radius R to form a torus.

The area of the circle is πr^2

The distance travelled by the centroid (= the centre) is = $2\pi R$

Therefore, by Pappus's second theorem, the volume of the torus is $\pi r^2 \times 2\pi R$
Hence:

Volume of the doughnut = $2\pi^2 r^2 R$

==============================

Exercise 8

[1] Use Pappus's second theorem to prove that the volume of a cone of height h and base radius r is $\frac{1}{3}\pi r^2 h$

(Hint: the centroid of the triangle is distance $\frac{1}{3}r$ from the axis)

[2] **The metal ring**

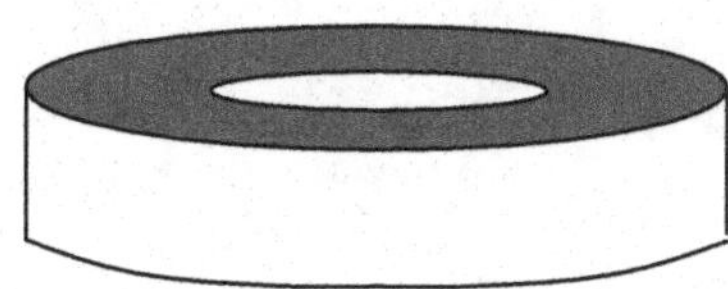

The cross section of a metal ring is an **a x a** square.
The radius of the ring from the centre of the square is **R.**

What is the volume of the ring?

==

Answers

Exercise 2

[1] 2x6+2x2 = 16   [2] 4x5-2x2 = 16   [3] 3x7+2x2 = 25   [4] 5x5 + 2 = 27

Exercise 4   [1] $2 + \frac{1}{4}\pi$ [2] $2\pi + \frac{1}{4}\pi^2$
(note: the second figure is an enlargement of the first by a scale factor π)

Exercise 5 [1] $\frac{1}{8}\pi$ [2] $\frac{1}{8}\pi^3$ [3] $A = \frac{1}{2}$ $B = \frac{1}{4}\pi - \frac{1}{2}$ $C = \frac{1}{2}$

(area [2] is an enlargement of [1] by a scale factor π)

Exercise 6 about 6 feet

====================================

CHAPTER 21

Ceva, Menelaus and Morley

Duality: Names and Shapes

Most people would call a three-sided figure a triangle and a four-sided figure a quadrilateral. The word triangle refers to the fact that the three-sided figure has three angles but the word quadrilateral refers to the fact that the figure has four sides. We could equally refer to the triangle as a trilateral and a quadrilateral as a quadrangle.

(Latin: tres=three; quattuor=four ; latus=side)

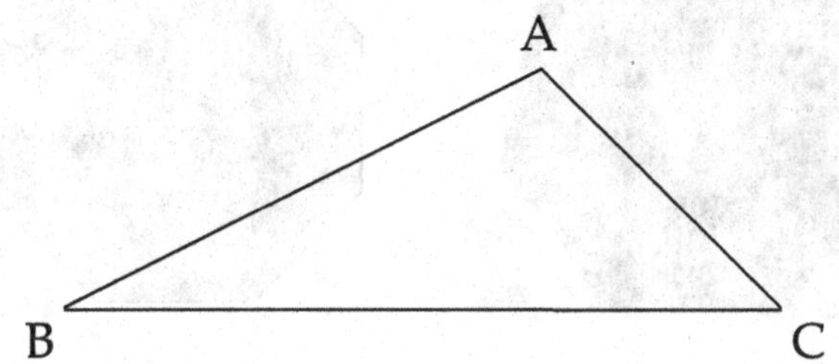

Triangle ABC or Trilateral ABC

According to the dictionary, a quadrangle is a four sided court in a college but in geometry, it is a figure with four angles whereas a quadrilateral is four sided figure.

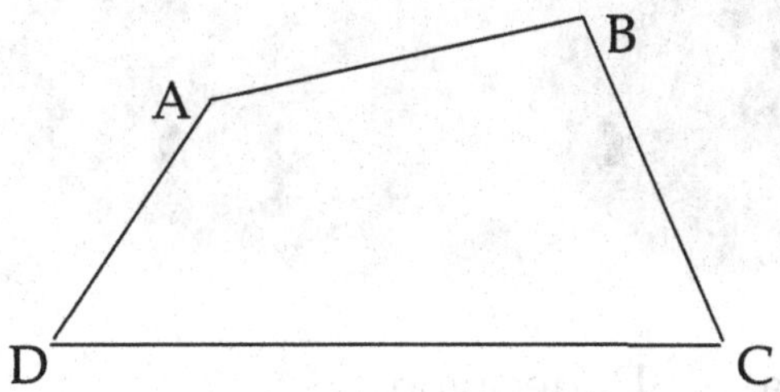

Quadrilateral ABCD has four sides

Quadrangle ABCD has four angles

From five sides onwards we switch to using Greek and refer to angles, "gonos" being the Greek for an angle, thus a polygon has many angles!

We do not call this a trigon 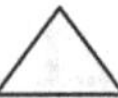 nor this a tetragon 

however, from five onwards we have

5 = penta
pentagon

6 = hexa
hexagon

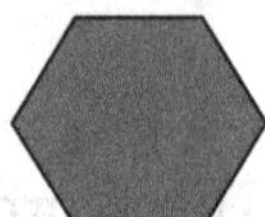

7 = hepta
heptagon

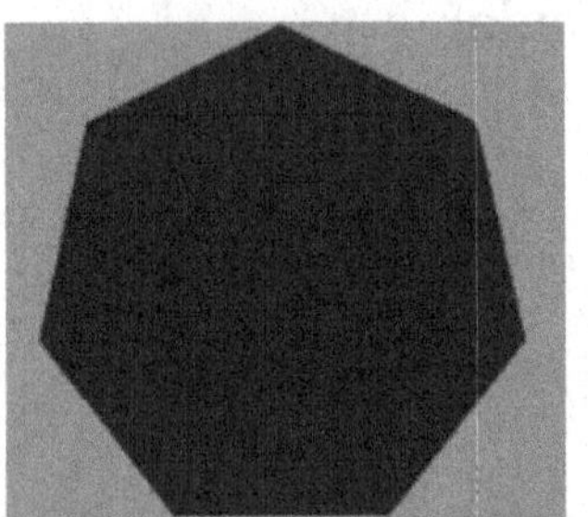

8 = okto
octagon

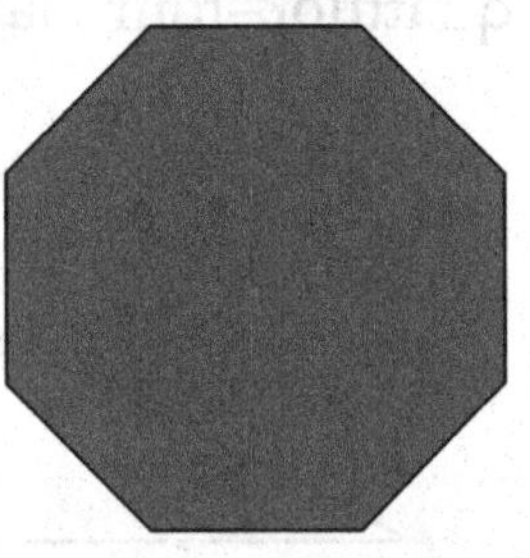

9 = nonus (Latin)
nonagon

10 = deka
decagon

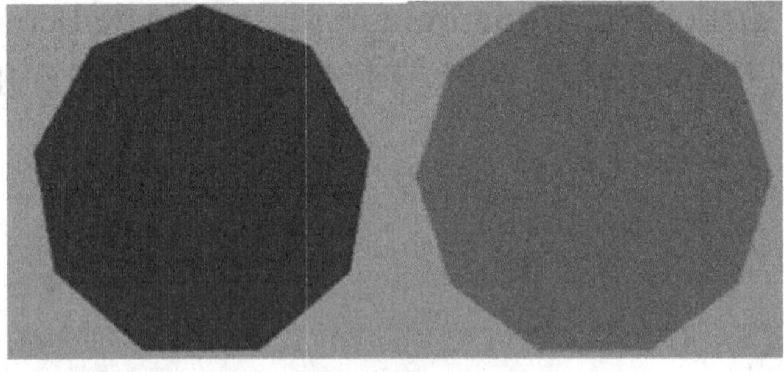

12 = dodeka Dodecagon

We have a similar use of Greek and Latin in naming familiar solid shapes

Tetrahedron = 4 faces

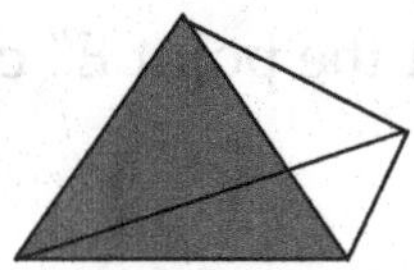

Octahedron = 8 faces

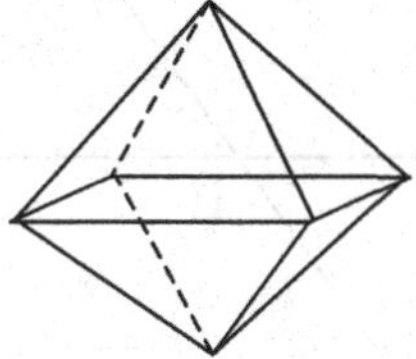

The Dodecahedron has 12 pentagonal faces

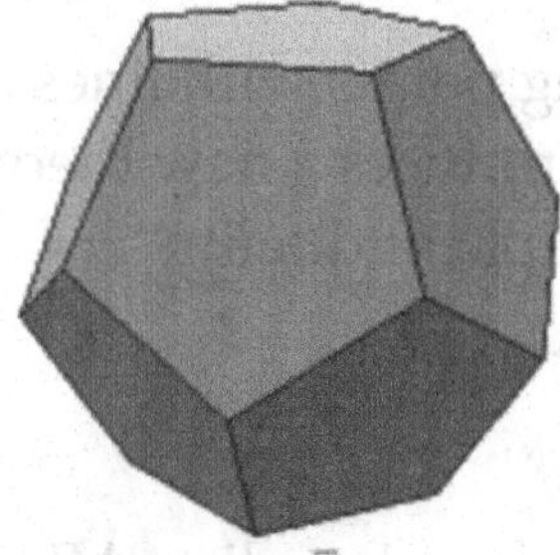

The Icosahedron has 20 triangular faces

(Greek: eikos=20)

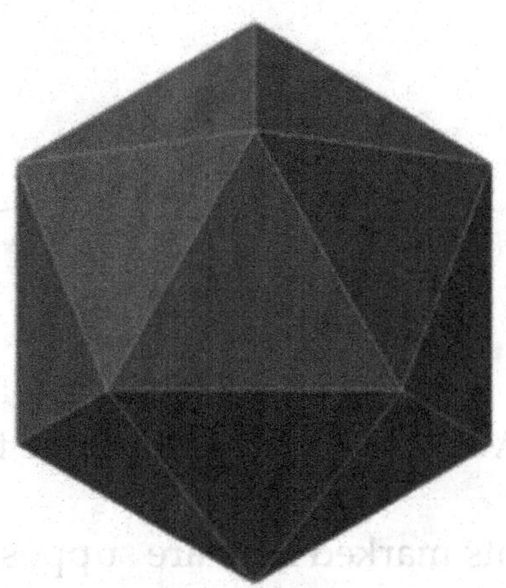

Notation
If two lines AB and CD intersect in a point E, then we can use the "intersection" symbol from set theory, that is the symbol ∩, to express this in a neat and concise way.
"The line AB meets the line CD at the point E" can be expressed as

$$AB \cap CD = E$$

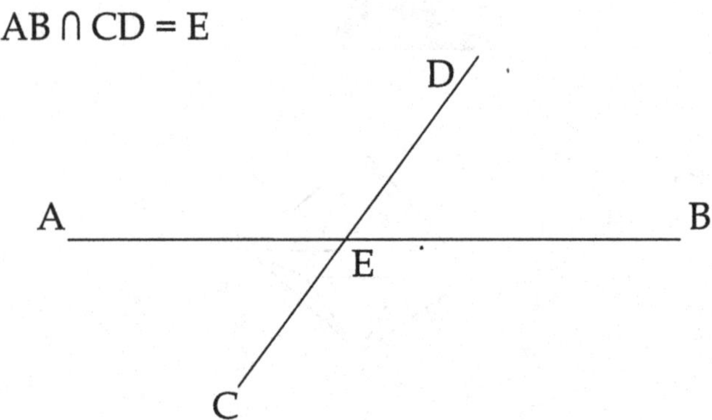

It is often convenient to denote points by capital letters and whole lines with lower case letters.
We will find that interchanging points with lines and lines with points, in the wording of a theorem, often produces a new theorem that is also true. This idea is described as "the principle of duality".

Duality:

Two points A and B determine a unique line AB.
Two lines a and b determine a unique point ab.

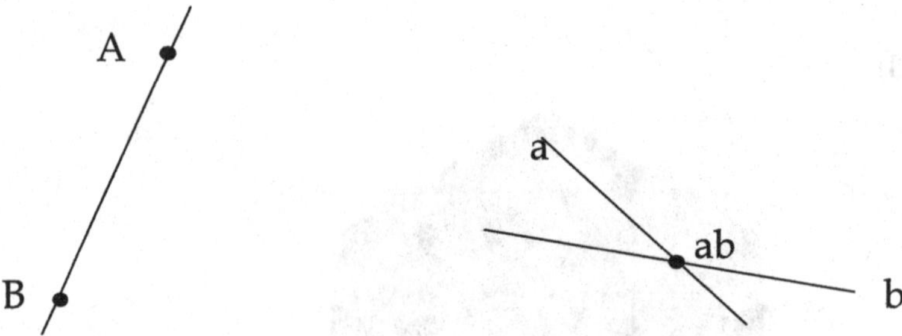

Two point A and B
Determine the unique line AB

Two lines a and b
determine the unique point ab

(Note however that the points marked here are supposed to have no dimension. They give position without size)

The use of the words triangle, trilateral, quadrilateral, quadrangle illustrates a duality in the way that we view and understand these plane figures.

Triangles and Trilaterals

The triangle formed by the three points A,B and C gives three sides named AB, BC and CA.

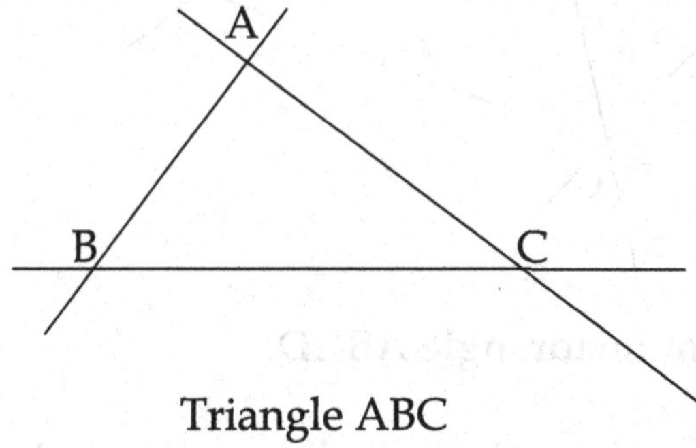

Triangle ABC

A trilateral formed by the three lines **a,b** and **c** gives three points named **ab, bc** and **ca.**

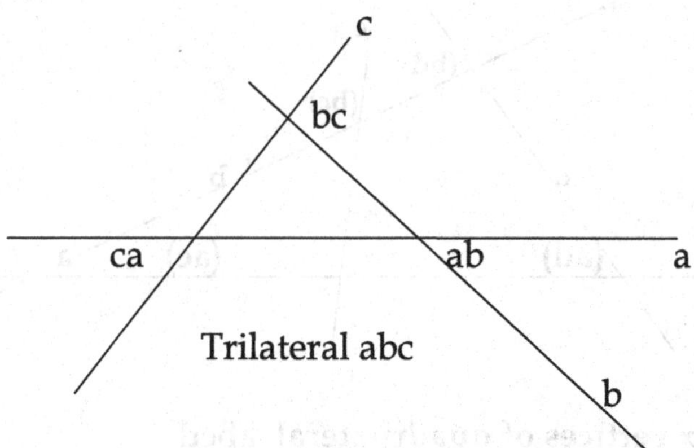

Trilateral abc

Using the intersection symbol ∩ we have:

$A = b \cap c$	$B = c \cap a$	$C = a \cap b$
a = BC,	b = CA,	c = AB

Quadrangles and Quadrilaterals

A quadrangle formed by the four points A,B,C and D gives six sides named AB, AC, AD, BC, BD and CD.

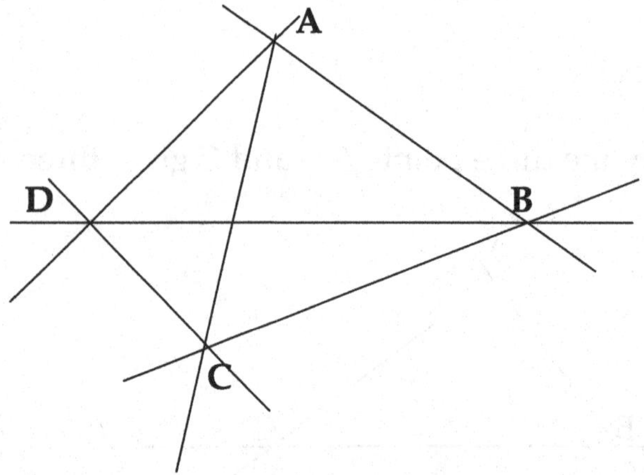

Six sides of quadrangle ABCD

The quadrilateral formed by the four lines **a,b,c** and **d** has six vertices named **ab, ac, ad, bc, bd** and **cd.**

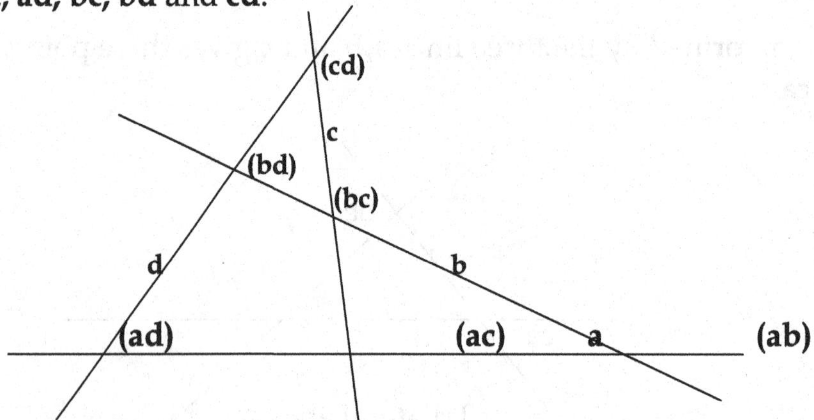

Six vertices of quadrilateral abcd

The Principle of Duality

The way in which we switch lines for points and points for lines in a statement concerning a geometrical figure composed of points and straight lines is called **Duality**

The principle of duality sometimes gives us two theorems for only one proof.

Examples of Duality

Two points A and B define a unique line AB called the join of A and B.

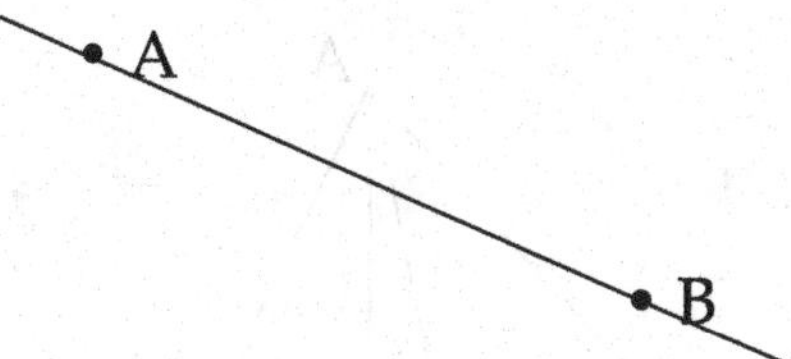

Two lines **a** and **b** define a unique point **ab** called the intersection of **a** and **b**.

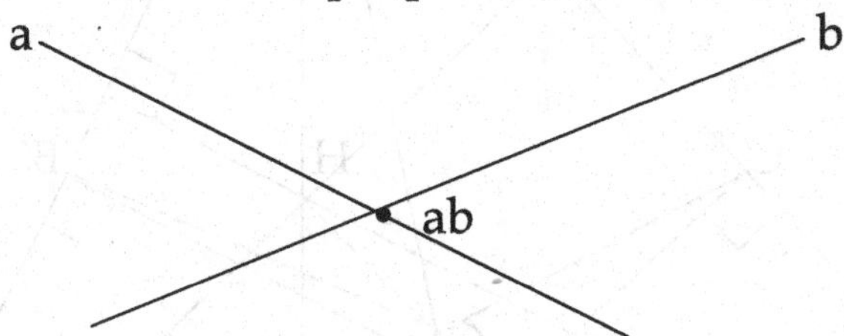

The points A,B and C are collinear, lying on the line **l**.

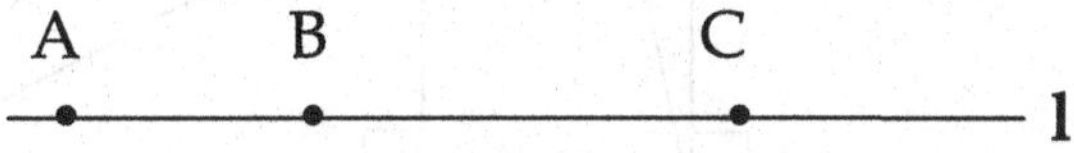

The lines **a,b** and **c** are concurrent, intersecting at the point L.

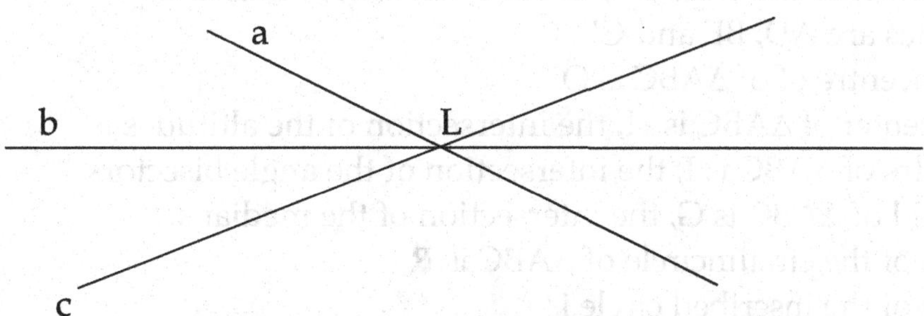

However, in three dimensions we have a different kind of duality. The dual of a plane is a point and the dual of a point is a plane but the line is self dual

Two plane intersect in a line

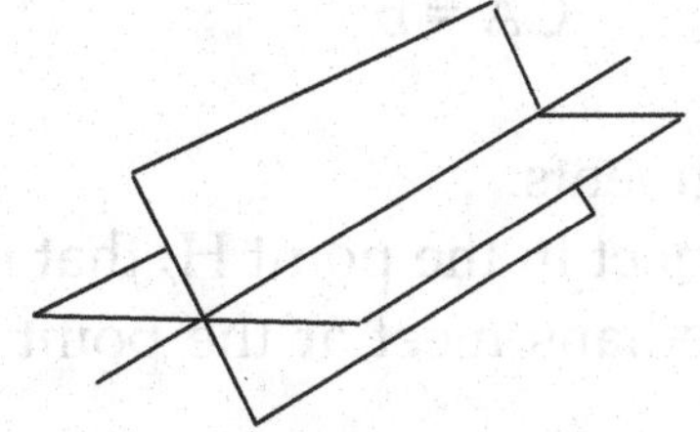

Two points lie on a line

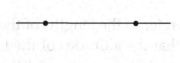

Triangle Notation

The next figure shows the usual way of labelling the important points of a general triangle ABC.

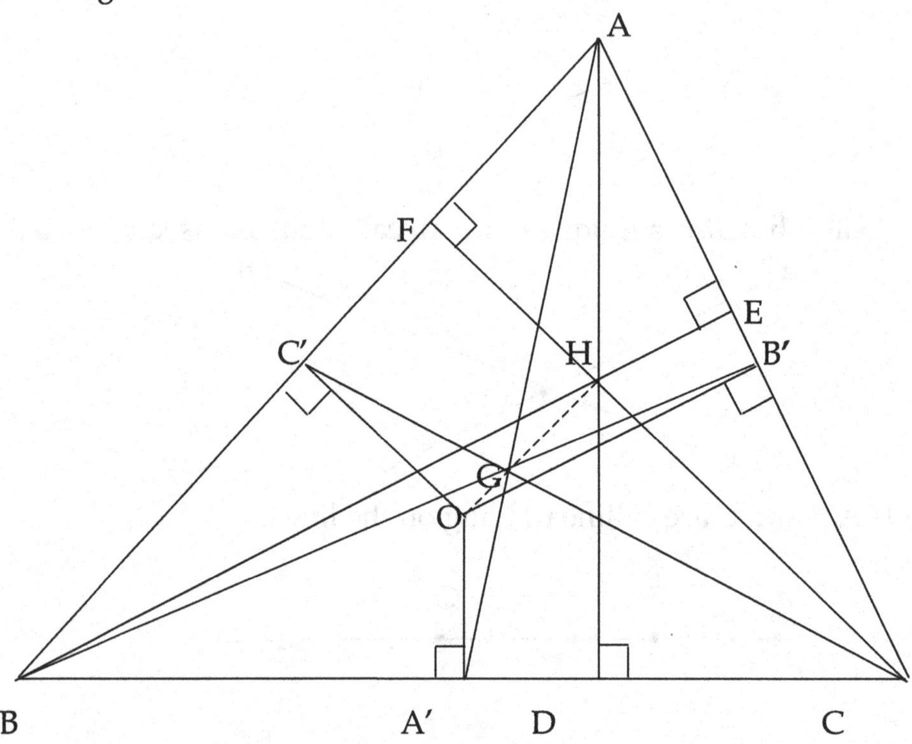

The mid points of the sides BC, CA and AB are A′, B′ and C′
The altitudes are AD, BE and CF
The circumcentre of of ΔABC is O
The orthocenter of ΔABC is H, the intersection of the altitudes
The in-centre of ΔABC is I, the intersection of the angle bisectors
The centroid of ΔABC is G, the intersection of the medians.
The radius of the circumcircle of ΔABC is **R**
The radius of the inscribed circle is **r**
The area of the triangle is denoted by Δ
O,G and H will always lie on a straight line (dotted in the figure)
The lengths of the sides of ΔABC are usually denoted by small letters **a,b** and **c** so that

AB = c BC = a CA = b

These equations refer to the lengths of the line segments.
We have stated that the altitudes of the triangle meet in the point H, that the angle bisectors meet in a point I and that the medians meet at the point G.

These facts of course, need to be proved. We will use Ceva's theorem to prove these results.

Giovani Ceva

Giovanni Ceva (1647-1734) was professor of mathematics at the University of Mantua (now Mántova) in northern Italy.
Before we can prove Ceva's theorem we need an algebraic result. In many books this would be referred to as a Lemma. A Lemma is a useful result that can be used to prove a theorem but which does not quite merit the status of a theorem itself.

Lemma 1.
If $a/b = c/d$ then $a/b = (a-c)/(b-d) = (a+c)/(b+d)$ assuming that $b \neq d$

Proof Let $a/b = c/d = k$
Then $a = bk$ and $c = dk$

$$(a-c)/(b-d) = (bk-dk)/(b-d) = k(b-d)/(b-d) = k$$

and

$$(a+c)/(b+d) = (bk+dk)/(b+d) = k(b+d)/(b+d) = k$$

$\therefore a/b = c/d = (a-c)/(b-d) = (a+c)/(b+d)$ because they are all equal to k

■

We now prove a geometrical Lemma that will help in the proof of Ceva's theorem:

Lemma 2

In the figure below, $\dfrac{BP}{PC} = \dfrac{\Delta BOA}{\Delta COA}$

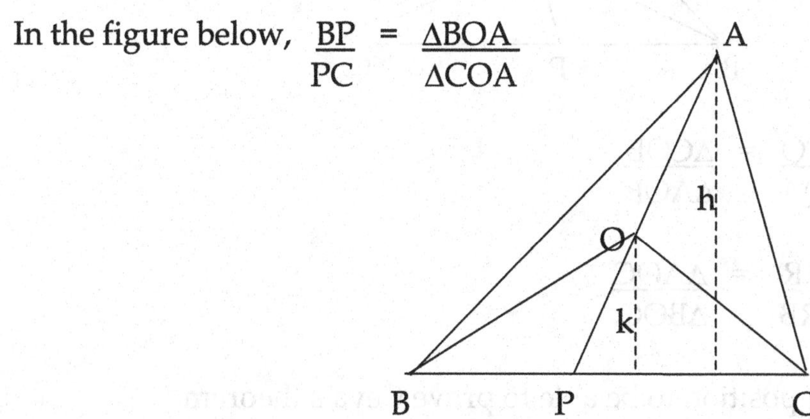

Proof

Let the height of ΔABC from base BC be h.

Then $$\frac{\Delta BAP}{\Delta CAP} = \frac{\frac{1}{2} BP.h}{\frac{1}{2} PC.h} = \frac{BP}{PC}$$

Similarly, if k is the height of triangle BOC from base BC, we have

$$\frac{\Delta BOP}{\Delta COP} = \frac{\frac{1}{2} BP.k}{\frac{1}{2} PC.k} = \frac{BP}{PC}$$

Therefore $$\frac{\Delta BAP}{\Delta CAP} = \frac{\Delta BOP}{\Delta COP} = \frac{BP}{PC}$$

Now using Lemma 1

$$\frac{BP}{PC} = \frac{\Delta BAP - \Delta BOP}{\Delta CAP - \Delta COP} = \frac{\Delta BOA}{\Delta COA}$$

Therefore

$$\frac{BP}{PC} = \frac{\Delta BOA}{\Delta COA}$$

Similarly in this figure,

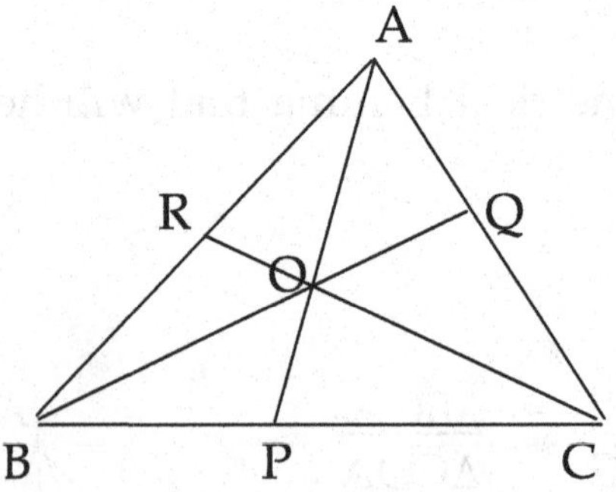

we have $$\frac{CQ}{QA} = \frac{\Delta COB}{\Delta AOB}$$

and $$\frac{AR}{RB} = \frac{\Delta AOC}{\Delta BOC}$$

We are now in a good position to be able to prove Ceva's theorem

Theorem 1 **Ceva' Theorem**

P,Q and R are points on the sides BC,CA and AB of triangle ABC
If the lines AP, BQ and CR in the triangle ABC are concurrent then

$$\frac{AR.BP.CQ}{RB.PC.QA} = 1$$

Proof
The point where AP, BQ and CR intersect could be either inside the triangle or outside the triangle, so there are two cases to consider:

Case 1: O is inside the triangle

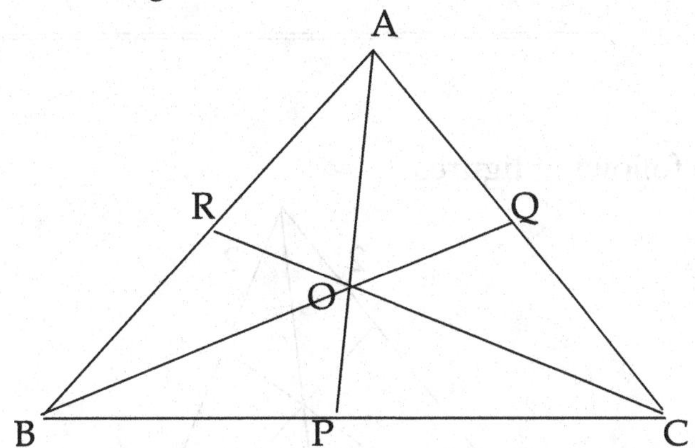

Using Lemma 2, we have

$$\frac{AR.BP.CQ}{RB.PC.QA} = \frac{\Delta AOC.\Delta BOA.\Delta COB}{\Delta BOC.\Delta COA.\Delta AOB} = 1$$

Case 2: O is outside the triangle

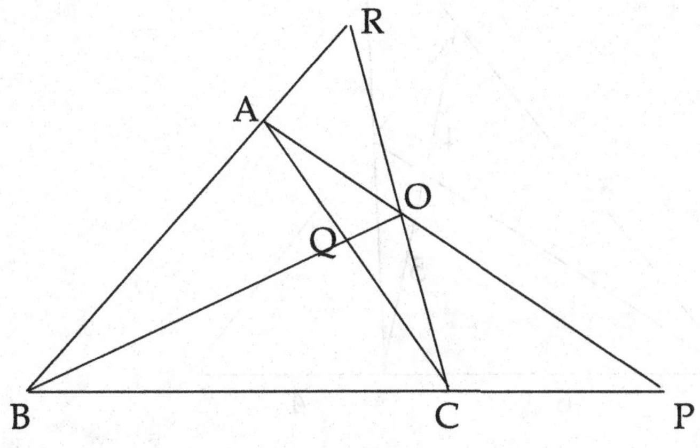

$$\frac{AR}{RB} = \frac{\Delta ARC}{\Delta RBC} = \frac{\Delta ARO}{\Delta RBO} = \frac{\Delta ARC - \Delta ARO}{\Delta RBC - \Delta RBO} = \frac{\Delta AOC}{\Delta BOC}$$

$$\frac{BP}{PC} = \frac{\Delta BPA}{\Delta PCA} = \frac{\Delta BPO}{\Delta PCO} = \frac{\Delta BPA - \Delta BPO}{\Delta PCA - \Delta PCO} = \frac{\Delta BOA}{\Delta COA}$$

$$\frac{CQ}{QA} = \frac{\Delta CQB}{\Delta QAB} = \frac{\Delta CQO}{\Delta QAO} = \frac{\Delta CQB + \Delta CQO}{\Delta QAB + \Delta QAO} = \frac{\Delta BOC}{\Delta BOA}$$

$$\therefore \frac{AR.BP.CQ}{RB\ PC\ QA} = \frac{\Delta AOC.\Delta BOA.\Delta BOC}{\Delta BOC\ \Delta COA\ \Delta BOA} = 1$$

■

Exercise 1

Find x in the following figures

1.

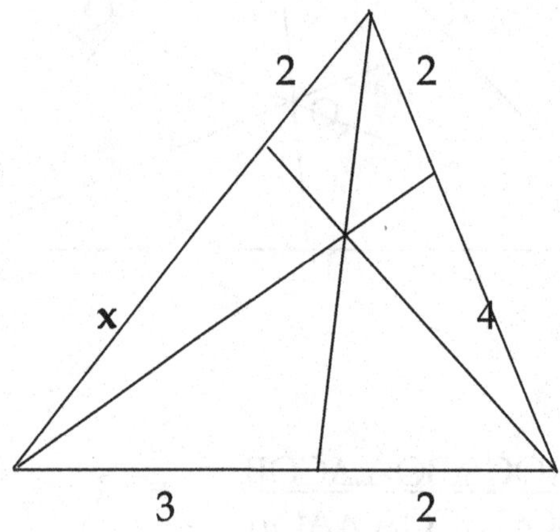

2.

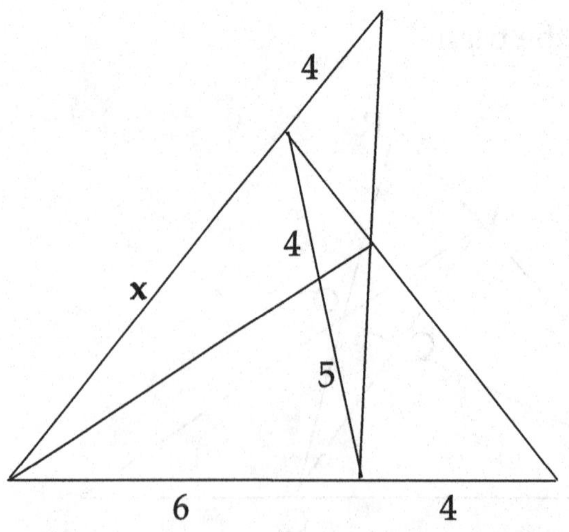

Theorem 2 The Converse of Ceva's Theorem

Case 1

P,Q and R are points on the sides BC, CA and AB of triangle ABC with P between Band C, Q between C and A and R between A and B.

If $\frac{AR.BP.CQ}{RB\,PC\,QA} = 1$ then the lines AP, BQ and CR are concurrent.

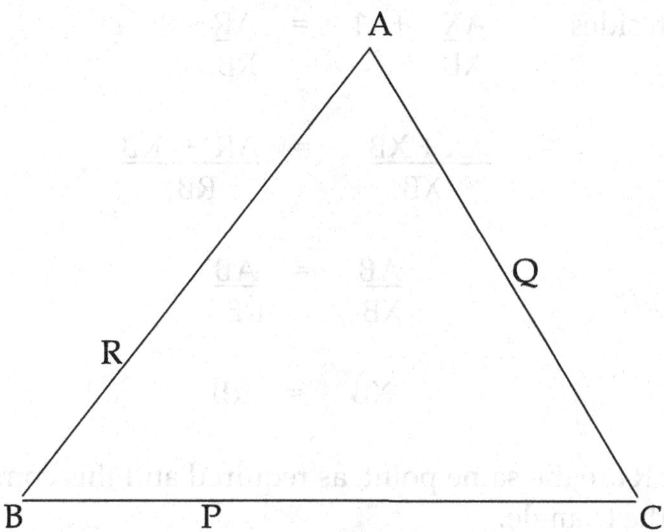

Proof

Let AP and BQ meet at O and let CO meet AB at X then we need to show that R and X are the same point.

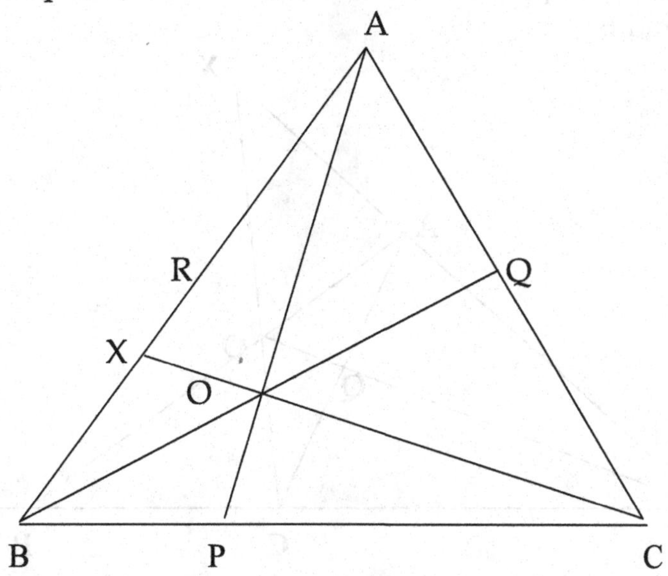

By Ceva's theorem, $\frac{AX}{XB}\cdot\frac{BP}{PC}\cdot\frac{CQ}{QA} = 1$

And we are given that $\frac{AR}{RB}\cdot\frac{BP}{PC}\cdot\frac{CQ}{QA} = 1$

Therefore $\frac{AX}{XB} = \frac{AR}{RB}$

Add 1 to both sides $\frac{AX}{XB} + 1 = \frac{AR}{RB} + 1$

$$\frac{AX + XB}{XB} = \frac{AR + RB}{RB}$$

$$\frac{AB}{XB} = \frac{AB}{RB}$$

Therefore $XB = RB$

Hence X and R are the same point, as required and this completes the proof for O inside the triangle.

Note that it is important to take into account the **direction** of any line segments. XB=RB alone does not prove that X and R are the same point. They could be on opposite sides of B but we know that both X and R are between A and B.

Case 2

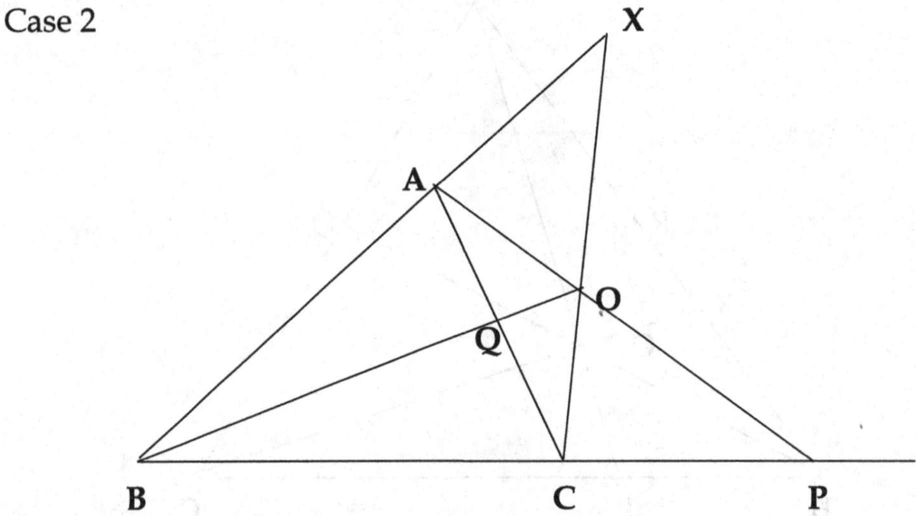

Given $\frac{BP}{PC}.\frac{CQ}{QA}.\frac{AR}{RB} = 1$ then two of the ratios could be negative

In the diagram shown, BP/PC and AR/RB are negative with P on BC produced and R on BA produced.

Let BQ and AP meet at O and suppose that CO meets BA produced at X.

Then by Ceva's theorem $\frac{BP}{PC}.\frac{CQ}{QA}.\frac{BX}{XA} = 1$

Therefore $$\frac{BR}{RA} = \frac{BX}{XA}$$

Subtract 1 from both sides to get
$$\frac{BR}{RA} - 1 = \frac{BX}{XA} - 1$$

$$\frac{BR - RA}{RA} = \frac{BX - XA}{XA}$$

Leading to $$\frac{BA}{RA} = \frac{BA}{XA}$$

Therefore RA = XA so that R and X must be the same point.

■

The converse of Ceva's Theorem is often used to prove some important results in geometry, for example (i) the medians of a triangle are concurrent (ii) the altitudes of a triangle are concurrent etc....

When proving the converse of Ceva's Theorem we should be aware that there are many diagrams for which the converse can apply and in theory, for completeness a proof of the converse should be justified for each possible figure. We have given a proof of the converse for (i) O being inside the triangle and (ii) O being between BA produced and BC produced. The medians of a triangle are always concurrent at a point inside the triangle but this is not the case for the altitudes of a triangle.
We display here some of the cases that we should bear in mind when considering Ceva's theorem and its converse.

Diagrams for Ceva's theorem.

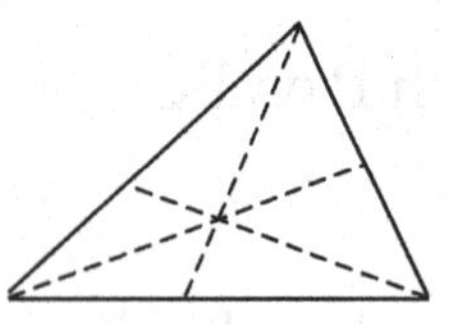

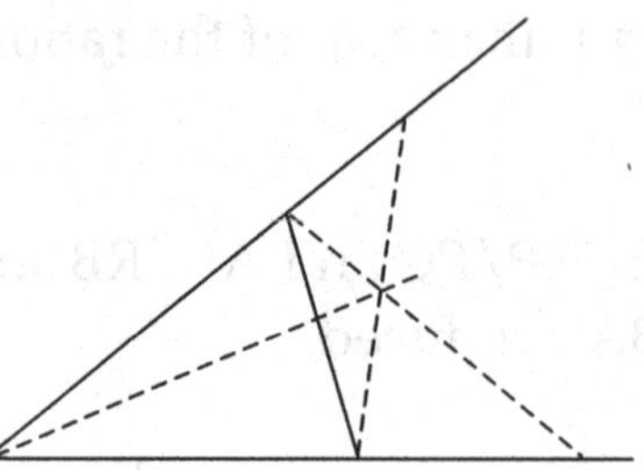

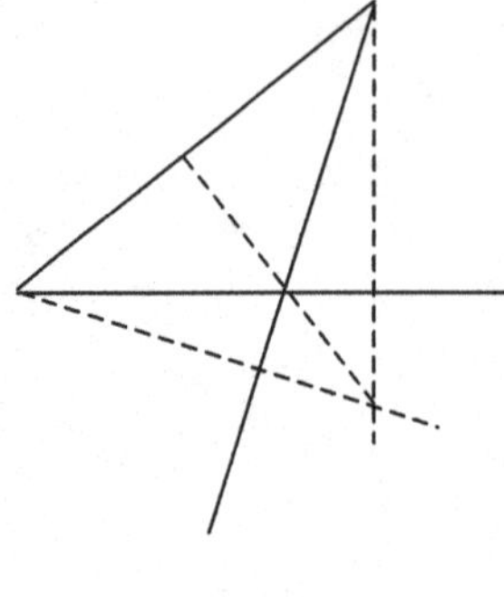

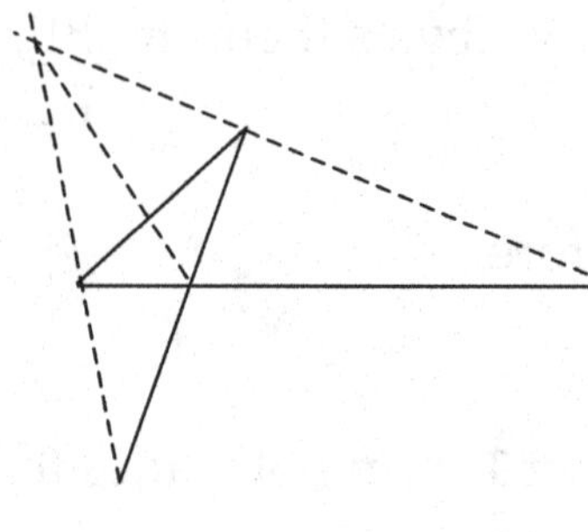

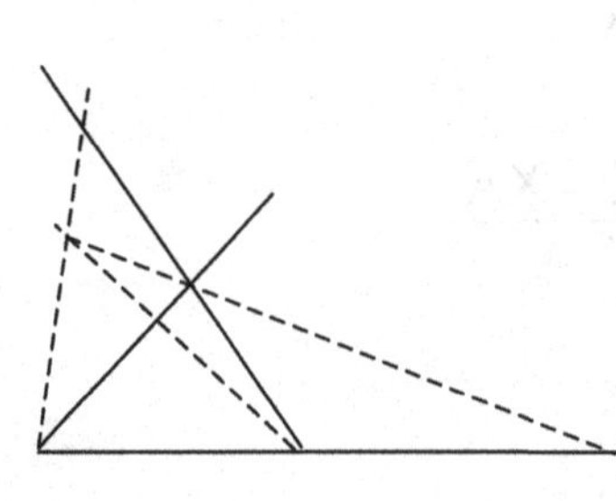

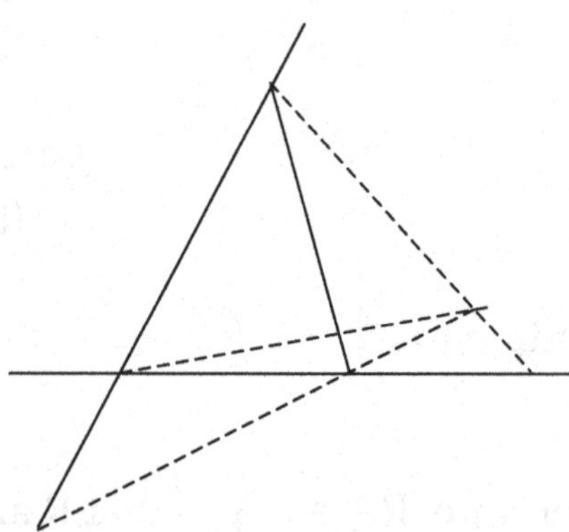

We now use the converse of Ceva's theorem to deduce some important results on the geometry of the triangle.
The first result concerns the medians of the triangle, which are the three lines joining the vertices A, B and C to the mid points of the opposite sides, called A′, B′ and C′.

Theorem 3 The medians of a triangle are concurrent.
Proof

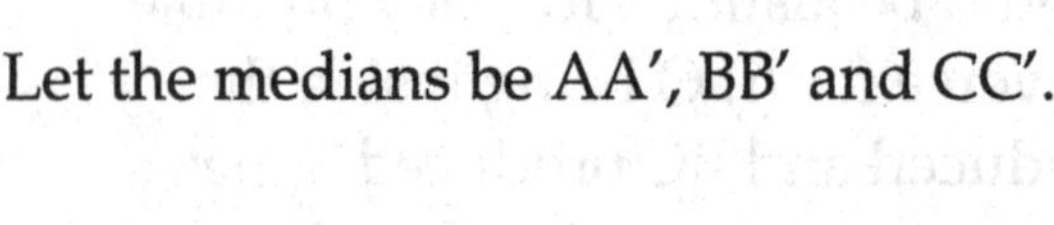

Let the medians be AA′, BB′ and CC′.

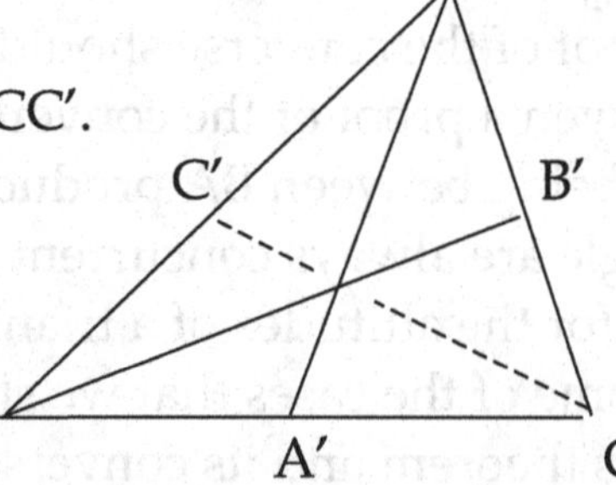

Then $\dfrac{AC'}{C'B} = \dfrac{BA'}{A'C} = \dfrac{CB'}{B'A} = 1$

So that $\dfrac{AC'}{C'B}.\dfrac{BA'}{A'C}.\dfrac{CB'}{B'A} = 1$

Therefore, by the converse of Ceva's theorem AA′, BB′ and CC′ are concurrent.
Thus, the medians of a triangle meet at a point (called G). ■

Another look at G

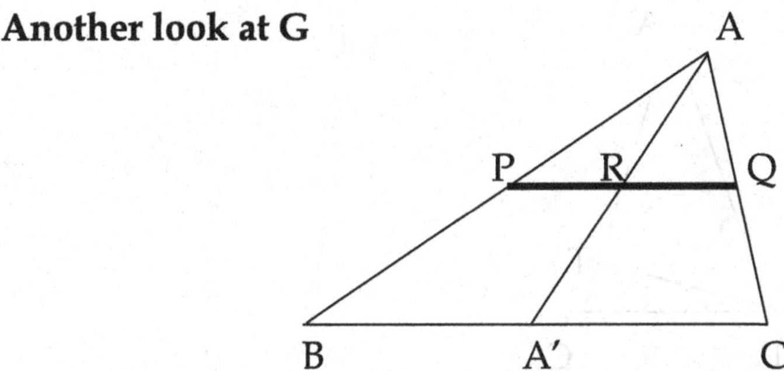

In the above figure, AA′ is the median of the triangle passing through vertex A. The line PQ is drawn parallel to the base BC and meets the median AA′ at R.
Since PQ is parallel to the base BC, ABC will be an enlargement of triangle APQ from centre A.
Further, since A′ is the mid point of BC, R will be the mid point of PQ.

Now consider triangle ABC to be made up of numerous parallel rods like PQ.

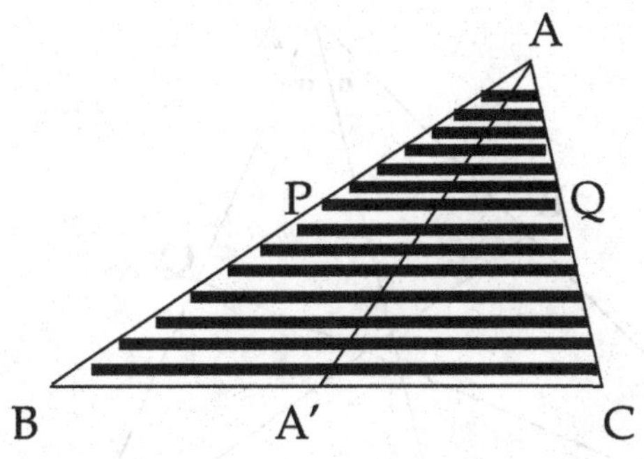

By the same argument, each of these parallel rods will be bisected by the median.

This means that if the triangle was to be cut out of a piece of cardboard, the whole triangle would balance on the median AA′ because each of the rods balances on its mid point which lies on AA′.
From an engineering view point we would say that the centre of mass of the triangle must lie on AA′.
Similarly, the centre of mass could be argued to lie on BB′ and also on CC′.
The centre of mass, or centre of Gravity, is called G. This is the point where the medians intersect.

Theorem 4 **The angle bisectors of a triangle are concurrent.**

Proof

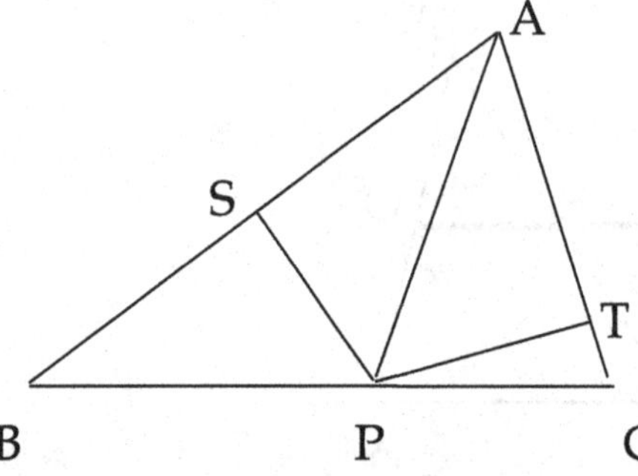

Let bisector of angle A meet BC at the point P.
Let the height of the triangle from A to BC be h.
Draw the perpendiculars PS and PT from P to the sides AB and AC.
These perpendiculars are equal in length. Let the length of each perpendicular be k then

$$\frac{BP}{PC} = \frac{\frac{1}{2}BP.h}{\frac{1}{2}PC.h} = \frac{\Delta ABP}{\Delta ACP} = \frac{\frac{1}{2}AB.k}{\frac{1}{2}AC.k} = \frac{AB}{AC}$$

Similarly, if the bisector of angle B meets CA at Q, we can prove that $\frac{CQ}{QA} = \frac{BC}{BA}$

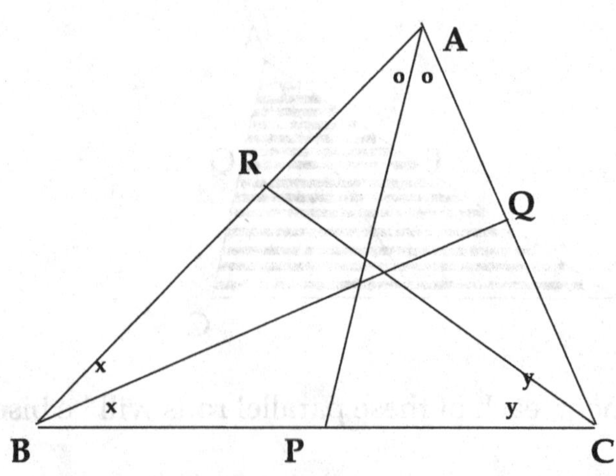

and if the bisector of angle C meets AB at R, then

$$\frac{AR}{RB} = \frac{CA}{CB}$$

Thus $$\frac{AR.BP.CQ}{RB\ PC\ QA} = \frac{CA.AB.BC}{CB\ AC\ BA}$$

$$= 1$$

Hence, by the converse of Ceva's Theorem, AP, BQ and CR are concurrent.

==================================

Thus the angle bisectors of any triangle meet at a point (called I, the incentre of the triangle)

An alternative demonstration that the angle bisectors of a triangle are concurrent comes from the observation that a circle with its centre on AP, the angle bisector of angle A, can be drawn to touch AB and AC. This is because the perpendiculars from the centre will be of equal length and the tangents to a circle are always perpendicular to the radius through the point of contact.

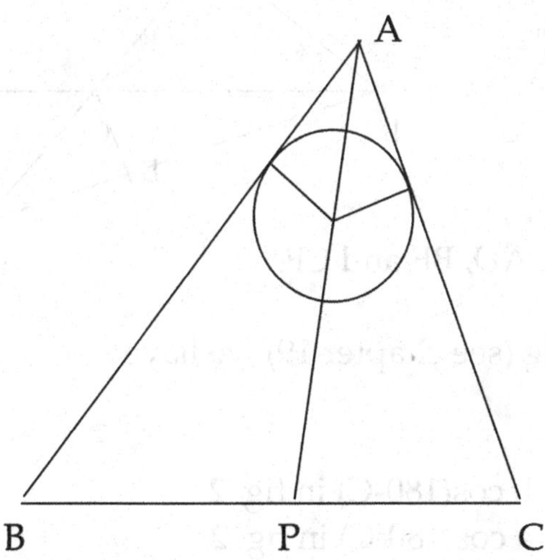

Suppose that the centre of this circle is initially close to A and that the radius is gradually increased but we keep the circle in contact with the sides AB and AC. Eventually, the circle will touch the base BC.
When the circle touches BC, we will have a circle that touches the three sides of the triangle. Further, the centre of this circle will lie on each of the angle

bisectors. Thus the angle bisectors will meet at the centre of this circle that can be drawn to touch the three sides of the triangle:

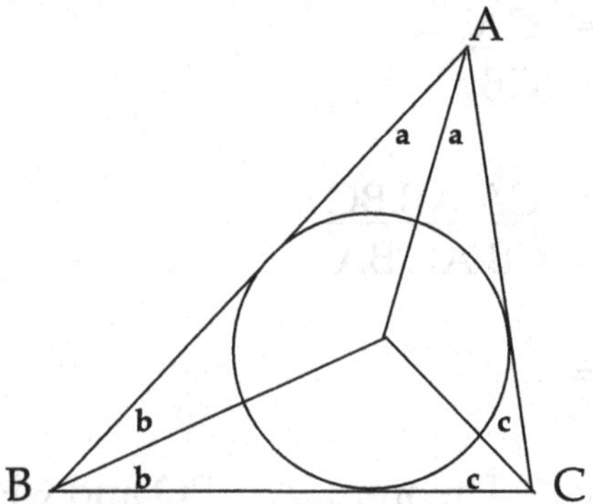

The centre of this circle is called I, the incentre of triangle ABC.

Theorem 5 **The altitudes of a triangle are concurrent**

Proof

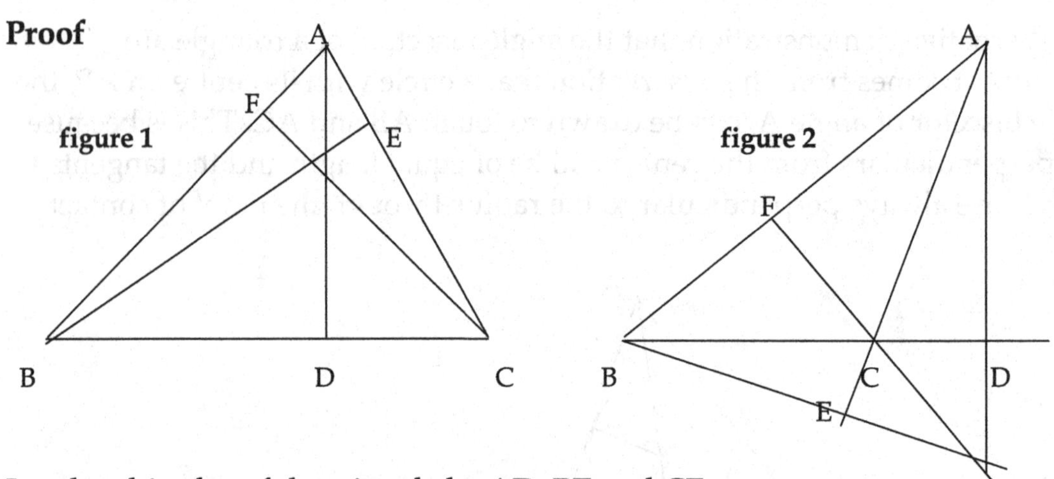

Let the altitudes of the triangle be AD, BE and CF.

Using results from elementary trig (see chapter 19) we have

$$
\begin{aligned}
BD &= c\cos B \\
DC &= b\cos C \qquad b\cos(180-C) \text{ in fig. 2} \\
CE &= a\cos C \qquad a\cos(180-C) \text{ in fig. 2} \\
EA &= c\cos A \\
AF &= b\cos A \\
FB &= a\cos B
\end{aligned}
$$

so that $\dfrac{AF.BD.CE}{FB\,DC\,EA} = \dfrac{b\cos A.c\cos B.\,a\cos C}{a\cos B\;b\cos C\;c\cos A} = \dfrac{abc\cos A\cos B\cos C}{abc\cos A\cos B\cos C} = 1$

and similarly for figure 2.
Hence, by the converse of Ceva's theorem, the altitudes of a triangle are concurrent.

The point of intersection is called H, the orthocentre of the triangle.

(the prefix ortho- stands for upright or rectangular so orthocentre could be interpreted as "the centre of the uprights")

Menelaus

Menelaus was born in Alexandria in Egypt. It is not certain when, but most writers and historians agree that it was in the first century around the year 70 AD. Little is known of his life but he appears to have lived and worked both in Alexandria and in Rome. He was famous for his books on astronomy and Geometry and was the first mathematician to develop the study of spherical geometry (the study of shapes drawn on the surface of a sphere). He wrote three books on the elements of Geometry and our Theorem 6 appears in his Book 3.

Theorem 6 **Menelaus's Theorem**

If the line PQR meets the sides BC, CA, AB at P, Q and R then

$$\frac{AR.BP.CQ}{RB\ PC\ QA} = -1$$

There are two basic figures:
(The line PQR must not pass through any of the vertices A, B or C)

figure 1

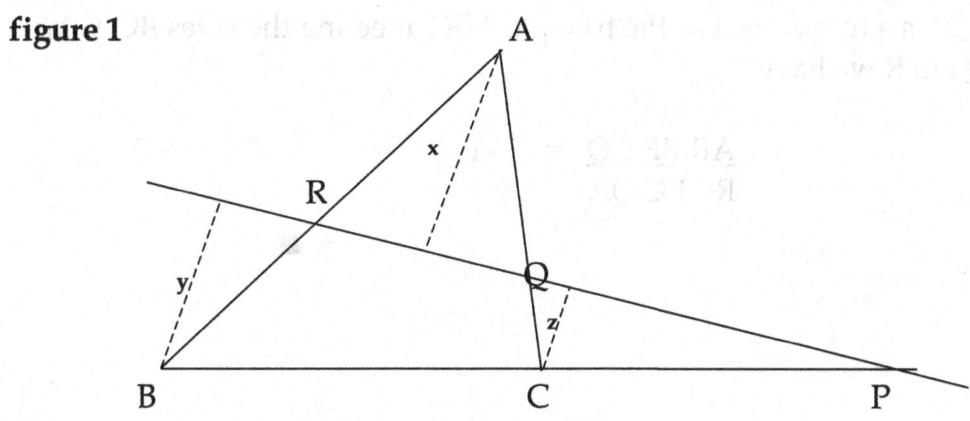

figure 2

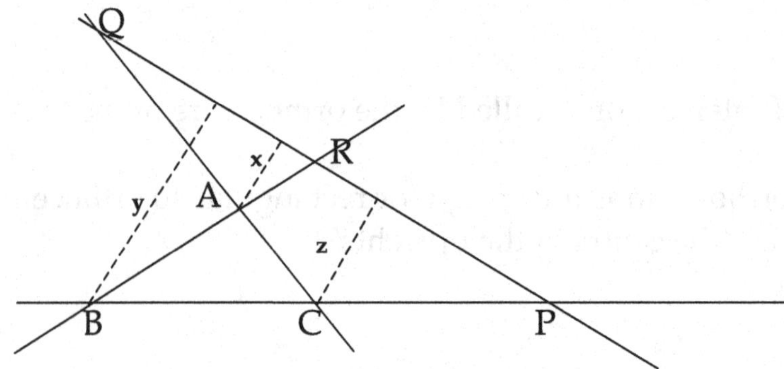

Proof

Let the lengths of the perpendiculars from A, B, C to the line PQR be x, y and z.
Then, using similar triangles,

$$\frac{AR}{RB} = \frac{x}{y} \qquad \frac{BP}{PC} = \frac{y}{z} \qquad \frac{CQ}{QA} = \frac{z}{x}$$

Thus, in magnitude, we have $\frac{AR.BP.CQ}{RB\ PC\ QA} = \frac{x.y.z}{y\ z\ x} = 1$

In figure 1, BP and PC are in opposite directions, providing the negative sign.

In figure 2, (BP, PC), (CQ, QA) and (AR,RB) are all in opposite directions providing three negative signs.

The negative sign distinguishes Menelaus's Theorem from Ceva's Theorem. Thus, if PQR is a transversal of the triangle ABC, meeting the sides BC, CA, AB at P, Q am R we have

$$\frac{AR.BP.CQ}{RB\ PC\ QA} = -1$$

■

Theorem 7 The Converse of Menelaus's Theorem

If P,Q and R are points on the sides BC, CA and AB of triangle ABC with

$$\frac{AR}{RB}.\frac{BP}{PC}.\frac{CQ}{QA} = -1$$

then the points P, Q and R lie on a straight line.
(note: the use of the minus sign is essential here, otherwise, this would be the converse of Ceva's theorem)

Proof

Because of the use of the sign, we know that at least one of P,Q or R lies outside of the corresponding side, BC, CA or AB. We can suppose P to be external to BC on BC produced, BC for if this is not the case then we can re-label the diagram. After a careful re-labelling of the diagram we will have one of these two possibilities:

Figure 1 : Q and R are internal to AC and AB and P is on BC produced and RQ meets BC produced at X.

Figure 2: Q and R are external to AC and AB and P is on BC produced and QR produced meets BC produced at X.

figure 1

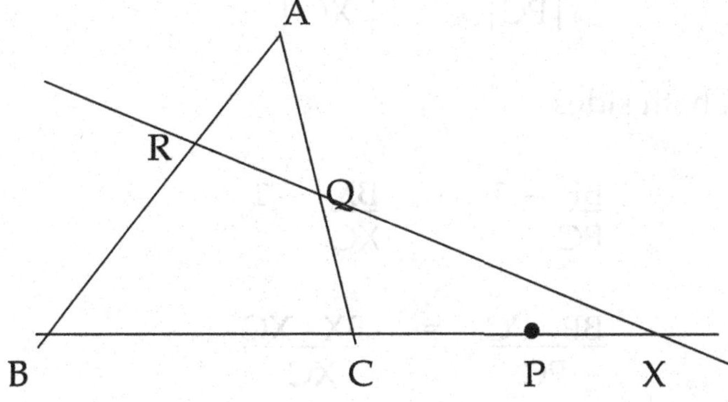

figure 2

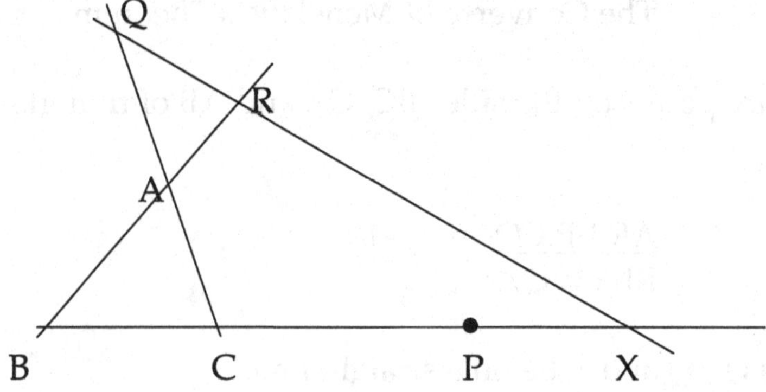

In both figures we are given

$$\frac{AR}{RB}\cdot\frac{BP}{PC}\cdot\frac{CQ}{QA} = -1$$

and we must prove that P and X coincide.
By Menelaus's Theorem we have

$$\frac{AR}{RB}\cdot\frac{BX}{XC}\cdot\frac{CQ}{QA} = -1$$

Therefore

$$\frac{BP}{PC} = \frac{BX}{XC} \quad \text{(sign being -1)}$$

*(see below) For clarity, at this point, we can switch to magnitudes, i.e. dispense with the minus signs:

$$\frac{|BP|}{|PC|} = \frac{|BX|}{|XC|}$$

subtract 1 from both sides

$$\frac{BP}{PC} - 1 = \frac{BX}{XC} - 1$$

$$\frac{BP - PC}{PC} = \frac{BX - XC}{XC}$$

$$\frac{BC}{PC} = \frac{BC}{XC}$$

Hence, PC = XC so that P and X are the same point.

(*) Alternatively we could stick to the convention of signs and proceed as follows:

$$\frac{BP}{PC} = \frac{BX}{XC} \qquad \text{(sign being -1)}$$

add 1 to both sides

$$\frac{BP}{PC} + 1 = \frac{BX}{XC} + 1$$

$$\frac{BP + PC}{PC} = \frac{BX + XC}{XC}$$

Using the sign convention, BP+PC = BC and BX+XC = BC (see chapter 11 or chapter 24)

$$\frac{BC}{PC} = \frac{BC}{XC}$$

Hence, PC = XC so that P and X are the same point.

Either way, P and X are the same point and this proves the converse of Menelaus's theorem:

If $\frac{AR.BP.CQ}{RB\ PC\ QA} = -1$ then PQR is a straight line. ■

Blaise Pascal

Pascal was born in France in 1623 and died at the age of 39 in 1662. The standard unit for pressure is called the Pascal, (=one Newton per square metre), to honour his work on hydrostatics. He wrote important works on conic sections, projective geometry and probability and Pascal's triangle for the coefficients in a binomial expansion is well known to any student of mathematics. Pascal's triangle for a binomial expansion with a positive integral index led Newton to discover the binomial expansion for negative and fractional indices. In his early twenties he invented a mechanical calculator and he proved a more advanced version of our next theorem that bears his name when he was only 16. In the advanced version, a theorem from Projective Geometry, the points A,B,C D,E and F are allowed to lie on any conic section for example a circle or an ellipse. In our version, the points lie on a degenerate hyperbola formed from two straight lines.

We use Menelaus's Theorem to prove our next theorem:

Theorem 8 **Pascals Theorem** (for straight lines)

A,B, C are points on one line.
D,E, F are points on a second line.

If AE ∩ DB = L, AF ∩ CD = M and BF ∩ CE = N then LMN will be a straight line.

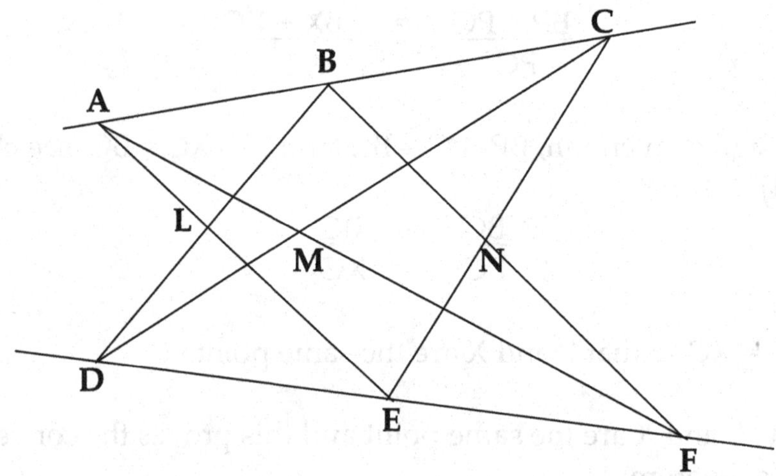

To Prove that LMN is a straight line

Proof
Let the lines AC, DF and LM form triangle PQR as shown

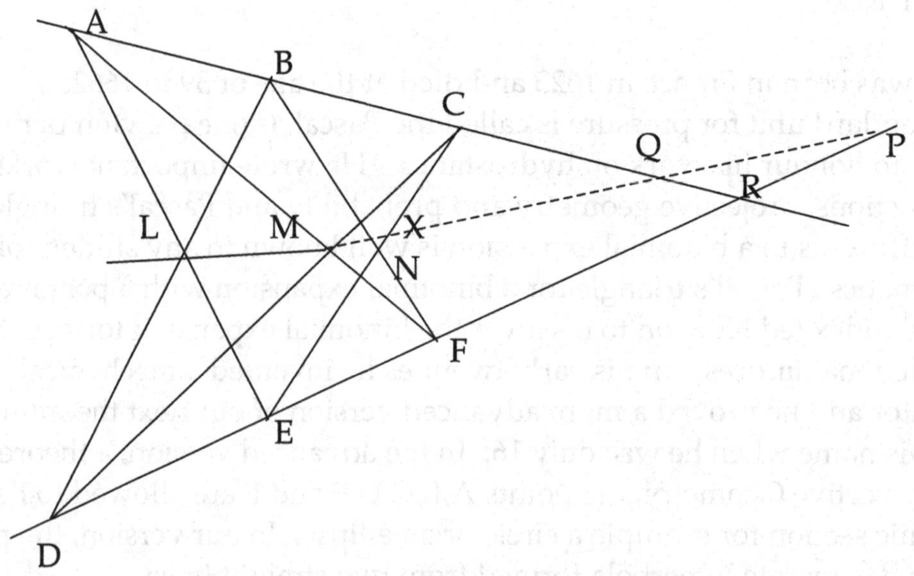

Let LM meet CE at X (LM produced is the broken line in the figure), then we must show that X and N coincide.

Using Menelaus's Theorem in triangle PQR

With transversal ALE we have $\frac{PL.QA.RE}{LQ\ AR\ EP} = -1$

With transversal BLD we have $\frac{PD.RB.QL}{DR\ BQ\ LP} = -1$

With transversal AMF we have $\frac{PF.RA.QM}{FR\ AQ\ MP} = -1$

With transversal CMD we have $\frac{PM.QC.RD}{MQ\ CR\ DP} = -1$

With transversal CXE we have $\frac{PE.RC.QX}{ER\ CQ\ XP} = -1$

Multiply these five equations together, and after cancelling, we have

$$\frac{PF.RB.QX}{FR\ BQ\ XP} = -1$$

Now by the converse of Menelaus's Theorem this shows that FBX is is a straight line. That is, FB cuts CE at X. But we know that FB cuts CE at N and therefore X and N must be the same point.
Since LMX is a straight line, by construction, and we have proved that N and X are the same point, we have shown that LMN is a straight line.

This conclude s the proof of Pascal's theorem.

■

The Dual of Pascal's theorem

Using the idea of duality, we can translate Pascal's Theorem into its "Dual" and find a new theorem. For the sake of symmetry, instead of using D,E and F we use here A′, B′ and C′.

Pascal	Dual
Points A,B,C lie on line p	Lines a,b,c meet at point P
Points A′,B′,C′ lie on line p′	Lines a′,b′,c′ meet at point P′
L is the intersection of BC′ and B′C	l is the join of bc′ and c′a
M is the intersection of CA′ and C′A	m is the join of ca′ and c′a
N is the intersection of AB′ and A′B	n is the join of ab′ and a′b
Pascal's theorem	Dual of Pascal
L,M,N are collinear	l,m,n are concurrent

Pascal's Theorem

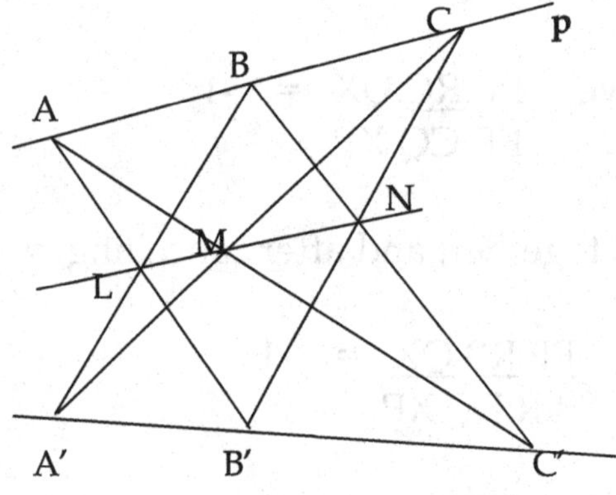

Points L, M and N are collinear

The Dual of Pascal's theorem

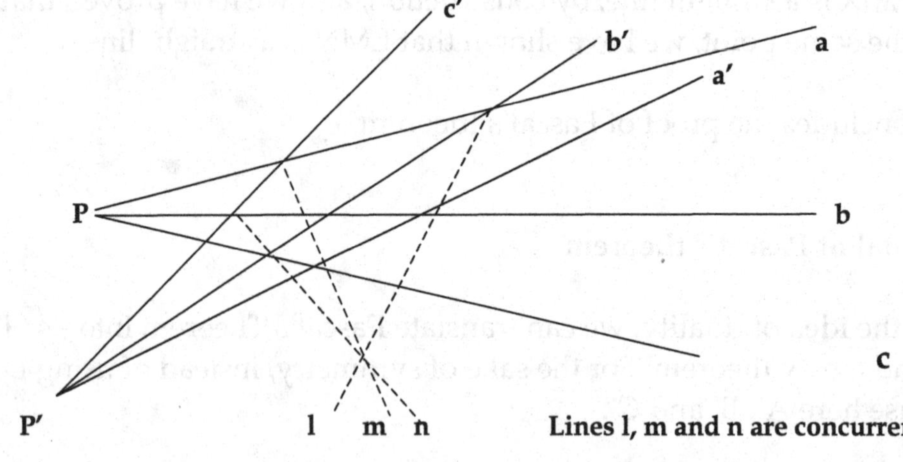

Lines l, m and n are concurrent

Girard Desargues was born in 1591 in France and died in 1661. He published important works on conic sections and was the discoverer of projective geometry. His famous theorem on perspective triangles was published in 1648.

Theorem 9 **Desargues Theorem**

ABC and A′B′C′ are two triangles in perspective from the point O.
That is, AA′, BB′ and CC′ are concurrent lines.
If $BC \cap B'C' = L$, $CA \cap C'A' = M$ and $AB \cap A'B' = N$
then LMN is a straight line.

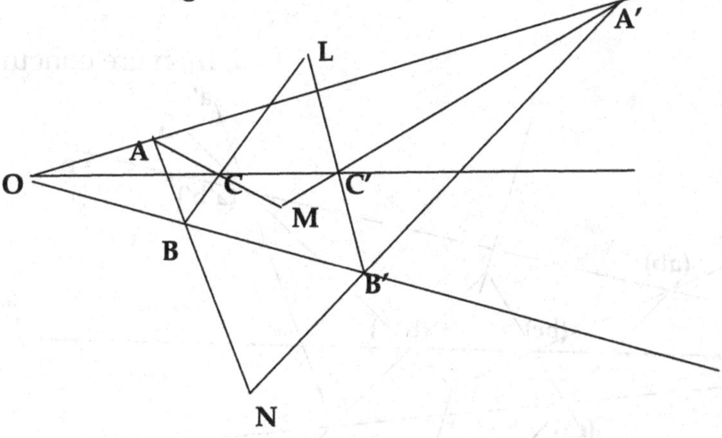

Proof
Using Menelaus
In Δ OAB with transversal A′B′N

$$\frac{OA'}{A'A}\cdot\frac{AN}{NB}\cdot\frac{BB'}{B'O} = -1$$

In Δ OBC with transversal B′C′L

$$\frac{OB'}{B'B}\cdot\frac{BL}{LC}\cdot\frac{CC'}{C'O} = -1$$

In Δ OCA with transversal C′A′M

$$\frac{OC'}{C'C}\cdot\frac{CM}{MA}\cdot\frac{AA'}{A'O} = -1$$

Multiplying these equations together, after some canceling, we have

$$\frac{AN}{NB}\cdot\frac{BL}{LC}\cdot\frac{CM}{MA} = -1$$

Hence, by the converse of Menelaus's Theorem, LMN is a transversal for triangle ABC. Therefore, LMN is a straight line. ■

Desargues theorem and its Dual

Desargues theorem	The Dual of Desargues
A,A′ and O are collinear	a, a′ and o are concurrent
B,B′ and O are collinear	b, b′ and o are concurrent
C,C′ and O are collinear	c, c′ and o are concurrent
BC meets B′C′ at L	l is the join of (bc) and (b′c′)
CA meets C′A′ at M	m is the join of (ca) and (c′a′)
AB meets A′B′ at N	n is the join of (ab) and (a′b′)
L, M, N are collinear	l, m, n are concurrent

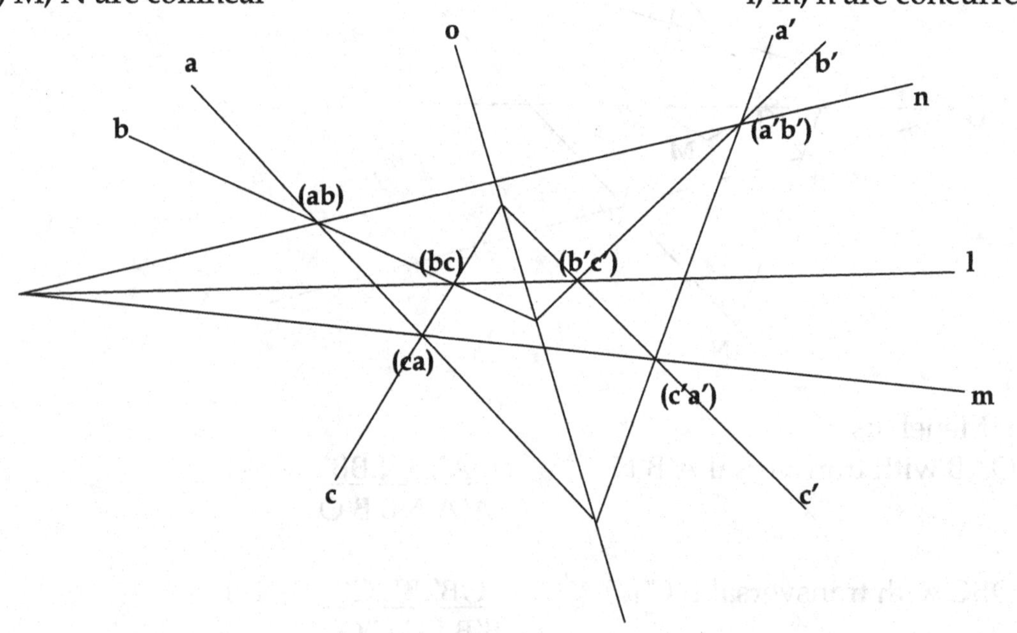

The points (aa′), (bb′) and (cc′) lie on the line o (dotted).

Line l is the join of (bc) and (b′c′)

Line m is the join of (ca) and (c′a′)

Line n is the join of (ab) and (a′b′)

The result of the Dual theorem is that the lines l, m and n are concurrent.

Now Desargues theorem itself, in this diagram would state that if lines l, m and n are concurrent then the points (aa′), (bb′) and (cc′) must lie on a straight lie, the line o.
The Dual Theorem states that if the points (aa′), (bb′) and (cc′) lie on the line o then the lines l, m and n must be concurrent.
The Dual theorem is therefore the Converse of Desargues theorem.

The Dual of Desargues theorem is the Converse of Desargues theorem.

Desargues Theorem in Three Dimensions

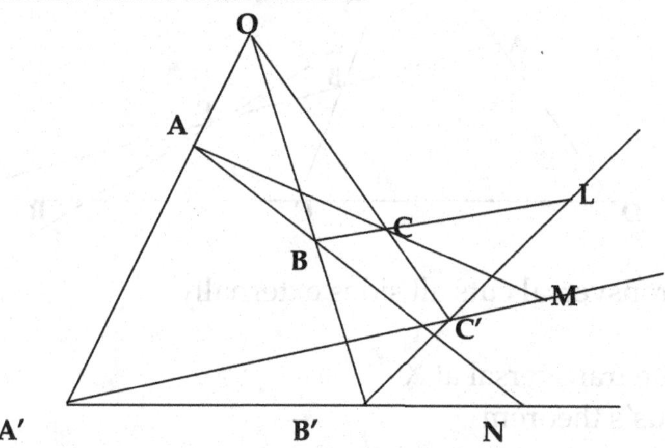

The diagram represents two triangles in perspective from O, but in three dimensions so that OA′B′C′ is a triangular pyramid on base A′B′C′.
Triangle ABC is a slice through the pyramid.

Lines AB and A′B′ lie in the plane OAA′B′B and intersect at N

Lines BC and B′C′ lie in the plane OBB′C′C and intersect at L

Lines AC and A′C′ lie in the plane OAA′C′C and intersect at M

The points L, M and N lie in the plane of triangle ABC and also lie in the plane of triangle A′B′C′ and so, LMN must be a straight line, the intersection of the two planes.

This proves Desargues Theorem in three dimensions.

■

Generalisations of Menelaus's theorem

Theorem 10 **Menelaus's theorem for a Quadrilateral**

If ABCD is a quadrilateral and a transversal meets AB, BC, CD, DA at P,Q,R and S respectively, then

$$\frac{AP}{PB}\cdot\frac{BQ}{QC}\cdot\frac{CR}{RD}\cdot\frac{DS}{SA} = +1$$

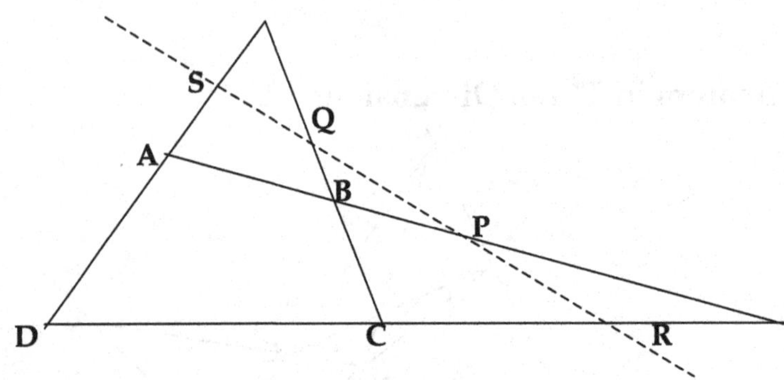

Proof

Case (i); The transversal cuts all sides externally

Let DB meet the transversal at X

Using Menelaus's theorem

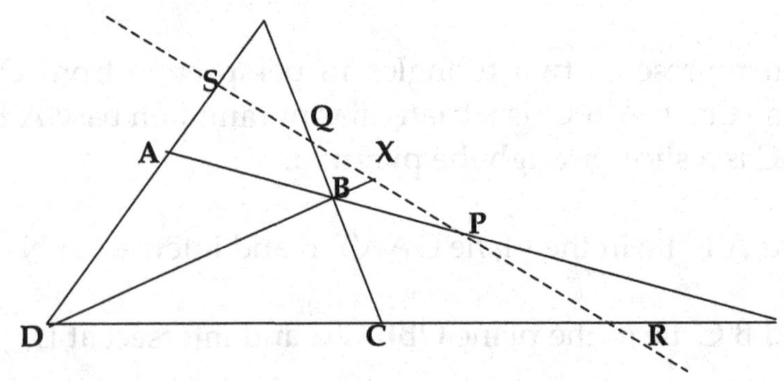

In triangle ABD $$\frac{AP}{PB}\cdot\frac{BX}{XD}\cdot\frac{DS}{SA} = -1$$

In triangle DBC $$\frac{DX}{XB}\cdot\frac{BQ}{QC}\cdot\frac{CR}{RD} = -1$$

Multiplying $$\frac{AP}{PB}\cdot\frac{BX}{XD}\cdot\frac{DS}{SA}\cdot\frac{DX}{XB}\cdot\frac{BQ}{QC}\cdot\frac{CR}{RD} = +1$$

Giving $$\frac{AP.BQ.CR.DS}{PB\,QC\,RD\,SA} = +1$$

as required

Case (ii): The transversal cuts sides internally

If the transversal "shifts over " a vertex so that it then cuts two sides internally, then two of the ratios will change sign from minus to plus. The sign of the result therefore remains the same.
Using this argument, the result follows for all transversals that do not pass through a vertex. ∎

Theorem 11 Extension of Menelaus's Theorem for a Pentagon

If a transversal of the pentagon ABCDE meets the sides AB, BC, CD, DE, EA at the points P,Q,R,S,T respectively.

then $$\frac{AP.BQ.CR.DS.ET}{PB\,QC\,RD\,SE\,TA} = -1$$

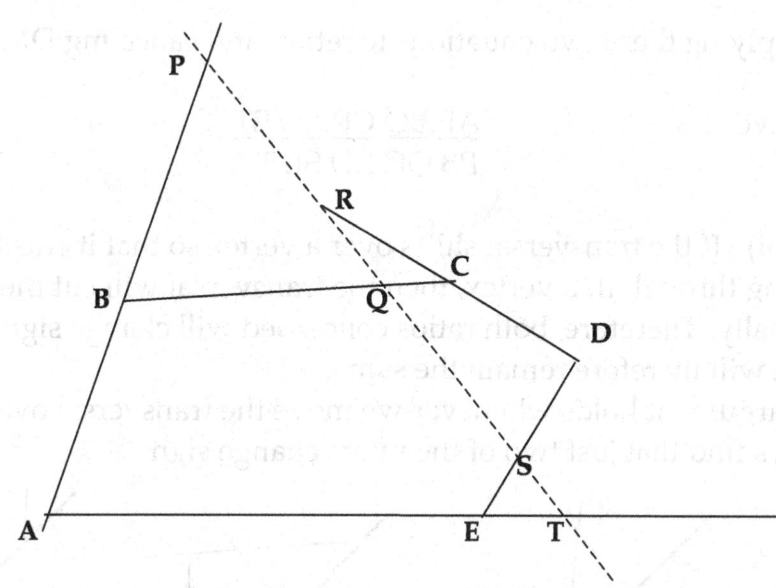

Proof

Case (i) The transversal meets all sides externally

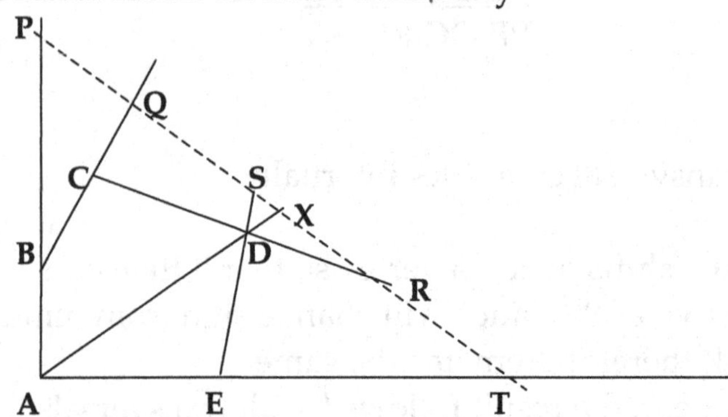

Let AD meet the transversal at X

Using Theorem 10 in Quad ABCD

We have $\frac{AP.BQ.CR.DX}{PB\ QC\ RD\ XA} = +1$

Using Menelaus's Theorem in triangle ADE

we have $\frac{AX.DS.ET}{XD\ SE\ TA} = -1$

Multiplying these two equations together and canceling DX/XA

we have $\frac{AP.BQ.CR.DS.ET}{PB\ QC\ RD\ SE\ TA} = -1$

Case(ii) If the transversal shifts over a vertex so that it cuts the two sides passing through that vertex, then the transversal will cut these two sides internally. Therefore, both ratios concerned will change sign. The sign of the result will therefore remain the same.

This argument holds whenever we move the transversal over a vertex. We always find that just two of the ratios change sign:

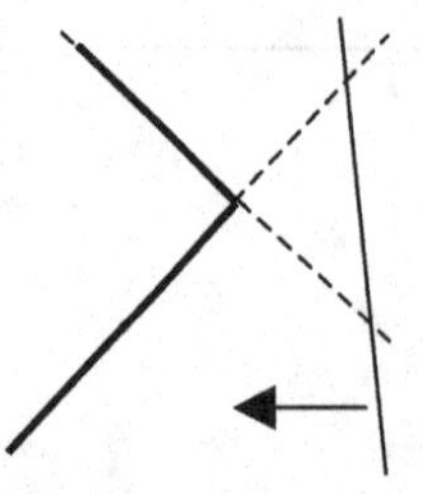

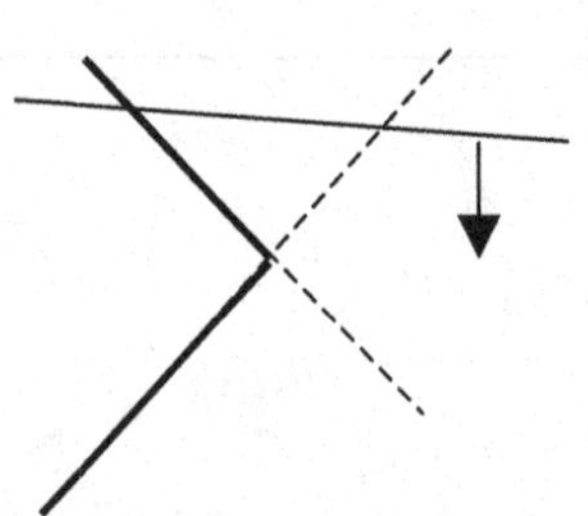

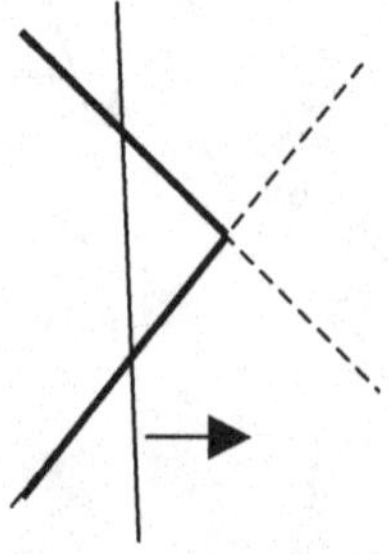

Transversal shifts to the left over a vertex Both ratios change from from minus to plus.	Transversal shifts down over a vertex. One ratio changes from plus to minus and the other from minus to plus.	Transversal shifts to Right over a vertex. Both ratios change plus to minus.

In every case, two just two ratios change sign and so the result of the product of the ratios remains the same.

This proves Menelaus's theorem for a pentagon.

■

This method of proof of Menelaus's Theorem for polygons clearly extends to polygons of any number of sides. The method is illustrated using a convenient notation, in Theorem 11 which is the extension of Menelaus's theorem to a transversal of a hexagon. In Theorem 12, Menelaus's Theorem is formally proved for a transversal of an n sided polygon.

Theorem 12. Menelaus's Theorem for a transversal of a hexagon.

If a straight line meets the sides A1A2, A2A3, A3A4, A4A5, A5A6 and A6A1 of hexagon A1A2A3A4A5A6 at points P1, P2, P3, P4, P5 and P6

then
$$\frac{A1P1}{P1A2}\cdot\frac{A2P2}{P2A3}\cdot\frac{A3P3}{P3A4}\cdot\frac{A4P5}{P5A6}\cdot\frac{A6P6}{P6A1} = (-1)^6$$

Proof

Let A1A5 meet the transversal at X.

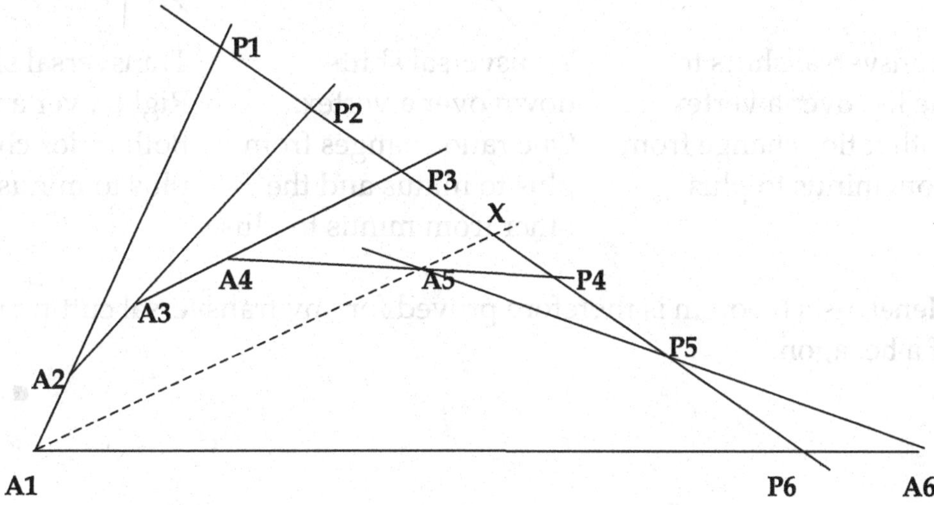

By theorem 11 for pentagon A1A2A3A4A5

We have then $\frac{A1P1.A2P2.A3P3.A4P4.A5X}{P1A2\ P2A3\ P3A4\ P4A5\ XA1} = (-1)^5$

Using Menelaus's Theorem in triangle A1A5A6

We have then $\frac{A1X.A5P5.A6P6}{XA5\ P5A6\ P6A1} = -1$

Multiplying these two equations together and canceling A1X/XA5 we have

The desired result $\frac{A1P1.A2P2.A3P3.A4P5.A6P6}{P1A2\ P2A3\ P3A4\ P5A6\ P6A1} = (-1)^6$

Now we use the same argument that we used for the pentagon to show that the same result hold if the transversal intersect any of the sides internally:

Whenever we move the transversal over a vertex we always find that just two of the ratios change sign:

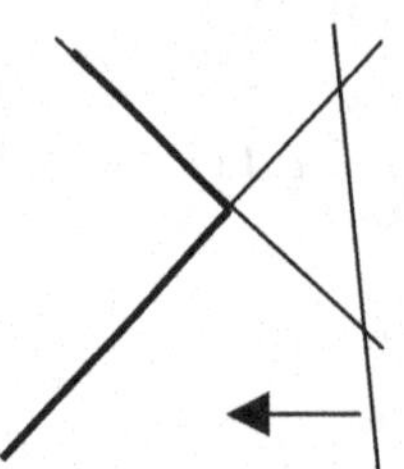

Transversal shifts to the left over a vertex Both ratios change from from minus to plus.

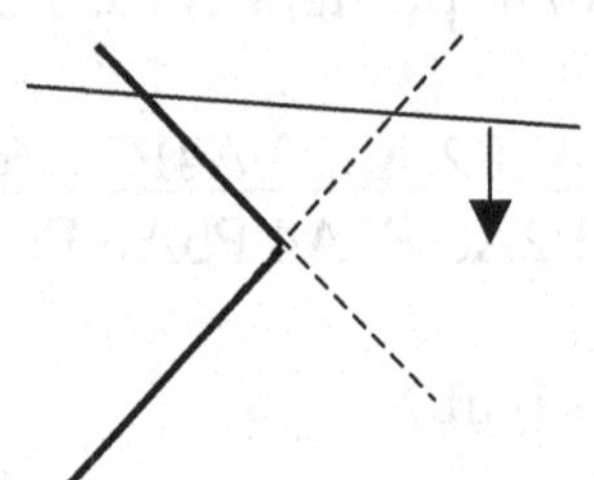

Transversal shifts down over a vertex. One ratio changes from plus to minus and the other from minus to plus.

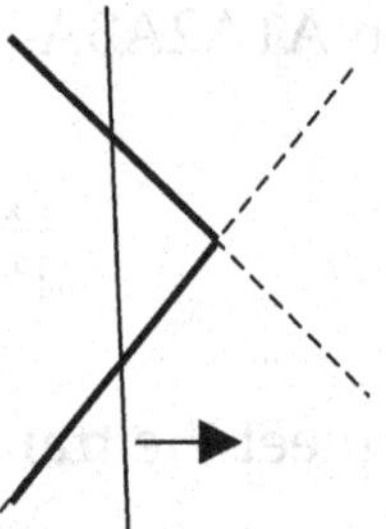

Transversal shifts to Right over a vertex. Both ratios change plus to minus.

Menelaus's theorem is therefore proved for any transversal cutting the sides of a hexagon.

■

Theorem 12 Menelaus's Theorem for a transversal of an n-sided polygon

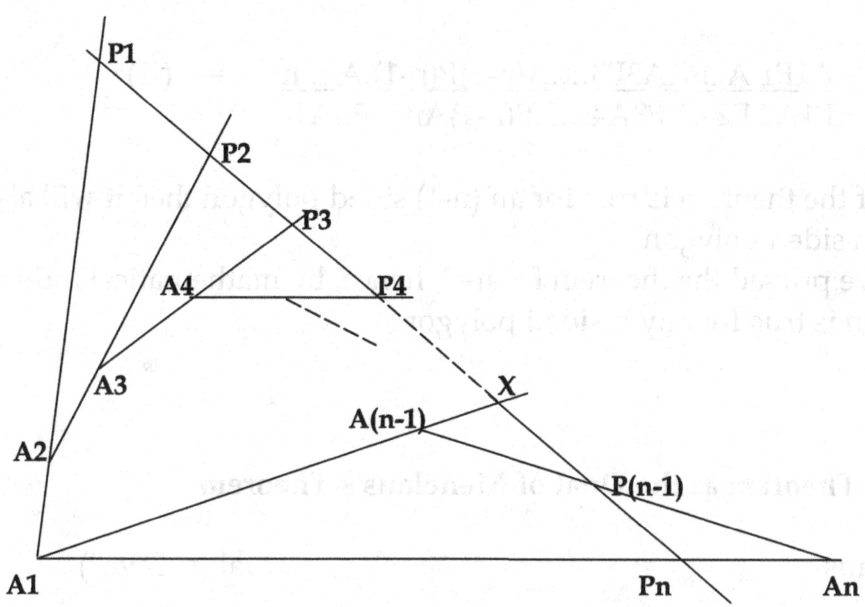

A1A2A3A4…A(n-1)An is an n-sided polygon. P1P2P3P4…P(n-1)Pn is a transversal that cuts the n sides of the polygon at points P1, P2,……
The line P1P2P3P4…P(n-1)Pn meets the sides A1A2, A2A3, A3A4,…A(n-1)An, AnA1 at points P1, P2, P3,….P(n-1), Pn.

Prove that $$\frac{A1P1}{P1A2}\cdot\frac{A2P2}{P2A3}\cdot\frac{A3P3}{P3A4}\cdots\frac{A(n-1)P(n-1)}{P(n-1)An}\cdot\frac{AnPn}{PnA1} = (-1)^n$$

The proof is by induction on n.
Suppose that the theorem holds for any (n-1) sided polygon.
Let A1A(n-1) meet the transversal at X then A1A2….A(n-1) in an (n-1) sided polygon so that the supposition gives

$$\frac{A1P1}{P1A2}\cdot\frac{A2P2}{P2A3}\cdot\frac{A3P3}{P3A4}\cdots\frac{A(n-1)X}{XA1} = (-1)^{n-1}$$

Use Menelaus's Theorem in triangle A1A(n-1)An to get

$$\frac{A1X}{XA(n-1)}\cdot\frac{A(n-1)P(n-1)}{P(n-1)An}\cdot\frac{AnPn}{PnA1} = -1$$

Multiplying these two equations together and cancelling A1X/XA(n-1) we have

$$\frac{A1P1.A2P2.A3P3....A(n\text{-}1)P(n\text{-}1).AnPn}{P1A2\ P2A3\ P3A4....\ P(n\text{-}1)An\ \ PnA1} = (-1)^n$$

Thus, if the theorem is true for an (n-1) sided polygon then it will also be true for an n sided polygon.
We have proved the theorem for n=3, hence, by mathematical induction, the theorem is true for any n-sided polygon.

■

Ceva's Theorem as the Dual of Menelaus's Theorem

Menelaus	Dual (≈ Ceva?)
Given triangle ABC and transversal l	Given trilateral abc and point L
BC meets l at P CA meets l at Q AB meets l at R	the join of (bc) and L is line p the join of (ca) and L is line q the join of (ab) and L is line r
$\frac{\lvert BP\rvert}{\lvert PC\rvert}.\frac{\lvert CQ\rvert}{\lvert QA\rvert}.\frac{\lvert AR\rvert}{\lvert RB\rvert} = 1$	$\frac{\lvert bp\rvert}{\lvert pc\rvert}.\frac{\lvert cq\rvert}{\lvert qa\rvert}.\frac{\lvert ar\rvert}{\lvert rb\rvert} = 1$

Where we have used the symbol |BP| for the distance between point B and point P. The meaning of this is clear, but in the dual it is not clear what should be use for the "distance" between the line b and the line p. Clearly, it should have something to do with the angle between the two lines b and p.

If we refer back to Ceva's Theorem to see if it can be rewritten introducing the angles between the lines b and p, p and c etc.

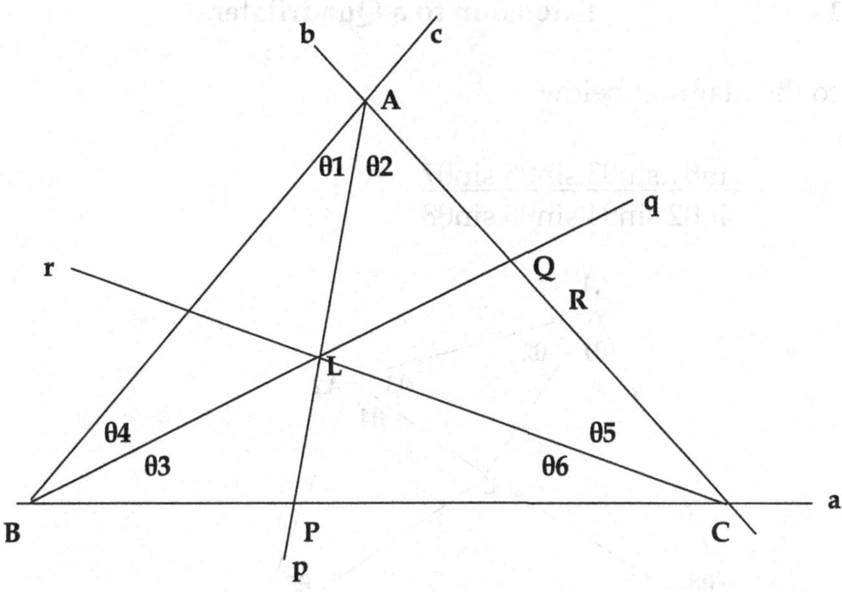

Now $\frac{BP}{PB} = \frac{\Delta BAP}{\Delta CAP} = \frac{\frac{1}{2}\ AP.AB.\sin\theta 1}{\frac{1}{2}\ AP.AC.\sin\theta 2} = \frac{AB.\sin\theta 1}{AC.\sin\theta 2}$

And similarly for the other two ratios

so we can rewrite $\frac{|BP|}{|PC|}.\frac{|CQ|}{|QA|}.\frac{|AR|}{|RB|} = \frac{AB\sin\theta 1}{AC\sin\theta 2}.\frac{BC\sin\theta 3}{BA\sin\theta 4}.\frac{CA\sin\theta 5}{CB\sin\theta 6}$

$$= \frac{\sin\theta 1}{\sin\theta 2}.\frac{\sin\theta 3}{\sin\theta 4}.\frac{\sin\theta 5}{\sin\theta 6}$$

thus we can almost regard sinθ1 as a distance measure between the line b and the line p. If we write |bp| = sinθ1, |pc| = sinθ2 etc then we have Ceva's Theorem looking like a Dual of Menelaus's Theorem.

Generalising Ceva's Theorem

The following theorems are not really extensions of Ceva's Theorem but relate to the version of what we have referred to as the dual theorem of Menelaus's theorem.

Theorem 14 **Extension to a Quadrilateral**

Referring to the diagram below

$$\frac{\sin\theta1.\sin\theta3.\sin\theta5.\sin\theta7}{\sin\theta2\ \sin\theta4\ \sin\theta6\ \sin\theta8} = 1$$

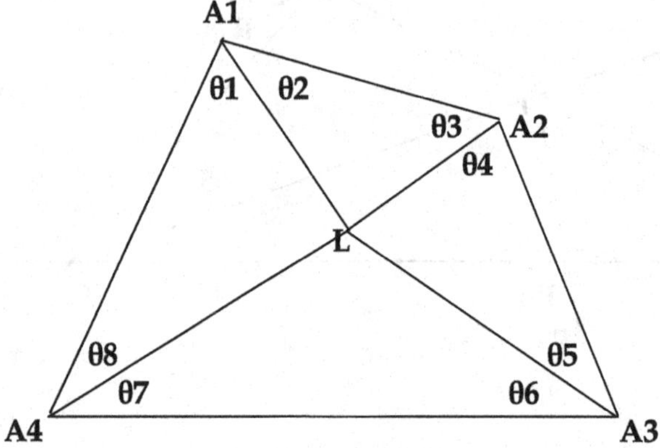

Proof
Let the lengths of the perpendiculars from L to the sides of the quadrilateral be p1, p2, p3 and p4 as shown.

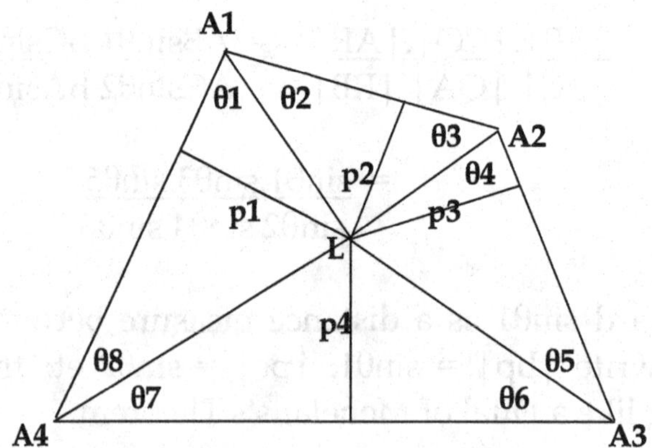

Then
$\sin\theta1 = p1/A1L$, $\sin\theta2 = p2/A1L$, $\sin\theta3 = p2/A2L$, $\sin\theta4 = p3/A2L$ etc

thus $\dfrac{\sin\theta1}{\sin\theta2} = \dfrac{p1}{p2}$, $\dfrac{\sin\theta3}{\sin\theta4} = \dfrac{p2}{p3}$ etc.

hence $$\frac{\sin\theta1.\sin\theta3.\sin\theta5.\sin\theta7}{\sin\theta2\ \sin\theta4\ \sin\theta6\ \sin\theta8} = \frac{p1.p2.p3.p4}{p2\ p3\ p4\ p1} = 1$$

■

This theorem can be generalized to a polygon of any number (n) of sides:

Theorem 15: Ceva for the n-agon

For any polygon A1A2A3....An and any point L inside, let the angles

between A1L and the sides AnA1 and A1A2 be θ1 and θ2

between A2L and the sides A1A2 and A2A3 be θ3 and θ4

between A3L and the sides A2A3 and A3A4 be θ5 and θ6
etc.

then $$\frac{\sin\theta1.\sin\theta3.......\sin\theta(2n-1)}{\sin\theta2\ \sin\theta4.......\ \sin\theta(2n)} = 1$$

Proof

Using a similar notation to theorem 14,

$$\frac{\sin\theta1.\sin\theta3.......\sin\theta(2n-1)}{\sin\theta2\ \sin\theta4.......\ \sin\theta(2n)} = \frac{p1.p2.p3\ .....\ pn}{p2\ p3\ p4\ .....\ p1} = 1$$

■

Brocard's Angle

Henri Brocard was a French army officer who, in 1875 stated that there was a unique position for L in Ceva's diagram that would make θ1, θ3 and θ5 equal.

The next discussion attempts to show why there is just one position for L, the intersection of AP, BQ and CR, that makes the three angles θ1 = θ3 = θ5.

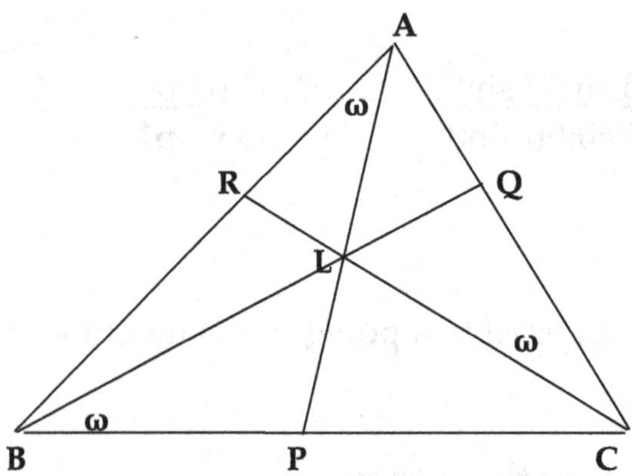

Ceva's diagram with θ1 = θ3 = θ5 (= ω)

To prove that there is just one position for L that makes the three angles equal.

First, we need to arrange the triangle so that angle A is greater (or equal) to angle B. This might involve relabelling the triangle, but does not affect the validity of the proof.
Draw the circle that touches BC at B and passes through the point A.
Since angle A is not less than angle B, the circle will not intersect AC between A and C. If A=B then the circle touches the triangle at both A and B.

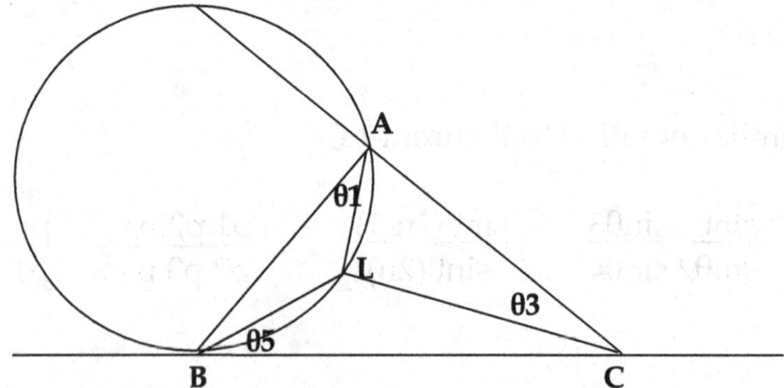

We examine what happens when L starts from B and moves round the arc of the circle from B to A.
As L moves round the arc of the circle from B to A, we will always have θ5 = θ1 (alternate segment)
Now θ5 starts with the value zero and increases to the value of angle B when L reaches A.
Also, as L moves from B to A, θ3 starts with the value of angle C and decreases to the value zero.

Consider the value of $\theta 5 - \theta 3$ and we have:

Initially $\theta 5 - \theta 3 = 0 - C = -C$

Finally $\theta 5 - \theta 3 = B - 0 = B$

So $\theta 5 - \theta 3$ starts with the value $-C$ and increases to the value B

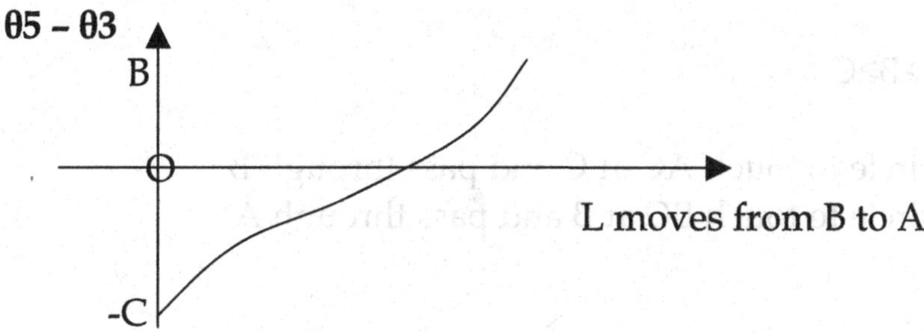

Thus **$\theta 5 - \theta 3$** increases from a negative value to a positive value and therefore must reach a value zero for just one position of L.
At this point **$\theta 1 = \theta 3 = \theta 5$.** This point is called a Brocard point of the triangle.

There is also a second Brocard point for which **$\theta 2 = \theta 4 = \theta 6$.**

Alternative argument (i)

Another way of dealing with the Brocard points is to keep all three angles equal , say **$\theta 1 = \theta 3 = \theta 5 = \omega$** and consider what happens in the diagram as ω increases from zero.

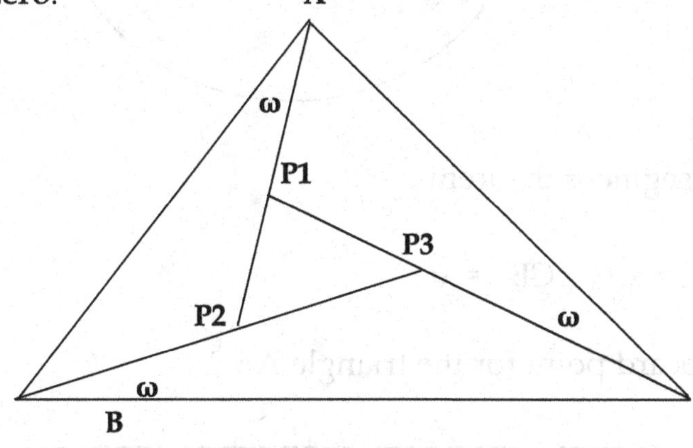

The lines through the vertices form a triangle P1P2P3. (When ω=0, P1, P2 and P3 are at A,B and C).

As ω increases from zero, the points P1, P2 and P3 get closer. They will always lie inside the triangle, for example, P1 starts A and moves away from A towards P2. P2 starts at B and moves away from B towards P3. Similarly P3 starts at C and moves away from C towards P1. If P2 were to reach P3 then P1 would also be at the same place (being the intersection of BP2 and CP1) and we would have found the point we were looking for.

Alternative argument (ii)

Suppose A≥B≥C

Draw the circle to touch AC at C and pass through B
Draw the circle to touch BC at B and pass through A

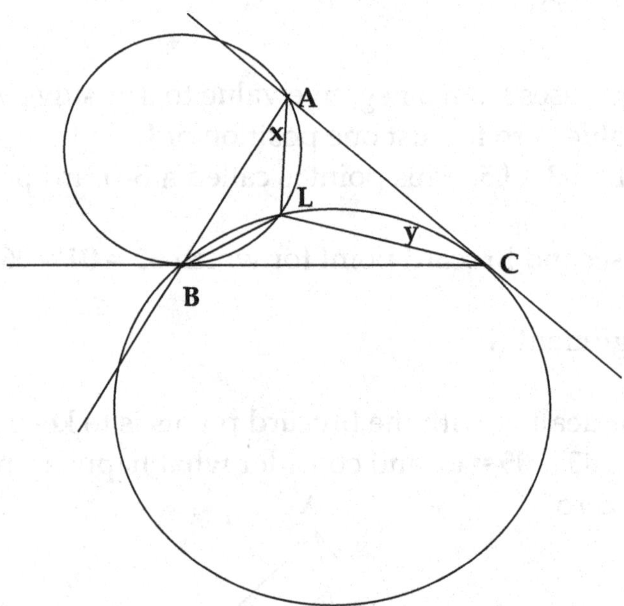

By the alternate segment theorem

$$\angle CBL = x, \quad \angle CBL = y$$

Thus L is the Brocard point for the triangle ABC.

==============================

Morley's triangle

Frank Morley was professor of mathematics at Haverford College in Philadelphia USA. In 1899 he found what is now referred to as Morley's Triangle. It is an intriguing result, easy to understand but not so easy to prove. If the trisectors of the angles of a triangle are drawn, we find they intersect to form a triangle which is always equilateral.

We end this chapter with a proof of Morley's triangle.

This proof is from "Elementary Geometry" by John Roe (Oxford) 1993.
The trisectors of the angles A, B and C are drawn.
Pairs of trisectors meet at P, Q and R to form a triangle.

Theorem 16 (Morley) **Triangle PQR is an equilateral triangle.**

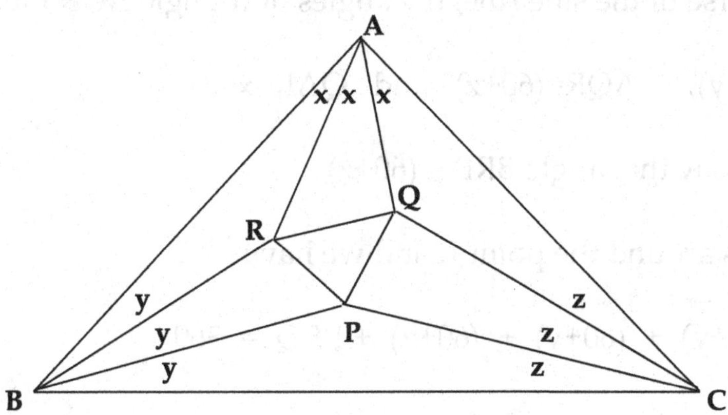

Proof

Let the angles A =3x, B=3y, C=3z so that $3x + 3y + 3z = 180$

$$x+y+z = 60$$

using the sine rule $\dfrac{a}{\sin A} = \dfrac{b}{\sin B} = \dfrac{c}{\sin C} = 2R$

Therefore $c = 2R.\sin C = 2R\sin.(3x+3y)$

also $AR = \dfrac{AB.\sin y}{\sin(x+y)} = \dfrac{2R.\sin(3x+3y).\sin y}{\sin(x+y)}$

Now $\frac{\sin 3\theta}{\sin \theta} = 4 \sin(60-\theta).\sin(60+\theta)$

Therefore $AR = 2R. 4\sin(60-[x+y]).\sin(60+[x+y]).\sin y$

$= 8R\sin y.\sin z.\sin(60+z)$

Similarly, $AQ = 8R\sin y.\sin z.\sin(60+y)$

Therefore $\frac{AQ}{\sin(60+y)} = \frac{AR}{\sin(60+z)}$

we also have $(60+y) + (60 + z) + x = 180$

hence, by the converse of the sine rule, the angles of triangle AQR must be

$ARQ=(60+y), \quad AQR=(60+z) \quad \text{and} \quad QAR=x$

Similarly, we can show that angle $BRP = (60+z)$

Now add the angles around the point R and we have

$$180-(x+y) + (60+y) + (60+z) + PRQ = 360$$

From which we have $\angle PRQ = 60$

Similarly, we can show that $\angle P = \angle Q = \angle R = 60$

This concludes the proof. ∎

Answers to Exercise 1

[1] $x = 6$ [2] $x = 8\frac{1}{2}$

==================================

CHAPTER 22

Circles

All points on the rim (circumference) of a circle are the same distance (the radius) from the centre of the circle. Thus, two radii and a chord form an isosceles triangle and we know that the base angles of this isosceles will be equal. This simple fact leads to a number of basic circle theorems. These theorems are then used to prove more interesting results involving two or more circles.

Theorem 1. **The angle in a semi circle is 90°**

AB is a diameter of the circle, centre O.
P is any point on the circumference.

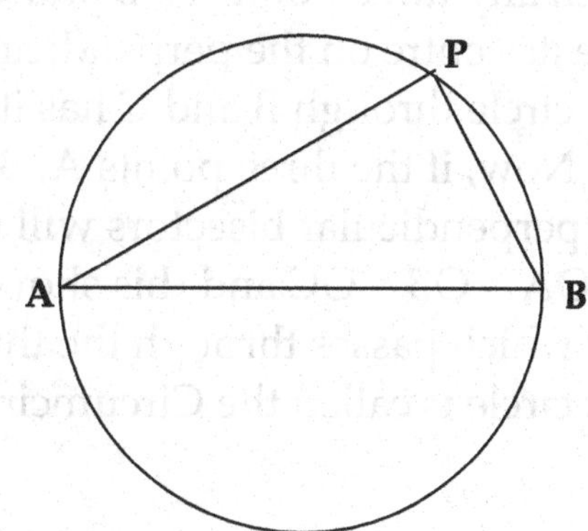

We must prove that $\angle APB = 90°$

Proof

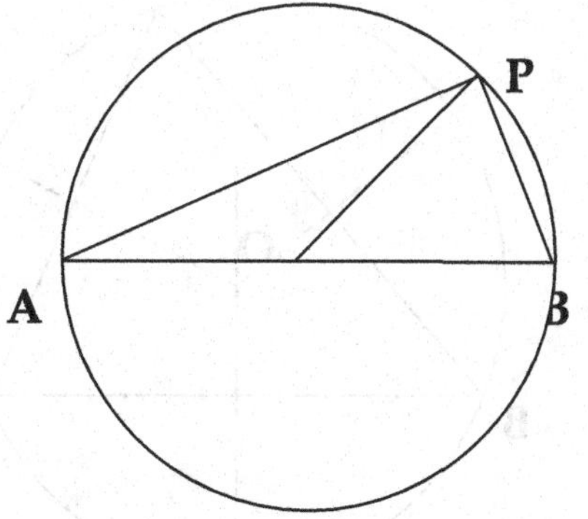

The angles x are equal (isosceles triangle OAP)
The angles y are equal (isosceles triangle OBA)

From ΔAPB, $2x + 2y = 180$

$\therefore \quad x+y = 90$

$\therefore \quad \angle APB = 90$ as required ■

The Circumcircle

If a circle passes through the two points A and B, then because the radii through these two points are equal, the centre of the circle will be on the perpendicular bisector of AB.

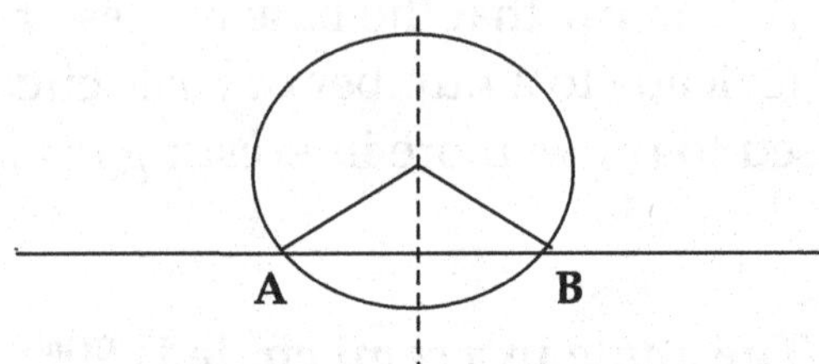

Given any three points A, B and C any circle passing through A and B will have its centre on the perpendicular bisector of AB.
Any circle through B and C has its centre on the perpendicular bisector of BC. Now, if the three points A, B and C are not in a straight line, then these two perpendicular bisectors will meet at a point O say. The point O, then has OA = OB = OC and this shows that a circle can be drawn with its centre at O which passes through the three points A, B and C.
This circle is called the Circumcircle of triangle ABC.

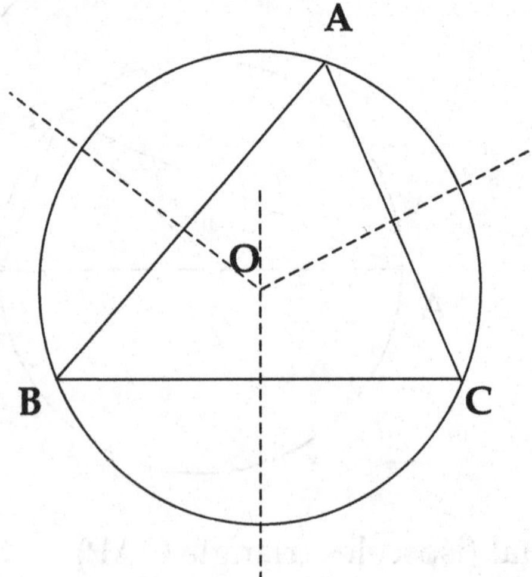

The perpendicular bisectors of the sides of a triangle ABC meet at the centre of the circle ABC.

==============================

Reductio Ad Absurdum

It is often the case that the converses of theorems lend themselves to proof by "reduction ad absurdum".
Suppose that we wish to prove the truth of a statement called X then the process of reduction ad absurdum (reduce to something absurd) is as follows:

Assume that X is false
Deduce from this assumption, the truth of another statement called Y (say)
Now show that Y must be false
Conclusion: X must have been true after all.

The proof of the converse of theorem 1, is a neat example of this method

Theorem 2 If $\angle APB = 90$ then the circle on AB as diameter passes through B.

Proof (by reduction ad absurdum)

Suppose that the circle on AB as diameter does not pass through P but meets AP at some other point Q

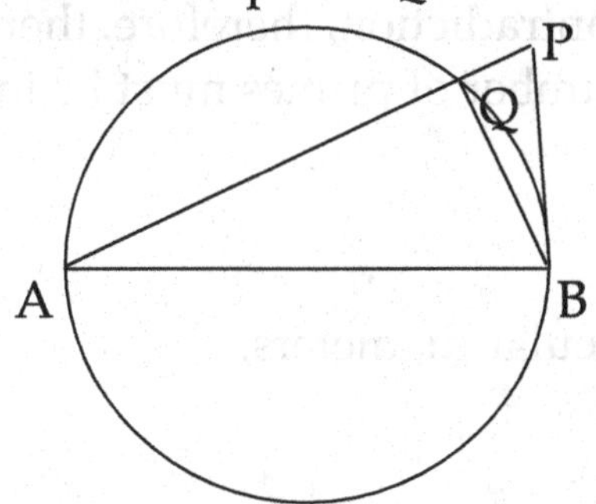

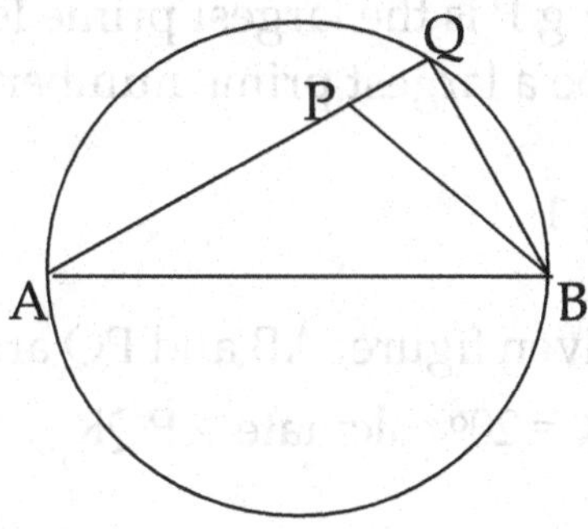

Then $\angle APB = 90$ (given)
and $\angle AQB = 90$ (by theorem 1)
Therefore, in triangle PQB, the angles at P and Q are both 90°
Therefore $\angle PBQ = 0$ which is absurd if Q and P are different points.

Therefore Q and P are the same point and the circle does pass through P.

■

An Infinity of primes

As a further illustration of the method of "reduction ad absurdum" we give here a proof that there is an infinite number of prime numbers.

Proof

Suppose that there is only a finite number of primes. Then there will be a largest prime number P say.

Now consider $P! + 1 = P(P-1)(P-2)(P-3)\ldots\ldots 3.2.1 + 1$

This is not divisible by any number less than or equal to P (because of the odd 1 that is added on).

Therefore, it is either a prime number, or, if not, must be divisible by a prime number that is greater than P.

In either case, there must be a prime number that is bigger than P.

Therefore, P cannot be the largest prime.

Assuming P is the largest prime leads to a contradiction, therefore, there cannot be a largest prime number and the number of primes must be infinite.

■

Exercise 1

In the given figure, AB and PQ are perpendicular diameters.
If $\angle BAX = 20^\circ$ calculate $\angle PQR$.

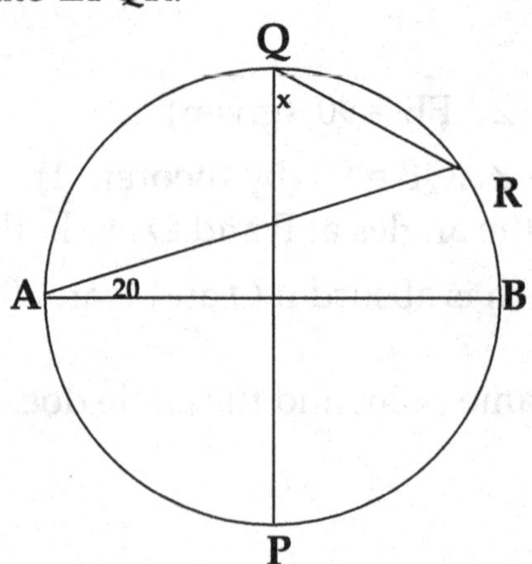

Theorem 3

The angle at the centre of a circle is twice the angle at the circumference.
There are three different cases to consider:

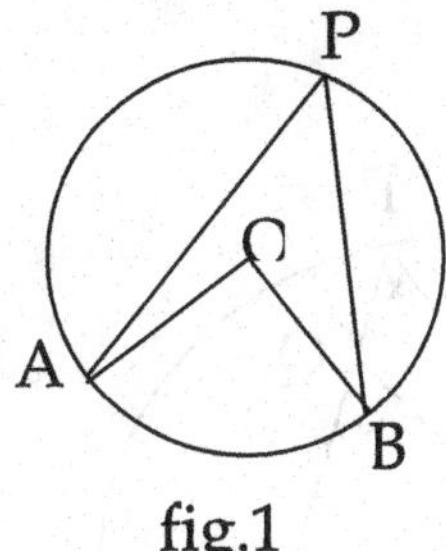

fig.1

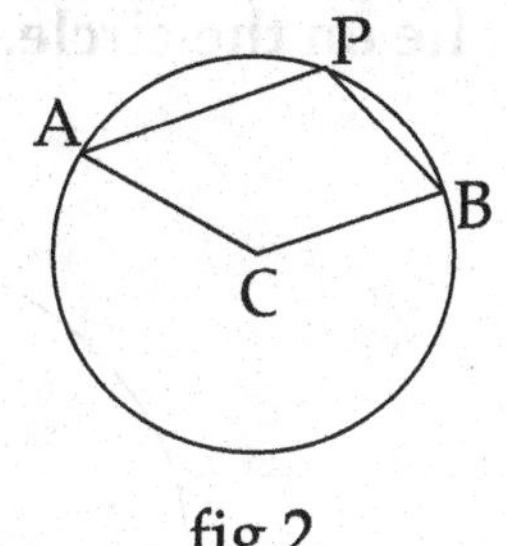

fig.2

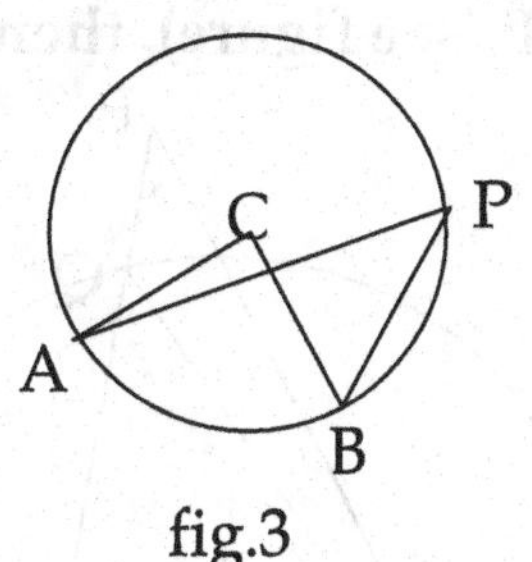

fig.3

In each figure we will show that ∠AOB = 2∠APB (reflex ∠AOB in fig.2)

Proof

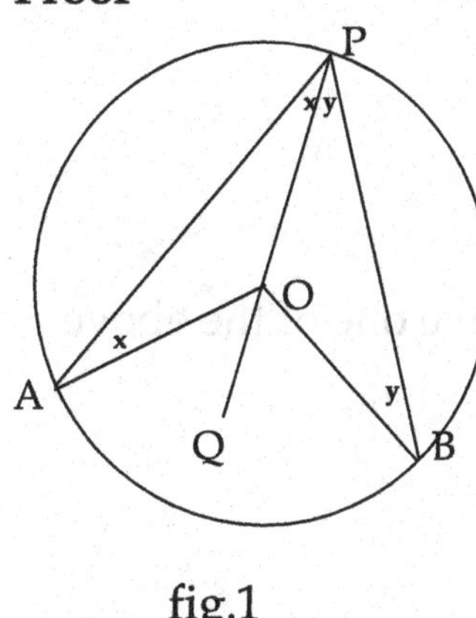

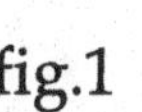

fig.1

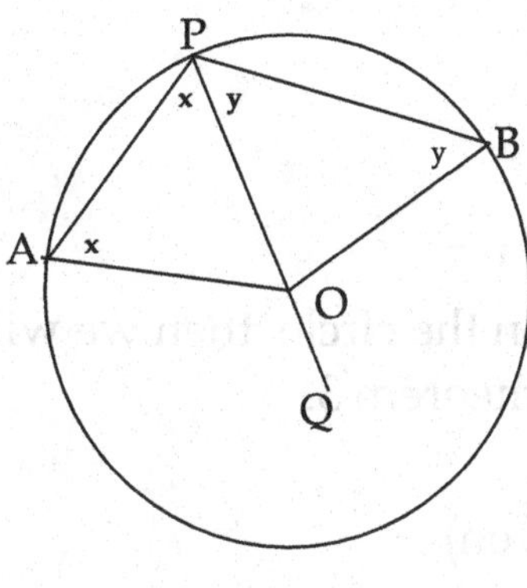

fig.2

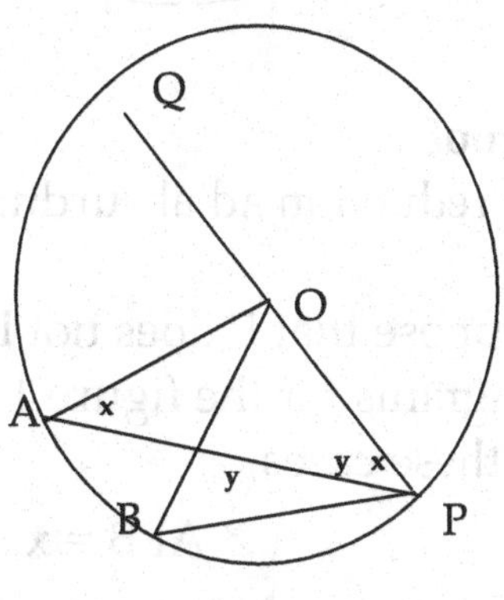

fig.3

Join PO and produce the line to Q

The angles x are equal (isosceles triangle OAP)

The angles y are equal (isosceles triangle OBP)

In (i) and (ii) ∠APB = x+y In (iii) ∠APB = y-x

But ∠AOQ = 2x and ∠BOQ = 2y (exterior angle of triangle)

In (i) ∠AOB = ∠AOQ + ∠BOQ = 2x+2y = 2(x+y) = 2∠APB

In (ii) reflex∠AOB=∠AOQ+∠BOQ = 2x+2y = 2(x+y) = 2∠APB

In(iii) ∠AOB = ∠BOQ - ∠AOQ = 2y – 2x = 2(y-x) = 2∠APB

Hence, in each case, the theorem is proved ■

The Converse of Theorem 3:

If an arc AB of a circle subtends 2x at the centre of the circle and x at a point P (see figure), then P must lie on the circle.

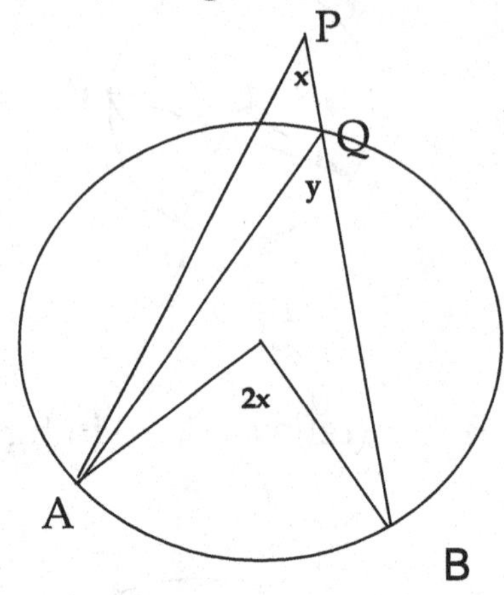

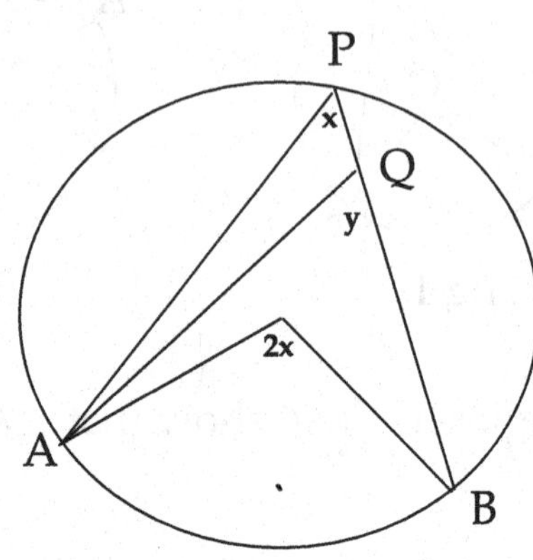

Proof

By reduction ad absurdum:

Suppose that P does not lie on the circle, then we will have one of the above diagrams for the figure 1 of theorem 3.

In these cases

$$\angle APB = x \text{ (given)}$$

and $y = x$ (since $\angle AQB = \frac{1}{2} \angle AOB$ by theorem 3)

$$\therefore \; y = x$$

but $y = x + \angle PAQ$ (ext angle of triangle)

therefore $\angle PAQ = 0$ which means P and Q are the same point. ■

This completes the proof for figure 1.

The reader is invited to consider the proofs for figures 2 and 3:

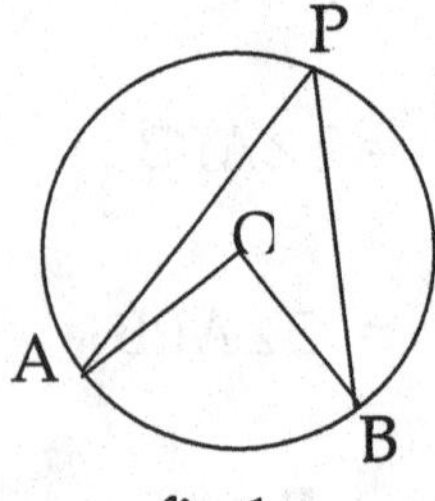

fig.1

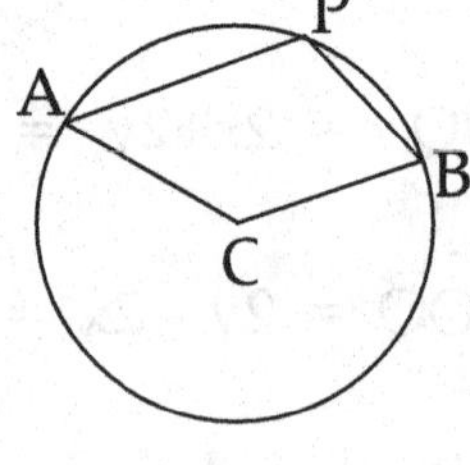

fig.2

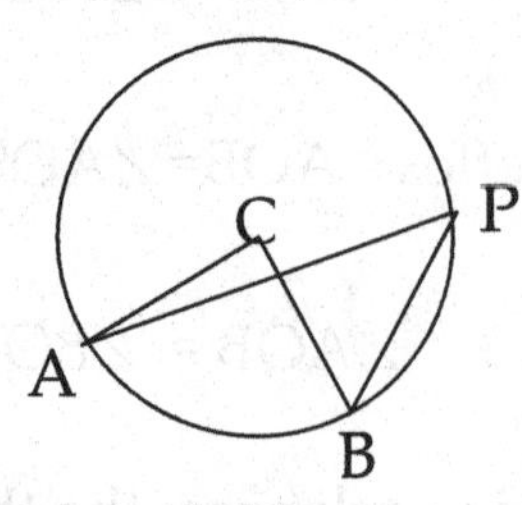

fig.3

Theorem 4 The Same Segment Theorem

In the following diagrams, the angles **a** and **b** are standing on the same arc AB of the circle centre O. The same **segment theorem** states that these two angles are equal.

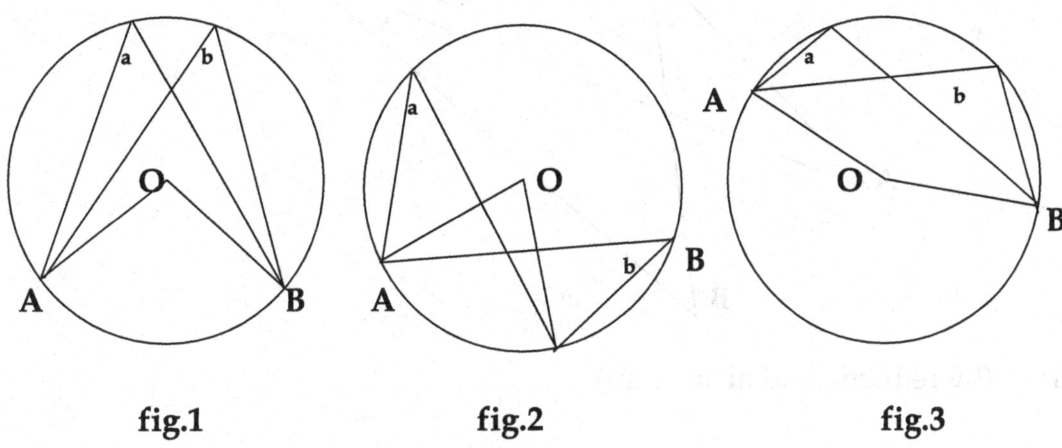

fig.1 fig.2 fig.3

Proof

In figures 1 and 2, by theorem 3, ∠ AOB = 2a,

and also ∠AOB = 2b therefore a=b

In figure 3 reflex ∠ AOB = 2a = 2b therefore a=b

■

Exercise 2

Find the angles marked **x** in the following figures

(i)

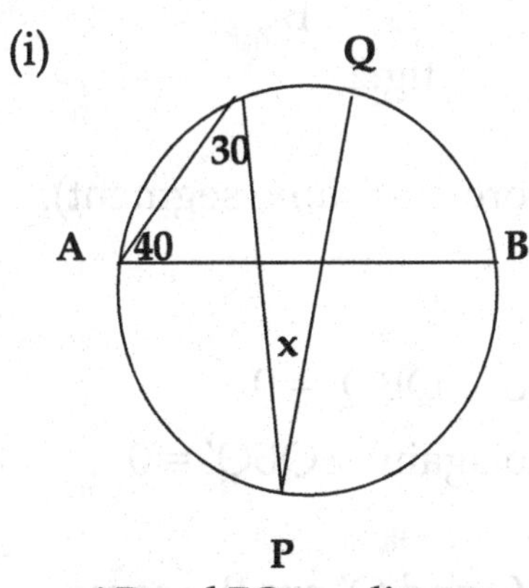

AB and PQ are diameters
Angles given are 40 and 30

(ii)

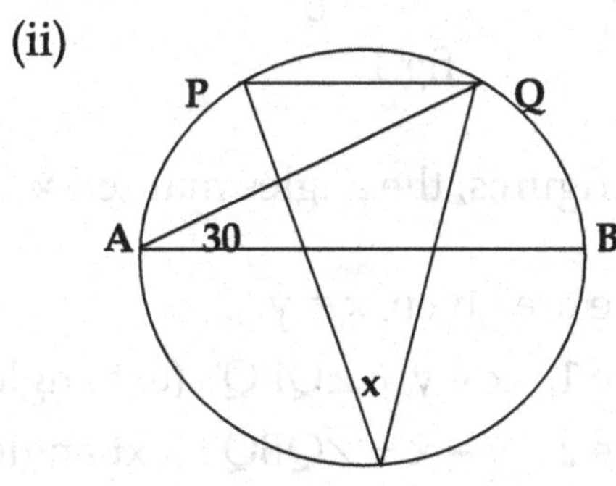

PQ is parallel to diameter AB
Given angle is 30

Theorem 5 **The Converse of Theorem 4**

In the following diagram, if x=y then the four points A,B, P and Q lie on a circle.

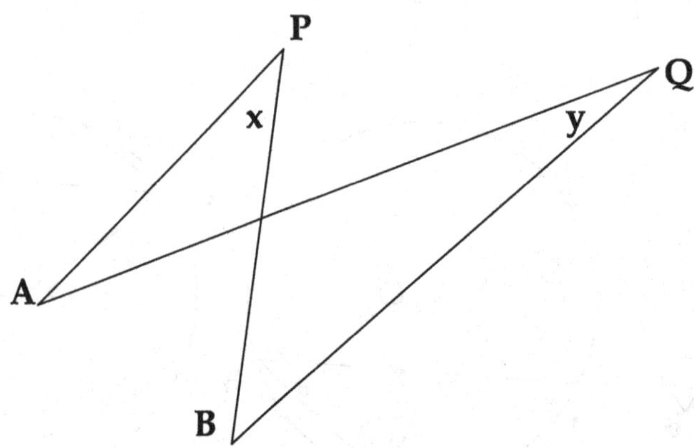

Proof (by reduction ad absurdum)

Suppose not and draw circle APB. Let AQ meet the circle at Q′.
There are two possible cases. In figure 1, Q′ lies on AQ produced. In Figure 2, Q′ lies between A and Q.

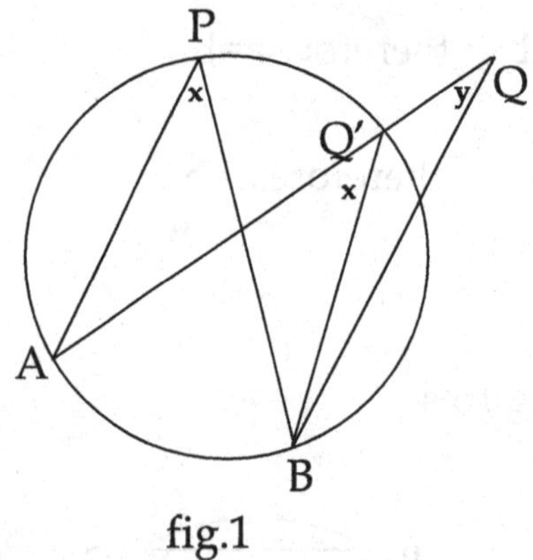

fig.1

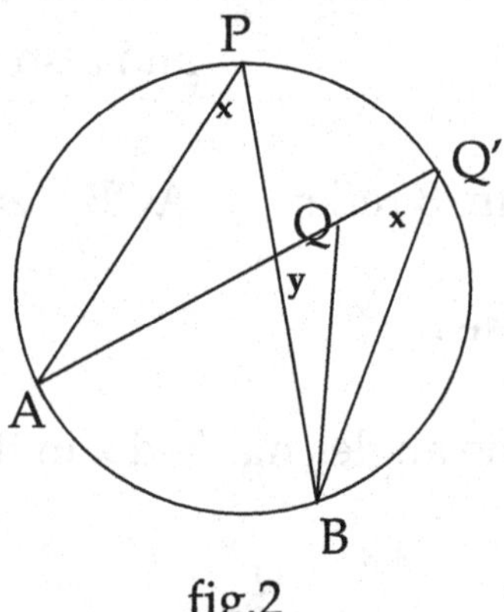

fig.2

In both figures, the angles marked x are equal by theorem 4 (same segment).

Also, we are given, x = y.
In figure 1, $x = y + \angle QBQ'$ (ext angle of triangle), so $\angle QBQ' = 0$
In figure 2, $y = x + \angle QBQ'$ (ext angle of triangle) , so again, $\angle QBQ' = 0$

In both cases, we have $\angle QBQ' = 0$ which means that Q and Q′ are the same point and this proves the result. ■

Theorem 6. Opposite Angles of a Cyclic Quad add up to 180 Degrees.

There are three cases to consider, one where the centre of the circle lies inside the quad, (fig.1), one where the centre lies on one side of the quad, (fig.2) and one where the centre of the circle lies outside the quad (fig.3).

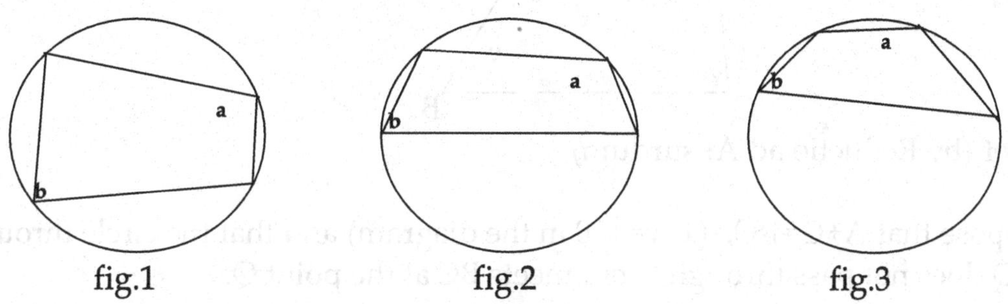

fig.1 fig.2 fig.3

In each of these figures we prove that a+b = 180°

Proof

Join the other two vertices to the centre as shown:
The angles at the centre will be 2a and 2b.

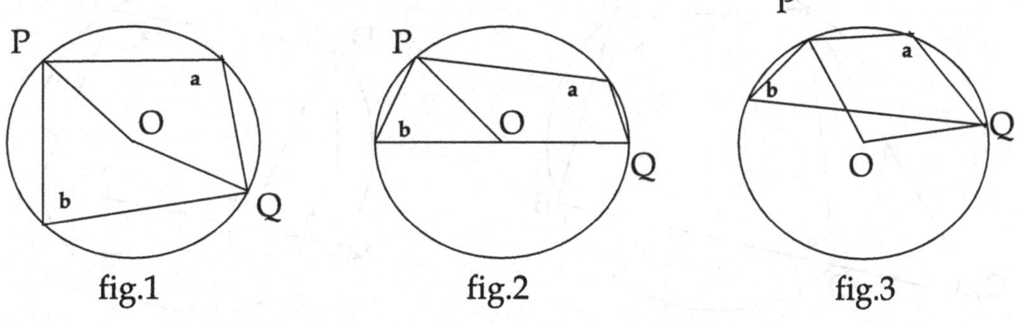

fig.1 fig.2 fig.3

In each on of these diagrams, by theorem 3 (angle at centre=2xangle at circ) we have

POQ reflex = 2a, POQ = 2b

∠ 2a + 2b = 360

∠ a+b = 180 ■

Exercise 3

Find the angle x in this figure:

AB is a diameter

O is the centre. Angles 50°, 100°

Theorem 7. **The Converse of Theorem 6**

For any quadrlateral ABCD, if A+C = 180 then a circle ABCD can be drawn.

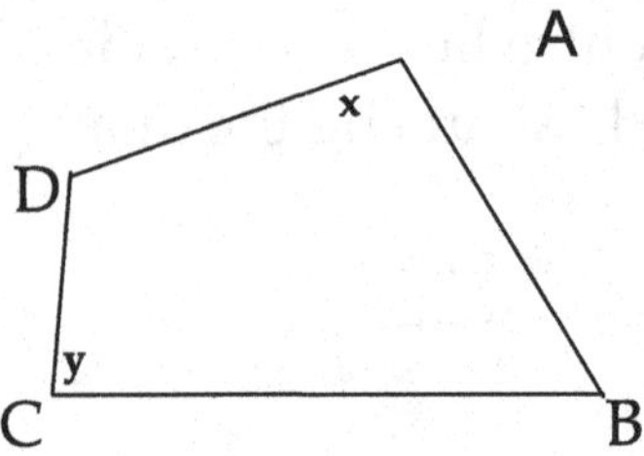

Proof (by Reductio ad Absurdum)

Suppose that A+C=180, (x+y=180 in the diagram) and that the circle through ABD does not pass through C but meets BC at the point Q.

There are two cases to consider, in figure 1 BC meets the circle at Q between B and C, and in figure 2, the circle meets BC at Q on BC produced.

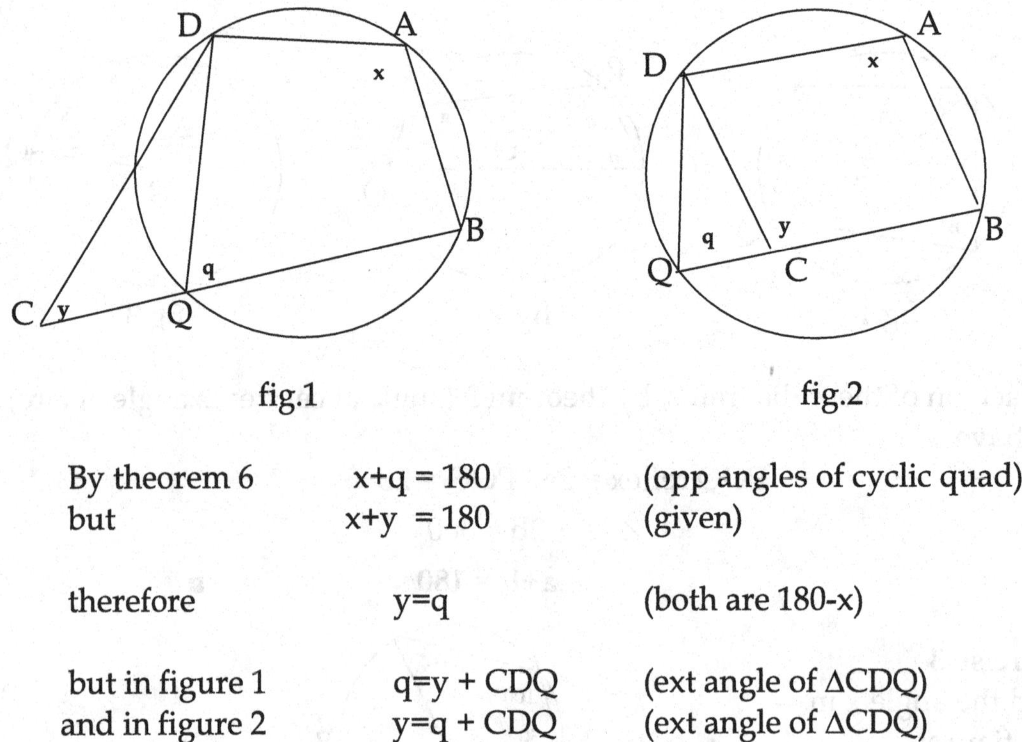

fig.1 fig.2

By theorem 6	x+q = 180	(opp angles of cyclic quad)
but	x+y = 180	(given)
therefore	y=q	(both are 180-x)
but in figure 1	q=y + CDQ	(ext angle of ΔCDQ)
and in figure 2	y=q + CDQ	(ext angle of ΔCDQ)

In both cases, we conclude that angle CDQ = 0 which means that C and Q must be the same point and this concludes the proof. ■

Tangents

A tangent to a circle is a line that touches the circle at a single point. The word derives from the Latin: tango = I touch. Other words from the same origin include contact, tangible and tactile.

The question arises; "Can a line touch a circle in just one point?"

Theorem 8
A Line can meet the Circle at just one Point only and it is then the perpendicular to radius through that point.
Such a line is called a tangent to the circle and the point is called the point of contact.

Proof(by reduction ad absurdum)

OP is a radius of the circle.
The line ST passes through P and is drawn perpendicular to OP.
Suppose that the line ST meets the circle again at a point Q.

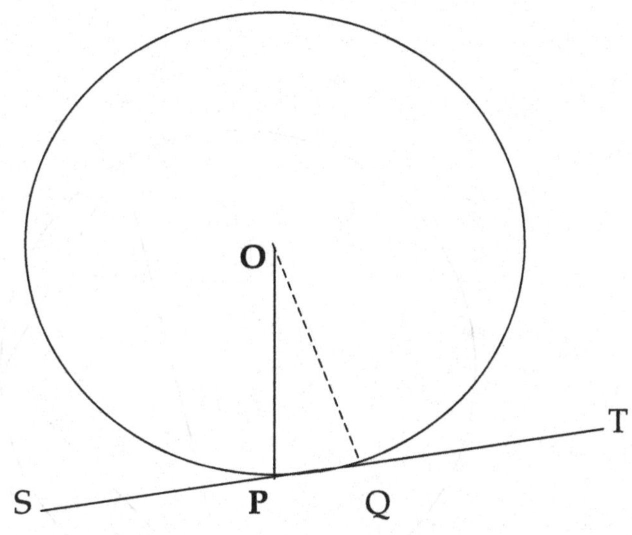

Join OQ, then OP = OQ (both radii),
so angles OPQ and OQP are equal (isos. Triangle)
But, $\angle OPQ = 90$ so, using the angle sum of triangle OPQ, we find angle POQ=0 so that P and Q are the same point. This concludes the proof. ■

Theorem 9. The Alternate Segment Theorem.

The angle between a tangent and a chord is equal to the angle standing on the chord in the alternate (i.e. the other side) segment.

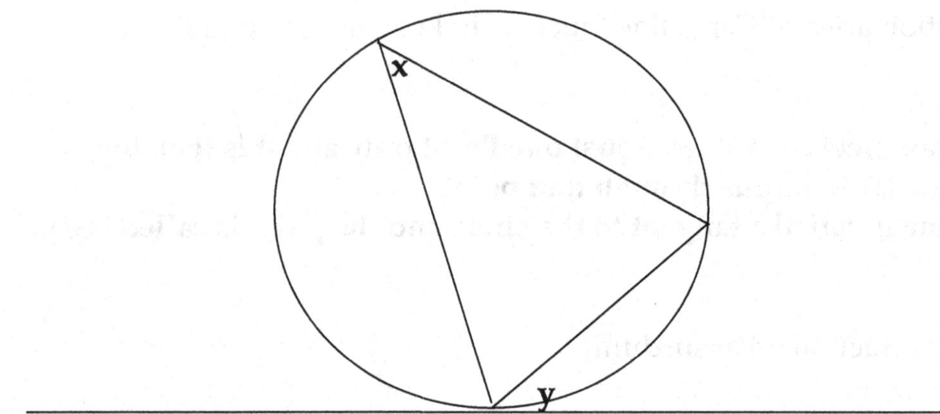

figure 1

figure 2

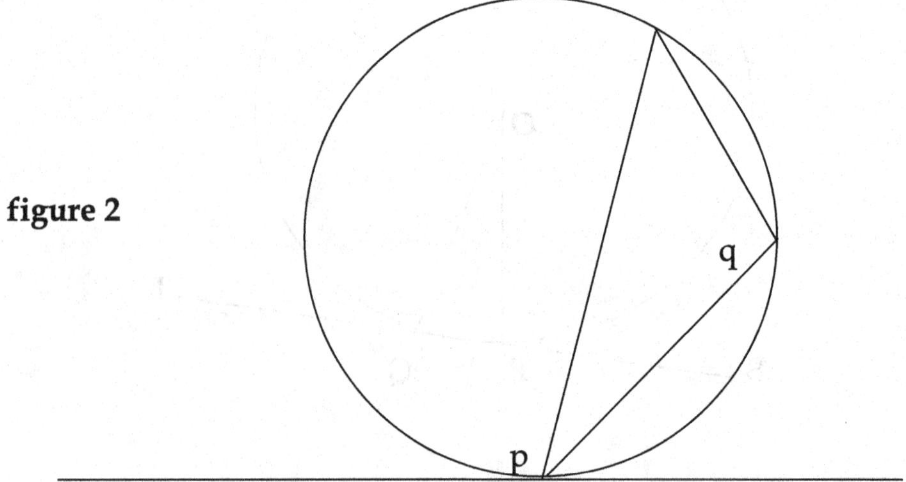

In figure 1, the angles x and y are equal.
In figure 2 the angles p and q are equal.

Proof
for figure 1

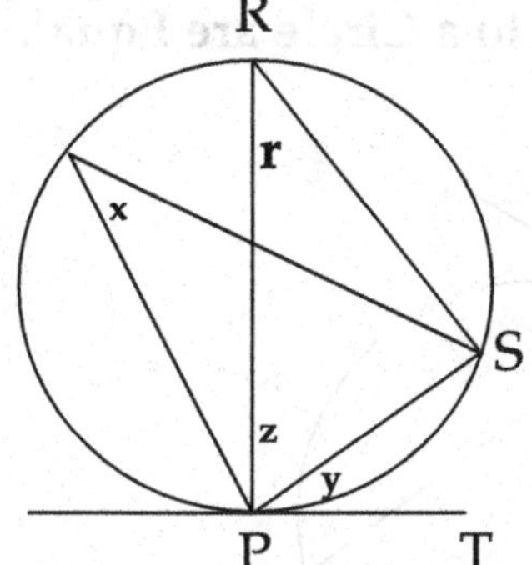

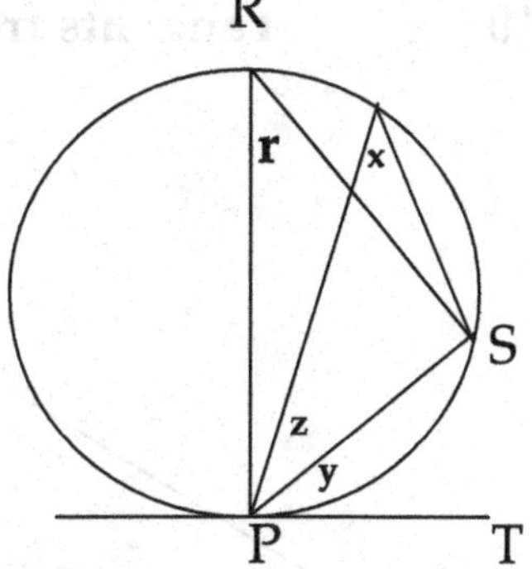

Draw the diameter PR then there are two possibilities. In both cases we have

	angle PSR=90	(angle in semi circle..Theorem 1)
therefore	r+z=90	(angle sum of triangle)
but	x=r	(same segment..Theorem 4)
therefore	x+z=90	
also	y+z=90	(tangent perp. radius..Theorem 7)
therefore	x=y	(both are 90-z)

■

figure 2

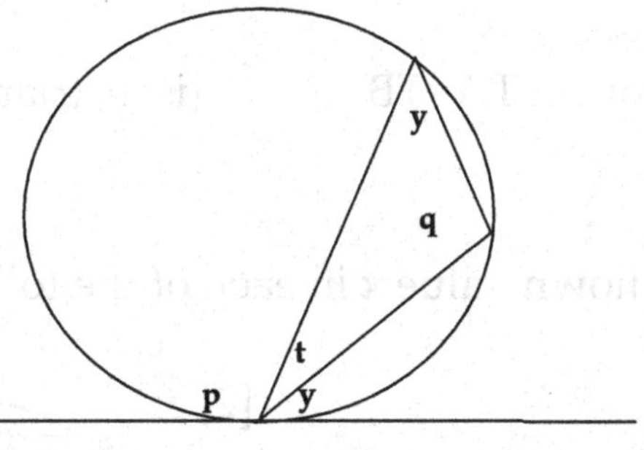

In figure 2, the angles marked y are equal, by the first part.

now	q+y+t=180	(angle sum of a triangle)
and also	p+y+t=180	(straight line PT)
therefore	p=q	(both are 180-y-t) ■

Theorem 10 **Tangents from a Point to a Circle are Equal.**

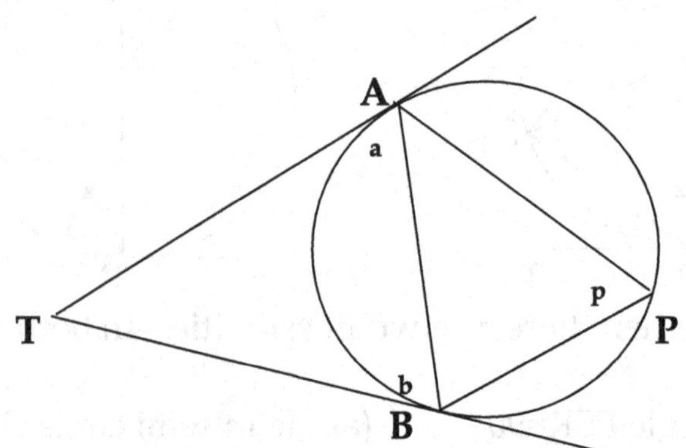

The tangents TA and TB to the circle are equal.

Proof

Let P be any point on the circle as shown,

then a=p (alt. seg Theorem 9)

b=p (alt. seg Theorem 9)

hence a=b

therefore TA=TB (isos. triangle)

■

Exercise 4 Find the unknown value **x** in each of the following diagrams:

[1]

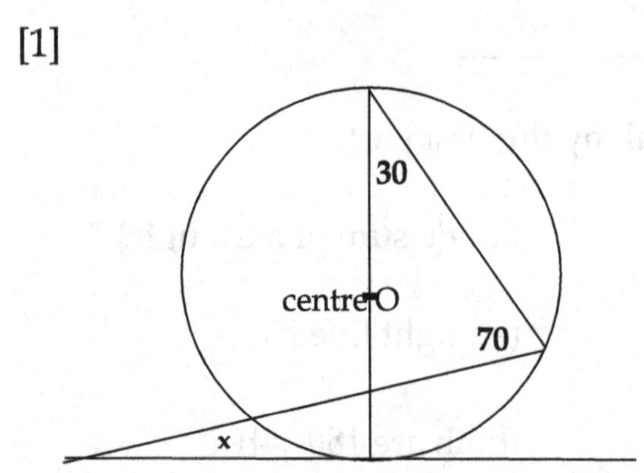

[2]

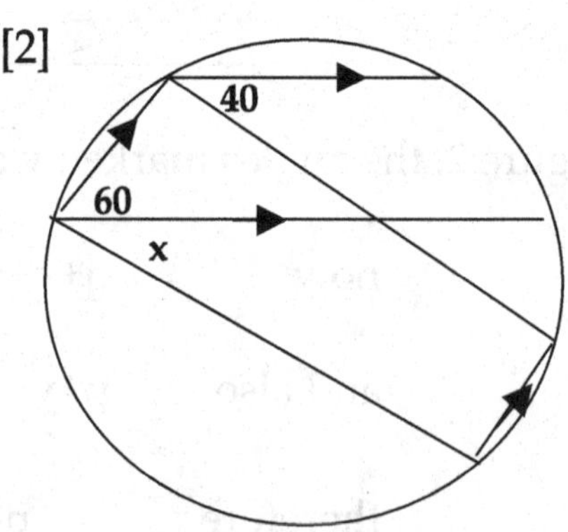

[3]

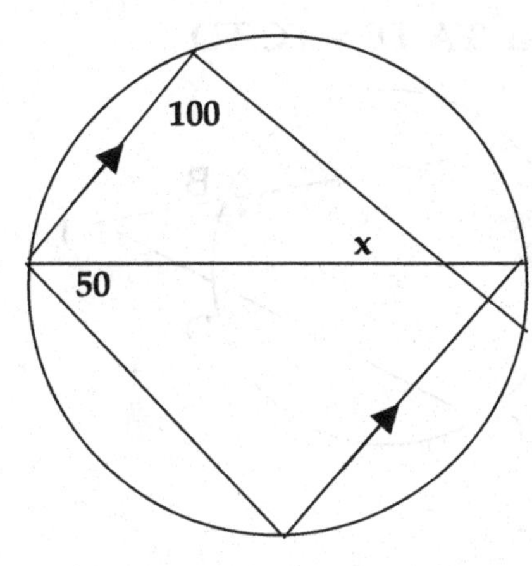

[4]

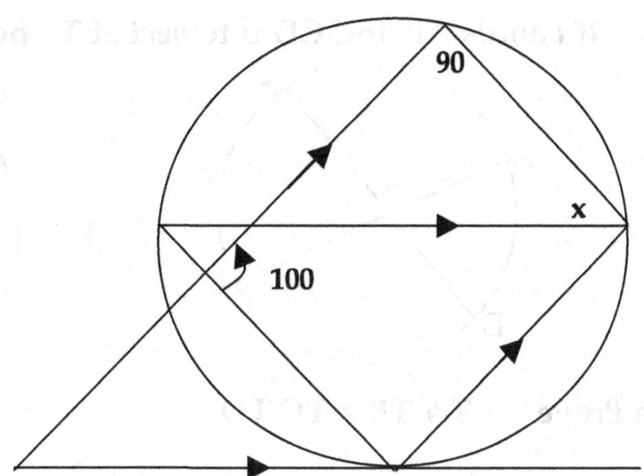

[5]

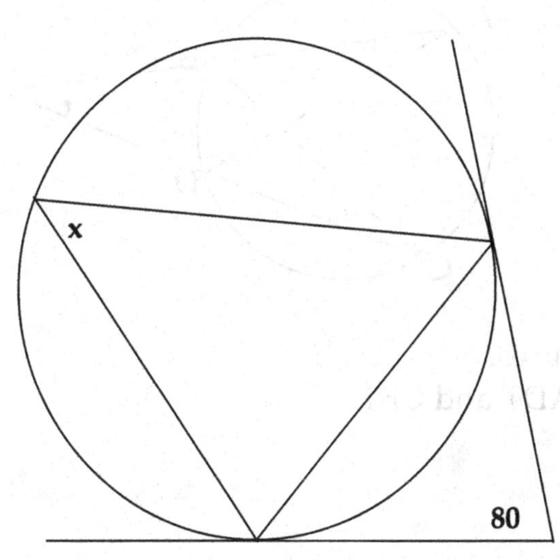

[6]

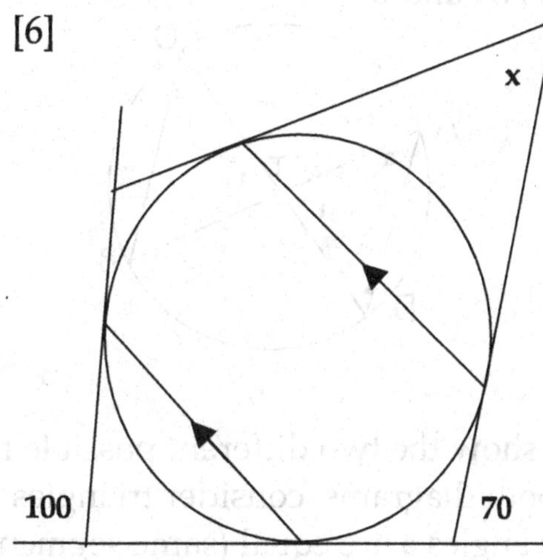

[7]

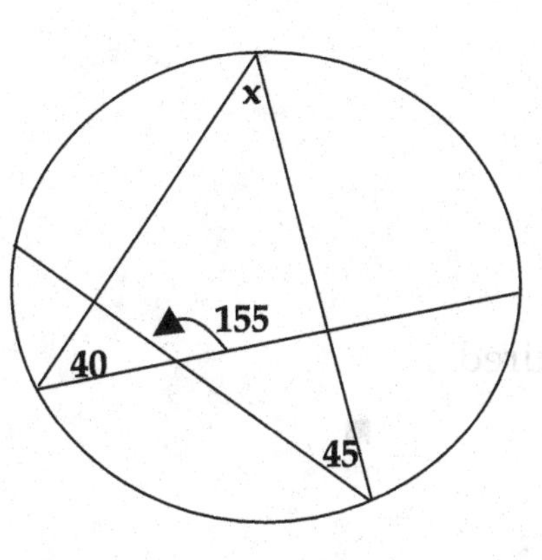

[8]

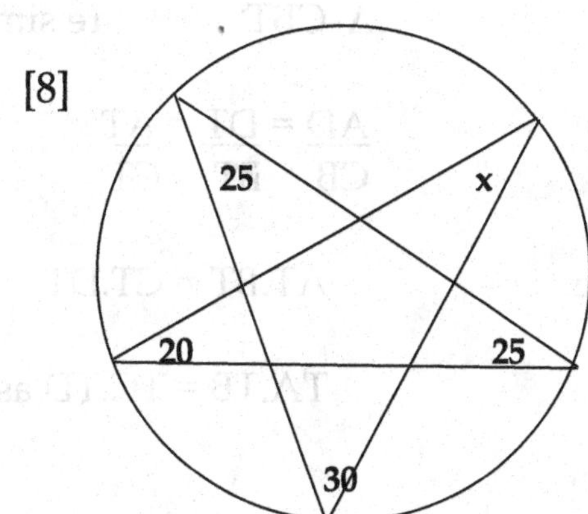

Theorem 11 **Intersecting Chords**

If chords AB and CD intersect at T then TA.TB = TC.TD

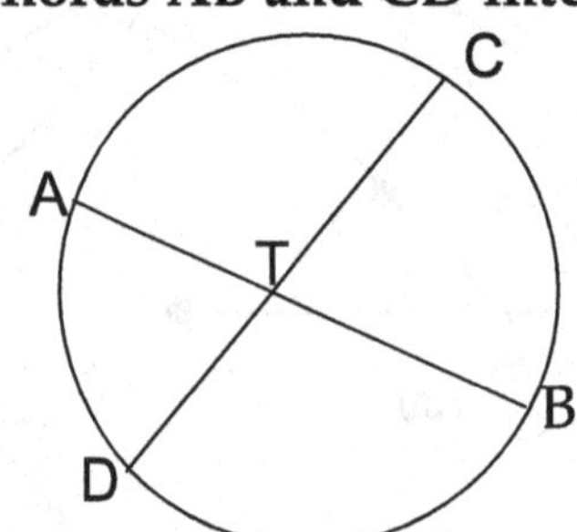

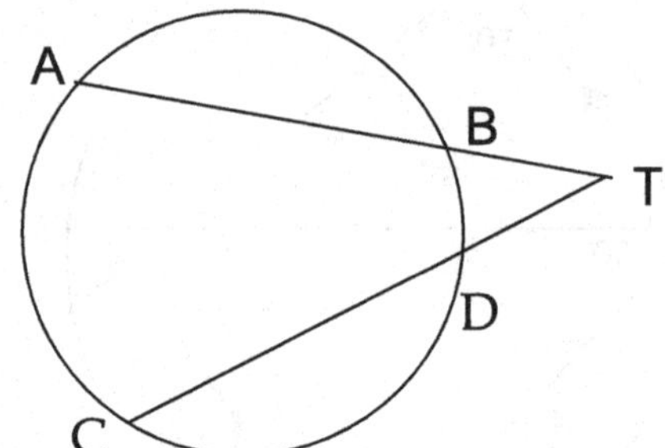

To Prove **TA.TB = TC.TD**

Proof
Join AD and BC

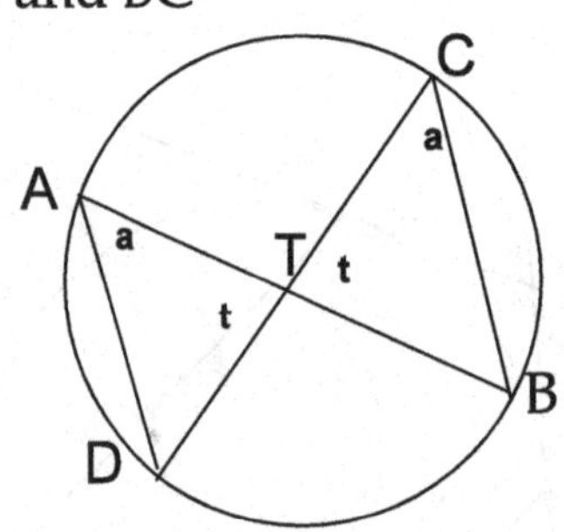

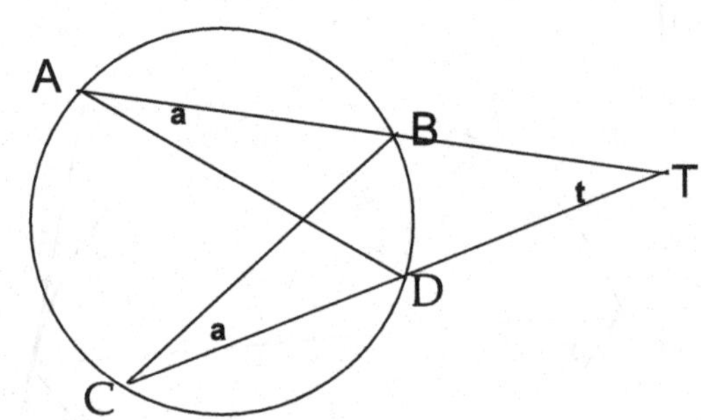

We show the two different possible figures.
In both diagrams, consider triangles ADT and CBT.
The angles **a** are equal (same segment)
And angles **t** occur in both triangles,

Therefore Δ ADT
and Δ CBT are similar triangles

therefore $\frac{AD}{CB} = \frac{DT}{BT} = \frac{AT}{CT}$

therefore AT.BT = CT.DT

or TA.TB = TC.TD as required

■

Theorem 12 $TA.TB = TC^2$

(This theorem is a special case of theorem 11.
When C and D coincide, then the chord CD becomes a tangent to the circle)

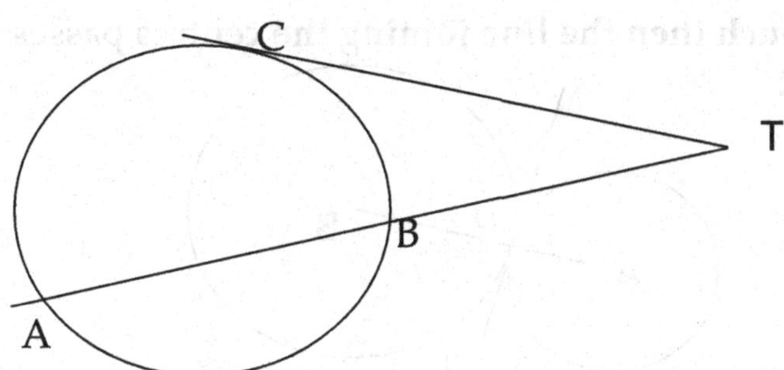

To Prove that

$$TA.TB = TC^2$$

Proof

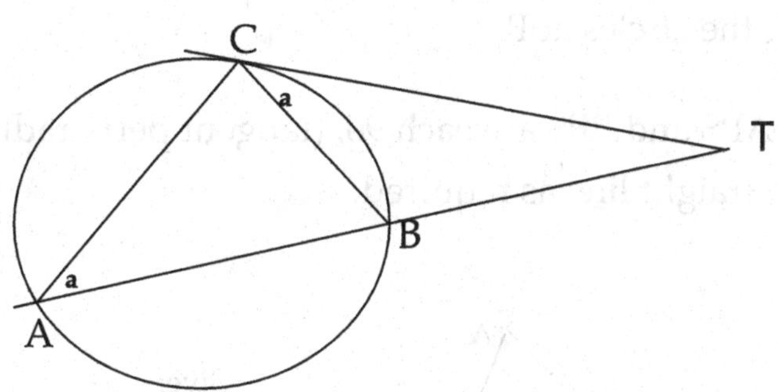

Consider triangles ATC and CTB.
The angles marked a are equal (alternate segment)
and angle b is common to both triangles.

Therefore Δ ATC
and Δ CTB are similar triangles

therefore $$\frac{AT}{CT} = \frac{TC}{TB} = \frac{AC}{CB}$$

giving $$AT.TB = CT.CT$$

or $TA.TB = TC^2$ as required

■

Touching circles If two circles touch, then they will have a common tangent at the point of contact.

Theorem 13

If two circles touch then the line joining the centres passes through the point of contact.

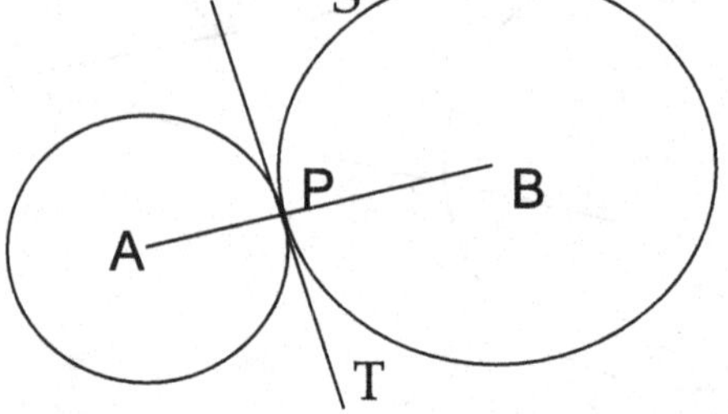

Proof

Let A and B be the centres of two touching circles and let ST be the common tangent touching the circles at P.

Then the angles APS and BPS are each 90, (tangent perp. radius)

Hence, APB is a straight line as required. ■

Exercise 5

[1]

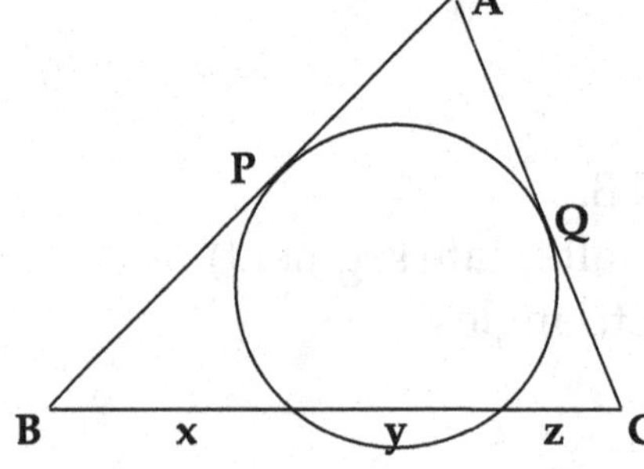

Given

BP = √15 , CQ = √8

BC = 7

Find the lengths **x, y and z**

[2]

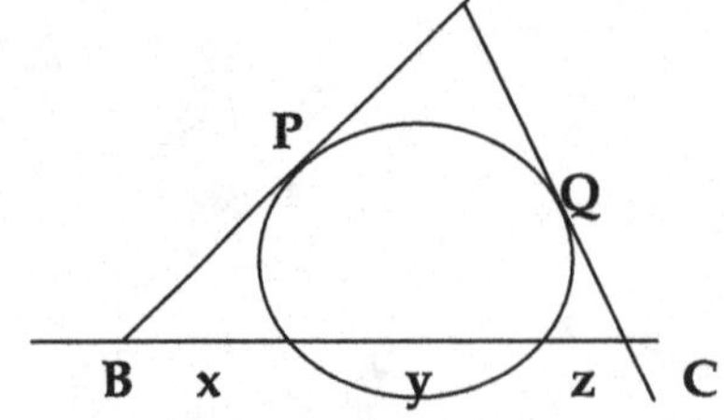

Given **BP = √15 , CQ = √15 and BC = 10**
Find **x, y and z**

[3}

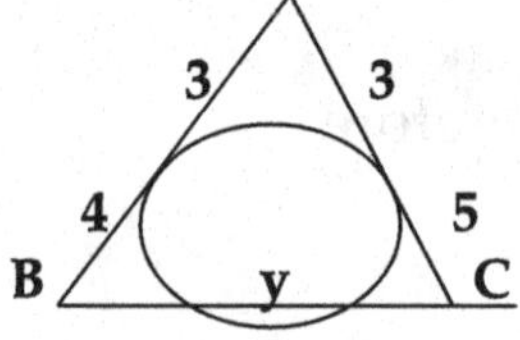

Given **BC = 9**
prove that **y = 0** so that the circle must be the in-circle of ΔABC

This chapter on circles continues with a number of problems that I hope you would agree, are often curious and surprising.

Problem 1 **Three Touching Circles**

(i) If three circles touch at the points P, Q and R prove that the common tangents are concurrent.
(ii) If O is the point of intersection of the three tangents prove that O is the centre of the circle PQR.

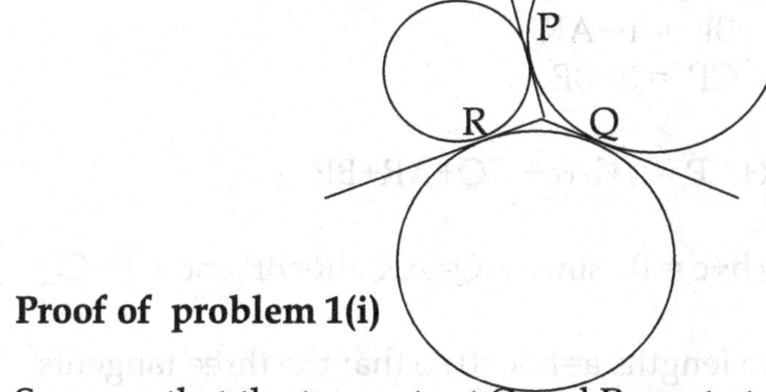

Proof of problem 1(i)

Suppose that the tangents at Q and R meet at A
Suppose that the tangents at R and P meet at B
Suppose that the tangents at P and Q meet at C

Then AQ=AR, BR=BP and CP=CQ (tangents to a circle...Theorem 10)

There are now two possible configurations:

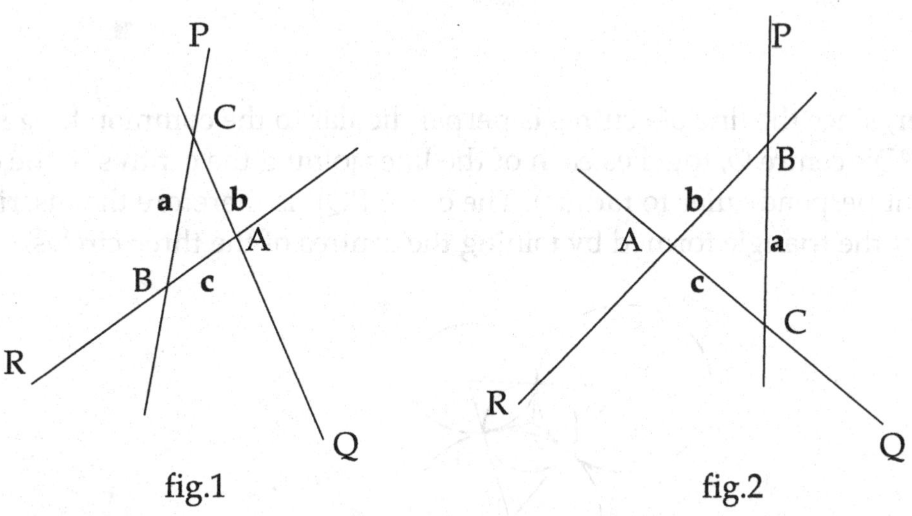

fig.1 fig.2

Let AB=c, BC=a and CA=b then

In figure 1 we have

$$AR = c+BR$$
$$BP = a+CP$$
$$CQ = b+AQ$$

Add to get $AR+BP+CQ = a+b+c+ BR+CP+AQ$

Hence $a+b+c = 0$ since AR=AQ, BP=BR and CP=CQ

In figure 2 we have

$$AQ = c+CQ$$
$$BR = b+AR$$
$$CP = a+BP$$

Add to get $AQ+BR+CP = a+b+c+ CQ+AR+BP$

Hence $a+b+c = 0$ since AQ=AR, BR=BP and CP=CQ

This means that in either case, the lengths a=b=c=0 so that the three tangents meet in a point.

■

Proof of problem 1(ii)

if the point of intersection of the tangents is O, then we have shown that PO=QO=RO so that O is the centre of triangle PQR

■

Further, since the line of centres is perpendicular to the common tangent, the circle PQR centre O, touches each of the lines joining the centres of the circles (tangent perpendicular to radius). The circle PQR is therefore the inscribed circle of the triangle formed by joining the centres of the three circles.

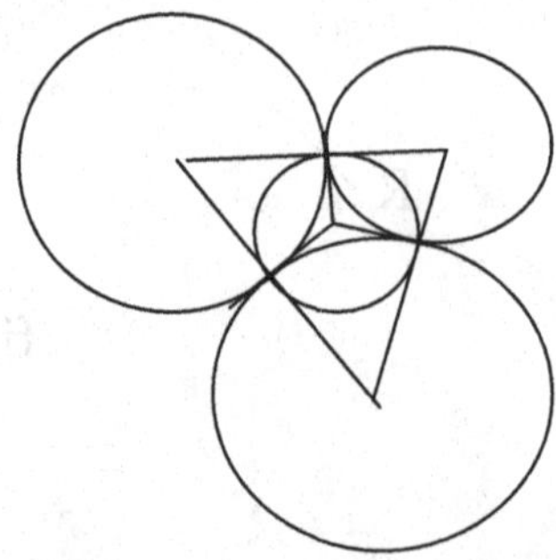

Problem 2 **Touching Circles Touch a Line**

Two circles, centre A and B, touch each other at C and touch a line at P and Q (see figure)

Prove that PCR is a straight line.

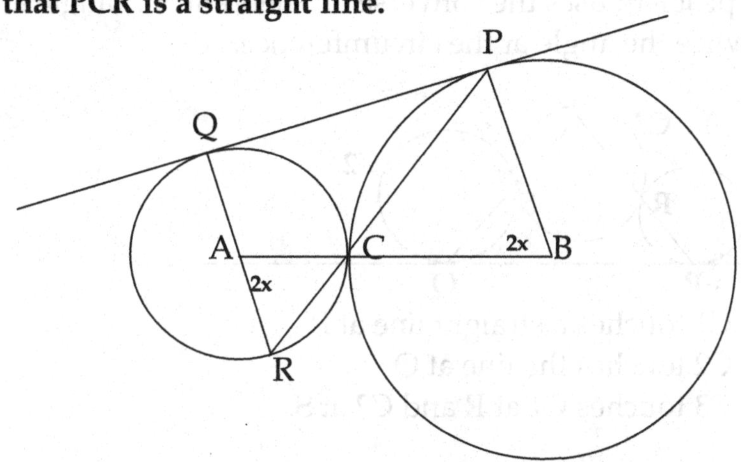

Proof

ACB is the line of centres and is a straight line.
The angles marked 2x are equal since QR // PB (tangent perp radius)
Therefore using the angle sum of the isosceles triangles ACR and PCB, angles ACR and PCB will both be (90-x).
Since ACB is a straight line, ∠ BCR will be 180-(90-x) = 90+x

Therefore ∠ PCR= ∠PCB+∠BCR = (90-x) + (90+x) = 180

Therefore PCR is a straight line.

■

Problem 3

Problem 3 involves three circles C1, C2 and C3. C1 touches C2 and C2 touches C3. C1 and C3 also touch a given line.

When we join the points of contact, we find that the two lines produced intersect on circle C3.

The solution to this problem uses the converse of Theorem 3, ...angle at the centre of a circle = twice the angle at the circumference.

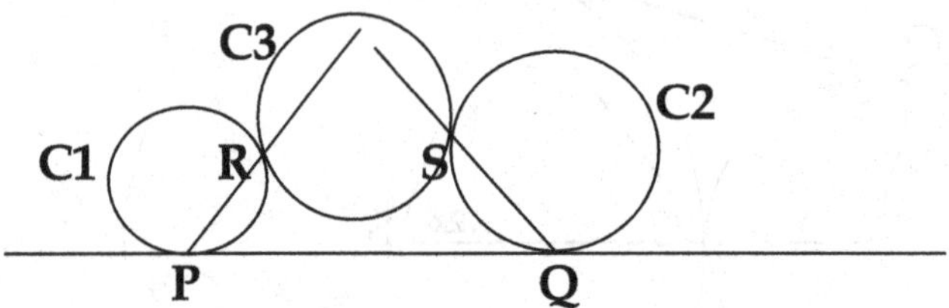

Given Circle C1 touches a straight line at P
Circle C2 touches the line at Q
Circle C3 touches C1 at R and C2 at S.

Prove

The chords PR and QS intersect at a point T that lies on circle C3.

Proof

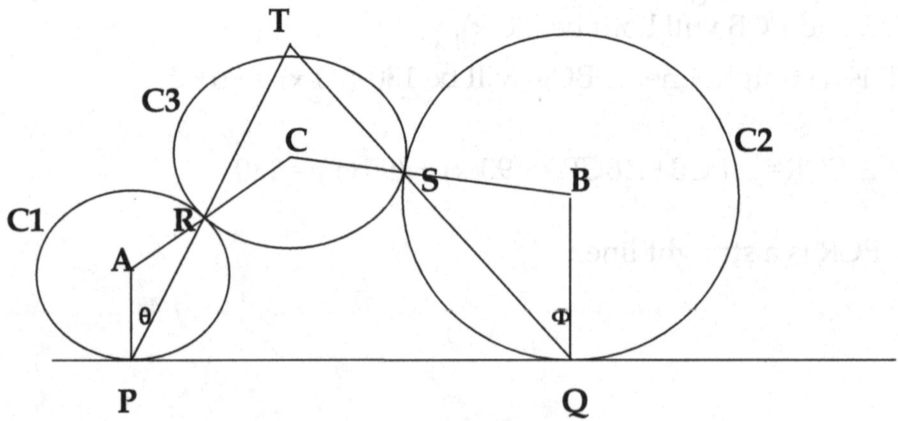

Let the centres of the circles be A, B and C.

ARC and BSC are straight lines since the circles touch on the line of centres.

Let the angles at P and Q be θ and Φ, as shown,

Then	∠RTS = ∠PTQ = θ+Φ	(PA//QB)
Now	∠CRT = θ and ∠CST = Φ	(opp ∠s + isos triangle)
But	∠RCS = CRT + RTS + CST	(ext angles of triangle)

$= \theta+(\theta+\Phi)+\Phi$

$= 2(\theta+\Phi)$

$\angle RCS = 2\angle RTS$

Therefore T lies on C3 (Converse of ∠at centre = 2∠at circ.)

■

But, there is more to this problem than meets the eye! Suppose that we draw a third circle:

Chord PR meets chord QS at T on C3.
But, by the same theorem, chord XY meets chord PR at T on C3. (PR only meets C3 at one other point, so there is only one T!)
Chord XY also meets chord QS at T on C3.

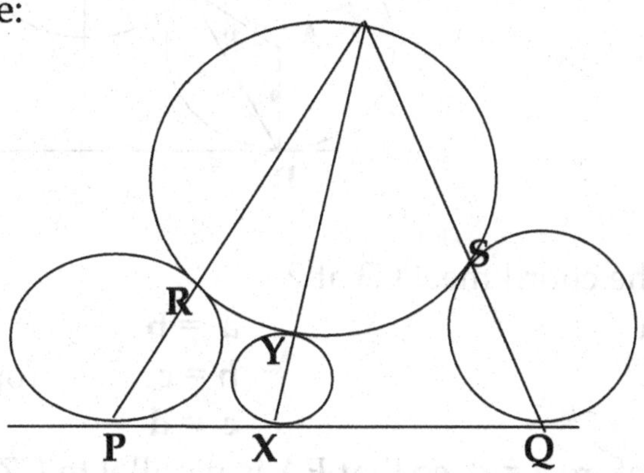

T is therefore a rather special point. To find out more about T, draw a symmetrical diagram:

From symmetry alone, we can see that T must be the point at the end of the diameter which is perpendicular to the line.

Instead of starting with problem 3 an easier approach would have been:

Problem 3*
If circle C1 touches a line and a circle C3, prove that the chord of contact passes through our point T.

Proof

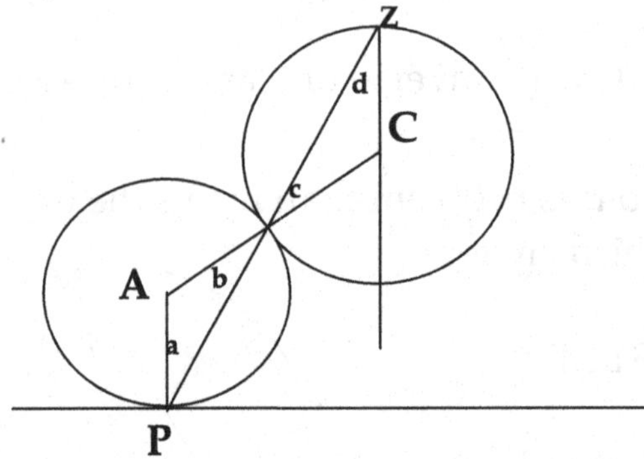

Let the chord meet C3 at Z

Then	**a** = **b**	(isos triangle)
	b = **c**	(opposite angles)
	c = **d**	(isos triangle)

Therefore **a** = **d** so that PA is parallel to CZ. Therefore CZ is perpendicular to the line and so Z must be the point T.
Problem 2 is now a special case of problem 3* in which C3 touches the line:

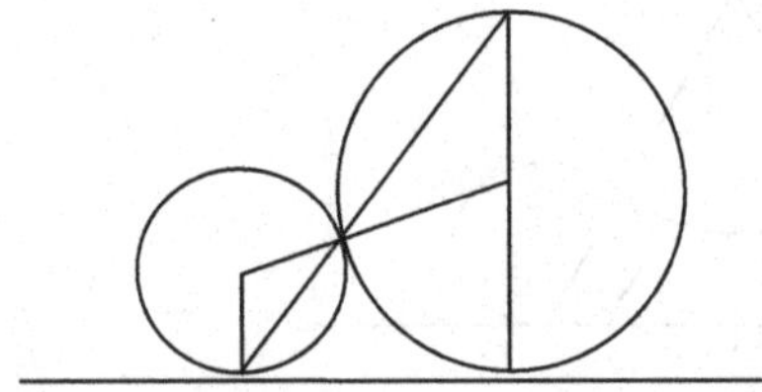

Problem 3* shows that PR always passes through T wherever C1 happens to be, as long as C1 touches the line and touches C3:

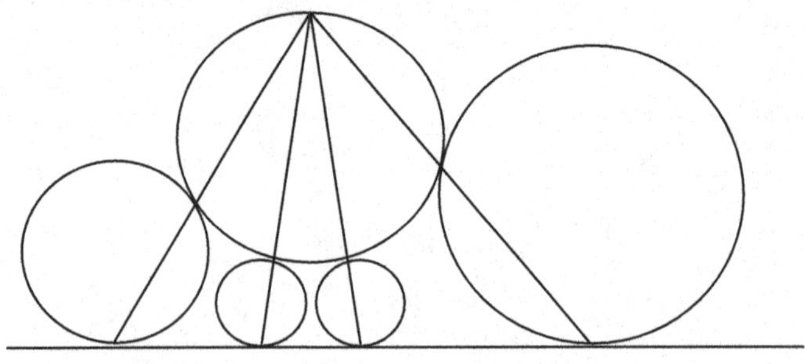

=======================================

Problem 4 A sequence of inscribed circles drawn as shown, producing a sequence of triangles labeled ABC, A1B1C2, A2B2C2, A3B3C3.....
What happens to the triangle AnBnCn as the value of n gets larger?

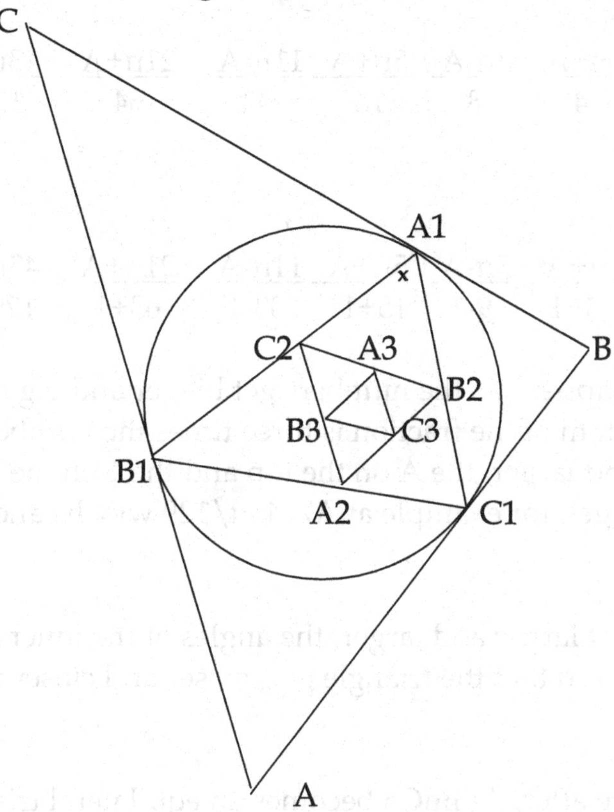

Solution

(Note: for convenience, in this section we use π for a half turn rather than 180..radian measure which is covered in chapter 20)

Using isosceles triangles and the alternate segment theorem, a short calculation shows that the angle x at A1 is given by

$$x = \frac{\pi - A}{2} \qquad \text{where A is the angle BAC}$$

Using the same calculation on triangle A2B2C2 we will have $\frac{\pi - x}{2} = \frac{\pi + A}{4}$

for the next angle in triangle A2B2C2.

These calculations give us a sequence of values for the angles in the triangles:

Angle at	A	A1	A2	A3	A4	A5	A6	A7
Value	A	$\frac{\pi-A}{2}$	$\frac{\pi+A}{4}$	$\frac{3\pi-A}{8}$	$\frac{5\pi+A}{16}$	$\frac{11\pi-A}{32}$	$\frac{21\pi+A}{64}$	$\frac{43\pi-A}{128}$

Write these values as:

Value	A	$\frac{\pi-A}{3-1}$	$\frac{\pi+A}{3+1}$	$\frac{3\pi-A}{9-1}$	$\frac{5\pi+A}{15+1}$	$\frac{11\pi-A}{33-1}$	$\frac{21\pi+A}{63+1}$	$\frac{43\pi-A}{129-1}$

and we can see what happens as the numbers get bigger and bigger. The number in the bottom of the fraction is three times the number in the top. As n gets larger and larger, the A on the top and the 1 on the bottom can be ignored, so that we get, for example at A7, $43\pi/129$ which cancels down to $\pi/3$.

As the n in AnBnCn get larger and larger, the angles of the inner triangle approach the value $\pi/3$ so that the triangle gets closer and closer to being an equilateral triangle.

This is the result we are after, AnBnCn becomes an equilateral triangle.

■

Problem 5 **Three Equal Circles**

Three equal circles intersect at P, Q and R and have a common point T.

Prove (i) that T is the orthocenter of triangle PQR

(ii) that the circle PQR has the same radius as the given three circles.

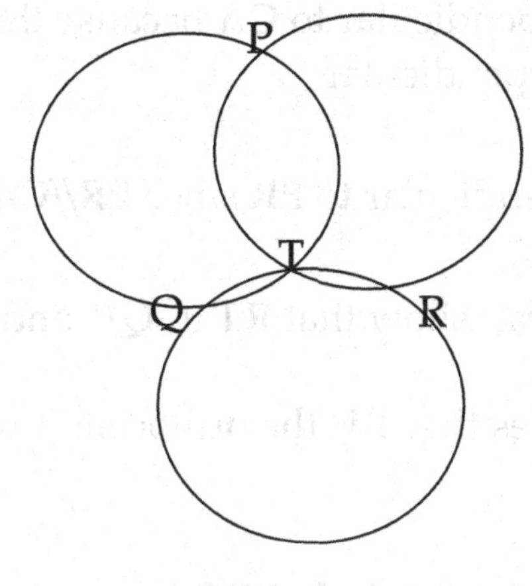

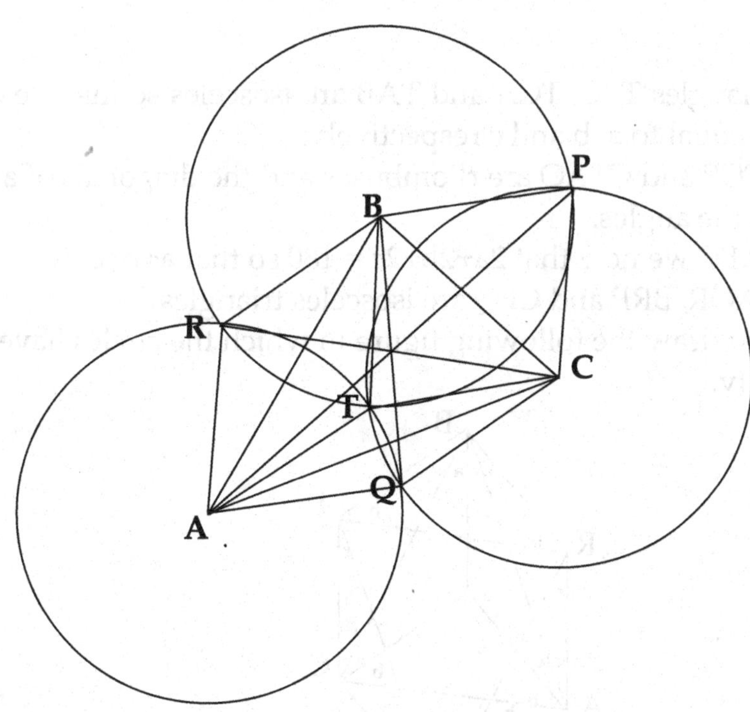

(i) To prove that T is the orthocentre of triangle PQR

Proof

BTCP and BTAR are both rhombuses (all sides are equal)

∴ PC = // BT = // RA

∴ PCAR is a parallelogram with **PR // CA**

but **QT** is perpendicular to **CA** because the diagonals of rhombus CTAQ are perpendicular

∴ **QT** is perpendicular to **PR** since **PR//CA**

Similarly we can show that **RT ⊥ QP** and **PT ⊥ RQ**

And this proves that T is the orthocenter of triangle PQR

■

(ii) To prove that radius of circle PQR is equal to the radii of the other circles

We note that triangles TBC, TCA and TAB are isosceles so that we can mark the base angles equal to a, b and c respectively.
Also, ATBR, BTCP and CTAQ are rhombuses and the diagonals of a rhombus bisect the angles.
From triangle ABC we note that 2a+2b+2c = 180 so that a+b+c=90
Also note that AQR, BRP and CPQ are isosceles triangles.
This allows us to draw the following figure in which the circles have been left out for clarity.

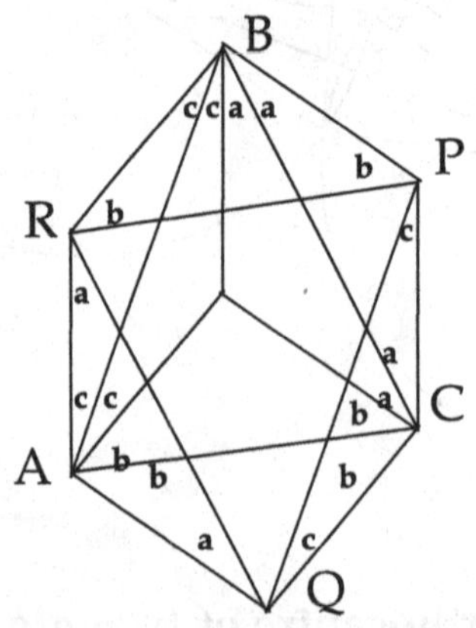

The angle sum of triangle BPC gives

$$\angle P = 180-(2a+b+c)$$

$$= 180-(a+b+c) - a$$

$$= 90-a \qquad \text{(since } 2a+2b+2c=180)$$

Using the sine rule in triangle PQR

$$\text{Radius of circle PQR} = \frac{QR}{2\sin P} = \frac{BC}{2\sin(90-a)} = \frac{BC}{2\cos a} = \frac{BC}{2\text{x } \tfrac{1}{2} BC/R} = R$$

where R is the radius for each of the given circles and this proves the result.

■

Pascal's Theorem

In chapter 21, we proved Pascal's theorem for points on straight lines. The diagram is shown below:

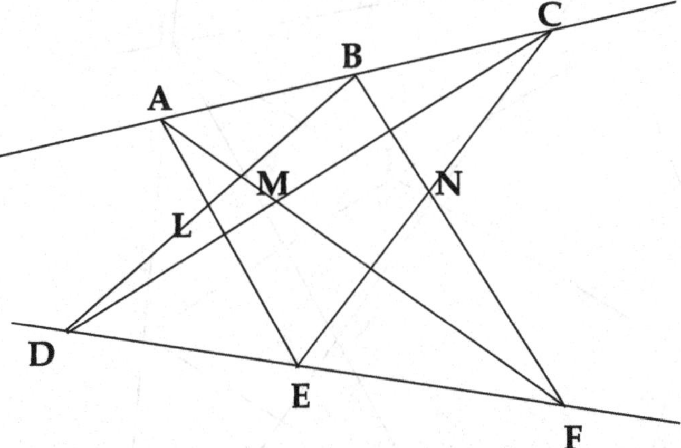

ABC and DEF are any two straight lines. L,M and N are the intersections shown. Pascal's theorem proves that LMN is always a straight line.

Students of projective geometry will know that Pascal's theorem holds for any conic section, that is, a parabola, an ellipse, or a hyperbola.
The two straight lines in the diagram above can be regarded as a degenerate hyperbola so that the theorem of chapter 21 is really a special case of a much

more general result. The conic sections referred to are the curves that we can find if we slice a cone by a plane.

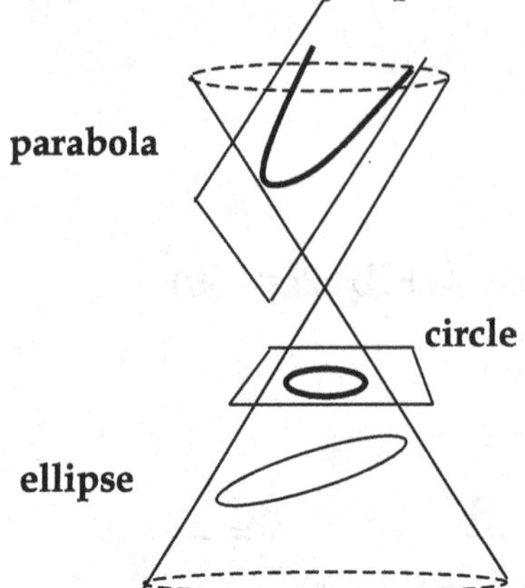

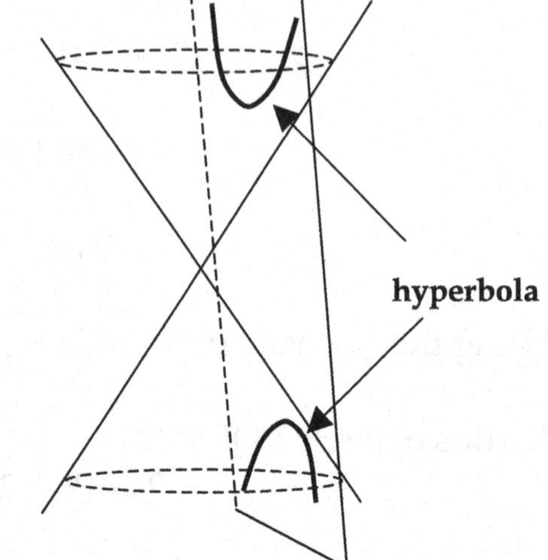

Exercise 6

As an exercise on Pascal's theorem, see if you can prove that the lines AC, BD, PQ and LM in this diagram, must be concurrent:

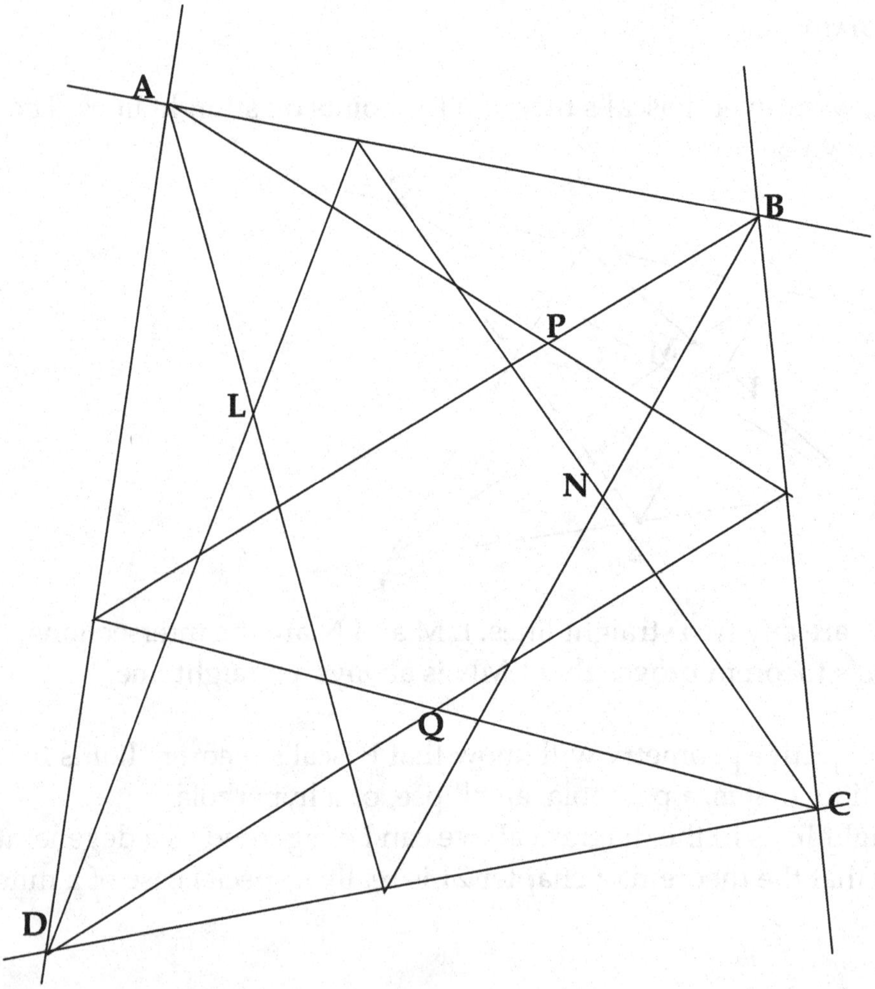

Pascal's Theorem for a Circle

A,B,C and D,E,F are six points on a circle.

If AE∩BD = L, AF∩CD = M, BF∩CD = N then LMN is a straight line.

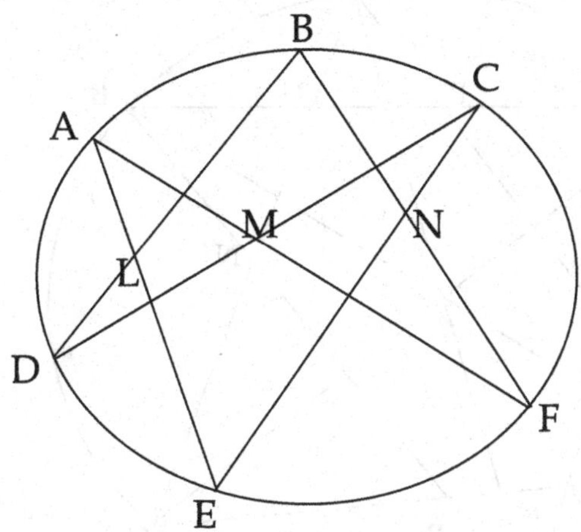

Proof (uses Menelaus's Theorem…chapter 21)

Let AE∩DC = Z, AE∩BF = Y, BF∩DC = X

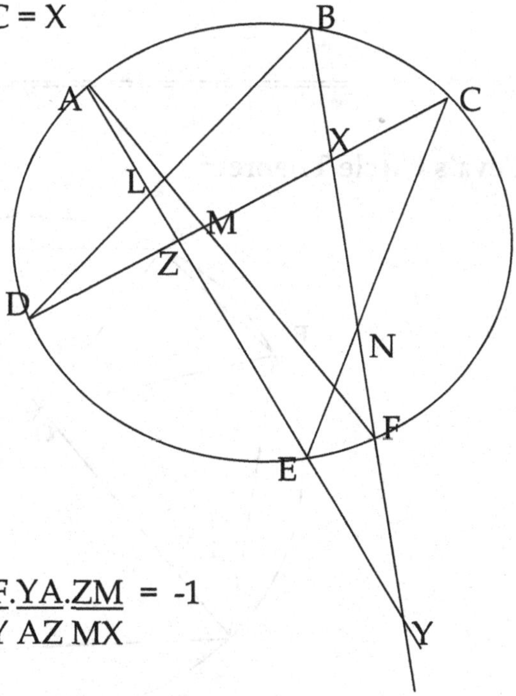

Using Menelaus's Theorem in triangle XYZ with

transversal BLD $$\frac{XB.YL.ZD}{BY\ LZ\ DX} = -1$$

transversal CNE $$\frac{XN.YE.ZC}{NY\ EZ\ CX} = -1$$

transversal AMF $$\frac{XF.YA.ZM}{FY\ AZ\ MX} = -1$$

Multiply: $$\frac{XB.YL.ZD.XN.YE.ZC.XF.YA.ZM}{BY\ LZ\ DX\ NY\ EZ\ CX\ FY\ AZ\ MX} = -1$$

Now use XB.XF=XC.XD, ZD.ZC=ZA.ZE, YE.YA=YF.YB

And after canceling, we are left with

$$\frac{XN.YL.ZM}{NY\ LZ\ MX} = -1$$

Thus, by the converse of Menelaus's theorem, LMN is a straight line. ■

Exercise 7

In this figure, prove that the lines AD, CF, LN, PR and UW are concurrent.

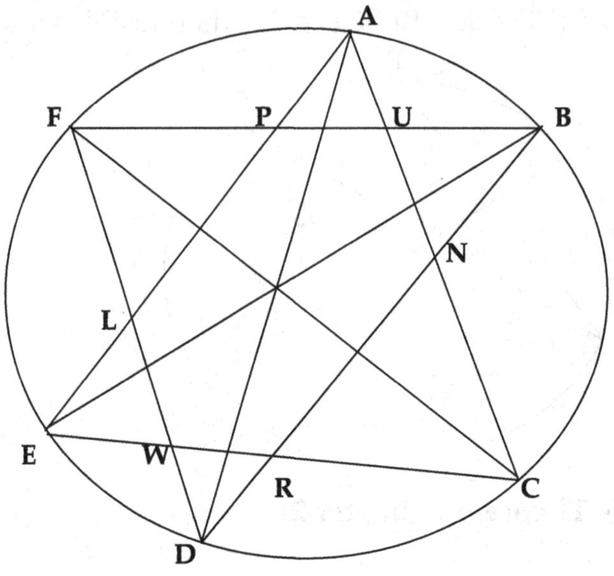

===================================

Ceva's Circle Theorem

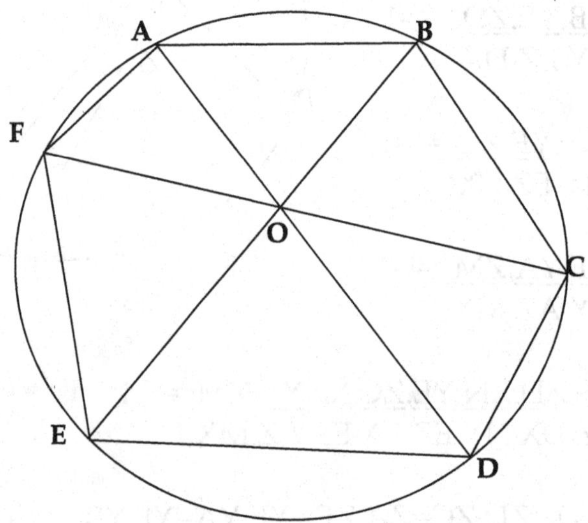

A,B,C,D,E,F are six points in order on a circle.
If AD, BE and CF are concurrent then

$$\frac{AB.CD.EF}{BC\ DE\ FA} = 1$$

Proof
Using the same segment theorem, we have

Triangles ABO and EDO are similar $\therefore \frac{AB}{ED} = \frac{AO}{EO}$

Triangles CDO and AFO are similar $\therefore \frac{CD}{AF} = \frac{DO}{FO}$

Triangles EFO and CBO are similar $\therefore \frac{EF}{CB} = \frac{FO}{BO}$

Multiply to get $$\frac{AB.CD.EF}{DE\ AF\ CB} = \frac{AO.DO.FO}{EO\ FO\ BO}$$

$$\therefore \frac{AB.CD.EF}{BC\ DE\ FA} = 1$$

■

The Converse of Ceva's Circle Theorem

Given $$\frac{AB.CD.EF}{BC\ DE\ FA} = 1$$

then AD, BE and CF are concurrent

Proof

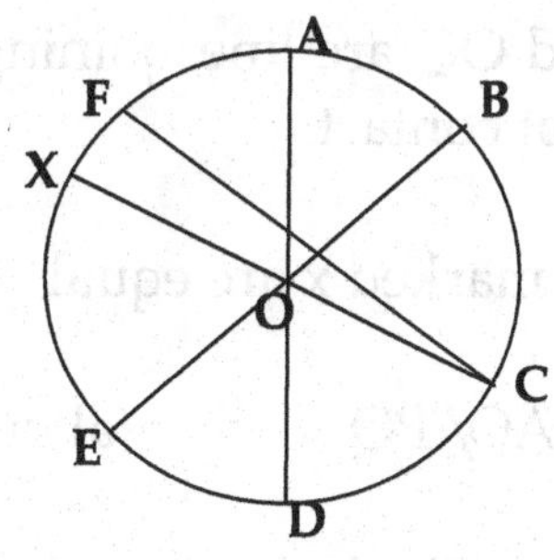

Let AD meet BE at O
Let CO meet the circle at X

Then by Ceva's Circle Theorem

$$\frac{\mathbf{AB.CD.EX}}{\mathbf{BC\ DE\ XA}} = \mathbf{1}$$

$\therefore \frac{\mathbf{EF}}{\mathbf{FA}} = \frac{\mathbf{EX}}{\mathbf{XA}}$ now add 1 to both sides to get $\frac{\mathbf{EA}}{\mathbf{FA}} = \frac{\mathbf{EA}}{\mathbf{XA}}$

Therefore **FA = XA** so that F and X must be the same point. ■

The Three Touching Circles Theorem

When three circles touch each other, then the chords joining the points of contact meet the circles again at the ends of a diameter.

To Prove AB, CD and EF will be diameters

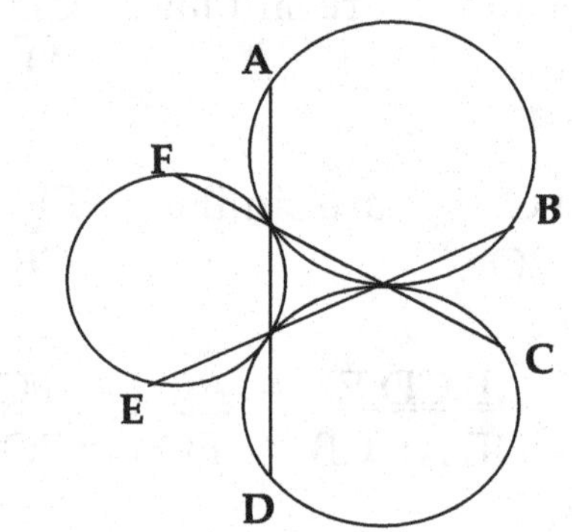

Proof

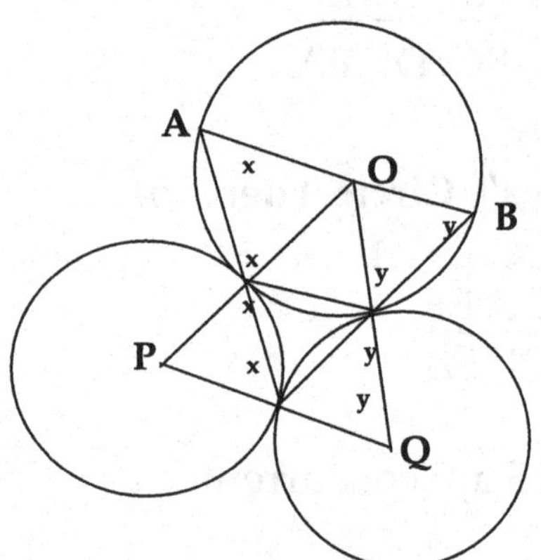

PQ, OP and OQ are lines joining the centres of the circles that pass through the points of contact.

The angles marked x are equal (isos. triangles and opposite angles)

Therefore AO//PQ (alternate angles equal)

The angles marked y are equal (isos. triangles and opposite angles)

Therefore BO//PQ (alternate angles equal)

Therefore AOB is a straight line and hence AB is a diameter. ■

The Seven Circles Theorem
(The seven circle theorem was discovered in 1974 by Evelyn, Money-Coutts and Tyrell "The Seven Circles Theorem and Other New Theorems..Stacey International London. This more elementary proof is after Rabinowitz, Pi Mu Epsilon Journal 8, 1987)

Six circles are drawn around a given circle so as to touch it and also to touch each of the two neighbouring circles.

If the points of contact are A,B,C,D,E,F then AD, BE and CF are concurrent.

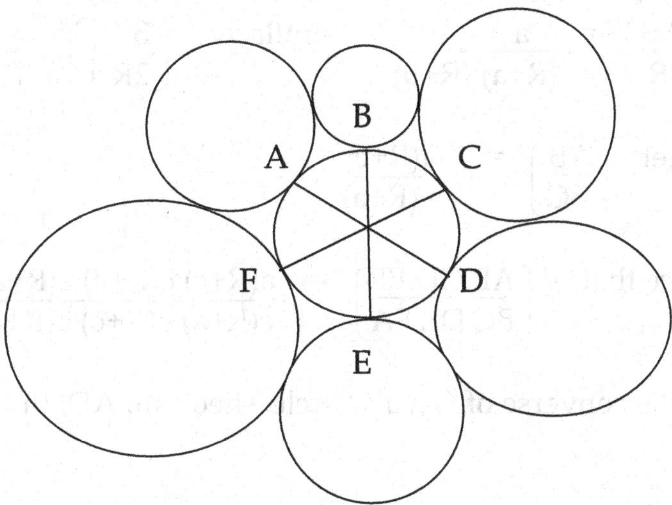

Proof

In the following discussion, it is convenient to refer to the circles by their point of contact so the circle that touches at A will be called "circle A".
Consider circles A and B. Let the centre of the middle circle be O.
PMQ is the line of centres of circles A and B.
By the **Three touching circles Theorem** MA and MB meet the middle circle at the ends of a diameter ROS and ROS//PMQ

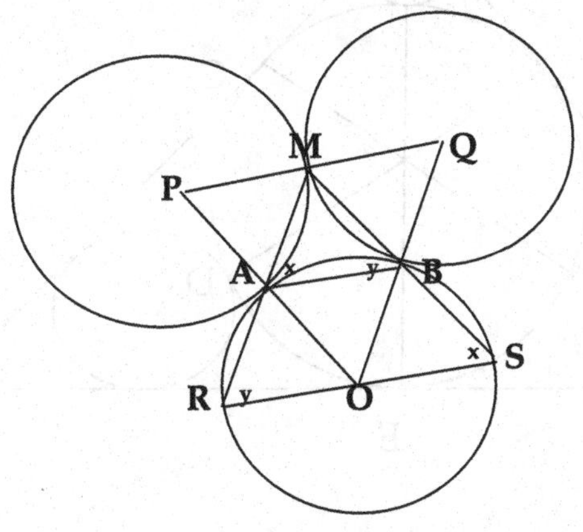

$$x = x \quad (\angle MAB = \angle BSR \text{ ext. } \angle \text{ cyclic quad})$$

$$y = y \quad (\angle MBA = \angle ARS \text{ ext. } \angle \text{ cyclic quad})$$

$$\therefore \Delta s \ \frac{ABM}{SRM} \text{ are similar and } \frac{AB}{SR} = \frac{AM}{SM} = \frac{BM}{RM}$$

$$\therefore \left(\frac{AB}{SR}\right)^2 = \frac{AM.BM}{SM.RM} = \frac{AM.BM}{RM.SM} = \frac{AP.\,BQ}{OP.OQ} \qquad (PQ//RS)$$

Therefore $\left(\frac{AB}{2R}\right)^2 = \frac{a}{(R+a)}\frac{b}{(R+b)}$ similarly $\left(\frac{BC}{2R}\right)^2 = \frac{b}{(R+b)}\frac{c}{(R+c)}$

Dividing, we get $\left(\frac{AB}{BC}\right)^2 = \frac{a(R+c)}{c(R+a)}$

Thus we deduce that $\left(\frac{AB.CD.EF}{BC\ DE\ FA}\right)^2 = \frac{a(R+c).c(R+e).e(R+a)}{c(R+a)\ e(R+c)\ a(R+e)} = 1$

Therefore, by the converse of Ceva's Circle Theorem, AD, BE and CF are concurrent.

■

Circles of Infinite Radius

A circle of infinite radius is, of course, a straight line but we can allow some of the seven circles to become infinite and deduce further results:

(i) Suppose that circles A, C and E have infinite) radii then, instead of being circles, they become straight lines and the figure then reduces this:

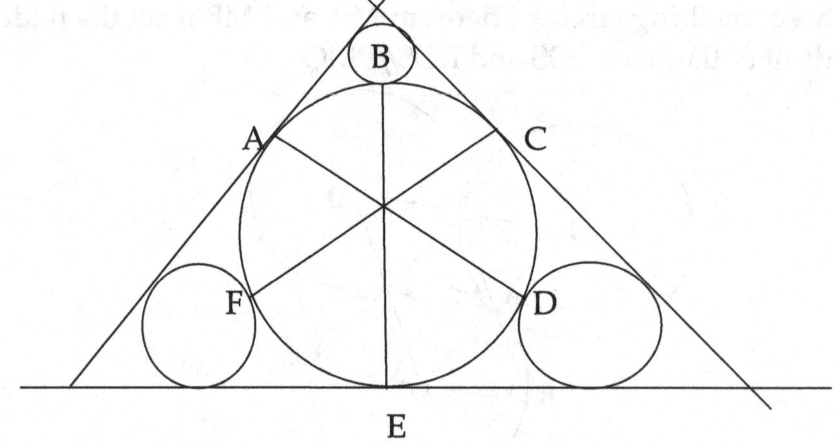

(ii) If circles A and C in the seven circles diagram have infinitely large radii then the diagram becomes:

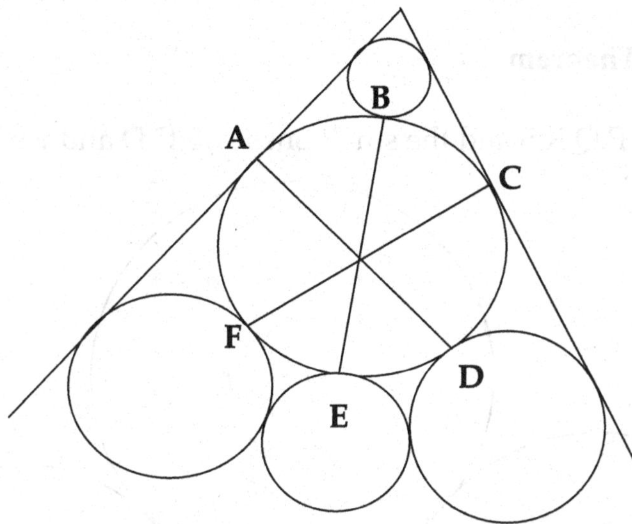

Exercise 8

(i) Deduce a theorem for five touching circles and a line by making the radius of circle A become infinite.

(ii) Draw the diagram for which circle A and circle D have infinite radii.

==============================

Pascals Line and the Touching Circles
There is a nice link between Pascal's theorem for the circle and the six touching circles theorem.

Pascal's Theorem
For the circle

p is the Pascal line
for the points ABC
DEF

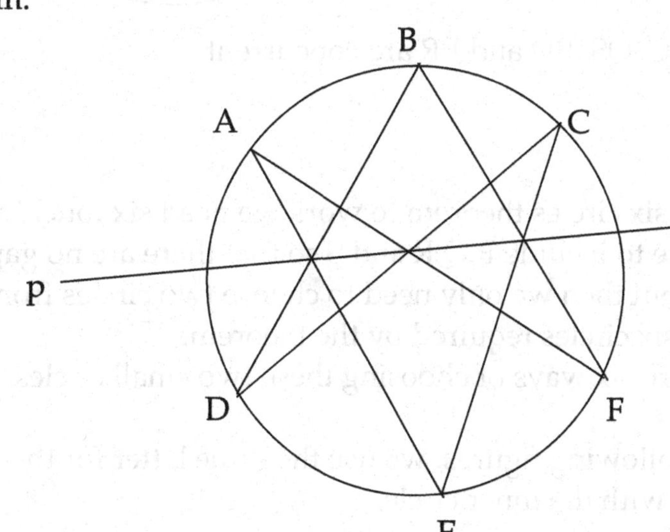

We now change our six circles diagram by expanding four of them until they touch and slotting in four smaller circles in the gaps.

The Nine Circles Theorem

Call the big circles P,Q,R,S and the small ones A,B,C,D and we have the following diagram:

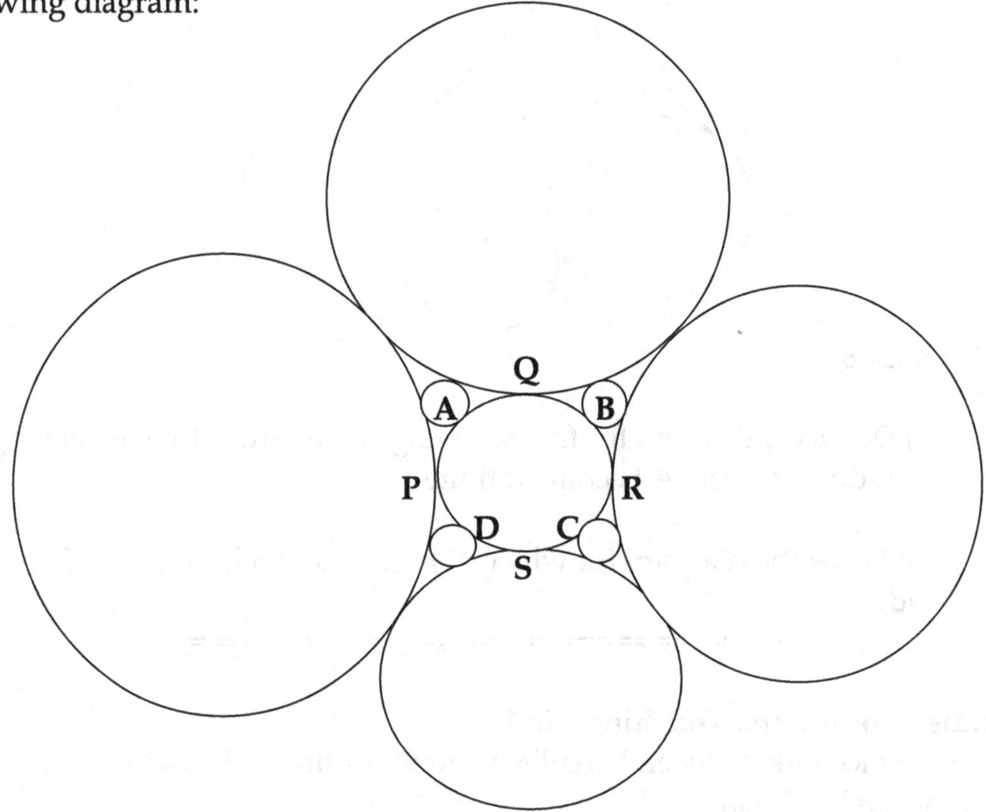

Then AC, QS, BD and PR are concurrent

Proof

For the six circles theorem to work we need six touching circles.
We have to include P,Q,R and S so that there are no gaps in the touching circles but then we only need to choose two circles from A,B,C or D to make up the six circles required by the theorem.
There are six ways of choosing these two small circles.

In the following figures, we use the same letter for the circle and the point of contact with the inner circle.

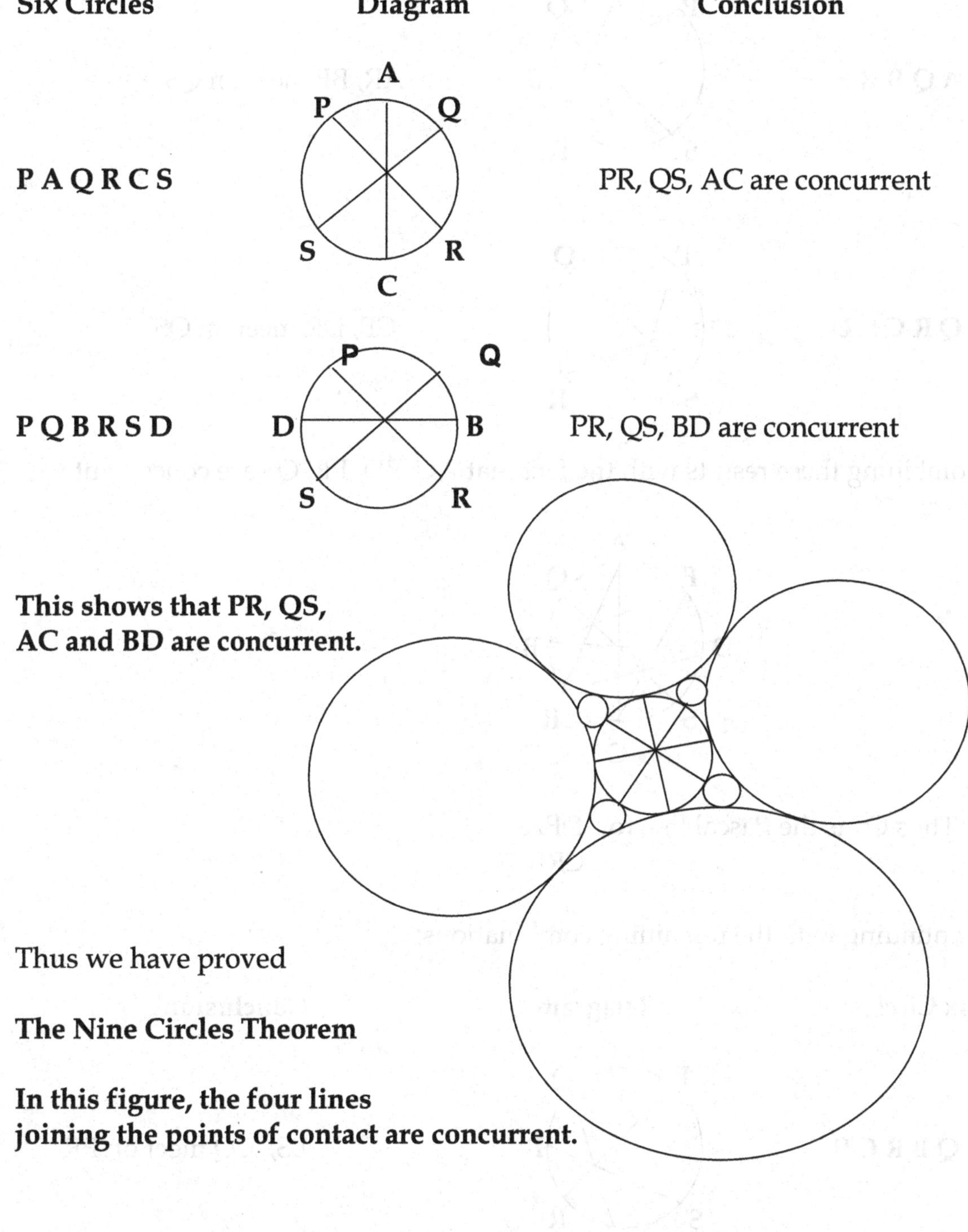

Six Circles	Diagram	Conclusion
P A Q R C S		PR, QS, AC are concurrent
P Q B R S D		PR, QS, BD are concurrent

This shows that PR, QS, AC and BD are concurrent.

Thus we have proved

The Nine Circles Theorem

In this figure, the four lines joining the points of contact are concurrent.

==

Further we also have:

Six Circles	**Diagram**	**Conclusion**
P A Q B R S	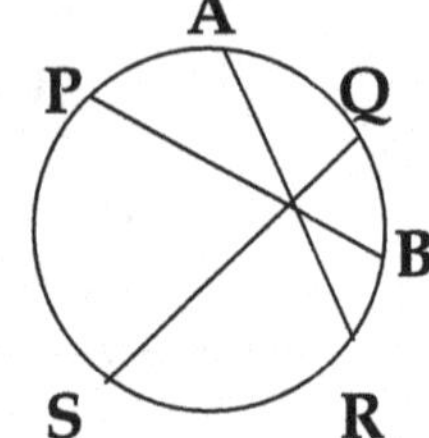	AR, BP meet on QS
P Q R C S D	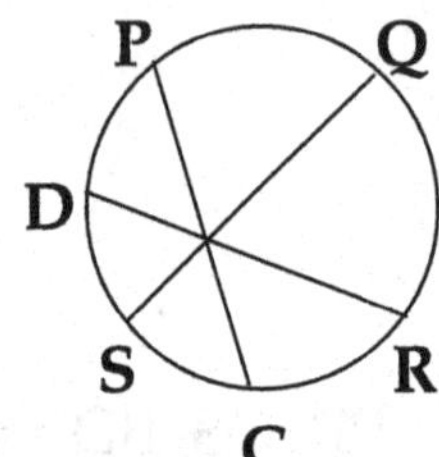	CP, DR meet on QS

Combining these results with the fact that AC, BD, PR, QS are concurrent we have:

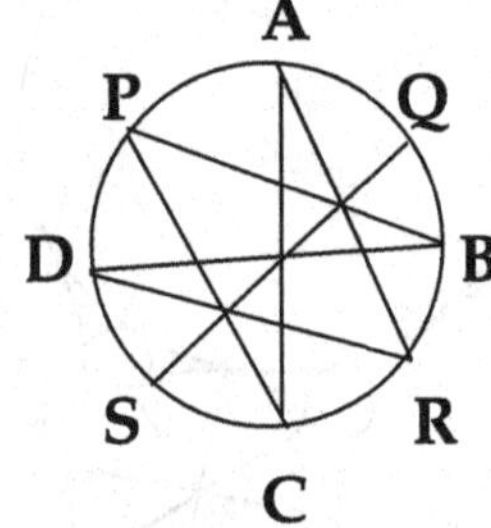

Thus QS is the Pascal line for DPA
CRB

Continuing with the remaining combinations:

Six Circles	**Diagram**	**Conclusion**
P Q B R C S	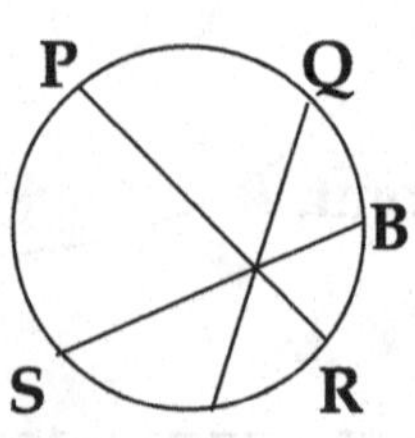	BS, CQ meet on PR

P A Q R S D

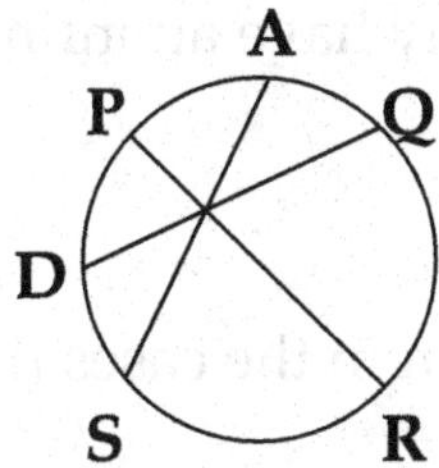

AS, DQ meet on PR

so that AC and DB must also meet on PR

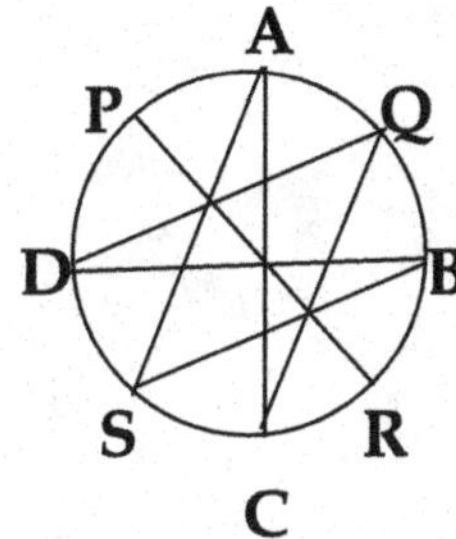

because must be PR is the Pascal line for AQB
DSC

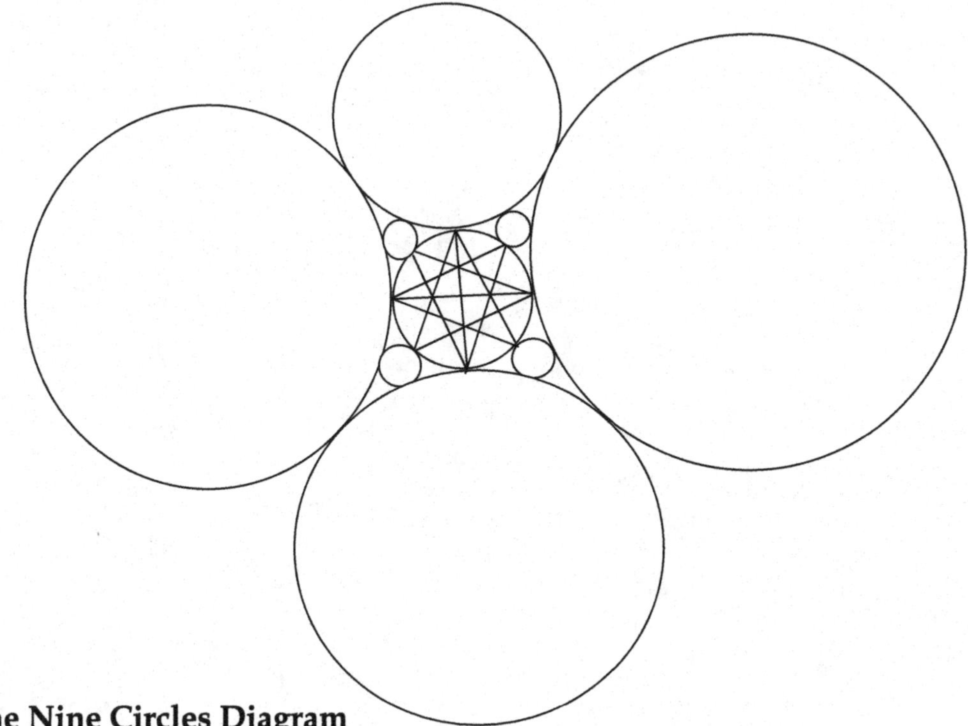

The Nine Circles Diagram

Of course, some of the circles P, Q, R, S may have an infinite radius.

Exercise 9

Draw diagrams for the nine circles theorem in the cases (i) $P=\infty$ (ii) $P=Q=\infty$ (iii) $P=R=\infty$

================================

Answers

Exercise 1 65°

Exercise 2 (i) x=10 (ii) x=30

Exercise 3 x = 50

Exercise 4 10, 20, 30, 40, 50, 60, 70, 80

Exercise 5 [1] x = 3, y = 2, z = 2

[2] x = 3, y = 5, z = 2

Exercise 6

LN is the Pascal line for A . B ∴ LN, AC and BD are concurrent
D . C

PQ is the Pascal line for A . D ∴ PQ, AC and BD are concurrent
D . C

Exercise 7

POR is the Pascal line for ABC
FED

UOW is the Pascal line for BCD
AFE

LON is the Pascal line for FAB
ECD

Therefore the lines meet at the central point O ■

CHAPTER 23

Space and Spacetime

Sound and Light

If you were in Parliament Square in London just before 12 noon looking at the face of Big Ben you would expect soon to hear the chimes striking the hour. If however, just before twelve, you were to jump into a rocket and speed away from Big Ben at the speed of sound or more, then you would not be surprised if you did not hear the chimes.

We have to assume here that

1. the rocket was infinitely quiet so that it did not generate any sound of its own
2. the rocket was infinitely smooth so that it did not set up any pressure waves of its own as it sped through the air

The sound from Big Ben, travelling at 340 ms^{-1} would not reach you although the light from the clock face traveling at 300 000 000 ms^{-1} would enable you to see the clock.

On the other hand, if you were to travel at twice the speed of sound, you could eventually, catch up with the pressure wave set up by the chimes for 11 o'clock and hear the chimes for 11 o'clock played backwards.

Now imagine that your rocket is travelling at the speed of light. If you leave at 12 o'clock would you be travelling along with the light that left the clock face at 12 noon? Does this mean that time would stand still and that your view of Big Ben would be frozen at 12 noon?

If your rocket could travel faster than the speed of light, would you catch up with the light that left the clock face just before 12 noon? Would you see time running backwards as you overtook the light that left Big Ben before you started?

We shall answer these questions at the end of the chapter.

Similar question occupied the thoughts of Albert Einstein in 1905.

Albert Einstein (1879-1955)

Einstein was born in Ulm in the south of Germany near Munich in 1879. In 1896, the family had moved to Switzerland and Albert was sent to the Swiss Federal Polytechnic School in Zurich to train as a mathematics and physics teacher. In 1901, having tried in vain to find a teaching post, he accepted a position in the Swiss Patent Office in Bern where he worked until 1909. It was during this period that that he earned his doctorate from the University of Zurich for a thesis on "the determination of molecular dimensions". He also worked on the foundations of quantum theory and in 1905 published a paper on Special Relativity that led to him finding the most famous of all equations in physics: $E=mc^2$.
In 1909 he resigned from the patent office and became professor of physics at the University of Zurich. In 1911 he was appointed professor of theoretical physics in Prague and in 1912 moved back to Zurich. In 1914 he became director of the Kaiser Wilhelm Physical Institute in Berlin.
In 1915 he published his General Theory of Relativity that equated the effects the force of gravity to the effects of acceleration and explained gravitation as a warp of spacetime produced by the presence of matter.

Einstein married Mileva Maric in 1903 and had a daughter and two sons but they separated in 1919 when he married his cousin Elsa Löwenthal.
In 1921 he received the Nobel Prize for Physics for work on the photoelectric effect whereby light could knock electrons from the surface of a metal as though they had been hit by small particles. Einstein postulated that the light beam was composed of small particles that he called photons.

In 1932 Einstein visited the United States and while there, the Nazi Party came to power and he accepted a post at Princeton University, New Jersey, never to return to Germany. He became a permanent resident in the United States in 1935 until his death in 1955.

Up until 1887 it was believed that light travelled through a fixed background substance called the ether that permeated the whole of the universe, its matter, solid, liquid and gas and was at absolute rest.
It was thought that light rippled through this ether at a constant speed of 340 000 km/second as an electromagnetic radiation.
The ether was supposed to be the canvas against which the speed of any moving object could be seen. In particular, it was the grid against which the speed of light and the speed of the Earth through space could be measured.

In 1887, Albert Michelson and Edward Morley, two scientists working in Cleveland Ohio, set up an experiment to measure the speed of the Earth relative to the ether. They planned to do this by measuring the speed of light on the Earth in two perpendicular directions. Since the Earth is moving round the Sun in its orbit at a speed of about 67000 mph, the speed of light relative to the ether should have been found to be different when measured in these two perpendicular directions.
From the two different answers, they expected to be able to determine the speed of the Earth through the ether.

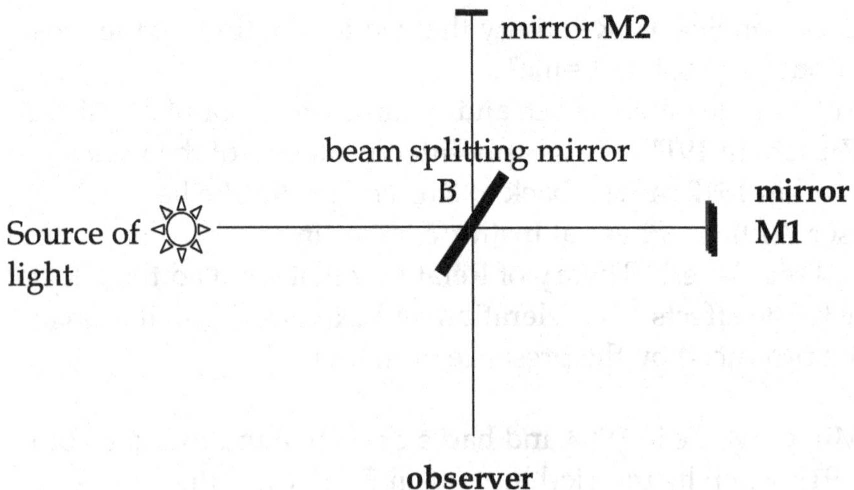

The Michelson Morley Experiment

The experiment used a beam splitting "half silvered" mirror **B** that let some of the light from the source straight through to be reflected from mirror **M1** abd reflected some to the mirror **M2.** The observer saw some light reflected for **M1** and some light reflected from **M2.**
So one path for the light from the source was

Source----------B---------------M1---------------B------------------observer

Some of the light from the source was reflected at **B** at right angles, to the mirror **M2** so that a second path for the light was

Source---------B----------------M2----------------B------------------observer

The distances from B to each of the mirrors was exactly the same.

The observer uses a telescope with cross hairs to view an interference pattern, light and dark fringes, produced by interference between the two beams of light. Now if the Earth is really travelling through the background ether and if light travels at the speed $c=3x10^8$ m/sec relative to the ether, first measured by Armand Fizeau in 1849, then the speed of light should be different when measured (i) in the direction of the Earth's motion and (ii) perpendicular to the Earth's direction. The Michelson interferometer reveals slight differences in the lengths of the path travelled by the two light beams. The difference is revealed to the observer as light and dark fringe patterns that move as the distance is changed.

The whole apparatus could be rotated through 90° in which case the fringe shifts in the interference pattern should move in the opposite direction. The interference pattern was expected to shift by 0.37 wavelengths due to the motion of the Earth through the ether but the observed value for the shift was always found to be virtually zero. The speed of light seemed to be exactly the same in each of the two perpendicular directions.

Michelson and Morley repeated the experiment at different times of the year when, supposedly, the Earth would be moving in a different direction relative to the background ether. However, they always came up with the same result for the speed of light. Despite the motion of the Earth round the sun, the earth always appeared to be at rest relative to the ether.

Two explanations were suggested to account for the "null result".

1. Light always travels at the same speed with respect to its source

2. Objects moving through the ether with speed v experience a length contraction by a factor $v(1 - v^2/c^2)$.

This was suggested both by George Fitzgerald, lecturer at Trinity College Dublin and by Hendrik Lorentz, professor of mathematics at Arnheim University, Holland and is referred to as the Fitzgerald-Lorentz contraction.

We will see shortly, that Einstein was able to explain the null results of the Michelson-Morley experiment using his Special Theory of Relativity but that not only would lengths be contracted but also that time would have to be stretched. An idea that was difficult to swallow in the early part of the 20th Century.

In order to understand Einstein's theory of special relativity, we need to think about pairs of coordinate axes that may be "at rest" or may be moving relative to each other.

Inertial Frames

An inertial frame of reference, or Galilean frame, is a set of coordinate axes for which Newton's first law of motion holds. That is

> Any mass continues in a state of rest or uniform motion in a straight line, unless acted on by an external force.

Consider a 2-D frame of reference with x-y axes placed flat on the surface of the Earth. Ignoring the fact that the Earth is following a curved orbit round the sun and the fact that the Earth has a curved surface, the 2-D x-y axes will define a good approximation to an inertial frame.

A snooker table is a good example of an inertial frame of reference.

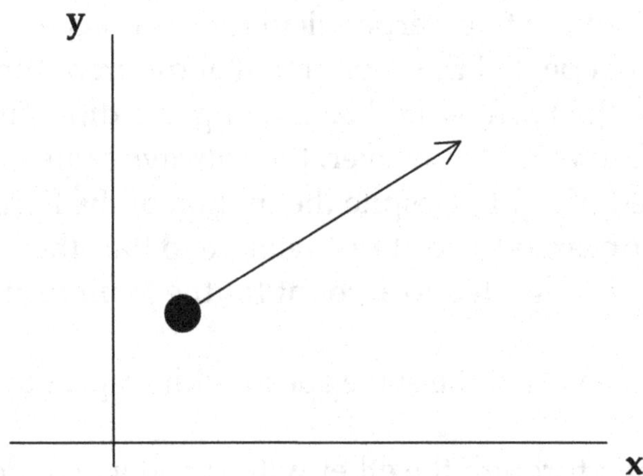

The snooker ball follows a straight line path.

Suppose that you are in the dining saloon of a train travelling at 60 m.p.h. along a straight track but there is absolutely no vibration and no noise from the tracks or the engine. The carriage does not rock from side to side. There is a snooker table in the dining saloon. You would be able to play snooker just as if the train was standing at the station platform.

Suppose that another train is approaching from the opposite direction also at 60 m.p.h. and just as noiseless and vibration free as the first train. Suppose that a certain Dr Fred is in the dining saloon of the other train, also playing snooker. Then both you and Dr Fred would be able to play a normal game of snooker in spite of the fact that relative speed of approach is 120 m.p.h.

If the two trains were running absolutely smooth and noise free, then you would have difficulty deciding, from the run of the balls on the snooker table, if your train was moving or not.

If however, your train went round a bend, then the frame of reference would cease to be an inertial frame.

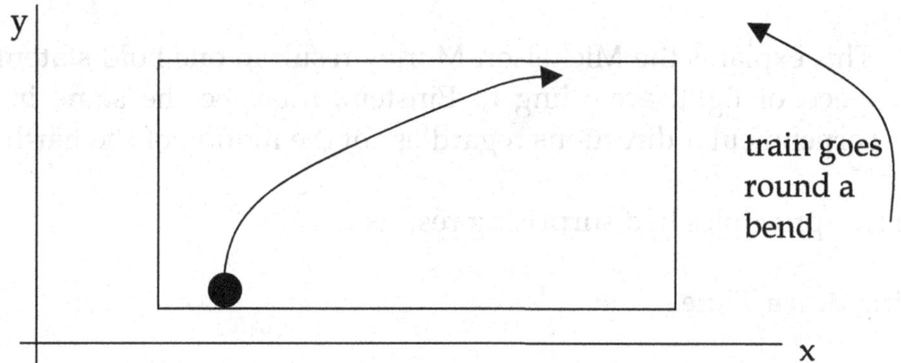

The ball deviates from the straight path when the train goes round a bend. When the train goes round a bend x-y frame of reference is no longer an inertial frame.

Notation

S(x,y,z) represents a set of 3-D Cartesian axes. **S′(x′,y′,z′)** also represents a set of 3-D Cartesian axes but not necessarily with axes parallel to the axes of the frame **S**.
If **S** is an inertial frame of reference and **S′** is moving at a constant velocity, without rotation, relative to **S**, then **S′** will also be an inertial frame.

Einstein and The Theory of Special Relativity.

In 1905, Einstein published his theory of special relativity which is based on just two simple postulates:

1. The Principle of Relativity. The laws of Physics are the same in all inertial frames.

This means, in effect, that you cannot detect if you are moving at a constant speed.

If two space ships float past each other the occupants of each ship would not be able to determine if one was at rest and the other moving or if they were both moving.

2. The speed of light through empty space is the same in all inertial frames and is independent of the velocity of the source.

This explains the Michelson-Morley result in one bold statement. The speed of light, according to Einstein, must be the same in the two perpendicular directions regardless of the motion of the Earth.

These two principles had surprising results.

Slowing down Time

While enjoying his train journey, our scientist Dr Fred sets up an experiment to measure the speed of light **c**, while travelling along the straight railway track at speed **v**.

To measure the speed of light, he flashes a torchlight at 12 noon, from the floor of his carriage to a mirror fixed to the roof of the carriage and measures the time $\mathbf{t_F}$ that it takes to go up to the roof of the carriage and back. The distance from the lamp to the roof is **L.**

Dr Fred measures the speed of light

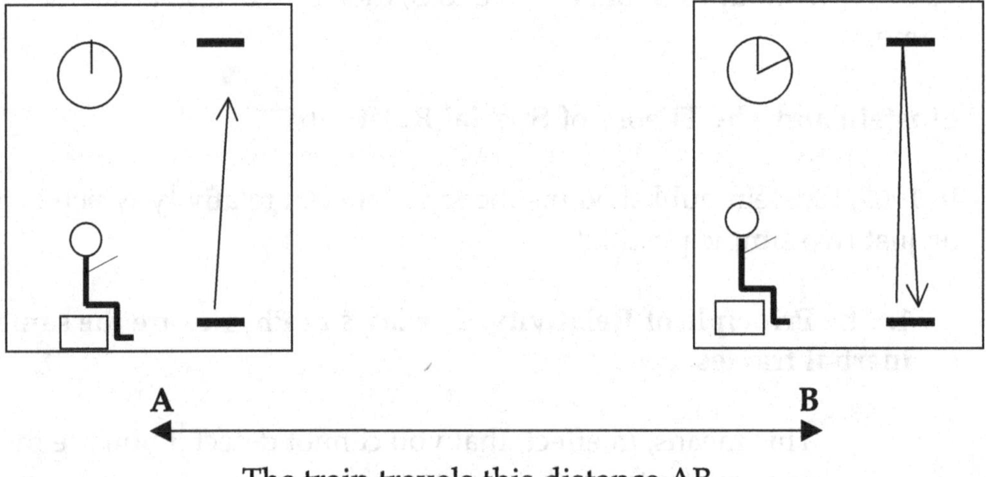

The train travels this distance AB

According to Dr Fred, the light travels a distance **2L** from the floor to the roof of the carriage and back, in time t_F and so Dr Fred calculates the speed of light, using **distance=speed x time,** and derives the equation

$$2L = c \times t_F$$

But as we, on the station platform, observe Dr Fred in his train carriage, we see a different picture:

What we see from the platform

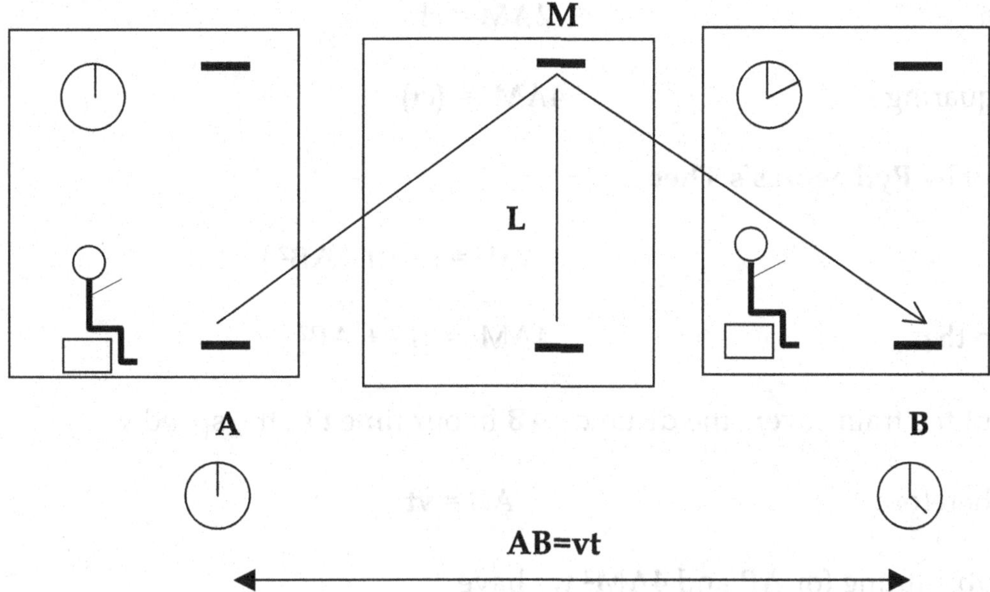

We see the flash of light leave the lamp at **A** and hit the mirror half way along at **M,** bounce back to the lamp again at **B.**

Remember, Einstein's second principle of relativity states that the speed of light must be the same for us on the station platform as for Dr Fred in the train. The light from the lamp has travelled a distance **AM+MB** which is greater than the distance **2L** that Dr Fred uses in his equation.

This means that the time we measure for the flash of light has to be greater the Dr Fred's time t_F . In order to measure the time we see for the flash to leave the lamp at **A** , hit the mirror at **M** and bounce back to **B,** we place two clocks at the points **A** and **B.** The clocks are synchronized with each other and read 12 noon when Dr Fred passes the point **A.** As soon as Dr Fred reaches **B** the second clock stops and our time **t** for the path of the light flash can be recorded.

When we inspect the clock at the point **B** we find that the time for the flash of light to travel the distance **AM+MB** is different from Dr Fred's time and we get the time for the flash to travel from **A** to **B** to be **t.**
The speed of travel for the light flash must be **c,** according to Einsteins 2nd principle.

so using **distance = speed x time** we have

$$AM + MB = ct$$

Or $$2AM = ct$$

Squaring $$4AM^2 = (ct)^2$$

But by **Pythagoras's Theorem**

$$AM^2 = L^2 + (\tfrac{1}{2}AB^2)$$

So that $$4AM^2 = 4L^2 + AB^2$$

But the train covers the distance **AB** in our time **t** at the speed **v**

Therefore $$AB = vt$$

Substituting for **AB** and **4AM²** we have

$$(ct)^2 = 4L^2 + (vt)^2$$

But from Dr Fred's equation we have

$$2L = c \times t_F$$

so, substituting $(c \times t_F)^2$ for $4L^2$ we have

$$(ct)^2 = (c \times t_F)^2 + (vt)^2$$

Therefore $$t^2(c^2 - v^2) = c^2 t_F{}^2$$

This gives us the formula for what is called the **dilation of time.** Dr Fred's time for the path of the flash is less than the time we get as we see the flash travel from **A** to **M** and to **B**. As far as Dr Fred is concerned, the flash of light leaves and comes back to the **same point** in his frame of reference. Dr Fred's time, t_F is called the **proper time** for the time between the events **"flash leaves lamp"** and the event **"flash arrives back at lamp".**

The proper time is the time between two events that occur at the same place in an inertial frame as measured by an observer in that frame.

The proper time is the least time that can occur between two events.

Rearranging the formula we have

$$t_F = t\sqrt{(1 - v^2/c^2)}$$

This formula shows that $1 - v^2/c^2$ must be greater than zero, otherwise we would be taking the square root of a negative quantity.

Hence $v^2/c^2 < 1$, so that $v < c$

Showing that **we cannot travel faster than light.** The speed **v** must be less than the speed of light.

(This answers one of the questions posed at the Big Ben experiment: you cannot catch up with the light that left earlier: you cannot travel faster than light)

The Dilation of Time

If an observer measures the time interval between two events that occur **at the same point** in his rest frame as time t_0, (this is the t_f in Dr Fred's experiment) then an observer whose frame is moving with a constant speed **v** relative to the first observer will measure the time interval as **t** where

$$t_0 = t\sqrt{(1 - v^2/c^2)}$$

t_0 is called the **proper time** between the two events.

Note that in our example of Dr Fred on the train, Dr Fred's time is the proper time. The two events "send Flash" and "receive flash" occur at the same point in his rest frame. The two events "send flash" and "receive flash" occur at different points in the frame of the station platform, and as far as Dr Fred is concerned, the station platform is moving back, while he is at rest in the frame of the train.
If we were to carry out the experiment on the station platform and we to be observed by Dr Fred as he sped by in the train, then our time would be the least time and Dr Fred's time would be longer. Dr Fred would see the two events "send" and "receive" occurring at different points.

Worked examples on time dilation

(In these exercises we take the speed of light as $\mathbf{c=3x10^8\ ms^{-1}}$)

Example 1. Natasha on Earth, sees Dr Fred approaching in his spaceship and flashes a signal to him for exactly 0.5 seconds. Dr Fred passes overhead in his spaceship at $\mathbf{1.8x10^8\ ms^{-1}}$.
For how long is the light on according to Dr Fred?

Solution

The two events to be considered here are (1) light flashes on and (2) light turns off. These two events occur at the same place in the Earth frame of reference and therefore, the time 0.5 seconds is the **proper time** $\mathbf{t_0}$ for the time of the flash of light.

Using $\mathbf{t_0 = t(1-v^2/c^2)^{1/2}}$

$\mathbf{t_0 = 0.5}$ and is the **proper time** for the flash in the rest frame of the Earth

t is the time as seen by Dr Fred and will be longer than the proper time.
$\mathbf{v = (1.8/3)\ c\ = 0.6\ c}$
hence

$$t = t_0/(1-v^2/c^2)^{1/2}$$

$$= 0.5/(1 - 0.6^2)^{1/2}$$

$$= 0.5/(0.64)^{1/2}$$

$$= 0.5/0.8 = \mathbf{6.25\ seconds}$$

=============

Example 2 A spaceship passes the Earth at **0.8c ms^{-1}** and the pilot flashes a light for **0.6** seconds. Dr Fred watches from the ground. What is the duration of the flash of light according to Dr Fred?

Solution

Two events to be considered are (1) light flashes on and (2) light turns off. These two events occur at the same place in the pilots frame of reference and therefore, the proper time for the flash, is that recorded by the pilot of the spaceship and is

$$t_0 = 0.6 \text{ seconds}$$

Dr Fred sees these two events occur at different points in his frame of reference and therefore has a longer time for the flash.

$$v = 0.8\,c$$

Using $t_0 = t(1-v^2/c^2)^{1/2}$

We have $$t = t_0/(1-v^2/c^2)^{1/2}$$

$$= 0.6/(1 - 0.8^2)^{1/2}$$

$$= 0.6/(1 - 0.64)^{1/2}$$

$$= 0.6/0.6 = 1 \text{ second}$$

Dr Fred sees the flash last for 1 second.

======================

Example 3 Dr Fred takes a round the world trip in an aircraft and flies a distance of 40000 km at a speed of 400 ms^{-1}.
How long does the flight take (i) according to an observer on the ground, (ii) according to Dr Fred?

Solution

For the observer on the ground, distance = 40000km, speed = 400 ms^{-1}

Therefore the time is 40 000 000/ 400 = 100 000 seconds (27.8 hours)

For the rest frame of Dr Fred on the plane, we have

$$t_0 = t(1-v^2/c^2)^{1/2}$$

t = 100 000 sec, v = 400, c = 300 000 000

Therefore $t_0 = 100\,000(1 - [133\text{x}10^{-8}]^2)^{1/2}$

To calculate this figure we need to use the approximation $\mathbf{(1-n)^{1/2} = 1 - \frac{1}{2} n}$

This approximation comes from the first two terms in Newton's binomial expansion. We thus have:

$$t_0 = 100\,000(1 - \tfrac{1}{2}[133\text{x}10^{-8}]^2)$$

$$= 100\,000(1 - 88445\text{x}10^{-16})$$

$$= 99999.99999912 \text{ seconds}$$

So, at the end of his trip, Dr Fred is about **0.9** millionths of a second younger than he would have been if he had not taken the flight.

========================

Example 4 Proxima Centauri, our nearest star, is nearly 4 light years from the sun.
How fast would an astronaut have to travel to reach Proxima Centauri in 3 years?

Solution

For galactic distances it is convenient to use the light year for the unit of distance. That is, the distance that light travels in one year. The speed of light is then one light year per year so that **c=1.**

Suppose that the speed of the spaceship is **k** light years per year.

Then the Earth time for the journey to Proxima Centauri will be **4/k** years.

The ship time for the journey is **3** years. This is the proper time, being the time in the rest frame of the astronaut.

$$t_0 = 3, \qquad t = 4/k, \qquad v = k$$

Using $\mathbf{t_0 = t(1-v^2/c^2)^{1/2}}$

We have $$3 = 4/k\,(1 - k^2)^{1/2}$$

$$9k^2 = 16(1 - k^2)$$

$$25k^2 = 16$$

$$k = 4/5 = 0.8$$

The speed of the spaceship is therefore **0.8** of the speed of light.

============================

Example 5. Dr Fred boards a spacecraft for Betelgeuse, a star that is approximately 650 light years from Earth. The spacecraft travels at 99% of the speed of light. How long does it take him to reach the star?
When he arrives at Betelgeuse, he immediately radios back to Earth. How much older would his friends be when they receive the news of his safe arrival?

Solution

The Earth time for the journey is **650/0.99 = 656.6 years**

Dr Fred's time for the journey gives time t_0, the **proper time** for the journey.

Using $t_0 = t(1-v^2/c^2)^{1/2}$

We have t = 656.6 years, v = 0.99c light years per year

Hence $t_0 = 656.6\,(1- 0.99^2)^{1/2}$

$= 92.6$ years

Dr Fred will therefore be **92.6 years** older on arrival at Betelgeuse.

If he immediately radios back to Earth then his friends back home will have aged

656.6 years (the Earth time for the journey)
+650 years (the time for the radio signal to reach Earth)

Thus, those on Earth would be **1306.6 years** older when they hear of Dr Fred's arrival at Betelgeuse

==================================

Example 6 A 20 year old spacepilot sets out on a journey to a star that is 30 light years away traveling at a speed of 0.95c. (95% of the speed of light) leaving his twin brother back on Earth.
When he reaches the star, he immediately returns to Earth at the same speed.

How old will he be when he returns to Earth and how old will his twin brother be?

Solution

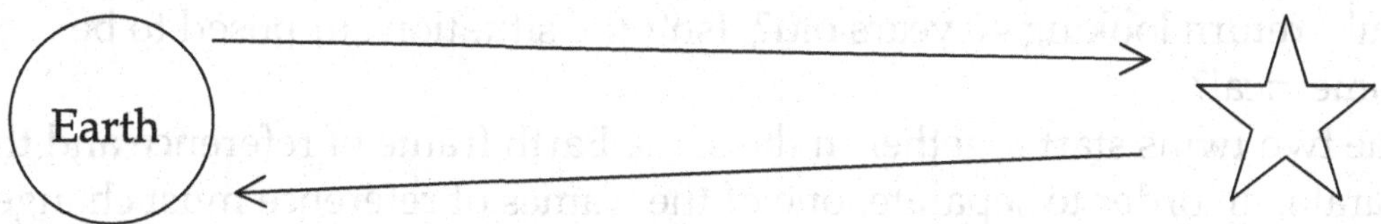

There are two events to be considered: (1) the spacecraft leaves the Earth and (2) the spacecraft arrives at the star.
(We will have to ignore the mechanics of how the spaceship can immediately reverse its direction while travelling at 95% of the speed of light!)

The Earthbound twin sees these events at two different points in his frame of reference. The distance 30 light years at a speed of 0.95c gives a time for the journey of 30/0.95 = 31.6 years.
For the spaceman, the two events "leave Earth" and "arrive at the star" occur at the same point in his frame of reference and so he will measure the **proper time** for the journey.

Using $\mathbf{t_0 = t(1-v^2/c^2)^{1/2}}$ gives

$$t_0 = 31.6\text{x}(1 - (0.95)^2)^{1/2} = 9.9 \text{ years}$$

If we multiply by 2, the Earthbound twin will be **31.6 x 2 = 63.2** years older.
The spaceman will be **9.9 x 2 = 19.8 years** older.

The Earthbound twin will therefore be **83.2 years old**
The spaceman will be **39.8 years old.**

=================================

The Twins Paradox

Suppose that Fred is the space pilot who sets out for the distant star at 0.95c, while his twin sister Freda remains on Earth. We have calculated that Fred will be nearly 40 when he returns but that he will find that Freda is now 83 years old.

However, from Fred's point of view, could we not say that he sees the Earth and Freda zoom away at 0.95 of the velocity of light and then return. If we did the calculation from Fred's point of view, he should be 83 and Freda should return looking 40 years old? Isn't the situation supposed to be symmetrical?

The two twins start together in the same Earth frame of reference and then separate. In order to separate, one of the frames of reference must change its motion, accelerate and feel the force of acceleration. Further, if two frames of reference are in uniform relative motion, we cannot bring the two frames of reference together, to compare ages and clocks, unless one of the frames is accelerated or decelerated.

In practice, it is the twin that is accelerated whose clocks run slow and ends up younger.

Atomic clocks have been flown around the world and found to have lost time compared with identical clocks that have remained "at rest".
(See "Around the World Atomic Clocks": Science vol 77 July 1972)

Exercise 1

1. A spaceship flies past Earth at $\mathbf{1.8 \times 10^8\ ms^{-1}}$ and sends a short pulse of light. Dr Fred observes from the Earth and measures the duration of the pulse as 0.1 seconds. What it the duration of the pulse of light as measured on the spaceship?

2. The ninth planet Pluto is about 20 000 light seconds from Earth. How long would it take an astronaut, travelling at one tenth of the speed of light, to reach Pluto

(i) as measured on Earth

(ii) as measured by the astronaut

3. Spaceman Fred sets off from Earth on his 20th birthday on a journey to Alpha Centauri, a star that is 4.3 light years away. He travels in his spaceship at 0.9c where c is the velocity of light. How old will he be when he gets there?

4. Dr Fred needs to visit Rigel in the constellation of Orion but Rigel is 770 light years from Earth so Dr Fred will be put in a state of suspended animation so that he can be woken up when he gets there. His spaceship will leave Earth at a speed of 0.8c where c is the speed of light.

   At what time, to the nearest half year, should the timer on his capsule be set so that he can be woken up as soon as he arrives?

================================

As a result of Einstein's two principles of relativity, we have thus seen that the faster you go the slower your time goes, as seen by someone who is not moving with you. Since $t > t_0$ the time between two events in the moving frame appears to be getting longer to an observer at rest, the faster it moves. This result that was hard to swallow in 1905. The second consequence that we examine here, length contraction, had been anticipated by two scientists, independently of each other, George Fitzgerald and Hendrik Lorentz. George Fitzgerald (1851-1901) was an Irish physicist working at Trinity College, Dublin. Hendrik Lorentz (1853-1928) was a Dutch physicist at Leiden, who is now best known for developing the Lorentz Transformation equations that describe the changes in time and length that occur between two inertial frames. Both Fitzgerald and Lorentz postulated independently, what is now known as the Fitzgerald-Lorentz contraction, in order to explain the null result obtained by the Michelson-Morley experiment. We will see in this section that Einstein's two postulates for special relativity lead to the same result: "the faster you go, the shorter you seem get". We shall give two derivations of the Fitzgerald-Lorentz contraction here and later, derive the Lorentz transformation equations. Having derived the Lorentz transformation equations we will then be able to prove time dilation and length contraction for more general cases.

Derivations of the Fitzgerald-Lorentz Contraction from Einstein's Principles of Special Relativity.

First Derivation:

AB is a rod of length L_0 that is at rest in an inertial frame **S′**.
A flash of light leaves a source at **A** and is bounced back from a mirror placed at **B**.

The frame **S′** is moving at a speed **v** relative to the frame **S** of an observer. The observer in frame **S** sees the flash of light leave **A** and travel to a point **B′** (where the mirror has moved to) and bounce back to a point **A″** (where **A** has got to when the flash of light gets to the start of the rod).

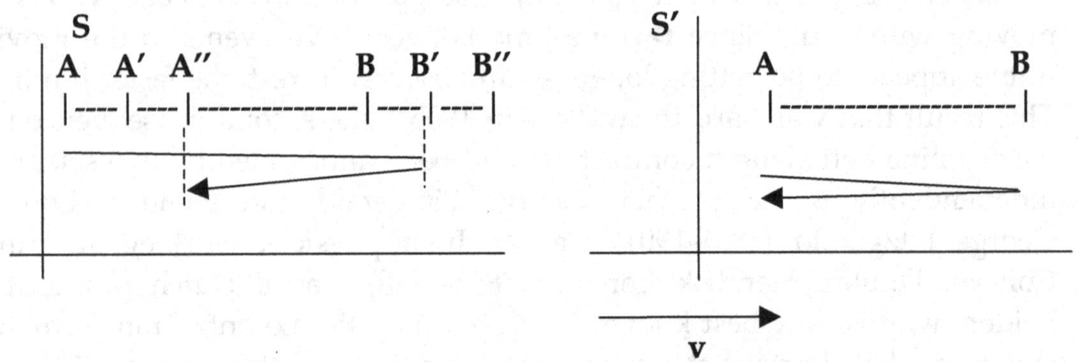

S′ is the rest frame for the rod, length L_0. Let the flash of light cover the distance $2L_0$ in time t_0.

Then using distance = speed x time we have $2L_0 = c.t_0$ (equation 1)

To the observer in **S**, suppose that the length of the rod is **AB** = **L** and the time for the flash to reach the mirror at the end of the rod is t_1. In the time t_1, the rod has moved from **AB** to **A′B′**
Hence **AA′** = **BB′** = $v.t_1$

and since the light flash travels from **A** to **B′** in time t_1 we have **AB′** = $c.\ t_1$

Now **AB′** = **AB** + **BB′** therefore $L + v.\ t_1 = c.t_1$ (equation 2)

Now suppose that the light flash travels back from the mirror at **B′** to **A″** (where **A** has got to) in time t_2, then **A″B′ = c.** t_2 (equation 3)

The rod moves from **A′B′** to **A″B″** in time t_2 therefore **A′A″ = B′B″ = v.** t_2

Also, **A″B′ = A″B″ – B′B″** therefore $c.\ t_2 = L - v.\ t_2$ (equation 4)

From equation 2 we have $t_1(c - v) = L$

From equation 4 we have $t_2(c + v) = L$

Hence $t_1 + t_2 = \frac{L}{(c - v)} + \frac{L}{(c + v)} = \frac{2cL}{(c^2 - v^2)}$ (equation 5)

Now t_0 is the time between two events ("send flash" and "get flash back") that occur at the same point in **S′** and $t = t_1 + t_2$ is the time between these two events in S and therefore we can use the time dilation formula to get

$$t_0 = t(1 - v^2/c^2)^{1/2}$$

thus using equations 1 and 5 we have

$$\frac{2L_0}{c} = \frac{2cL}{(c^2 - v^2)}(1 - v^2/c^2)^{1/2}$$

$$\frac{2L_0}{c} = \frac{2L}{(c^2 - v^2)}(c^2 - v^2)^{1/2}$$

$$L = \frac{L_0(c^2 - v^2)^{1/2}}{c}$$

$$L = L_0(1 - v^2/c^2)^{1/2}$$

This is the Fitzgerald-Lorentz contraction formula. It is the same formula that Fitzgerald and Lorentz postulated explain the Michelson-Morley experiment.

Second Derivation:

For comparison, we give here a shorter derivation of the same formula. Dr Fred's rocket passes the Earth at a speed **v** on the way to a star a distance x_0 away.

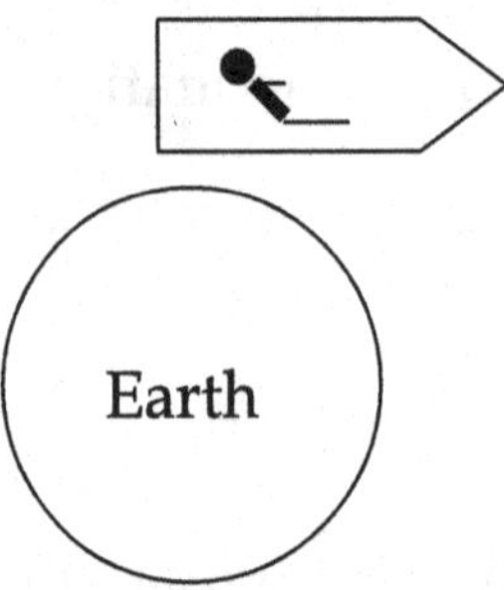

Distance $x_0 = vt$ in the Earth Star frame

The time taken according to an observer in the Earth – Star frame is $t = x_0/v$

In the ships frame, the events, "leave Earth" and "arrive at star" take place at the same point, i.e. in the ship, so the time between the events "leave" and "arrive" in the ship is the proper time t_0 hence

$$t_0 = t(1 - v^2/c^2)^{1/2}$$

therefore

$$t_0 = \frac{x_0}{v}(1 - v^2/c^2)^{1/2}$$

$$v\,t_0 = x_0(1 - v^2/c^2)^{1/2}$$

Now $v\,t_0$ is the distance between the Earth and the Star as calculated in moving frame of the ship (distance = speed x time). If we call this distance **x** then the distance calculated in the frame of the ship is

$$x = x_0(1 - v^2/c^2)^{1/2}$$

This corresponds to the Fitzgerald-Lorentz formula $L = L_0(1 - v^2/c^2)^{1/2}$

Worked Examples on the Fitzgerald-Lorentz Contraction

Example 1

An observer on Earth watches a spaceship, travelling at speed **v,** pass over two points **A** and **B.**
How do the observer on Earth and the pilot of the spaceship measure the time taken to cover the distance **AB?**
Compare values they each get for the distance **AB.**

Solution

There are two events to be considered:
1. the ship passes over point **A**
and 2. the ship passes over the point **B.**

In the pilots rest frame, these two events occur at the same point and so, the time recorded by the pilot is the **proper time** for the time taken to cover the distance **AB.** Therefore, the pilots time is denoted by t_0.
The pilot of the spaceship observes the points **A** and **B** pass below, moving backwards at speed **v.**
Using the formula **distance = speed x time,** the pilot would calculate the distance between the two points from

$$AB_{pilot} = vt_0$$

The earthbound observer sees the two events "ship passes over **A**" and "ship passes over **B**" occurring at different points in his rest frame and therefore measure a longer time **t** for the interval between the two events.
The earthbound observer would calculate the distance **AB** from the equation

$$AB_{earth} = vt$$

Using the time dilation formula we have

$$t_0 = t\sqrt{(1 - v^2/c^2)}$$

thus we have

$$AB_{pilot} = vt_0 = vt\sqrt{(1 - v^2/c^2)} = AB_{earth}\sqrt{(1 - v^2/c^2)}$$

The pilot therefore calculates a shorter distance for **AB,** than the observer on Earth. The pilot's time is also shorter than the time measured by the earthbound observer.

===========================

An alternative solution using the Fitzgerald-Lorentz contraction:

The points **A** and **B** are at rest relative to the earthbound observer and so $\mathbf{AB_{earth}}$ can be taken as the rest length $\mathbf{L_0}$ in the length contraction formula.

The pilot sees the moving frame containing $\mathbf{AB_{pilot} = L}$ moving back at a speed **v** and so the pilot sees the contracted length given by

$$\mathbf{AB_{pilot} = L = L_0(1 - v^2/c^2)^{1/2}}$$

thus

$$\mathbf{AB_{pilot} = AB_{earth} \times (1 - v^2/c^2)^{1/2}}$$

==============================

Example 2 . How fast should a metre rule go in order to shrink to 99cm?

Solution

Let the speed be k times the velocity of light.

Using $L = L_0(1 - v^2/c^2)^{1/2}$

$L / L_0 = 0.99$ hence $0.99 = (1 - k^2)^{1/2}$

Square both sides:

$$0.99 \times 0.99 = (1 - k^2)$$

$$k^2 = 1 - 0.9801$$

$$k^2 = 0.0199$$

$$k = 0.14 \text{ (2 dp)}$$

The metre rule needs to travel at **0.14c** where **c** is the velocity of light.

===========================

Worked Examples on Special Relativity

Formulae:

The formula for Time Dilation is:

$$t_0 = t(1 - v^2/c^2)^{1/2}$$

Where t_0 is the time between two events that occur at the **same point** in an observers rest frame.

t is the time between those two events measured by an observer (who sees them at two different points) who has a constant velocity **v** with respect to the first observer. ($t_0 < t$)

t_0 is called the proper time between the two events. (t_0 is also the least possible time between the two events)

The formula for Length Contraction is:

$$L = L_0(1 - v^2/c^2)^{1/2}$$

Where L_0 is the distance between two points **A** and **B** as measured in their rest frame **F′**, moving at a speed **v** in the direction of L_0, relative to the frame **F**.

L is the distance **AB** measured in **F** $\qquad$ L_0 is **AB** measured in **F′**

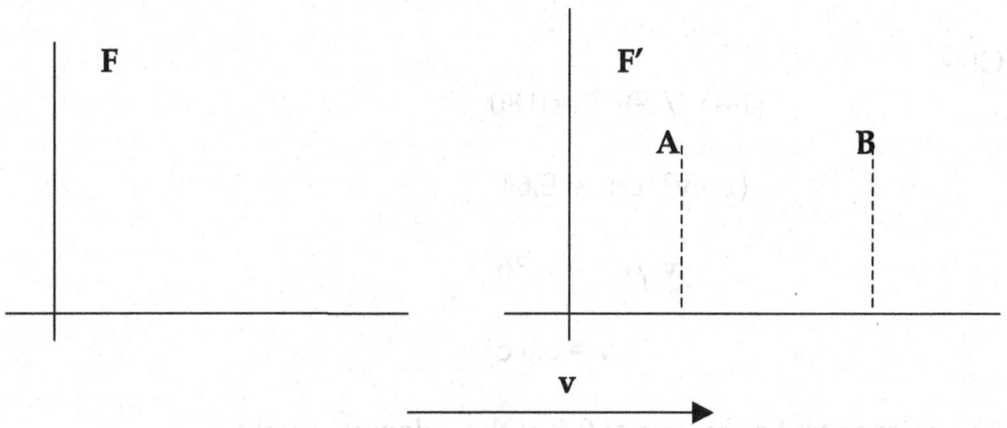

F

F′

A B

v

Example 1

A spaceship of rest length 60 m passes the Earth at a speed of 0.8c, where c is the velocity of light.
What is the length of the spaceship as seen by someone on Earth?

Solution

In the rest frame of the spaceship, we have $\mathbf{L_0 = 60}$

The speed of the spaceship relative to the earth is $\mathbf{v = 0.8c}$

Using $$\mathbf{L = L_0(1 - v^2/c^2)^{1/2}}$$

we have $$\mathbf{L = 60x(1 - 0.64)^{1/2}}$$

$$\mathbf{= 36\ m}$$

==============================

Example 2

An observer measures the length of a metre rule as **80 cm**. At what speed is the ruler moving relative to the observer?

Solution

The observer measures the length of the moving ruler as $\mathbf{L = 0.80\ m}$

The rest length of the ruler is $\mathbf{L_0 = 1}$

$$\mathbf{L = L_0(1 - v^2/c^2)^{1/2}}$$

Gives

$$(1 - v^2/c^2)^{1/2} = 0.80$$

$$(1 - v^2/c^2) = 0.64$$

$$v^2/c^2 = 0.36$$

$$v = 0.6\ c$$

The ruler must be moving at **0.6** of the velocity of light.

Example 3

A spaceship travels between two points **A** and **B,** on Earth, at a speed of **1.8 x 10^8** metres per second. The pilot of the spaceship measures the time of the flight between **A** and **B** as 4 seconds.
What is the distance **AB**

(i) as measured by the pilot
(ii) according to someone on earth?

Solution

(i) The pilot sees the Earth moving backwards, below him at a speed of **1.8 x 10^8** metres per second.
Using distance = speed x time, gives the distance as

$$AB = 1.8 \times 10^8 \times 4$$
$$= 7.2 \times 10^8 \text{ metres.}$$

==================================

(ii) The Earth observer measures a different time for the flight since 4 seconds was measured in the frame of the pilot that was moving relative to the Earth observer.

Using $t_0 = t(1 - v^2/c^2)^{1/2}$, since the time in the rest frame of the spaceship is $t_0 = 4$ **sec** we have

$$4 = t(1 - (1.8)^2/(3.0)^2)^{1/2} \qquad \text{(taking } c = 3.0 \times 10^8 \text{)}$$

$$4 = t(0.64)^{1/2}$$

thus $$t = 4 / 0.8 = 5 \text{ seconds}$$

Using distance = speed x time, the Earth observer therefore calculates the distance **AB** as

$$AB = 1.8 \times 10^8 \times 5 = 9.0 \times 10^8 \text{ metres}$$

=============================

Alternatively, we could solve part (ii) as follows:

If L_0 is the distance in the rest frame of the Earth, then, with the usual notation:

$$L = L_0(1 - v^2/c^2)^{1/2}$$

giving $7.2 \times 10^8 = L_0(1 - (0.6)^2)^{1/2}$

$$L_0 = (7.2 \times 10^8) / 0.8 = \mathbf{9 \times 10^8 \text{ metres}}$$

==============================

The Light Year

A **light year** is the distance travelled by light in one year.
It is a convenient measure of distance when dealing with astronomical problems.

The speed of light is **299 792 458 metres per second**

One solar year = **365 days 5 hours 48 min 45.5 sec**

= **31 556 925.5 seconds**

∴ **one light year = 299 729 458 x 31 556 925.5 = 9.460 528 263 x 10^{15} metres**

==========================

The nearest star to the **Earth** is **Proxima Centauri** which is about **4.2 light years** from us which means that whenever train our telescope on Proxima Centauri, we are seeing it as it was 4.2 years ago.

Example 4
Observers on Earth track a spaceship moving at 0.9c for 3 years.

(i) How far does the spaceship travel according to observers on Earth?

(ii) How far does the spaceship travel according to the crew of the spaceship?

(iii) What time elapses on the spaceship?

Solution
Suppose that observers on Earth start tracking with the spacecraft at point **A**, and finish tracking when the spacecraft has reached the point **B** three years later. There are two events to be considered: (1) the spacecraft passes point **A** and (2) the spacecraft passes point **B.**

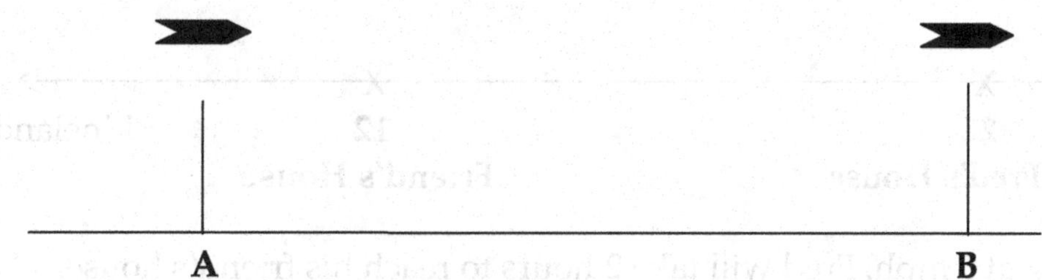

(i)
Observers on Earth see the spaceship moving at a speed = **0.9c,** for a time of **3** years. Hence, they will calculate the distance travelled from:

$$AB_{earth} = \text{distance} = \text{speed} \times \text{time} = 0.9c \times 3 = 2.7 \text{ light years}$$

(ii)
The crew of the spacecraft see the points **A** and **B** moving back at a speed 0.9c and so see the contracted length **L** where

$$L = L_0(1 - v^2/c^2)^{1/2}$$

or

$$AB_{crew} = AB_{earth} \times (1 - v^2/c^2)^{1/2}$$

$$= 2.7 \times (1 - 0.9^2)^{1/2} \quad = 2.7 \times 0.436 \quad = 1.177 \text{ light years}$$

===========

(iii)
The time for the journey according to the crew is the proper time for the period between the two events [1] spacecraft passes point **A** and [2] spacecraft passes point **B**

The proper time t_0 is given by $t_0 = t(1 - v^2/c^2)^{1/2}$

where **t=3** years
thus $t_0 = 3 \times (1 - 0.9^2)^{1/2}$

$$= 3 \times 0.436 = 1.308 \text{ years}$$

=======

World Lines in Lineland

One dimensional Fred decides to walk from his home in Lineland at **x=2,** leaving home at 12 noon, to go to a party at his friends house at **x=12.** Fred walks at a constant speed of 5mph.

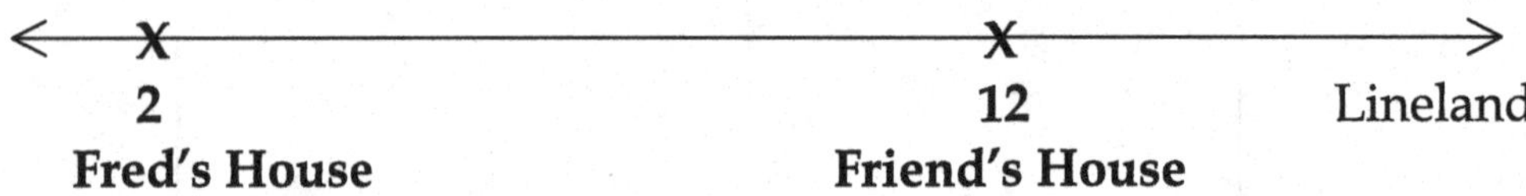

Walking at 5 mph, Fred will take 2 hours to reach his friend's house.
In each half hour, Fred will walk 2 ½ miles.

Fred's progress can be plotted as he makes his way to his friend's house:

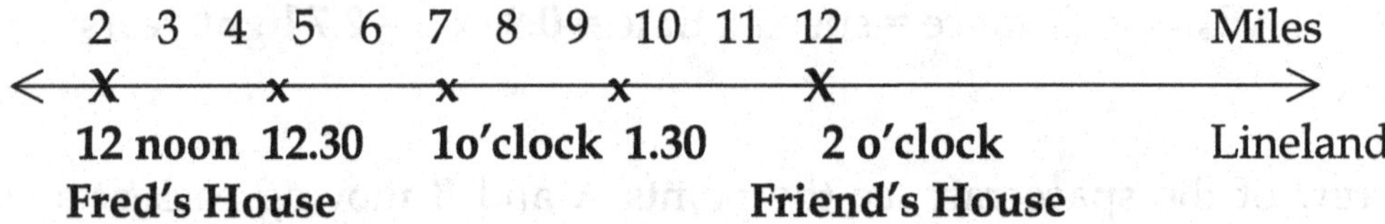

We can extend this diagram into 2 dimensional "space-time" by plotting the coordinates **x** against time **t** and then we can view Fred's progress through the day using a 2 D space-time "world line".

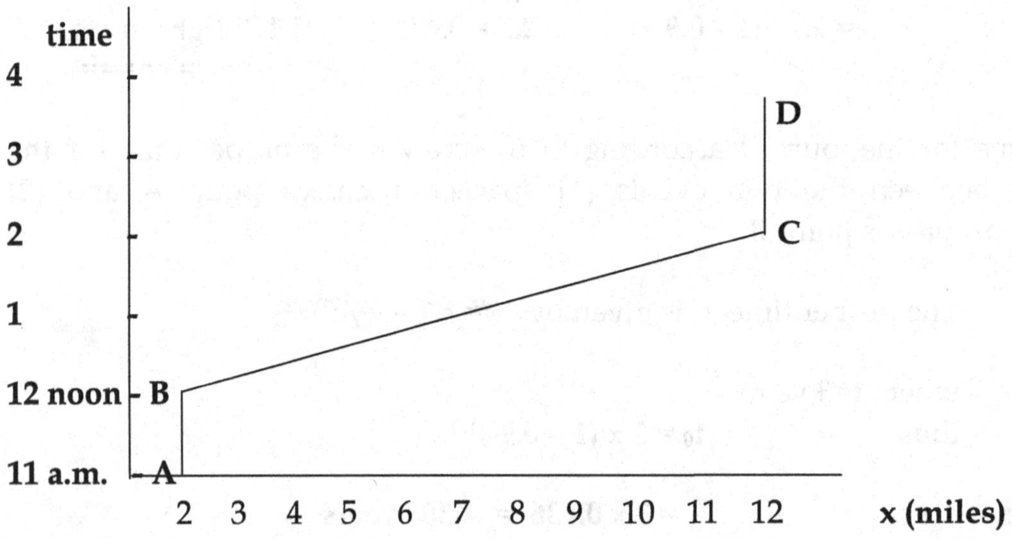

On Fred's world line, **AB** shows Fred at home before setting off for his friend's house at **12 noon.** **BC** plots Fred's walk, at a constant speed of 5 mph, to the party at his friend's house. The event **"arrive at friend's house"** is represented by the point **C** with space-time coordinates **(12,2).**
CD shows that Fred is at the party.
Fred stays at the party for 3 hours, but has too much to drink and has to hire a taxi to get back home. The taxi picks up Fred at 5 p.m. drops him home at 5.30. We suppose that the taxi drives at a constant speed...admittedly creating a problem for Fred in getting in and out of the taxi!
Fred gets home at 5.30 and immediately falls asleep.

The next day, Fred thinks about the party and carries out a "thought experiment". He cannot draw his world line for the previous day because he does not have two dimensions to draw his space-time diagram.
However, he can imagine what his world line would be.
This is the result of Fred's thought experiment:

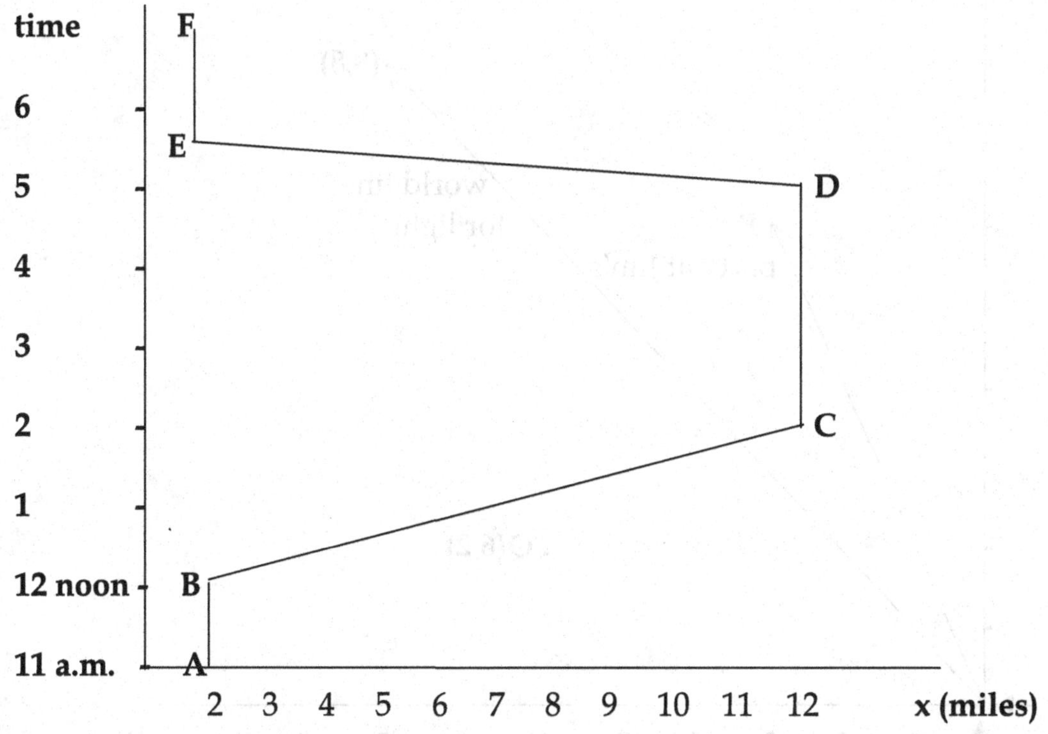

CD shows Fred at the party, **DE** represents the taxi ride back home and **EF** is Fred sleeping off his hangover.

Fred has another friend Jim who lives on a planet that orbits a star that is 3 light years away and Jim is throwing a party but not until 6 years time. Fred, of course, is a trained lineland astronaut and can pilot a very fast rocket but he needs to check if he can reach Jim's house in time for the party.
To do this, Fred imagines a world line using light years for distance and years for time.
Fred carries out another "thought experiment" in 2 dimensions.

The world line he imagines is in the next diagram.
The event "**party in 6 years time at Jim's"** is the event **P.**

To get there, Fred would have to travel 3 light years in 6 years, i.e. he would need to go at **0.5** of the speed of light.
The world line for a light ray leaving Fred at **(here,now)** is also drawn on the diagram.

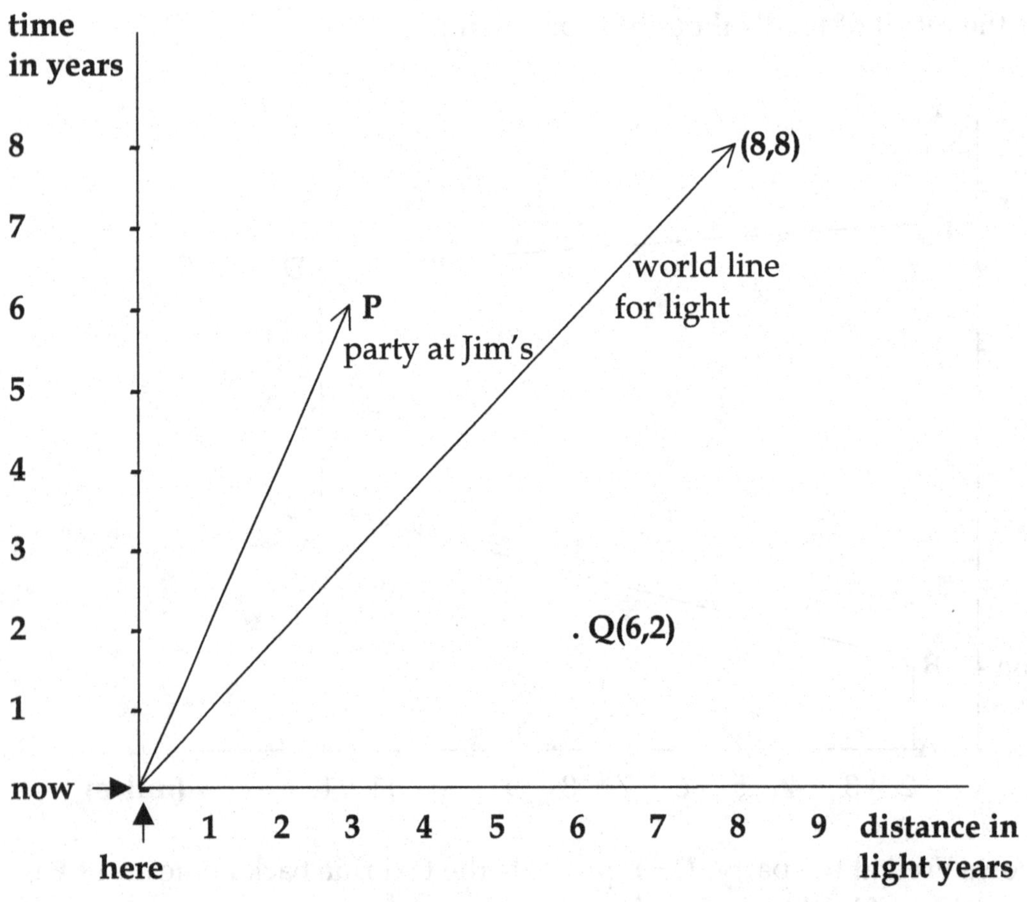

Fred's rocket can reach any party that has **2D** space-time coordinates above the world line for light, providing that his rocket can go fast enough!

The event **Q(6,2)**, represents a party that is to come on in 2 years time on a planet 6 light years away. In order to reach the party at Q, Fred would have to travel a distance of 6 light years in 2 years, i.e. at three times the speed of light- which he cannot do. The party at **Q** is out of reach!

Space-Time Distance

In ordinary three dimensional space, the distance between two points is fixed, in particular, the distance from the origin **O(0,0,0)** to the point **P(x,y,z)** is given by

$$OP^2 = x^2 + y^2 + z^2$$

However, in space-time coordinates, we have seen that distances are not fixed because of the Lorentz-Fitzgerald contraction. The distance measured between two points depends on the speed of the frame of reference relative to the observer.

In **4D** spacetime, the spacetime "distance" between the event **O(here, now)** which would be represented by the **4D** coordinates **x=0, y=0, z=0, t=0** and the event **P(x,y,z,t)** is given by

$$OP^2 = x^2 + y^2 + z^2 - c^2t^2$$

In Fred's case, since he is in one dimensional lineland, there are no y and z coordinates, and we are taking the velocity of light to be 1 light year per year, that is **c=1.**

The spacetime distance for Fred's one dimensional universe is therefore

$$OP^2 = x^2 - t^2 \quad \text{(since c=1)}$$

The spacetime distance between the event **O(here,now)** and Jim's party **P(3,6),** three light years away, in six years time is therefore

$$OP^2 = 3^2 - 6^2$$

$$= 9 - 36 = -27$$

It may seem odd that the spacetime distance squared turns out to be negative, so that the spacetime distance itself is imaginary. When this happens, the event can be reached, travelling at less than the speed of light. In ordinary **3** space, we can always get from one point to another but if there is a real "spacetime distance" between two events, we cannot reach one event from the other... its like having to be in two places at the same time. Using light years for distance and years for time, so that **c=1,** if the $x^2 + y^2 + z^2$ part is greater than the t^2 part, then to get to one event from the other, we would have to travel faster than the speed of light. If the Pythagoras part, i.e. the usual distance part, was 2 light years (so that $x^2 + y^2 + z^2 = 4$) and the time difference between the two events was one year **(t=1)**, you would need to travel 2 light years in one year, which means travelling at twice the speed of light!

If OP^2 is negative then Fred will be able to reach the event **P**. If the spacetime distance is positive, then it would not be possible for Fred to reach the event unless he was able to travel faster than the speed of light.

The spacetime distance between **(here, now)** and the event **Q(6,2),** the party 6 light years away in 2 years time is calculated from

$$OQ^2 = 6^2 - 2^2$$

$$= 36 - 4 = 32$$

Since OQ^2 is positive it indicates that it is impossible for Fred to reach the event **Q.** The distance is 6 light years and Fred has 2 years to get there so he would have to travel at three times the speed of light!

In ordinary **3D** space, the distance **OP** given by Pythagoras, that is $OP^2 = x^2 + y^2 + z^2$ is invariant, regardless of which set of coordinate axes is chosen.

In **4D** spacetime, it is

$$OP^2 = x^2 + y^2 + z^2 - c^2t^2$$

that is invariant.

In the rest frame coordinate system for Jim's party, we have **x=3** light years, **t=6** years and **c=1** light years per year so that

$$OP^2 = x^2 + y^2 + z^2 - c^2t^2 = 9 - 36 = -27$$

In Fred's coordinate system **x=0** because Fred does not move relative the his rocket, but due to the slowing down of Fred's time, as he travels at **v=0.5** of the speed of light, we have

$$\text{Fred's time} = t_f = t(1 - v^2/c^2)^{1/2} = 6\,x(1 - 0.5^2)^{1/2} = 6x(3/4)^{1/2}$$

So, according to Fred

$$OP^2 = x^2 + y^2 + z^2 - c^2t^2 = -36x3/4 = -9x3 = -27$$

So that $OP^2 = x^2 - c^2t^2$ has the same value for the stationary observer on Earth and for Fred in his rocket.

In spacetime, length, time and also mass are not fixed, but depend on the speed of the frame of reference, that is the set of coordinate axes being used. The speed of light however, remains constant for all observers, it is invariant. The spacetime distance between two events in **4D** spacetime is also invariant.

Example

A spaceship travels at **0.8** of the velocity of light to a star **10** light years from Earth.
Find the spacetime distance between the events (i) leaving Earth and (ii) arriving at the star using **(a)** the Earths frame of reference and **(b)** the spaceships frame of reference and show that the results are the same.

Solution

(a) In the Earths frame of reference, the distance is 10 light years and the speed is 0.8 light years per year,
hence the time is 10/0.8 = 12½ years
if **d** is the spacetime distance,
then

$$d^2 = x^2 - c^2t^2 = 10^2 - (12.5)^2 = -56.25$$

(b) since the spaceship is traveling at **0.8c** , the time for the journey on the spaceship is

$$t_{ship} = 12.5 \times (1 - 0.8^2)^{1/2} = 12.5 \times 0.6 = 7.5 \text{ years}$$

The events "leave Earth" and "arrive at the star" occur at the same place on the ship, i.e. at **(0,0,0)**.

Hence $$d^2 = x^2 - c^2t^2 = -7.5^2$$

$$= -56.25$$

=================

The Lorentz Transformation

The Lorentz equations show how the coordinates of an event (x,y,z,t) in a rest frame S, are related to the coordinates of the same event (x′,y′,z′,t′) as seen in a frame S′ that is moving at a speed v relative to S, in the direction of the positive x-axis in S. At time **t=0,** the two frames coincide.

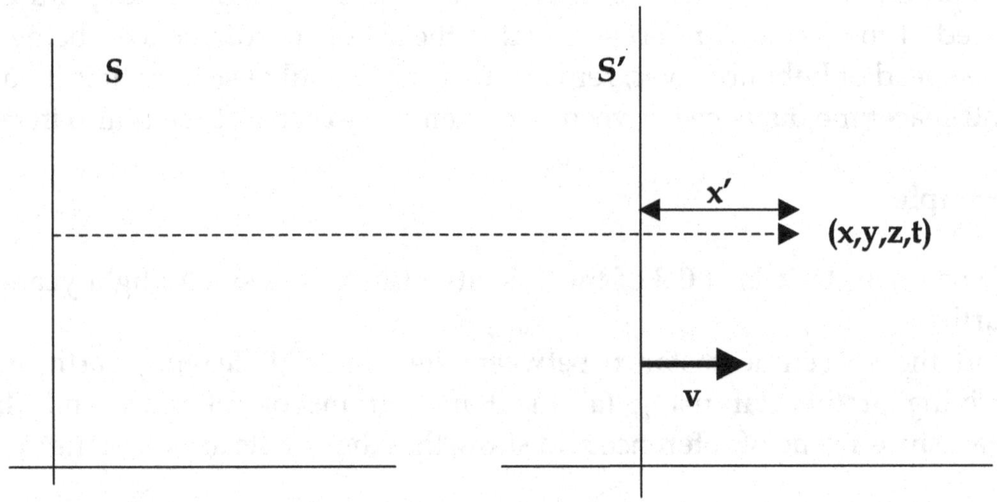

The reference frame **S′** moves at a speed **v** along the positive x-axis of frame **S.**
The origins of both frames coincide at **t=0.**
An event **(x,y,z,t)** is observed in the frame **S.**
An observer in the frame **S′** records the coordinates of the same event as **(x′,y′,z′,t′).**

At time **t** in **S** we would expect **S′** to have moved a distance **v.t** and might expect the observer in **S′** to calculate **$x' = x - vt$** but because of the effects of time dilation and length contraction for the moving frame of reference, this equation does not apply.

We suppose that **x′** is affected by a factor γ that depends on the velocity **v** so that

$$x' = \gamma(x - vt) \qquad (1)$$

Now this expression must be symmetrical since **S′** sees **S** moving the other way at speed **–v,** so we can also write

$$x = \gamma(x' + vt) \qquad (2)$$

We now use Einstein's principle that the speed of light is the same in both reference frames:

Let **E1** be the event "light flash at **t=0** at the origin"
Then in **S** we have **E1(0,0,0,0)**
And since the two frames coincide at time **t=0,** we also have in **S′ E1(0,0,0,0)**

Now consider an observer on the x axis in **S** at coordinate **(x,0,0)** receiving the light flash at time **t.**
Call this event **E2 (x,0,0,t).**
In **S** we have **x=ct** where **c** is the velocity of light so that is the frame **S** we have **E2(ct,0,0,t).**
In the frame **S′** the event **E2** has coordinates **(x′,0,0,t′)** but the speed of light in the frame **S′** must also be **c** and hence **x′ = ct′** so that in the frame **S′**, we have **E2(ct′,0,0,t′)**

Substituting for **x** and **x′** in equations **1** and **2** we have

$$ct' = \gamma(ct - vt)$$

$$ct' = \gamma t(c - v) \qquad (3)$$

and

$$ct = \gamma(ct' + vt')$$

$$ct = \gamma t'(c + v) \qquad (4)$$

Multiplying equations (3) and (4) we have

$$c^2tt' = \gamma^2tt'(c-v)(c+v)$$

$$c^2 = \gamma^2(c^2 - v^2)$$

so that $$\gamma = \{1/(1-v^2/c^2)\}^{1/2} \qquad (5)$$

To find the formula for time in the moving frame **S′** we solve equations **(1)** and **(2)** for **t′** in terms of **x** and **t.**

Since $$x = \gamma(x'+vt')$$

We have $$\gamma vt' = x-\gamma x'$$

$$= x - \gamma^2(x-vt) \qquad \text{using equation 1}$$

$$= x(1-\gamma^2) + \gamma^2vt$$

$$t' = x(1-\gamma^2)/\gamma v + \gamma t$$

Now from (5) $\gamma^2(1-v^2/c^2) = 1, \quad \therefore 1 - \gamma^2 = -\gamma^2 v^2/c^2$

$$t' = x(-\gamma v/c^2) + \gamma t$$

$$t' = \gamma(t - vx/c^2)$$

Changing the sign of **v** and swopping the dashes, we also have

$$t = \gamma(t' + vx'/c^2)$$

This gives us the set of Lorentz equations relating the coordinates of an event **(x,y,z,t)** in **S** to its coordinates **(x′,y′,z′,t′)** in the frame **S′**.

These equations show how to relate the coordinates of an event that is observed by two different observers that are in relative motion.

If **S′** is moving at a speed **v** along the positive x axis according to an observer in the frame **S** then an event that has coordinates **(x,y,z,t)** in the frame **S** and coordinates (x',y',z',t') in the frame S′ satisfies the following **Lorentz transformation equations.**

From S to S′ use

$$y' = y$$

$$z' = z$$

$$x' = \gamma(x - vt)$$

$$t' = \gamma(t - vx/c^2)$$

From S′ to S use

$$y = y'$$

$$z = z'$$

$$x = \gamma(x' + vt')$$

$$t = \gamma(t' + vx'/c^2)$$

where $\gamma = (1/(1- v^2/c^2))^{1/2}$

==================================

Dr Fred's Experiment and the Lorentz Equations

In Dr Fred's experiment whereby he discovers time dilation:
S′ is the rest frame for Dr Fred in the railway carriage and
S is the frame for the observer on the station platform

Dr Fred measures the speed of light in frame S′

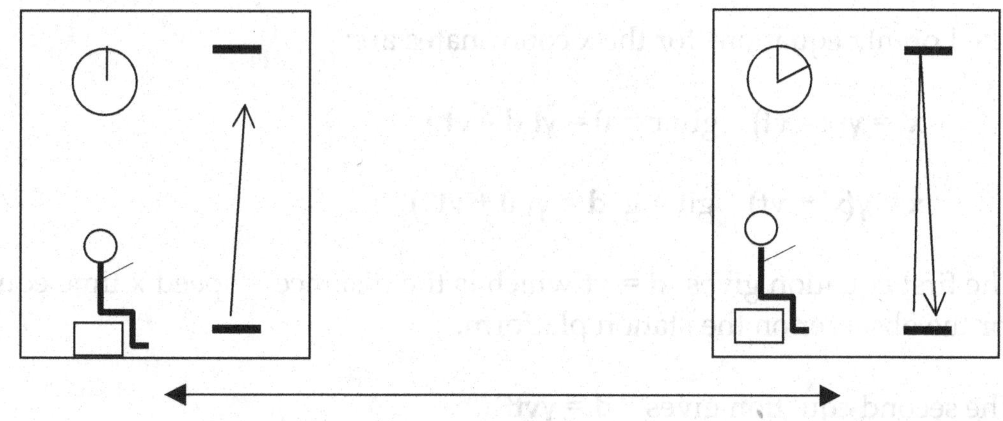

The train travels this distance

Initially, the two frames coincide, when the flash of light is set off towards the mirror in the roof of the carriage. After some time, the two observers record the event "flash returns to the floor".
In the frame **S′**, Dr Fred's frame, this event will have coordinates **(0,t′)**.

What we see from the platform

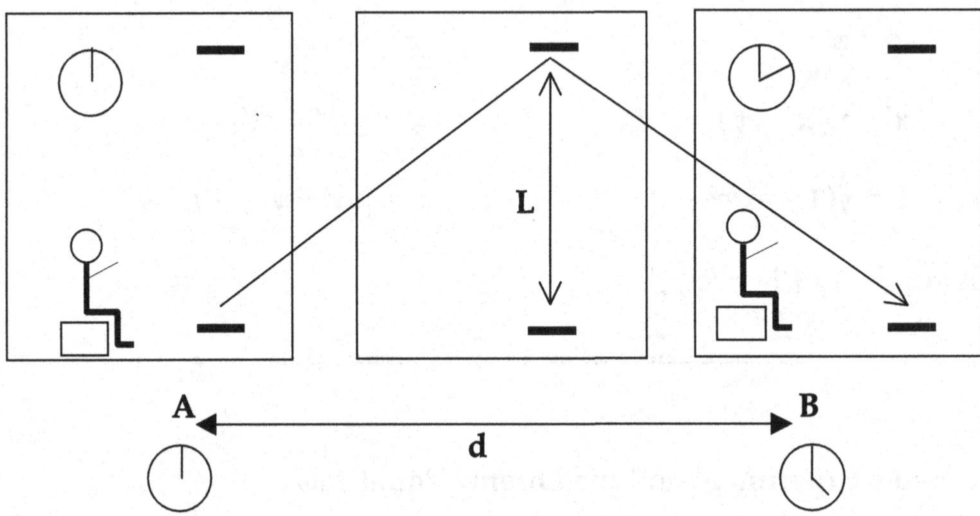

In the frame **S**, of the observer on the platform, the event "flash returns to floor" has coordinates **(d,t)**.

Using the x-coordinates

The Lorentz equations for the x coordinates are:

$$x' = \gamma(x - vt) \quad \text{giving} \quad 0 = \gamma(d - vt)$$

$$x = \gamma(x' + vt) \quad \text{giving} \quad d = \gamma(0 + vt')$$

The first equation gives **d = vt** which is the distance = speed x time equation for the observer on the station platform.

The second equation gives $d = \gamma vt'$

or $vt = \gamma vt'$

that is, $t = \gamma t'$

which we can write as $t' = t\,(1 - v^2/c^2)^{1/2}$ since $\gamma = (1/(1 - v^2/c^2))^{½}$

so that the Lorentz equations give the time dilation formula that Dr Fred discovered in his experiment.

Alternatively we can use the Time Equations:

The first Lorentz equation for time is

$$t' = \gamma(t - vx/c^2)$$

which gives $t'(1 - v^2/c^2)^{1/2} = t - vd/c^2$

$$= t - t\,v^2/c^2 \qquad \text{since } \mathbf{d = vt}$$

$$= t(1 - v^2/c^2)$$

so that we again have the time dilation equation

$$t' = t\,(1 - v^2/c^2)^{1/2}$$

The second Lorentz equation for time is

$$t = \gamma(t' + vx'/c^2)$$

and since $\mathbf{x'=0}$ this leads directly to

$$t = \gamma t'$$

again giving the time dilation equation

$$t' = t\,(1 - v^2/c^2)^{1/2}$$

==================================

A Formal Derivation of the Time Dilation Equation from the Lorentz Equations

The coordinate frame **S′** moves in the direction of the x axis in frame **S** with speed **v.**
The two frames of reference coincide at time **t=0** in **S** and time **t′=0** in **S′.**
Two events **E1** and **E2** are observed in frame **S.** The two events occur at different x coordinates but the **y** and **z** are the same and do not change. Let the events as seen in frame **S** be **E1**(x_1,y,z,t_1) and **E2**(x_2,y,z,t_2).
The same events are observed in the frame **S′** and are given the coordinates **E1**(x_1',y,z,t_1') and **E2**(x_2',y,z,t_2')
The Lorentz equations relating coordinates in **S** and **S′** are

$$x' = \gamma(x - vt) \qquad\qquad x = \gamma(x' + vt')$$

$$t' = \gamma(t - vx/c^2) \qquad\qquad t = \gamma(t' + vx'/c^2)$$

where $\gamma = (1/(1 - v^2/c^2))^{1/2}$

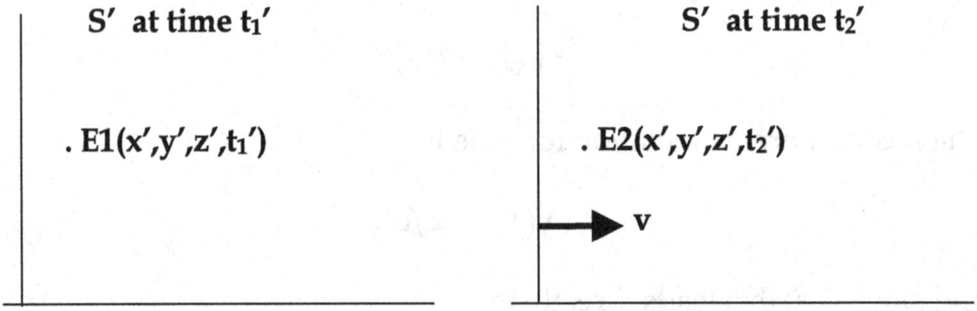

S′ at time t_1'

. E1(x',y',z',t_1')

S′ at time t_2'

. E2(x',y',z',t_2')

v

Now the time dilation formula gives the time between two events in **S′** that take place **at the same point** in **S′** which means, $x_1' = x_2'$ (=x say)

We wish to find the relation between the time interval between the two events as seen in **S′** (which is $t_2' - t_1'$) and the corresponding time interval in **S** (which is $t_2 - t_1$).

The second Lorentz time equation gives

$$t_1 = \gamma(t_1' - vx_1'/c^2)$$

and

$$t_2 = \gamma(t_2' - vx_2'/c^2)$$

and since $x_1'=x_2'$ (=x) we can write

$$t_1 = \gamma(t_1' - vx/c^2)$$

and

$$t_2 = \gamma(t_2' - vx/c^2)$$

subtracting the to get t_2-t_1 we have

$$t_2 - t_1 = \gamma(t_2' - t_1')$$

and since $\gamma = (1/(1 - v^2/c^2))^{1/2}$

this gives

$$t_2' - t_1' = (t_2-t_1)(1-v^2/c^2)^{1/2}$$

which is equivalent to the time dilation formula

$$t' = t\,(1 - v^2/c^2)^{1/2}$$

==================================

The Invariance of Spacetime Distance

Using the Lorentz equations

$$x' = \gamma(x - vt) \text{ and } t' = \gamma(t - vx/c^2)$$

by direct substitution, we have

$$x'^2 - c^2t'^2 = \gamma^2(x - vt)^2 - c^2\, \gamma^2(t - vx/c^2)^2$$

$$= \gamma^2(x^2 - 2xvt + v^2t^2) - c^2\, \gamma^2(t^2 - 2tvx/c^2 + v^2x^2/c^4)$$

$$= \gamma^2(x^2 - 2xvt + v^2t^2) - \gamma^2(c^2 t^2 - 2tvx + v^2x^2/c^2)$$

$$= \gamma^2(x^2 - 2xvt + v^2t^2 - c^2 t^2 + 2tvx - v^2x^2/c^2)$$

$$= \gamma^2(x^2 [1 - v^2/c^2] - c^2 t^2 [1 - v^2/c^2] - \cancel{2xvt} + \cancel{2tvx})$$

$$= \gamma^2 [1 - v^2/c^2] (x^2 - c^2 t^2)$$

$$= x^2 - c^2 t^2 \qquad \text{since } \gamma^2 = 1/(1 - v^2/c^2)$$

This proves the invariance of the spacetime distance for events on the x axis.

The Invariance of Spacetime Distance

Consider now, two frames of reference **S** and **S′** as shown:

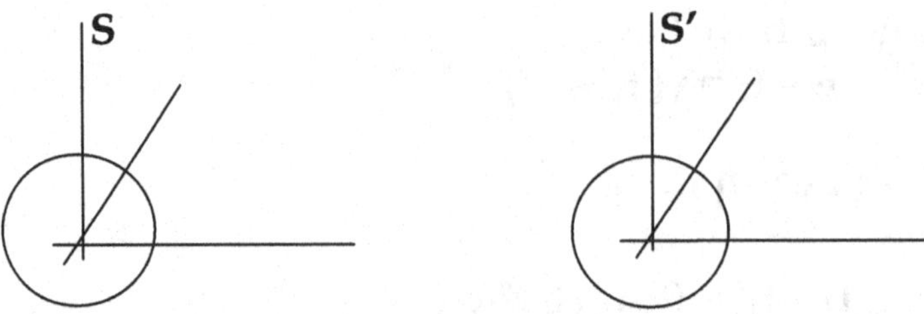

Suppose that **S** and **S′** initially coincide at **t=t′=0** and that a pulse of light leaves the common origin.
In the frame **S** at time **t**, the pulse forms a spherical wavefront of radius **r** where

$$r = ct$$

Now the speed of light is also **c** in the frame **S′** and so, in the frame **S′**, the pulse of light forms a spherical wavefront of radius **r′** say, where

$$r' = ct'$$

but $\quad r^2 = x^2 + y^2 + z^2$ and $r'^2 = x'^2 + y'^2 + z'^2$
hence, by squaring and transposing we have

$$x^2 + y^2 + z^2 - c^2t^2 = 0$$

and $$x'^2 + y'^2 + z'^2 - c^2t^2 = 0$$

therefore, we must have $x^2 + y^2 + z^2 - c^2t^2 = k(x'^2 + y'^2 + z'^2 - c^2t'^2)$ for some constant **k** since both sides have the dimension of (distance)^2.

Now for events on the x axis, we have **y=y′=z=z′=0**
so that $x^2 - c^2t^2 = k(x'^2 - c^2t^2)$ and from the first result of this section, we know that **k=1.** Therefore, for any event in **4D** spacetime with coordinates **(x,y,z,t)** in **S** and **(x′,y′,z′,t′)** in **S′**, the spacetime distance is invariant:

$$x^2 + y^2 + z^2 - c^2t^2 = x'^2 + y'^2 + z'^2 - c^2t'^2$$

====================================

Example

NASA launched their Voyager 2 spacecraft in 1977 and it took 12 years to reach Neptune.
It is now leaving the solar system in the direction of Sirius, the dog star, 8.6 light years away, at a speed of 10 km/sec.

(i) How long would it take to reach Sirius at this speed?
(ii) If NASA wanted to reach Sirius in 9 years, at what speed would Voyager 2 need to travel?
(iii) At this speed, what time would elapse on the ship?

Solution

(i) the velocity of light is 300 000 km/sec,
so the speed of Voyager 2 is 10/300 000 of the speed of light.

Time taken = distance/speed = **8.6 / (10/300 000) = 8.6x30000 = 258 000years**

(ii) 8.6 light years in 9 years = **0.96c**

(iii) we can us the invariance of spacetime distance to find the time that elapses on the spaceship

using $$x^2 - c^2 t^2 = x'^2 - c^2 t'^2$$

we have $x=8.6, c=1, t=9$ and $x'=0$

thus $$8.6^2 - 9^2 = 0^2 - t'^2$$

giving $$t'^2 = 9^2 - 8.6^2$$

$$t' = 2.65 \text{ years}$$

==============================

Alternatively we can solve (iii) using time dilation formula,

$$t' = t\,(1 - v^2/c^2)^{1/2}$$

$$= 9\,(1 - (8.6/9)^2)^{1/2}$$ since $c=1$ and $v=8.6/9$ from part (ii)

which gives the same calculation as before.

==============================

Relative Velocities

We live in an expanding universe in which galaxies are speeding away from us at huge speeds proportional to the distance away from us. The usual analogy that explains how this can be is to imagine two dimensional flatland as the surface of a balloon that is being inflated.

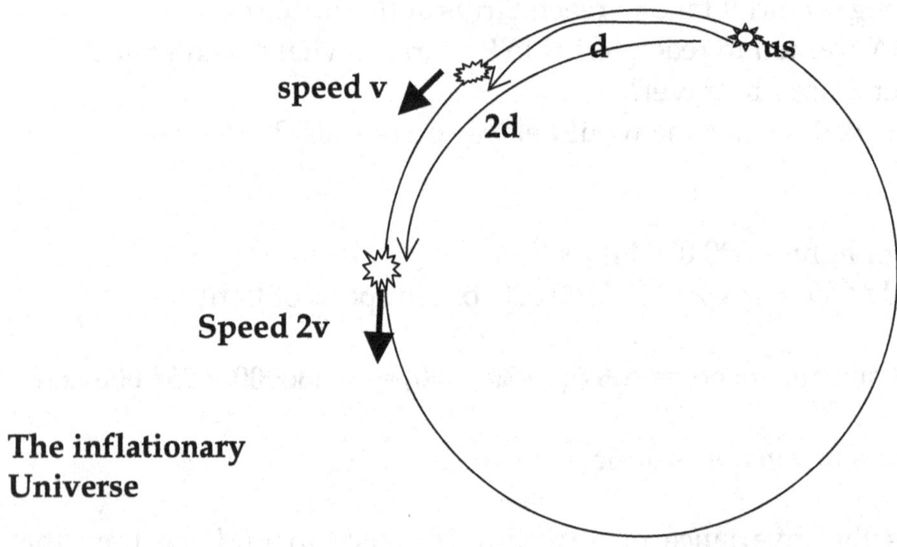

The inflationary Universe

If a galaxy at a distance **d** away is receding a at speed **v,** then another galaxy at a distance **2d** speeds away from us at a speed **2v.**

Note however, that gravity destroys the inflation of space. The galaxies themselves do not increase in size. Anything in the presence of a gravitational field does not inflate, it is the space between the galaxies, where there is no gravitation, that expands.

A paradox?

Suppose that we see a galaxy speeding away from us at **0.5c,** half the speed of light. Then we train our telescope in exactly the opposite direction and spy another galaxy, also speeding away from us at **0.5c.**
Then, are not the two galaxies separating from each other at the speed of light?

Clearly something is wrong because this cannot be.

To understand this paradox we need to examine how velocities are combined. The law of relative velocities that we learn in school applies to Galileian frames of reference:
For example, an ant walks along a sheet of paper at 2 cm per second:

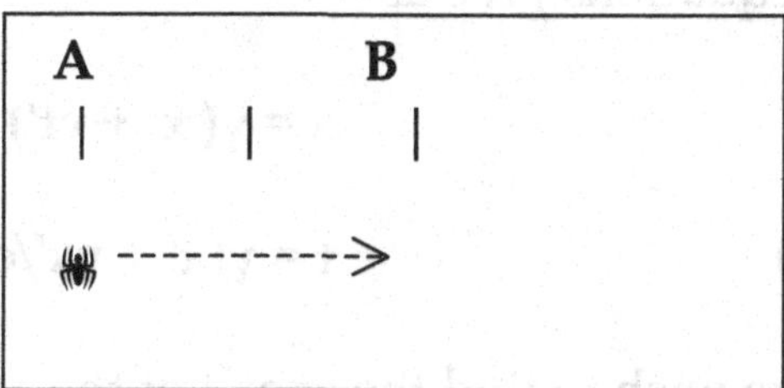

in one second, the ant moves from A to B on the paper, a distance of 2 cm.

We now play a trick on the ant and drag the paper over the table at 4 cm per second:

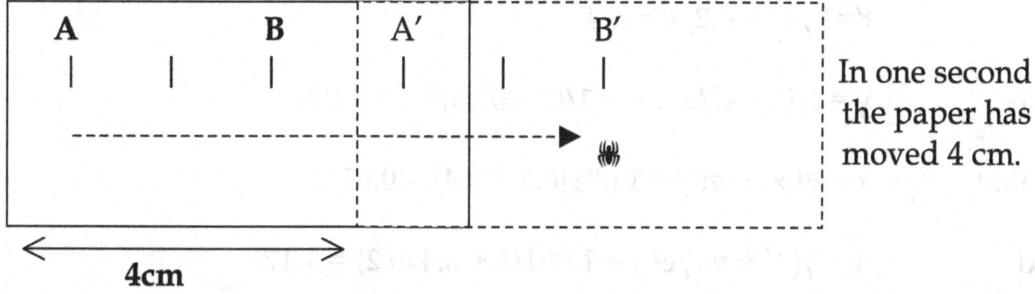

In one second the paper has moved 4 cm.

From A to A′ is 4 cm and from A′ to B′ is the 2 cm that the ant has crawled. Therefore, the distance from the start A to the finish at B′ is **AA′+A′B′ = 6 cm** Thus, relative to the table, it seems that the ant has moved 6cm in 2 seconds and is therefore traveling at 6 cm per second relative ti the table.

More formally, if **S′** is the frame of the paper, which moves at a constant speed **v** (=4 cm/sec) and the ant moves at a speed **u′** relative to **S′** (=2 cm/sec) then the speed of the ant relative to frame of the table **S,** is given by

$$\mathbf{u = u' + v}$$

However, when the speeds approach the speed of light, the rules are not the same: time dilates and lengths contract.

Suppose that we change the units from centimeters and seconds to light years and years so that the speed of the ant is **0.2** light years per year over the paper and the speed of the paper over the table is **0.4** light years per year.

(Note that the speed of light is **c=1** light year per year and the paper would need to be larger than A4)

What speed do we observe for the ant over the table?

The Lorentz transformation equations give us

$$x' = \gamma(x - vt) \qquad x = \gamma(x' + vt')$$

$$t' = \gamma(t - vx/c^2) \qquad t = \gamma(t' + vx'/c^2)$$

we have **u'=0.2, v=0.4** and we wish to find the speed of the ant over the table **(=u)**

Suppose that the ant moves for one year over the paper, then we have

$$t'=1\,,\ x' = 0.2,\ v = 0.4$$

Here, $\gamma = 1/(1 - v^2/c^2)^{1/2} = 1/(1 - 0.16)^{1/2} = 1.091$

So that $x = \gamma(x' + vt') = 1.091(0.2 + 0.4) = 0.6546$

and $t = \gamma(t' + vx'/c^2) = 1.091(1 + 0.4x0.2) = 1.178$

hence the speed of the ant according to the observer on the table, is

$$u = x/t = 0.6546/1.178 = 0.556 \text{ (3dp)}$$

The speed of the ant according to an observer on the table is therefore **0.556** light years per year and not 0.6.

Combining Velocities

Suppose that the frame **S′** is coincident with the frame **S** at time **t=0** and that the **S′** frame moves with a constant speed **v** in the direction of the x axis in **S.**

The Lorentz transformation equations then give

$$x' = \gamma(x - vt) \qquad x = \gamma(x' + vt')$$

$$t' = \gamma(t - vx/c^2) \qquad t = \gamma(t' + vx'/c^2)$$

thus

$$\frac{x}{t} = \frac{\gamma(x' + vt')}{\gamma(t' + vx'/c^2)}$$

$$= \frac{x'/t' + v}{1 + vx'/c^2t'}$$

$$= \frac{u' + v}{1 + u'v/c^2}$$

The law for combing velocities is therefore

$$u = \frac{u' + v}{1 + u'v/c^2}$$

where **v** is the speed of the moving frame, **u′** is the speed of an object in the moving frame and **u** is the speed of that object in the observers frame.

The Paradox of the Galaxies

An astronomer on Earth sees a galaxy **G1** receding at a speed **0.5c** where **c** is the speed of light.

When he turns his telescope round to the opposite direction he sees another galaxy **G2** that is also receding at half the speed of light.
Another astronomer residing somewhere in the galaxy **G1** observes the galaxy **G2.**
What will be the speed of recession of the galaxy **G2** according to the astronomer in G1?

Solution

Let **S** be the frame belonging to the astronomer in galaxy **G1** then **Earth** frame **S′** is moving away from the astronomer in the frame **S** with speed **0.5c.** **(v=0.5c)**

G2 is receding in the moving Earth frame **S′** with a speed **0.5c.** **(u′=0.5c)**

combining these velocities, we have $\mathbf{u} = \dfrac{\mathbf{u' + v}}{\mathbf{1+u'v/c^2}}$

$$= \frac{0.5c + 0.5c}{1+0.5\text{x}0.5}$$

$$= c/1.25 = 0.8c$$

The astronomer in **G1** sees the galaxy **G2** receding at **0.8** of the speed of light.

Catching up with Light

We can now answer the question posed at the start of the chapter when we stood in Parliament Square and aced away at 12 noon.
Could we catch up with the light that left Big Ben at 11 a.m. in the same way that, in theory, we could catch up with the sound waves for the ii o'clock chimes?
Suppose that we sped off from Big Ben at a speed v.
Let the frame of Parliament square be **S,** and our moving frame be **S′** speeding away at v.
The 11 am clock face is speeding away in the frame **S** at the speed of light.
Let **u′** be the speed of the 11 o'clock face in our frame **S′**.

The law for combining velocities gives

$$u = \frac{u' + v}{1+u'v/c^2}$$

and in our case, **u=c** so that

$$c(1+u'v/c^2) = u'+v$$

$$u'(v/c - 1) = v-c$$

multiplying by **c**

$$u'(v-c) = c(v-c)$$

hence $$u' = c$$

so we still see the 11 o'clock face of Big Ben flying away from us at the speed of light. Regardless of how fast e go to try and catch up, the clock face still appears to be flying away at the speed of light!
But we really knew this anyway. The speed of light is the same in all inertial frames.

Answers

Exercise 1

[1] 0.08 seconds

[2] (i) from Earth 55hours 33 minutes

(ii) by the astronaut 55 hours 17 minutes

[3] 22 years and 30 days old

[4] 577 ½ years

CHAPTER 24

Vectors

Two Biological Vectors---A Digression

There seem to be very few English words in the English language. My concise Oxford Dictionary of English states in the preface, under the subtitle "origins", that "It is difficult to be sure exactly what we mean by an 'English' word". If I scan through my English dictionary, I find words derived from 'Hawaiian', 'Africaans', 'Spanish', 'Latin', 'Hebrew', 'Old French', 'French', 'Arabic', 'Greek' and so on, in fact, out of the first 100 words, I only find three real English words! (by real English I mean derived from Old English). Those three were "abaft", "abide" and "abode". The only words you can usually guarantee to be derived from Old English seem to be swear words. But I suppose that's what we should expect if we allow ourselves to be conquered by Romans, French, Vikings, Saxons and so on and then later on, sail round the world trying to conquer most of the rest of the world.

This chapter is called **"Vectors"** a word derived from the Latin: **veho, vehere, vexi, vectum...** I carry, to carry, I carried, carried.

From this origin, we also have **vehicle**...for carrying people and things, and **vehement** ... carrying anger!

I have used the derivation of the word **vector** as an excuse for a digression into the world of biology. The word vector is used in biology to describe an agent that carries a parasite from one organism to another.
By the way, dear reader, if you are of a squeamish nature and would not like to read about blood and faeces, you can skip the sections on biological vectors and go to the section on mathematical vectors, five pages on. There is nothing mathematical about a biological vector but on the other hand, stories of human parasites seem to me to so weird that they are better than science fiction!

Over half of the world's population live in areas where tropical diseases undermine the health of large numbers of the people living there. Here is a list of some of the problems people in these areas have to deal with:

Problem (=disease)	parasite	vector
Sleeping sickness	trypansoma	tsetse fly
Elephantiasis	filarial worm	culex mosquito et al.
Yellow fever	virus	aedes mosquito
Dengue	virus	aedes mosquito
Bilharzia	flat worm (fluke)	fresh water snail
Guinea worm	guinea worm	water flea

We will discuss two of these problems and the men who discovered the life cycles of the corresponding parasite: Sir Ronald Ross and Theodor Bilharz

Sir Ronald Ross (1857-1932)

Ronald Ross was born in 1857 in Almoro, about 150 miles from Delhi and 50 miles from the border with Nepal, the son of a General in the British army. When he was eight years old he was sent to a boarding school in Southampton and in 1874, aged 17, he went to St Bartholomew's Hospital, London, to study medicine. He joined the British Indian Army Medical Corps in 1881 at a time when about one million people each year, were dying of malaria. Most of his time was taken up treating soldiers who had been stricken down by the fever. The name malaria derives from the Italian mal=bad, aria=air but it was not known what caused the disease.

In 1894-1895, while on leave in England, he met Dr Patrick Manson, a specialist in tropical diseases and he told Ross that he suspected that malaria was spread by mosquitoes. Also, a French doctor, Dr Laveran, had identified a parasite now called plasmodium, in the blood of malaria patients from North Africa. Ross then decided to specialize in malaria and on returning to India, while working in Calcutta, he began dissecting the many different kinds of candidate mosquito. He paid one of his patients, a Mr Husain Khan, one Anna per mosquito bite during his investigations. In 1897, then working in Secunderabad, Ross dissected what he called a "dapple winged" mosquito, and found cells 10 microns in diameter in tissue from the mosquito's stomach. The "dapple winged" mosquito was a species called anopheles.

In 1898 Ross proved that this mosquito was the vector for the malaria parasite. Using sparrows that were infected with bird malaria, he showed that the disease could be transmitted to a healthy bird when it was bitten by a mosquito that had fed on the infected sparrow.

In 1899, Ross returned to England and took up a post as a lecturer in tropical diseases at Liverpool School of Tropical Medicine. In 1902, Ross as awarded the Nobel Prize for his work on the life cycle of the malaria parasite. He became Sir Ronald Ross in 1911.

He returned to London in 1912 and founded the Ross Institute for Tropical Diseases in Putney.

He is buried in Putney Vale cemetery.

Malaria occurs widely throughout the tropics in Africa, Asia, the Pacific and the Americas. It is caused by the reproduction of one of four species of the Plasmodium parasite in the red blood cells.

Plasmodium Falciparum is common in tropical Africa and causes about 1 million deaths per year, mainly in children (2006). Plasmodium Ovale is common in Africa but is not so dangerous.

Plasmodium Vivax has been increasing in frequency since the 1970's in the Indian subcontinent but does not readily lead to death. Plasmodium Malariae is the least common of the four species.

To catch malaria, you need to provide an infected female anopheles mosquito with a meal of your blood.

When the female mosquito feeds on infected blood, the two sexual forms of the plasmodium parasite reach the stomach of the mosquito. (The sexual forms of the parasite are called gametocytes. A gamete is a mature cell that can unite with another in sexual reproduction producing many more gametocytes: Greek "Gamos=marriage" + Greek "kutos=cell")

The female form of the parasite is fertilized by the male form in the stomach of the mosquito. The fertilized cell then penetrates the lining of the mosquito's stomach wall and forms an oocyst, that is, a cyst containing an egg (Greek:"oion=egg"+"kustis=bladder").

In one or two weeks, sporozoites are released from the cyst .
(Greek: "spora=seed"+"zoion=animal").

The sporozoites travel through the blood stream of the mosquito until they arrive at the salivary gland.

When the infected mosquito feeds on you, the next victim of the malaria parasite, the sporozoites are introduced into your blood stream. The sporozoites are carried round your blood stream until they reach your liver.

They enter cells in the liver where they develop further and after 6 to 8 days, they rupture the liver cells releasing numerous merozoites into your blood stream (Greek: "Greek: "meros=part"+"zoion=animal")
These merozoites invade your red blood cells.

Inside the red blood cells, the merozoites grow and change into trophozoites (Greek: "tropheia=nourishment") which in turn, change into schizonts (Greek: "skhizo=split").

The schizonts multiply to form one or two dozen new merozoites which rupture the red blood cell and are released back into the blood stream, entering other red blood cells and developing into asexual or sexual forms. The sexual forms act as a reservoir of infection waiting for the next mosquito bite.

This is how the mosquito acts as a vector, or carrier for the malaria parasite. The cycle of events in the blood stream takes 2 days in falciparum, vivax and ovale, and 3 days in malariae, producing the recurrent fever.
Worldwide about 2 million deaths are caused by the various forms of malaria (2006).

Theodor Bilharz (1825-1862)

In 1851, a German physician, Theodor Bilharz, from the University of Tubingen, described a worm that he had found in the portal vein of a man during a routine autopsy in Egypt. (The portal vein is a large vein that carries blood to the liver). He had found a **schistosome.**

Schistosomiasis, or Bilharzia, affects about 300 million people (2006) in the tropics and subtropics. The schistosome is a parasitic tropical flatworm about 1 cm long that lives in the small veins around the intestines, colon or bladder, depending on the species, where they fed on blood and other liquid nutrients. There are at least 19 known species of schistosome worm of which the following are known parasites of man:

S. haematobium occurs in the East Mediterranean and Africa

S. mansoni is found in 52 countries in Africa, the Caribbean, East Mediterranean and South America (named after....guess who!)

S. Japonicum is found in 7 African countries and the Pacific region

S. intercalatum occurs in parts of central Africa

S. mekongi is found in Laos and Cambodia (found in the Mekong river)

S. malaysiensis is found in the Malasian peninsular

Other species of schistosome worm infect cattle, rodents and birds.

Eggs, about 100 to 200 microns in size, are passed in faeces or urine into 'fresh' water (hardly "fresh" but not salty). As soon as the eggs come into contact with the water, they hatch into a free swimming larva called a miracidium (pl. miracidia).

They look for a particular species of water snail, depending on the species of the worm and penetrate the soft tissue of the snail. The miracidium is about 100 microns in length and can survive for about 48 hours. It has both male and female forms and detects the snails secretions in the water

Inside the snail, over a period of 3 or 4 weeks, the miracidia change into sporocysts that divide to produce two generations of more sporocysts which then produce thousands of another form of the parasite called cercaria. 25 to 30 days after the snail was infected, the snail excretes the cercaria into the water. The cercaria do not feed and will run out of energy if they do not find a host within 5 to 8 hours. Once it finds a human host, its tail drops off and its head dissolves its way through the skin, becoming a schistosomulum larva that finds its way via the blood stream, to the liver, where it grows into an adult worm. The cercaria is about 1mm long.

Some forms have 2 to 3 days in which to find a human host before changing into a schistosomule and passes into the lungs before reaching the portal vein where they mature into an adult worm about 2 cm long. About 30-45 days after infecting the human host, the cercaria has changed into an adult worm, male or female.

Male and female worms pair off in the portal vein, where Bilharz made his first discovery. They live together with the female tucked into a groove along the length of the male worm. The adult worm has two suckers so that it can migrate around the veins. Mansoni and Japonicum aim for the colon and the intestines, while haematobium aims for the small blood vessels around the bladder. There they feed off of blood and other liquid nutrients. The male and female are in a continuous state of copulation.

The female can pass 200 to 2000 eggs per day and can survive for five years. About half of the eggs are excreted, in the urine or faeces, depending on the species of worm and the rest are carried around in the blood and can get blocked in various organs. Severe problems arise when many eggs get

stuck in the liver or kidneys. World wide, about 200 million people are infected and 600 million are at risk.
The Theodor Bilharz Research Institute was founded in 1962 near Giza in Egypt by Dr Ahmed Hafez Mousa. Calcified eggs of haematobium have been found in Egyptian mummies from 1184 BC so Bilharzia has been with us a long time. A survey in S.E. Nigeria examined the urine of 2071 children. 48% were found to be positive for the Bilharzia parasite.

The message surely is clear: "Don't pee in the river and don't crap in the lakes!"

Vectors in Mathematics

Vector quantities

Many quantities are associated with both a magnitude and a direction. If you are speeding at 100mph up the M1 going North, then that is different to speeding at 100mph going south. The speed is 100 mph in both cases, but the velocity, which has a direction, is 100mph North in the first case and 100mph South in the second case.
Velocity has both a magnitude (the speed) and a direction.
Forces too have both a magnitude and a direction.
In school, we learn that forces can be combined using a parallelogram law.
In the following diagram, two pulleys A and B have 30 gram and 40 gram weights hung over them and they are balanced in equilibrium by a 50 gram weight.

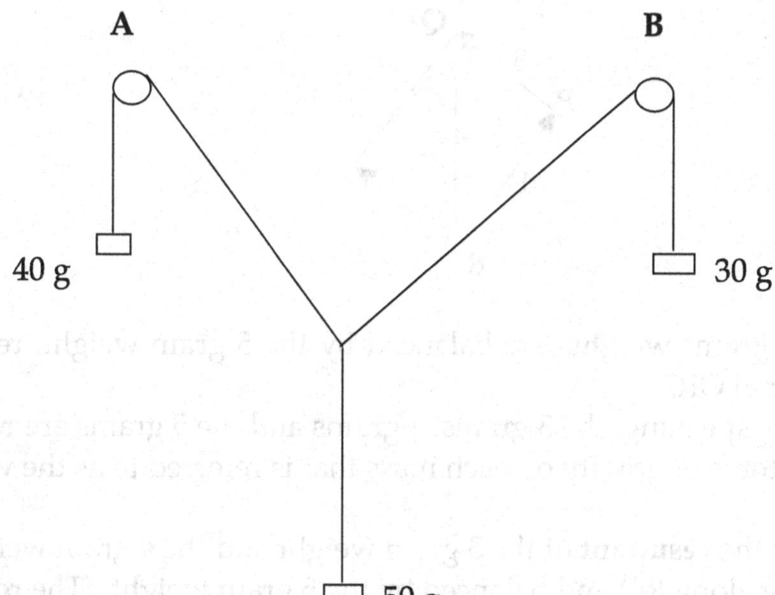

If we make a scale drawing representing the weights, we find that the two strings settle when they are at right angles:

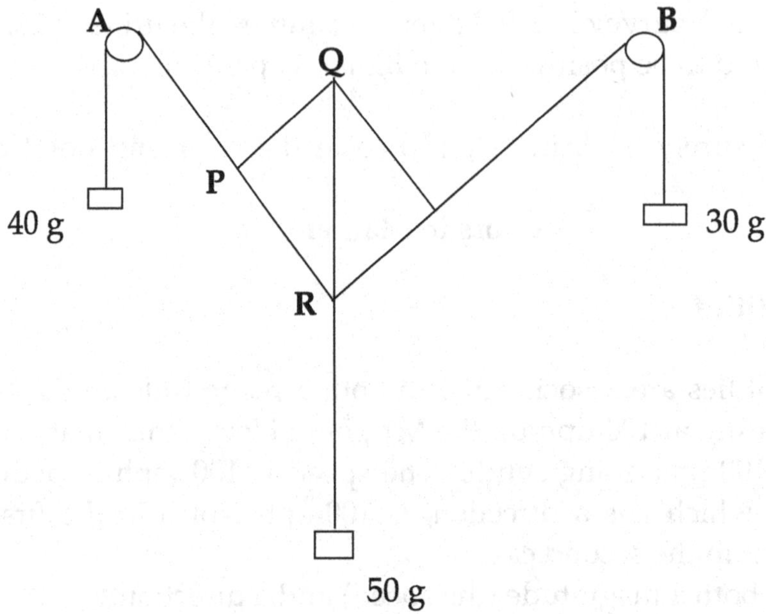

The triangle PQR, has sides proportional to the weights and the angle at P is 90°

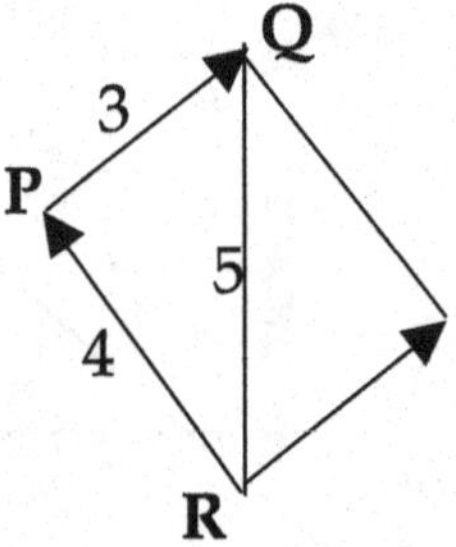

The 3 and 4 grams weights are balanced by the 5 gram weight, represented by the diagonal QR.
(Note: strictly speaking, the 3 grams, 4 grams and the 5 grams are **masses** and it is the force of gravity on each mass that is referred to as the **weight)**

We note that the **resultant** of the 3 gram weight and the 4 gram weight is 5 grams, acting along RQ and balanced by the 5 gram weight. (The **resultant** of two forces is a single force that has the same effect as the two forces acting together).

By experiment, we find that forces can be combined by a **parallelogram law of forces.** The resultant of two forces acting on a point is represented by the diagonal of the parallelogram in the scale drawing.

Example
Two forces 5 Newtons and 6 Newtons act at a point at an angle of 60° .
Calculate the magnitude of the resultant of these two forces.
(Note: one Newton is about the weight of an apple)

Solution

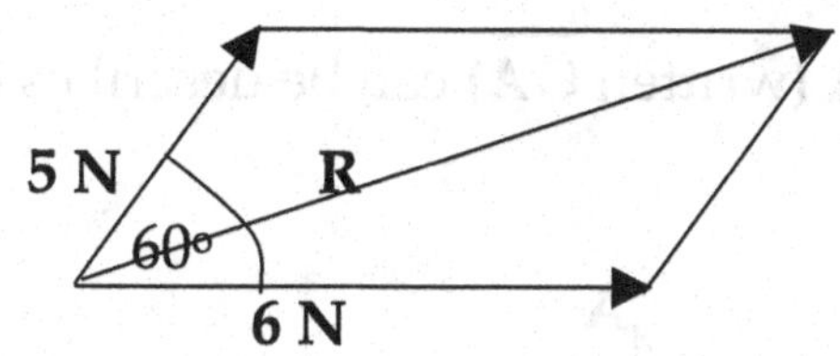

The length of the diagonal is **R** and is given by

$$R^2 = 5^2 + 6^2 + 2x5x6 \cos 60 = 91$$

$$R = 9.54 \text{ Newtons (2dp)}$$

Note that the "parallelogram law" is equivalent to a "triangle law" as far as the calculation of the resultant force is concerned:

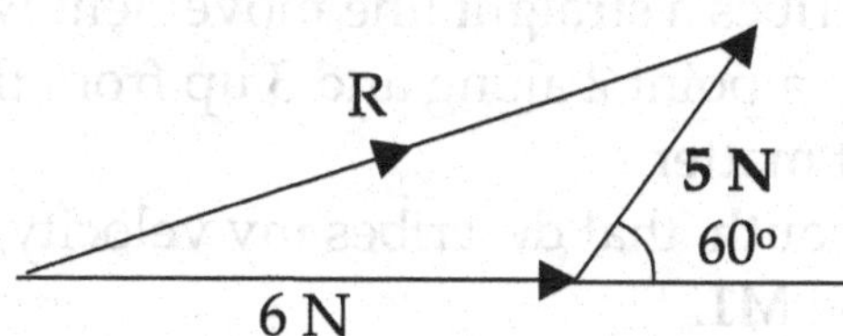

The cosine rule gives

$$R^2 = 5^2 + 6^2 - 2x5x6 \cos 120$$

Which is the same as

$$R^2 = 5^2 + 6^2 + 2x5x6 \cos 60$$

================================

Definition

Quantities that can be completely described by a magnitude and a direction and obey a triangle law of addition are called vectors.

The simplest vector quantity is a straight line movement from one point to another and is called a **displacement.**

Displacements

Let O be the origin for x-y coordinate axes and suppose that **A** is the point **A(2,3).**

The displacement from **O** to **A** (written $\overline{\mathbf{OA}}$) can be describes as "2 steps along, then 3 steps up".

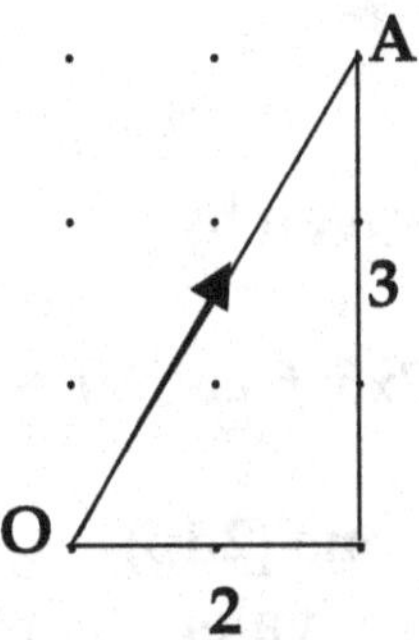

The displacement **OA** describes a straight line movement which starts a given point, and ends up at a point 2 along and 3 up from the starting point. The starting point does not matter.
If I am driving at 100mph south, that describes my velocity, whether I am driving down the **M6** or the **M1**.
We would write this displacement vector as:

$$\overline{\mathbf{OA}} = \begin{pmatrix} 2 \\ 3 \end{pmatrix}$$

If **P** and **Q** are two other points such that **Q** is "2 along and 3 up " from **P** then $\overline{\mathbf{PQ}}$ will describe the same displacement as $\overline{\mathbf{OA}}$ and we would write

$$\overline{\mathbf{PQ}} = \overline{\mathbf{OA}}$$

For example, if **P** is the point **P(4,1)** and **Q** is the point **Q(6,4)** we have

$$\overline{PQ} = \begin{bmatrix} 2 \\ 3 \end{bmatrix}$$

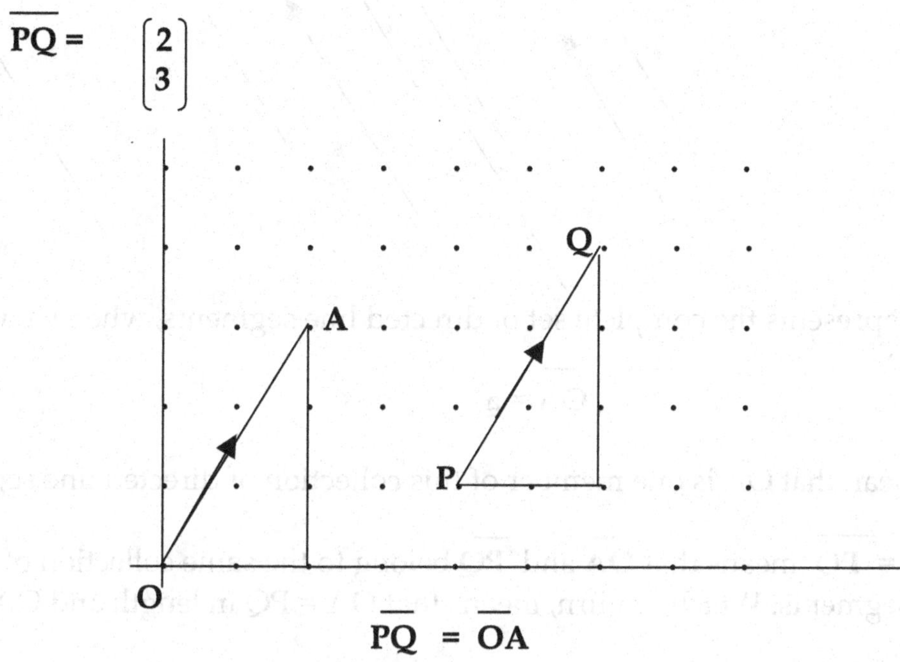

$$\overline{PQ} = \overline{OA}$$

We can regard the vector $\begin{bmatrix} 2 \\ 3 \end{bmatrix}$ as representing all of the directed line Segments equal and parallel to **OA** and pointing in the same direction. If we represent the whole class of equal parallel directed line segments by **a̲** then we would write

$$\underline{a} = \begin{bmatrix} 2 \\ 3 \end{bmatrix}$$

But, we must not think that a vector is tied to a particular set of coordinate axes. Axes are sometimes useful, but the concept of the geometrical vector is independent of any set of axes. For example, we will find that our theorems apply just as well in three dimensions as in two.

Definition

A mathematical vector is a complete class of **equal** directed line segments all pointing in the same **direction.**

A vector is often denoted using the underline and lower case, while a particular directed line segment is often denoted by an overline.

Diagram of the vector a

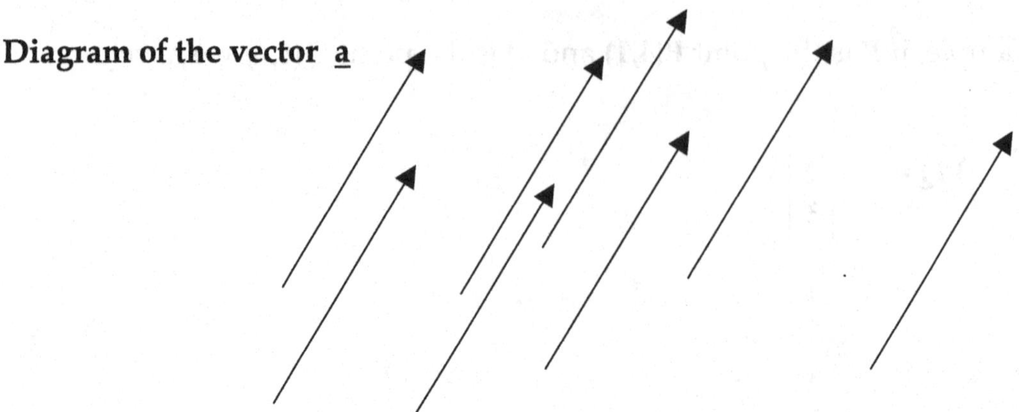

If a represents the complete set of directed line segments, when we write

$$\overline{\mathbf{OA}} = \underline{\mathbf{a}}$$

we mean that $\overline{\text{OA}}$ is one member of this collection of directed line segments.

$\overline{\mathbf{OA}} = \overline{\mathbf{PQ}}$ means that $\overline{\mathbf{OA}}$ and $\overline{\mathbf{PQ}}$ belong to the same collection of directed line segments. Which, in turn, means that OA = PQ in length and OA // PQ.

Addition of Vectors: The triangle law

Suppose that displacement vector a is " 4 along and 1 up" and that displacement vector b is "3 along and 6 up". Then it seems clear, that displacement a followed by displacement b would be "7 along and 7 up".

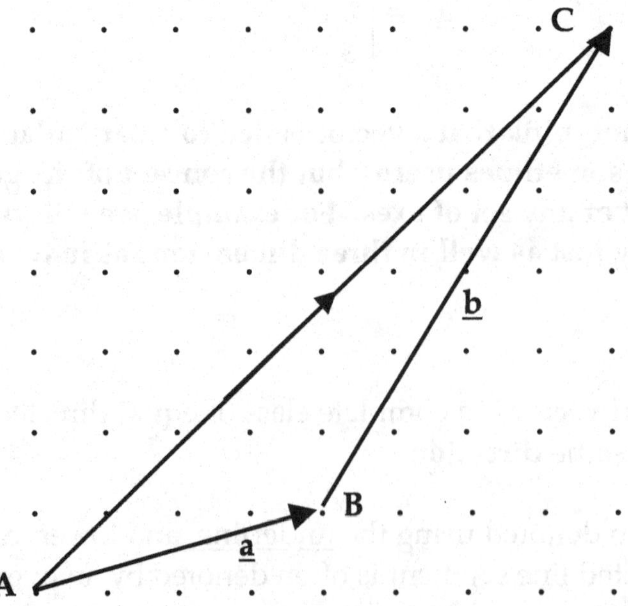

This is how we add vectors together. To add the two vectors, we would find the result of displacement $\underline{a}$ followed by the displacement $\underline{b}$ as illustrated in the diagram.

The vector $\underline{a}$ carries you from point **A** to point **B** and then vector $\underline{b}$ carries you from point **B** to point **C.**

The result of combining the two displacements is the same as the displacement from **A** to **C. A** to **B** followed by **B** to **C** is the same as **A** to **C.**

This is an example of the **triangle law** for the addition of two vectors. In general, we always have

$$\overline{AB} + \overline{BC} = \overline{AC}$$

In our example, we have $\underline{a} = \begin{pmatrix} 4 \\ 1 \end{pmatrix}$ **and** $\underline{b} = \begin{pmatrix} 3 \\ 6 \end{pmatrix}$

and the addition rule gives

$$\begin{pmatrix} 4 \\ 1 \end{pmatrix} + \begin{pmatrix} 3 \\ 6 \end{pmatrix} = \begin{pmatrix} 7 \\ 7 \end{pmatrix}$$

which illustrates the way in which column vectors are added.
The following three diagrams illustrate the triangle law of addition:

(1)

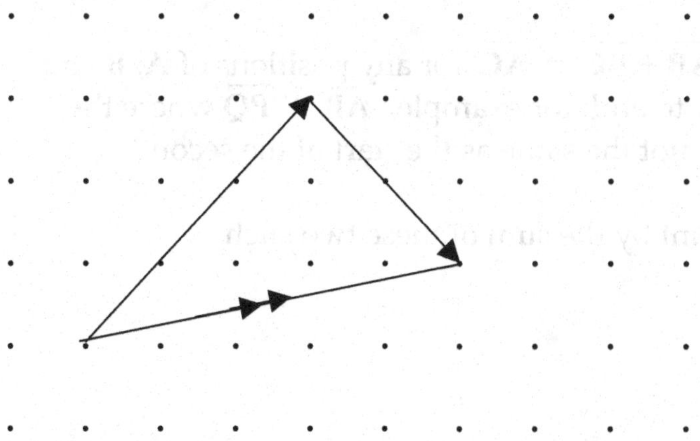

$$\begin{pmatrix} 3 \\ 3 \end{pmatrix} + \begin{pmatrix} 2 \\ -2 \end{pmatrix} = \begin{pmatrix} 5 \\ 1 \end{pmatrix}$$

(2)

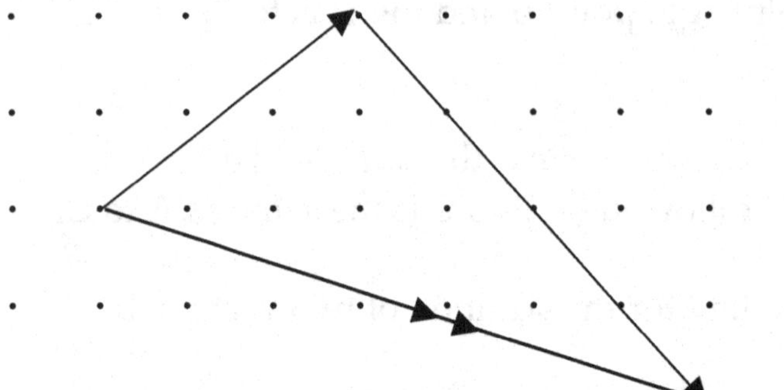

$$\begin{pmatrix} 3 \\ 2 \end{pmatrix} + \begin{pmatrix} 4 \\ -4 \end{pmatrix} = \begin{pmatrix} 7 \\ -2 \end{pmatrix}$$

(3)

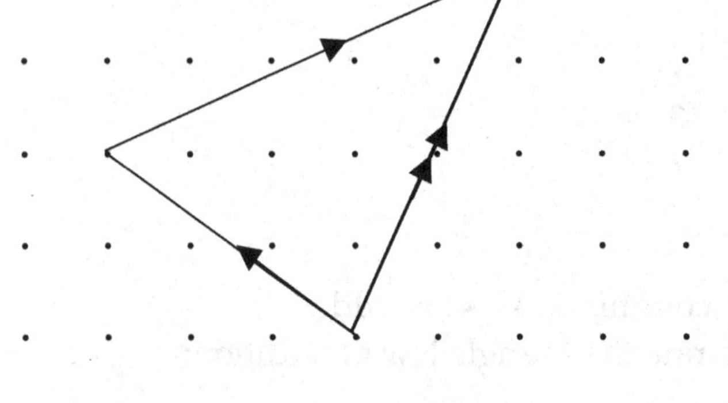

$$\begin{pmatrix} -3 \\ 2 \end{pmatrix} + \begin{pmatrix} 5 \\ 2 \end{pmatrix} = \begin{pmatrix} 2 \\ 4 \end{pmatrix}$$

Adding Displacement Vectors

The triangle law tells us that $\overline{\mathbf{AB}} + \overline{\mathbf{BC}} = \overline{\mathbf{AC}}$ for any positions of **A, B** and **C** but it does not show us how to add, for example, $\overline{\mathbf{AB}} + \overline{\mathbf{PQ}}$ where the end of the first displacement is not the same as the start of the second displacement.
We need to **define** what is meant by the sum of these two such displacements in general.

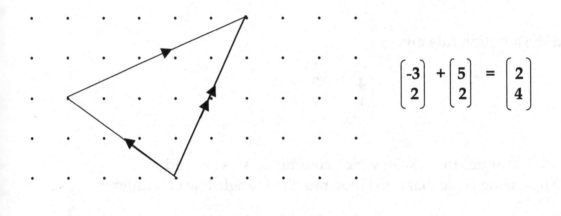

Remember $\overline{\mathbf{PQ}}$ is just one "representative" of an infinite class of directed line segments. If we call the whole class of directed line segments $\underline{\mathbf{p}}$ and then, we write

$$\overline{\mathbf{PQ}} = \underline{\mathbf{p}}$$

indicating that $\overline{\mathbf{PQ}}$ is one of the directed line segments in the whole class that is represented by $\underline{\mathbf{p}}$

Now there is just one of the directed line segments belonging to the class $\underline{\mathbf{p}}$ that starts on the point **B.** If we call this directed line segment $\overline{\mathbf{BX}}$, then $\overline{\mathbf{BX}}$ is also one of the class of directed line segments $\underline{\mathbf{p}}$ hence we can write

$$\overline{\mathbf{BX}} = \underline{\mathbf{p}}$$

and we can write

$$\overline{\mathbf{BX}} = \overline{\mathbf{PQ}}$$

indicating that $\overline{\mathbf{BX}}$ and $\overline{\mathbf{PQ}}$ each belong to the same class (i.e. $\underline{\mathbf{p}}$) with BX=PQ and BX//PQ.

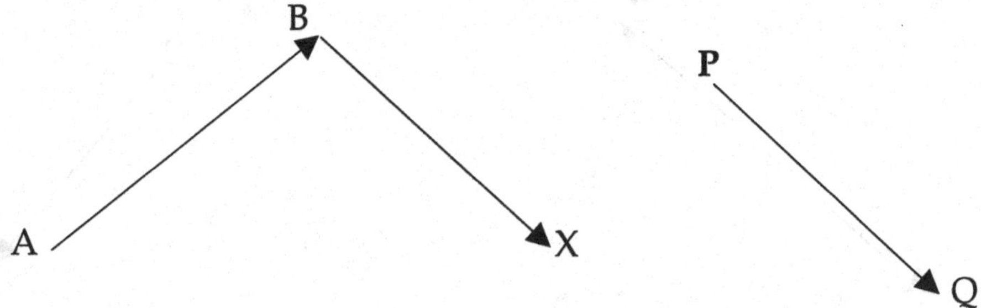

Hence, to find $\overline{\mathbf{AB}} + \overline{\mathbf{PQ}}$ we draw BX equal and parallel to PQ and then we have

$$\overline{\mathbf{AB}} + \overline{\mathbf{PQ}} = \overline{\mathbf{AB}} + \overline{\mathbf{BX}}$$

$$= \overline{\mathbf{AX}} \quad \text{by the triangle law.}$$

The Parallelogram Law

Suppose that we have two vectors (i.e. two different classes of directed line segments) called $\underline{\mathbf{b}}$ and $\underline{\mathbf{p}}$

Choose any point A and draw $\overline{\mathbf{AB}} = \underline{\mathbf{b}}$

Also draw $\overline{\mathbf{AP}} = \underline{\mathbf{p}}$

Complete parallelogram $\overline{\mathbf{ABXP}}$ then $\overline{\mathbf{BX}} = \overline{\mathbf{AD}}$ because they have the same direction and the same length (a property of the parallelogram).

Hence $\overline{\mathbf{BX}}$ belongs to the same class of directed line segments as does $\overline{\mathbf{AP}}$.
Therefore

$$\overline{\mathbf{BX}} = \underline{\mathbf{p}}$$

So that $\underline{\mathbf{b}} + \underline{\mathbf{p}}$, by the last section, is represented by $\overline{\mathbf{AX}}$

Therefore, the vector sum of the two vectors, $\underline{\mathbf{b}} + \underline{\mathbf{p}}$ is represented by the diagonal through A, of any parallelogram formed by AB and AP, where $\overline{\mathbf{AB}} = \underline{\mathbf{b}}$ and $\overline{\mathbf{AP}} = \underline{\mathbf{p}}$

The Commutative Law

To show that $\overline{\mathbf{AB}} + \overline{\mathbf{CD}} = \overline{\mathbf{CD}} + \overline{\mathbf{AB}}$

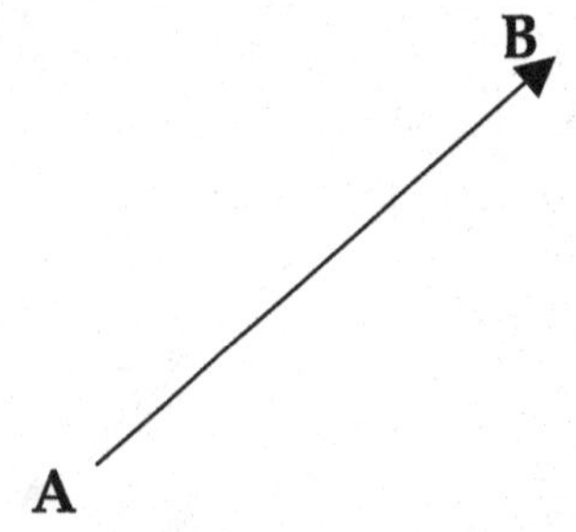

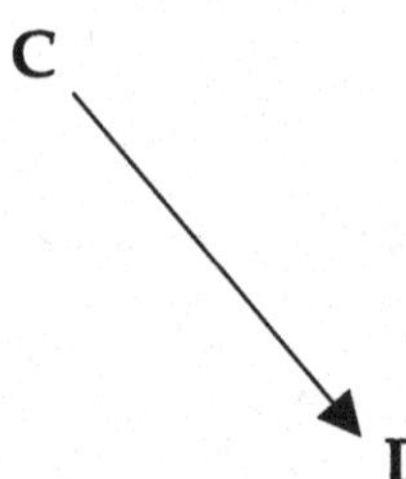

Solution

Complete the parallelogram ABXY, with $\overline{\mathbf{BX}} = \overline{\mathbf{CD}}$

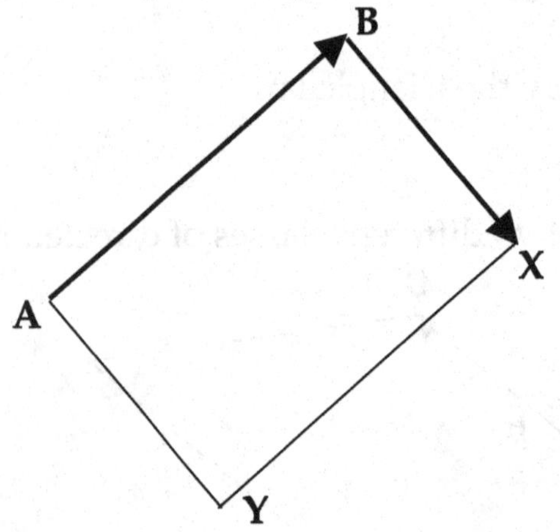

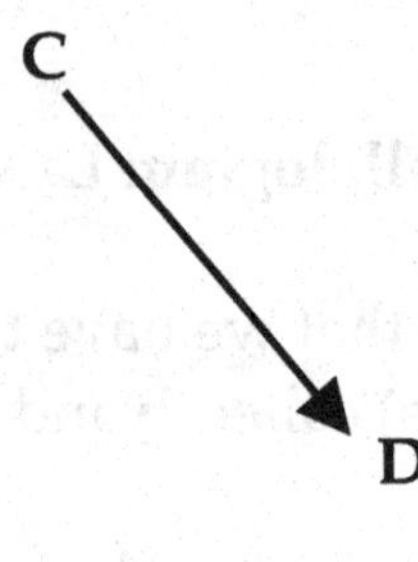

Then $\overline{AB} + \overline{CD} = \overline{AB} + \overline{BX}$

$= \overline{AX}$

$= \overline{AY} + \overline{YX}$ (triangle law)

$= \overline{BX} + \overline{AB}$ (// gram)

$= \overline{CD} + \overline{AB}$

Thus $\overline{AB} + \overline{CD} = \overline{CD} + \overline{AB}$

If we write vectors as column vectors then the commutative law reduces to the commutative law for adding numbers, for example

$$\begin{pmatrix}5\\2\end{pmatrix} + \begin{pmatrix}3\\4\end{pmatrix} = \begin{pmatrix}5+3\\2+4\end{pmatrix} = \begin{pmatrix}8\\6\end{pmatrix}$$

$$\begin{pmatrix}3\\4\end{pmatrix} + \begin{pmatrix}5\\2\end{pmatrix} = \begin{pmatrix}3+5\\4+2\end{pmatrix} = \begin{pmatrix}8\\6\end{pmatrix}$$

The Associative Law

The associative law for the addition of numbers allows us to add three numbers together regardless of the order that we carry out the addition, for example, 2+4+7 can be calculated either by adding 2+4 first or by adding 4+7 first. That is:

$$2+4+7 = (2+4) + 7 = 6+7 = 13$$

or

$$2+4+7 = 2 + (4+7) = 2+11 = 13$$

We ask, can we add vectors in a similar way?
can we write

$$\underline{a} + (\underline{b} + \underline{c}) = (\underline{a} + \underline{b}) + \underline{c}$$

Using column vectors the result follows immediately from the associative law for numbers, for example

$$\begin{pmatrix}5\\2\end{pmatrix} + \begin{pmatrix}3\\4\end{pmatrix} + \begin{pmatrix}6\\1\end{pmatrix} = \begin{pmatrix}5\\2\end{pmatrix} + \begin{pmatrix}3+6\\4+1\end{pmatrix} = \begin{pmatrix}5\\2\end{pmatrix} + \begin{pmatrix}9\\5\end{pmatrix} = \begin{pmatrix}14\\7\end{pmatrix}$$

$$\begin{pmatrix}5\\2\end{pmatrix} + \begin{pmatrix}3\\4\end{pmatrix} + \begin{pmatrix}6\\1\end{pmatrix} = \begin{pmatrix}5+3\\2+4\end{pmatrix} + \begin{pmatrix}6\\1\end{pmatrix} = \begin{pmatrix}8\\6\end{pmatrix} + \begin{pmatrix}6\\1\end{pmatrix} = \begin{pmatrix}14\\7\end{pmatrix}$$

but it is worth noting that our **geometrical vectors** are not tied to any particular axes, and in fact, they work just as well in three dimensions.

We can also demonstrate this associative law geometrically:

Let $\overline{\mathbf{AB}} = \underline{\mathbf{a}}$ $\overline{\mathbf{BC}} = \underline{\mathbf{b}}$ and $\overline{\mathbf{CD}} = \underline{\mathbf{c}}$

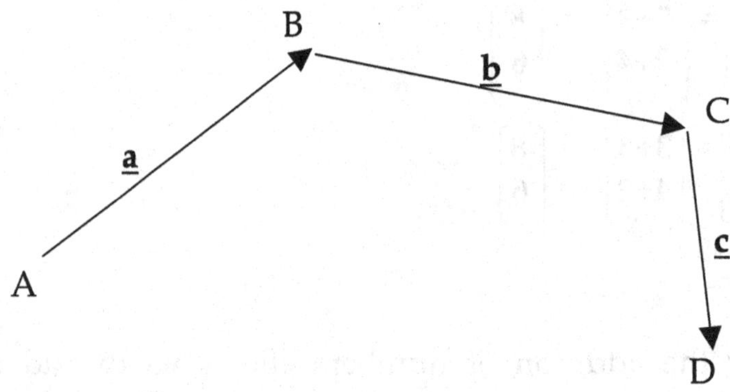

Then $\underline{\mathbf{a}} + (\underline{\mathbf{b}} + \underline{\mathbf{c}}) = \overline{\mathbf{AB}} + \overline{\mathbf{BD}} = \overline{\mathbf{AD}}$

and $(\underline{\mathbf{a}} + \underline{\mathbf{b}}) + \underline{\mathbf{c}} = \overline{\mathbf{AC}} + \overline{\mathbf{CD}} = \overline{\mathbf{AD}}$
directly from the application of the triangle law

thus we have the associative law for vectors:

$$\underline{\mathbf{a}} + (\underline{\mathbf{b}} + \underline{\mathbf{c}}) = (\underline{\mathbf{a}} + \underline{\mathbf{b}}) + \underline{\mathbf{c}}$$

Note on Notation:

At this point we adopt a notation for vectors that is more convenient to use when you are typing stuff in at the keyboard.

We dispense with the overline for displacement vectors so that **bold** and upper case will indicate a particular displacement vector. **Bold** and lower case underlined will represent a general vector regarded as a complete class of equal directed line segments.

The Zero Vector

If we denote a zero vector by the symbol $\underline{O}$ then it should have the property that

$$\underline{\mathbf{a}} + \underline{\mathbf{O}} = \underline{\mathbf{a}}$$

and

$$\underline{\mathbf{O}} + \underline{\mathbf{a}} = \underline{\mathbf{a}}$$

for any vector $\underline{\mathbf{a}}$

We would also like the triangle law, the commutative law and the associative laws to hold for the zero vector.

We use these properties to help us define $\underline{\mathbf{O}}$

If we write $\mathbf{AB} + \underline{\mathbf{O}} = \mathbf{AB}$

Then we can argue that a displacement from A to B followed by the zero vector, leaves us at B. Therefore, the zero vector will have zero length, otherwise we would move away from the point B. So, the zero vector starts at the point B and ends at the point B and is therefore represented by the displacement

$$\underline{\mathbf{O}} = \mathbf{BB}$$

Of course, we could use any other displacement vector and use the same argument, for example,

$$\mathbf{PQ} + \underline{\mathbf{O}} = \mathbf{PQ}$$

requires that the displacement $\underline{\mathbf{O}}$ starts at the point Q and ends at the point Q so that we have

$$\underline{\mathbf{O}} = \mathbf{QQ}$$

A zero displacement vector then, may be represented by **AA,** or **BB** or **PP** etc and so on for ever.

Now, if we assume the triangle law and the commutative laws hold, then we have

$$\mathbf{AA} = \mathbf{AB} + \mathbf{BA} \qquad \text{(triangle law)}$$

$$= \mathbf{BA} + \mathbf{AB} \qquad \text{(commutative law)}$$

$$= \mathbf{BB} \qquad \text{(triangle law)}$$

Thus **AA** and **BB** must be representatives of the same vector and if we require that these two laws hold for zero vectors we must have

$$\mathbf{AA} = \mathbf{BB} = \mathbf{CC} = \mathbf{PP} = \mathbf{QQ} = \text{etc...}$$

and conveniently, we represent all of these zero displacement vectors by the symbol $\underline{O}$.
It is now easy to check that this zero vector behaves as we would wish any self respecting zero vector to behave:

We want $\underline{a} + \underline{O} = \underline{a}$

and $\underline{O} + \underline{a} = \underline{a}$ for any vector $\underline{a}$

To verify this, let **AB** be a representative of the vector $\underline{a}$ then we have

$$\underline{a} + \underline{O} = \mathbf{AB} + \underline{O} = \mathbf{AB} + \mathbf{BB} = \mathbf{AB} = \underline{a} \qquad \text{(triangle law)}$$

and

$$\underline{O} + \underline{a} = \underline{O} + \mathbf{AB} = \mathbf{AA} + \mathbf{AB} = \mathbf{AB} = \underline{a} \qquad \text{(triangle law)}$$

The Negative of a Vector : Vector Subtraction

Given any vector $\underline{a}$ say, then the negative of $\underline{a}$, written $(-\underline{a})$, is any vector for which

$$\underline{a} + (-\underline{a}) = (-\underline{a}) + \underline{a} = \underline{O}$$

If we represent the vector $\underline{a}$ by the displacement **AB,** we have,

$$\mathbf{AB} + \mathbf{BA} = \mathbf{AA} = \underline{O} \text{ so that } \underline{a} + \mathbf{BA} = \underline{O}$$

and also

$$\mathbf{BA} + \mathbf{AB} = \mathbf{BB} = \underline{O} \text{ giving } \mathbf{BA} + \underline{a} = \underline{O}$$

so the vector represented by the displacement **BA** qualifies as the negative of the vector $\underline{a}$ = **AB**

It is natural to ask now, if this vector, for the negative of $\underline{a}$ is the only answer, .. is it unique?

Suppose that $\underline{a} + \underline{x} = \underline{O}$

Then $AB + \underline{x} = AA$

Add **BA** to both sides to get

$$BA + AB + \underline{x} = BA + AA$$

$$BB + \underline{x} = BA \qquad \text{(associative law)}$$

$$\underline{O} + \underline{x} = BA$$

$$\underline{x} = BA$$

so that the vector represented by **BA** is the unique negative of the vector represented by **AB.**

Subtraction of Vectors

If $\underline{a}$ and $\underline{b}$ are any two vectors then we define

$$\underline{a} - \underline{b} = \underline{a} + (-\underline{b})$$

For example, **AB - PQ = AB + QP**

Note that the difference of two vectors can be represented by "the other" diagonal of the parallelogram:

Let **OA** represent $\underline{a}$ and **OB** represent $\underline{b}$ then the sum $\underline{a} + \underline{b}$ is represented by the diagonal **OP** of the parallelogram **OAPB.**

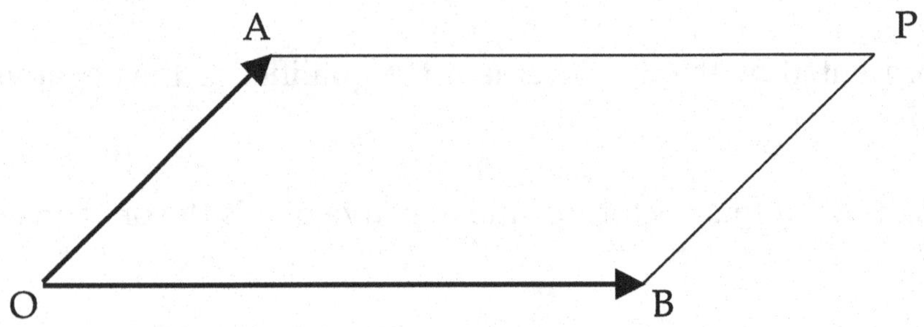

The difference $\underline{a} - \underline{b}$ determined as follows:

$$\underline{a} - \underline{b} = \mathbf{OA} - \mathbf{OB}$$

$$= \mathbf{OA} + \mathbf{BO}$$

$$= \mathbf{BP} + \mathbf{PA}$$

$$= \mathbf{BA} \qquad \text{(triangle law)}$$

Multiplication of a vector by a number

The vector sum **AB + AB** is drawn by constructing **BC = AB**

So that **AB + AB = AB + BC**

= AC

so that the vector sum **AB + AB** is represented by **AC** where **AC** is in the direction of **AB** and the length AC = 2AB.

We write **AB + AB = 2AB**

Similarly, 3**AB** is represented by a displacement in the direction of **AB** but three times the length of **AB.**

½**AB** is represented by a vector in the same direction as **AB** but half the length of **AB.**

It is not difficult to define multiplication by a negative number, for example, -3**AB** can be represented by a displacement in the direction of **BA** and three times the length of **AB**

Example

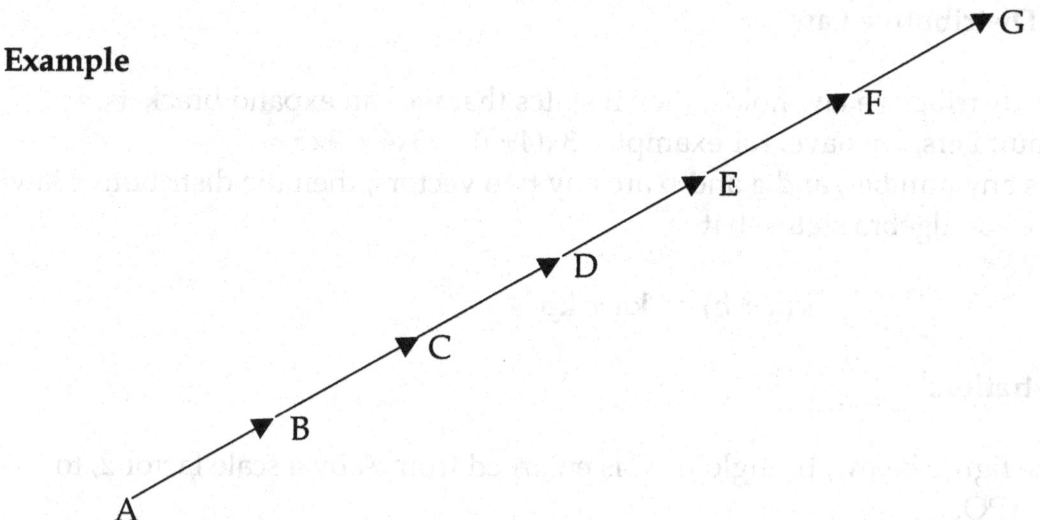

In the above figure, the line AG is divided into six equal parts at B, C, D, E and F.

Suppose that the displacement **DF** is represented by **p** then we have, for example,

DF = p

FD = -p

CG = 2p

CD = ½p

GA = -3p

DA = -1½p

Exercise 1

What are **BF, DG, EA, GB** in terms of **p** ?

The Distributive Law

If the distributive law holds, then it states that we can expand brackets.
For numbers, we have, for example, 3x(4+5) = 3x4 + 3x5.
If k is any number, and $\underline{a}$ and $\underline{b}$ are any two vectors, then the distributive law for vector algebra states that

$$k(\underline{a} + \underline{b}) = k\underline{a} + k\underline{b}$$

Illustration:

In the figure below, triangle ABC is enlarged from A by a scale factor 2, to give APQ.

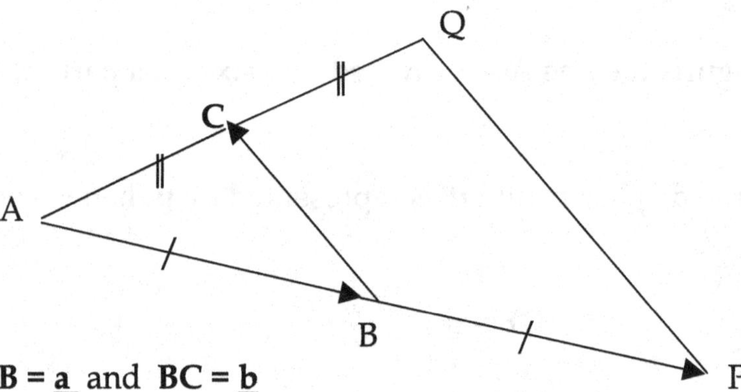

Let **AB = $\underline{a}$ and BC = $\underline{b}$**

Then AP = 2$\underline{a}$ **and** PQ = 2$\underline{b}$ **since ABC is enlarged by 2x to give APQ**

Also, **AC = $\underline{a}$ + $\underline{b}$** and **AQ = 2AC**

Hence **AQ = 2($\underline{a}$ + $\underline{b}$)**

But **AQ = AP + PQ**

= 2$\underline{a}$ + 2$\underline{b}$

therefore **2($\underline{a}$ + $\underline{b}$) = 2$\underline{a}$ + 2$\underline{b}$**

====================================

This is an illustration of the distributive law for a positive value of k that is greater than 1. We also have to show that the distributive law holds for values of k in the range $0<k<1$, for negative values of k in the range $-1<k<0$ and for negative values of k in the range $k<-1$. These cases can be dealt with in two separate treatments.

The Distributive Law for $k>0$

To Prove that $\quad k(\underline{a} + \underline{b}) = k\underline{a} + k\underline{b}$ where $k>0$

Proof

In the figure 1 below, triangle ABC is enlarged from A by a scale factor $k>1$, to give APQ. In figure 2, ΔABC is enlarged by a scale factor k in the range $0<k<1$.

figure 1

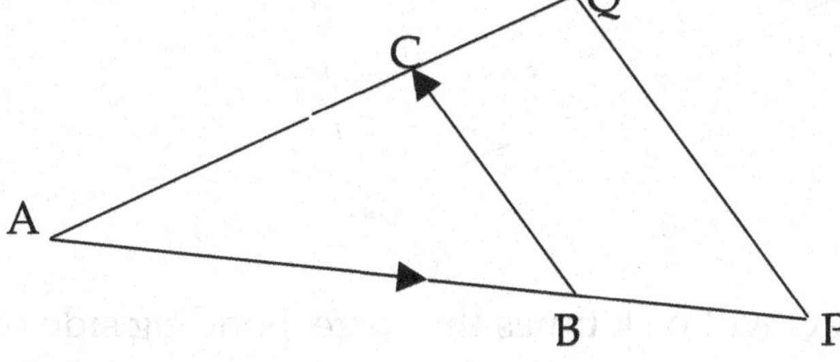

figure 2

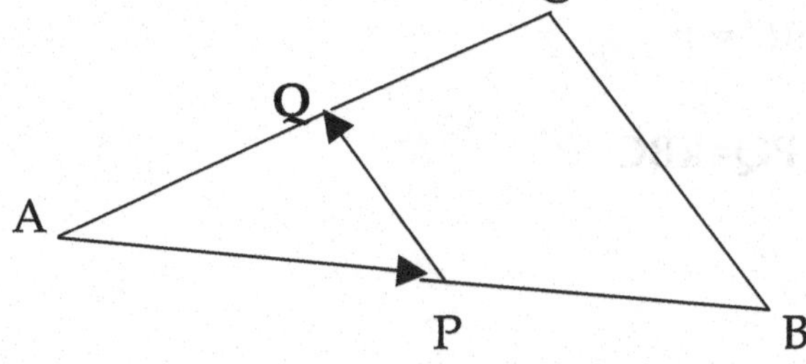

Let $\quad AB = \underline{a}$ and $BC = \underline{b}$

Then $\quad AP = k\underline{a}$ **and** $PQ = k\underline{b}$ **since ABC is enlarged by k times to give APQ**

Also, $\quad AC = \underline{a} + \underline{b}$ and $AQ = kAC$

Hence $\quad AQ = k(\underline{a} + \underline{b})$

But $\quad AQ = AP + PQ = k\underline{a} + k\underline{b}$

therefore $\quad k(\underline{a} + \underline{b}) = k\underline{a} + k\underline{b} \qquad (0<k)$

===

The Distributive law for k<0

In the following diagram, ABC is enlarged by a negative scale factor k to give APQ.

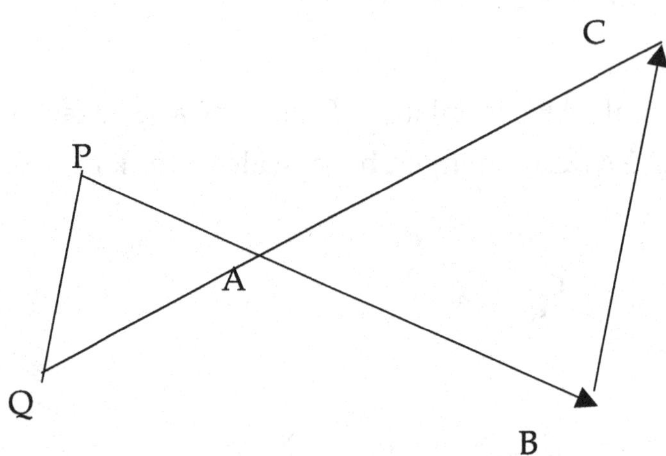

Each side of triangle APQ will be k times the corresponding side of triangle ABC and PQ//BC.

Let

$$\mathbf{AB} = \underline{a} \text{ and } \mathbf{BC} = \underline{b}$$

We have

$$\mathbf{AP} = k\mathbf{AB}, \quad \mathbf{PQ} = k\mathbf{BC}$$

Hence

$$\mathbf{AP} = k\underline{a}$$

$$\mathbf{PQ} = k\underline{b}$$

Add to get

$$\mathbf{AP} + \mathbf{PQ} = k\underline{a} + k\underline{b}$$

so that $\mathbf{AQ} = k\underline{a} + k\underline{b}$

but $\mathbf{AQ} = k\mathbf{AC}$ and $\mathbf{AC} = \underline{a} + \underline{b}$

Therefore $k(\underline{a} + \underline{b}) = k\underline{a} + k\underline{b} \qquad (k<0)$

■

====================================

Note:

Using components and column vectors, the distributive law shows up as the distributive law for numbers thus;

Let $\underline{a} = \begin{pmatrix} x \\ y \end{pmatrix}$ and $\underline{b} = \begin{pmatrix} p \\ q \end{pmatrix}$

then $k(\underline{a} + \underline{b}) = k\begin{pmatrix} x+p \\ y+q \end{pmatrix} = \begin{pmatrix} kx+kp \\ ky+kq \end{pmatrix} = \begin{pmatrix} kx \\ ky \end{pmatrix} + \begin{pmatrix} kp \\ kq \end{pmatrix} = k\begin{pmatrix} x \\ y \end{pmatrix} + k\begin{pmatrix} p \\ q \end{pmatrix}$

$= k\underline{a} + k\underline{b}$

==

A Geometrical example

The mid points of the sides of a quadrilateral form a parallelogram

Proof

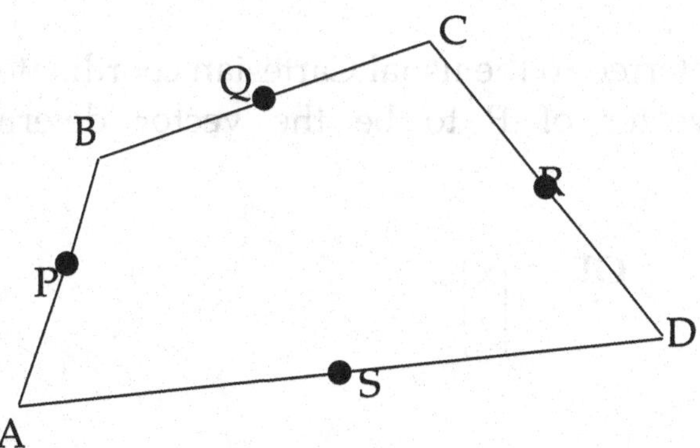

P,Q,R and S are the mid points of the sides of the quadrilateral ABCD.

We must show that PQRS is a parallelogram:

Now **PQ = PB + BQ**

= ½AB + ½BC

= ½(AB + BC)

= ½AC

= ½(AD + DC)

= ½AD + ½DC

= **SD + DR**

= **SR**

we have shown that **PQ = SR**

hence, PQ//SR

similarly, we can show that **PS = QR,** so that PS//SR

therefore PQRS is a parallelogram ■

=================================

Position Vectors

If P(x,y), is any point referred to the usual Cartesian coordinate axes, then we define the position vector of P to be the vector determined by the displacement

$$\mathbf{OP} = \begin{pmatrix} x \\ y \end{pmatrix}$$

Theorem 1

If A and B have position vectors $\underline{a}$ and $\underline{b}$ then **AB =** $\underline{b} - \underline{a}$

Proof

AB = AO + OB

= OB + AO

= OB – OA

= $\underline{b} - \underline{a}$ ■

Mid Points

Theorem 2 If A and B have position vectors $\underline{a}$ and $\underline{b}$

then the mid point of AB has position vector $\frac{1}{2}(\underline{a} + \underline{b})$

Proof

Since M is the mid point of AB,

$$\mathbf{AM} = \mathbf{MB}$$

Therefore $$\underline{m} - \underline{a} = \underline{b} - \underline{m}$$

Transpose $$2\underline{m} = \underline{b} + \underline{a}$$

$$\underline{m} = \tfrac{1}{2}(\underline{a} + \underline{b}) \quad \blacksquare$$

==

Point of Division

Theorem 3 If P divides AB in the ratio m : n then

$$\underline{p} = \frac{n\underline{a} + m\underline{b}}{m+n}$$

Proof

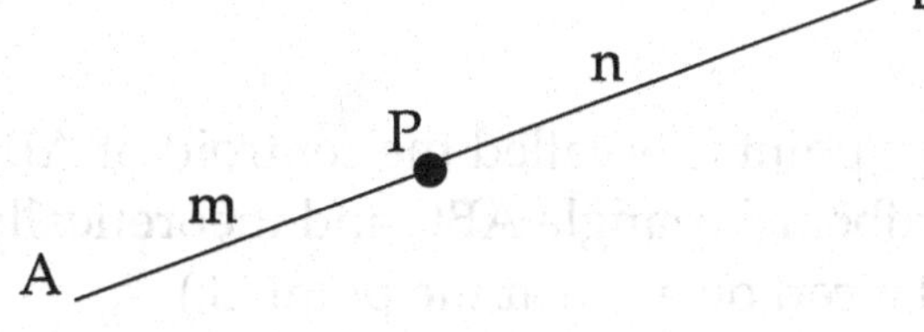

referring to the diagram, we have

$$n\mathbf{AP} = m\mathbf{PB}$$

$$n(\underline{p} - \underline{a}) = m(\underline{b} - \underline{p})$$

transpose $$(m+n)\underline{p} = m\underline{b} + n\underline{a}$$

$$\underline{p} = \frac{n\underline{a} + m\underline{b}}{m+n} \quad \blacksquare$$

Geometrical example

The medians of a triangle are concurrent (at a point called G)

Proof Let A′, B′, C′ be the mid points of BC, CA, AB

Then the medians are AA′, BB′ and CC′.

Consider the point dividing AA′ in the ratio 2:1 and call this point G

Now $\underline{a}' = \frac{1}{2}(\underline{b} + \underline{c})$ (Theorem 2)

Hence the position vector of the point dividing AA′ in the ration 2:1 is

$$\underline{g} = \frac{\underline{a} + 2\underline{a}'}{3} = \frac{\underline{a} + \underline{b} + \underline{c}}{3} \qquad \text{(Theorem 3)}$$

looking at the symmetry of the result, we deduce that the point dividing BB′ in the ratio 2:1 will have the same position vector.
Similarly, the point dividing CC′ in the ration 2:1 will have the same position vector.

Therefore, the medians of the triangle are concurrent at a point that lies two thirds of the way down the median from the vertex.

■

(The point G is called the centroid of ABC. It is the centre of mass of a cardboard triangle ABC and theoretically, the triangle ABC could be balanced on a pin at the point G.)

The centroid of four points A, B, C and D is the point with position vector

$$\frac{1}{4}(\underline{a} + \underline{b} + \underline{c} + \underline{d})$$

Exercise 2

[1] If P, Q, R and S are the mid points of AB, BC, CD and DA respectively, prove that PR and QS have the same mid points.

1.

[2] In the tetrahedron ABCD, if G_a, G_b, G_c and G_d are the centroids of BCD, ACD, ABD and ABC respectively, the AG_a, BG_b, CG_c and DG_d are each divided in the ration 3:1 by the centroid of ABCD.

[3] If A′, B′ and C′ are the mid points of BC, CA and AB respectively, prove that the centroid of triangle ABC is the same as the centroid of triangle A′B′C′.

[4] If G is the centroid of triangle ABC, prove that **GA + GB + GC = 0**

[5] If A′, B′, C′, D′, E′, F′ are the mid points of AB, BC, CD, DE, EF, FA respectively, prove that triangles A′C′E′ and B′D′F′ have the same centroid.

[6] If P, Q and R divide AB, BC and CA respectively in the same ratio m:n, prove that triangles ABC and PQR have the same centroid.

[7] Prove that the lines joining the mid points of opposite edges of a tetrahedron meet at a common point and also bisect each other.

Note on notation

For convenience when typing, position vectors will be denoted by **bold lower case,** with the same letter name as the point in question so that **a** will be understood to represent the position vector of A. We therefore know immediately that **a = OA.**

When writing however, it is better to use $\underline{a} = \overline{OA}$.

====================================

Unit Vectors in the x-y plane

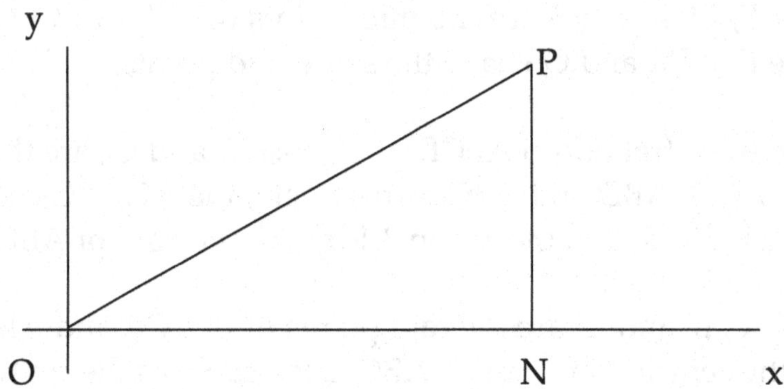

Let P(x,y) be any point referred to Cartesian x-y axes.
Let PN be the perpendicular from P onto the x axis.

Let **i** and **j** be unit vectors along Ox and Oy respectively.

Then **ON** = x**i** and **NP** = y**j**

And since **OP** = **ON** + **NP**

We have **OP** = x**i** + y**j** (1)

OP is called the position vector of P relative to O.

We can also write **OP** $= \begin{pmatrix} x \\ y \end{pmatrix}$ (2)

(1) and (2) mean the same thing.

The position vector of a general point P(x,y) is usually referred to by **r**, thus

r = **OP**

r = x**i** + y**j**

r $= \begin{pmatrix} x \\ y \end{pmatrix}$

are all familiar expressions for the same idea.

Example 1

Let A(3,4) and B(5,7)

Find the vector **AB**

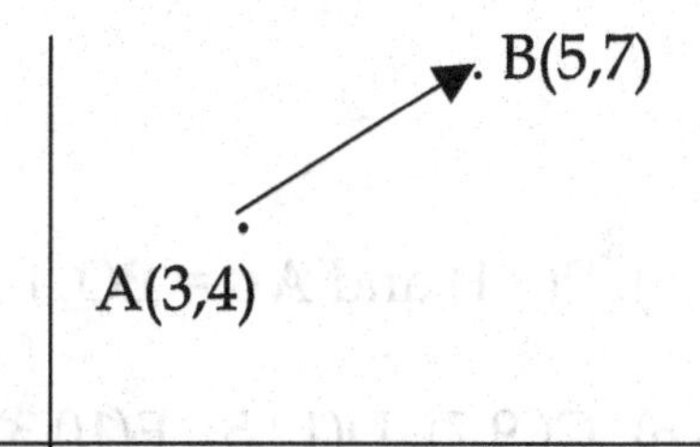

(Note: purely for illustrative purposes, we give solutions using the two notations)

Solution 1 $\quad \mathbf{OA} = 3\,\mathbf{i} + 4\,\mathbf{j}$ and $\mathbf{OB} = 5\,\mathbf{i} + 7\,\mathbf{j}$

$$\mathbf{AB} = \mathbf{b} - \mathbf{a} = (5\,\mathbf{i} + 7\,\mathbf{j}) - (3\,\mathbf{i} + 4\,\mathbf{j}) = 2\,\mathbf{i} + 3\,\mathbf{j}$$

Solution 2 $\quad \mathbf{OA} = \begin{pmatrix} 3 \\ 4 \end{pmatrix}$ and $\mathbf{OB} = \begin{pmatrix} 5 \\ 7 \end{pmatrix}$

$$\mathbf{AB} = \mathbf{b} - \mathbf{a} = \begin{pmatrix} 5 \\ 7 \end{pmatrix} - \begin{pmatrix} 3 \\ 4 \end{pmatrix} = \begin{pmatrix} 2 \\ 3 \end{pmatrix}$$

Example 2

A(5,7), B(2,4) and C(9,2) are three vertices of parallelogram ABCD. Find the position vector of D.

Solution

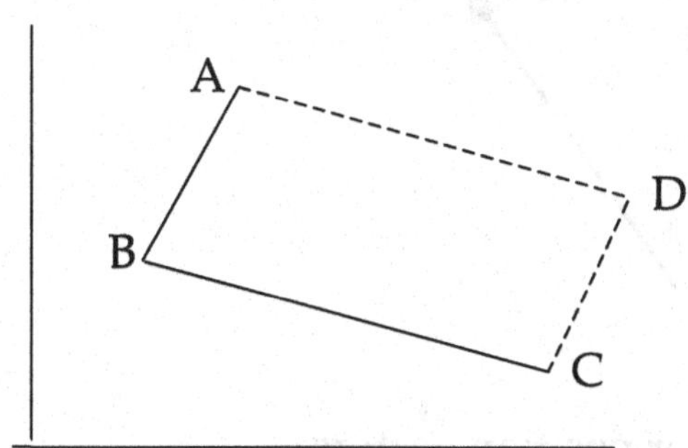

$$\mathbf{OD} = \mathbf{OA} + \mathbf{AD}$$

$$= \mathbf{a} + \mathbf{BC}$$

$$= \mathbf{a} + \mathbf{c} - \mathbf{b}$$

$$= \begin{pmatrix} 5 \\ 7 \end{pmatrix} + \begin{pmatrix} 9 \\ 2 \end{pmatrix} - \begin{pmatrix} 2 \\ 4 \end{pmatrix}$$

therefore $\mathbf{OD} = \begin{pmatrix} 12 \\ 5 \end{pmatrix}$

============

Exercise 3

1. Given A(2,2), B(4,5), P(3,1) and **AB** = **PQ**, find **OQ**

2. Given A(2,3), B(4,6), C(9,7), D(12,5), E(10,3)
Find **AB, BC, CD and DE** and verify that **AE** = **AB+BC+CD+DE**

==

The Vector equation of a Line

Let A be any given point and $\underline{\mathbf{b}}$ a given fixed vector.

We wish to find a vector equation that is satisfied by any point on the unique line through A and in the direction of $\underline{\mathbf{b}}$ (and only by points on this line). This will be called a vector equation for the line through A in the direction of $\underline{\mathbf{b}}$.

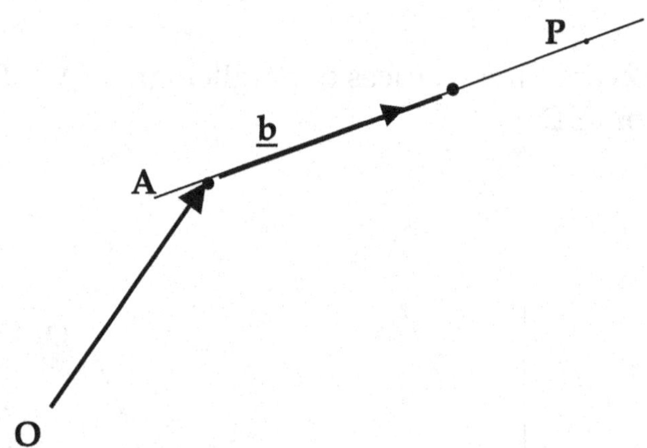

O is the origin for position vectors.

Since the vector **AP** is in the direction of the vector $\underline{\mathbf{b}}$, then **AP** must be some multiple of $\underline{\mathbf{b}}$ so we can write

$$\mathbf{AP} = t.\underline{\mathbf{b}}$$

where t is some number that will identify a particular point P on the line.
Now **OP** = **OA** + **AP**

So that $$\mathbf{OP} = \mathbf{OA} + t\underline{\mathbf{b}}$$

And if we use $\underline{r}$ for the position vector of the general point P and $\underline{a}$ for the position vector of A, then we have

$$\underline{r} = \underline{a} + t\underline{b}$$

for a vector equation of the line through A in the direction of $\underline{b}$.

(Note that if any point K say, has a position vector $\underline{k}$ satisfying $\underline{k} = \underline{a} + t\underline{b}$ for some value t, then $\underline{k} - \underline{a} = t\underline{b}$ so that **AK** = $t\underline{b}$ showing that the displacement **AK** is parallel to $\underline{b}$ so that K must be on the line)

Thus any point on the line through A in the direction of the vector $\underline{b}$ has a position vector $\underline{r}$ that satisfies the equation

$$\underline{r} = \underline{a} + t\underline{b}$$

for some value of the number t.

The number t is called the parameter for the point P. The parameter t is in effect a coordinate for points on the line with respect to the origin A. Values of t are indicated on the next diagram. Each value of t determines a unique point P.

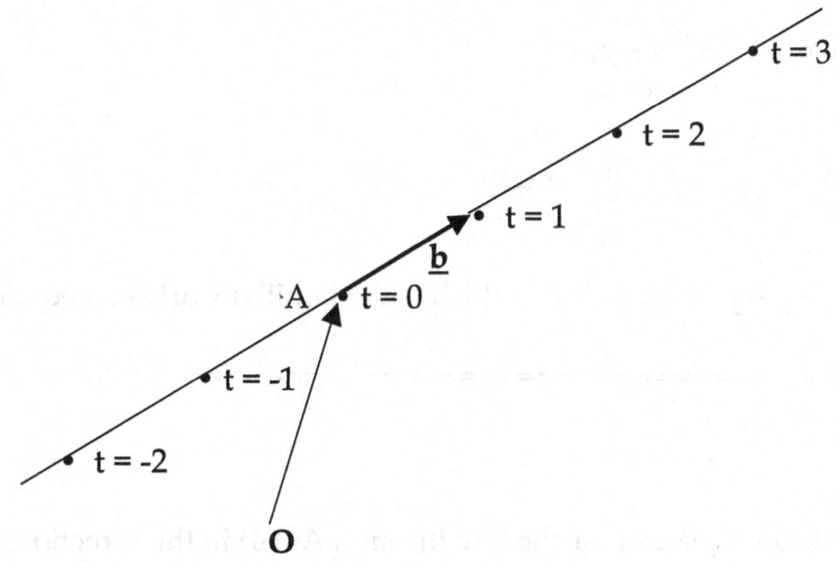

Comparison with the Cartesian Equation in 2 Dimensions

Suppose A is the point $A(x_1,y_1)$ and $\underline{b}$ is the vector $\underline{b} = \begin{pmatrix} r \\ s \end{pmatrix}$

If P is a general point P(x,y) then the vector equation becomes

$$\begin{pmatrix} x \\ y \end{pmatrix} = \begin{pmatrix} x_1 \\ y_1 \end{pmatrix} + t\begin{pmatrix} r \\ s \end{pmatrix}$$

in separate components this gives

$$x = x_1 + tr$$

and $$y = y_1 + ts$$

if the gradient of the line is m, then we have $m = s/r$ since the vector $\underline{b}$ is "r along and s up":

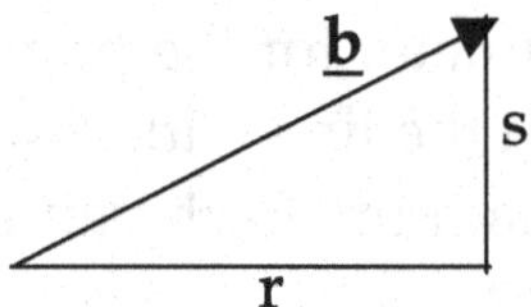

hence we have $ts = y - y_1$
and $tr = x - x_1$

dividing gives $$m = \frac{y - y_1}{x - x_1}$$

or $y - y_1 = m(x - x_1)$ which is the familiar Cartesian equation

==============================

Example 1
Find

(i) a vector equation for the line through A(3,4) in the direction of

$$\underline{b} = 2i + 3j$$

(ii) find an equivalent Cartesian equation.

Solution

(i) a vector equation is $\underline{r} = \begin{pmatrix} 3 \\ 4 \end{pmatrix} + t\begin{pmatrix} 2 \\ 3 \end{pmatrix}$

(ii) this gives $x = 3 + 2t$
$y = 4 + 3t$

so that $2t = x - 3$
$3t = y - 4$

$$2(y - 4) = 3(x - 3)$$

or $2y = 3x - 1$ ■

===================================

Example 2

Find the point of intersection of the lines

$$\mathbf{r} = 7\mathbf{i} + t(\mathbf{i} + \mathbf{j}) \quad \text{and} \quad \mathbf{r} = 5\mathbf{i} + 6\mathbf{j} + s(\mathbf{i} - \mathbf{j})$$

Solution

Note first, that we must not assume that the variable parameters for the two lines are the same at the point of intersection.
The values for t and s should be expected to be different at the point where the lines meet.
At the point of intersection, the position vectors **r** will be the same.
Equating components, we have

$7 + t = 5 + s$ (equating **i** components)
and $t = 6 - s$ (equating **j** components)

we solve these two equations for t and s.

adding, we get $7 + 2t = 11$

$t = 2$

Substituting t=2 in the equation of the first line we have

$$\mathbf{r} = 7\mathbf{i} + 2(\mathbf{i} + \mathbf{j})$$

$$\mathbf{r} = 9\mathbf{i} + 2\mathbf{j}$$

giving the point of intersection as (9,2)

■

==================================

Exercise 4

1. Find (i) a vector equation and (ii) a Cartesian equation for the line joining the points A(3,4) and B(7,2).

2. Find the point of intersection of the lines

$$\mathbf{r} = \begin{pmatrix} 2 \\ 3 \end{pmatrix} + t\begin{pmatrix} 4 \\ 5 \end{pmatrix} \quad \text{and} \quad \mathbf{r} = \begin{pmatrix} 8 \\ 2 \end{pmatrix} + t\begin{pmatrix} 2 \\ 11 \end{pmatrix}$$

(Hint: change the second t to an s)

====================================

Unit Vectors in 3D

Let P(x,y,z) be any point referred to coordinate axes Oxyz:

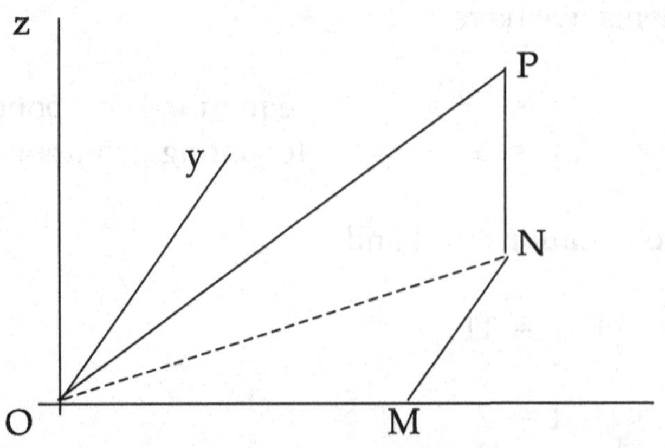

PN is the perpendicular from P onto the xy plane.
NM is the perpendicular from N onto the x axis.

i is the unit vector in the direction of Ox
j is the unit vector in the direction of Oy
k is the unit vector in the direction of Oz

Then $\mathbf{OM} = x\mathbf{i}, \quad \mathbf{MN} = y\mathbf{j}, \quad \mathbf{NP} = z\mathbf{k}$

But $\mathbf{OP} = \mathbf{OM} + \mathbf{MN} + \mathbf{NP}$

Hence we have $\mathbf{OP} = x\mathbf{i} + y\mathbf{j} + z\mathbf{k}$ or $\mathbf{r} = \begin{pmatrix} x \\ y \\ z \end{pmatrix}$

Vector methods can be used equally well in 2 or 3 dimensions.

==================================

Example 1

Find a vector equation for the line through A(1,2,3)

in the direction of the vector $\mathbf{b} = 3\mathbf{i} + 4\mathbf{j} + 5\mathbf{k}$

Solution

$$\mathbf{r} = \mathbf{OA} + t\mathbf{b}$$

gives
$$\mathbf{r} = \begin{pmatrix} 1 \\ 2 \\ 3 \end{pmatrix} + t\begin{pmatrix} 3 \\ 4 \\ 5 \end{pmatrix}$$

===========================

Note: solving this vector equation for t gives

$$t = \frac{x-1}{3} = \frac{y-2}{4} = \frac{z-3}{5}$$

The Cartesian form for a line in three dimensions requires two equations.
A linear equation in 2 dimensions represents a 1 dimensional line.
A linear equation in 3 dimensions represents a 2 dimensional plane.

The only way to specify a line in 3 dimensional Cartesian coordinates is to give the equations of two planes. The line is then the intersection of the two planes.

===================================

Example 2

Show that the lines $\mathbf{r} = \begin{pmatrix} 2 \\ 3 \\ 4 \end{pmatrix} + t\begin{pmatrix} 1 \\ 1 \\ 1 \end{pmatrix}$ and $\mathbf{r} = \begin{pmatrix} -2 \\ 0 \\ 2 \end{pmatrix} + s\begin{pmatrix} 6 \\ 5 \\ 4 \end{pmatrix}$ intersect

Solution

If the lines intersect then the position vectors **r** will be the same from each equation, so, try equating components to get

$$2 + t = -2 + 6s$$
$$3 + t = 0 + 5s$$
$$4 + t = 2 + 4s$$

solve the first two equations for s, by subtracting, to get

$$-1 = -2 + s$$

giving $s = 1$ and $t = 2$

we now find that the third equation is satisfied by these two numbers:

$$4+2=2+4.$$

Using t=2 in the first vector equation gives the point of intersection as (4,5,6)

==========================

Exercise 5

1. The four vertices of a tetrahedron are A(2,2,2), B(23,8,17), C(8,5,14) and D(11,17,23).

Write down

(i) the centroid of BCD called E
(ii)
(iii) the centroid of ADC called F

Find a vector equation for BF

Find a vector equation for AE

Show that BF intersects AE and find the point of intersection G.

Show that G lies on the line joining D to the centroid of ABC.

(Hint: show that **GD** is parallel to **HG**)

==================================

The Dot Product

If $\underline{\mathbf{a}}$ and $\underline{\mathbf{b}}$ are any two vectors then the dot product $\underline{\mathbf{a}}.\underline{\mathbf{b}}$ is defined to be

$$\underline{\mathbf{a}}.\underline{\mathbf{b}} = a\,b\cos\theta$$

where θ is the rotation from $\underline{\mathbf{a}}$ to $\underline{\mathbf{b}}$

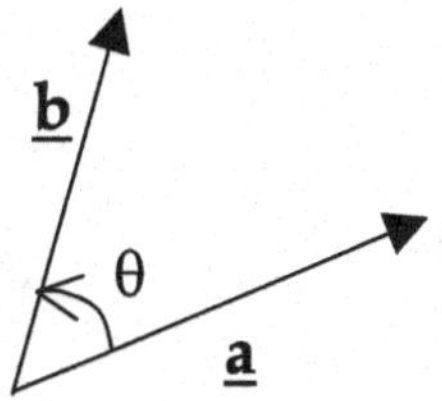

The following diagrams show that the dot product has the same value whether the angle between $\underline{\mathbf{a}}$ and $\underline{\mathbf{b}}$ is measured from $\underline{\mathbf{a}}$ to $\underline{\mathbf{b}}$ or from $\underline{\mathbf{b}}$ to $\underline{\mathbf{a}}$ and has the same value whether we choose to use the reflex angle between $\underline{\mathbf{a}}$ and $\underline{\mathbf{b}}$ or not.

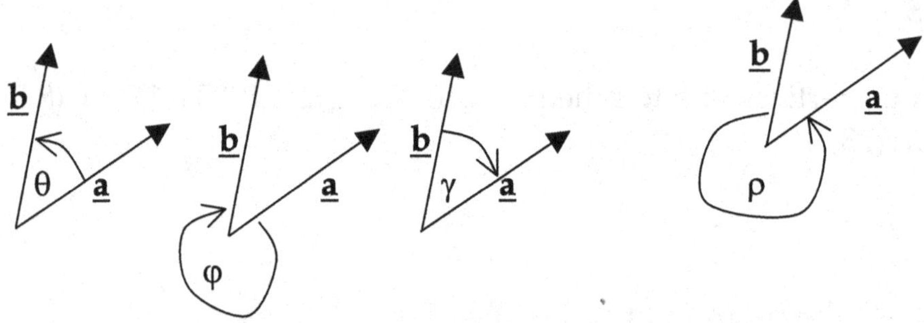

here, $\cos\theta = \cos\phi = \cos\gamma = \cos\rho$

The Commutative Law

$ab.\cos\theta = ba.\cos(360-\theta)$ which is enough to prove $\underline{a}.\underline{b} = \underline{b}.\underline{a}$ in all these cases

The Distributive Law

We can show diagrammatically that the distributive law holds for the dot product over vector addition. We illustrate with one example:

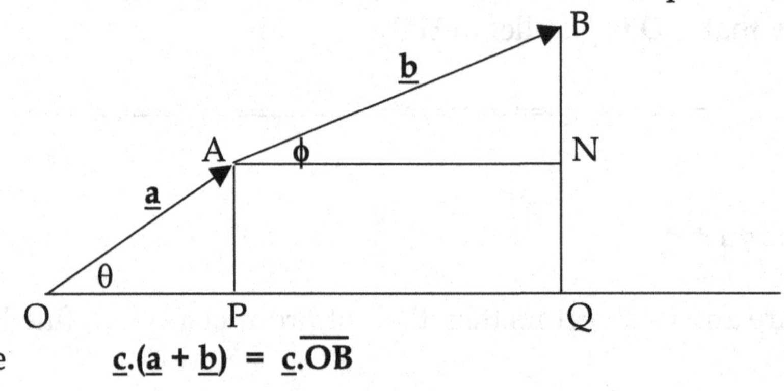

In this figure $\underline{c}.(\underline{a} + \underline{b}) = \underline{c}.\overline{\mathbf{OB}}$

$= c.OB.\cos BOQ$

$= c.OQ$

$= c(OP + OQ)$

$= c.OP + c.AN$

$= c.a \cos\theta \;\; c.b \cos\phi$

$= \underline{c}.\underline{a} + \underline{c}.\underline{b}$ ■

This result allows us to expand brackets (and we remind ourselves that $\underline{\mathbf{a}}.\underline{\mathbf{b}}$ is just a number, so that we can use the Laws of Arithmetic when manipulating quantities like $\underline{\mathbf{a}}.\underline{\mathbf{b}}$, $\underline{\mathbf{c}}.\underline{\mathbf{d}}$ etc), thus, for example:

$(\underline{\mathbf{a}} + \underline{\mathbf{b}})(\underline{\mathbf{c}} + \underline{\mathbf{d}}) = (\underline{\mathbf{a}} + \underline{\mathbf{b}}).\underline{\mathbf{c}} + (\underline{\mathbf{a}} + \underline{\mathbf{b}}).\underline{\mathbf{d}}$	Distributive Law
$= \underline{\mathbf{c}}.(\underline{\mathbf{a}} + \underline{\mathbf{b}}) + \underline{\mathbf{d}}.(\underline{\mathbf{a}} + \underline{\mathbf{b}})$	Commutative Law
$= \underline{\mathbf{c}}.\underline{\mathbf{a}} + \underline{\mathbf{c}}.\underline{\mathbf{b}} + \underline{\mathbf{d}}.\underline{\mathbf{a}} + \underline{\mathbf{d}}.\underline{\mathbf{b}}$	Distributive Law
$= \underline{\mathbf{a}}.\underline{\mathbf{c}} + \underline{\mathbf{b}}.\underline{\mathbf{c}} + \underline{\mathbf{a}}.\underline{\mathbf{c}} + \underline{\mathbf{b}}.\underline{\mathbf{c}}$	Commutative Law

Perpendicular Vectors

Note that if $\underline{\mathbf{a}}$ is perpendicular to $\underline{\mathbf{b}}$ then $\underline{\mathbf{a}}.\underline{\mathbf{b}} = \mathbf{0}$ (because $\cos 90^\circ = 0$)

Parallel Vectors

If $\underline{\mathbf{a}}$ is parallel to $\underline{\mathbf{b}}$ then $\underline{\mathbf{a}}.\underline{\mathbf{b}} = ab.\cos 0 = ab$

In particular $\underline{\mathbf{a}}.\underline{\mathbf{a}} = a^2$

======================================

Pythagoras

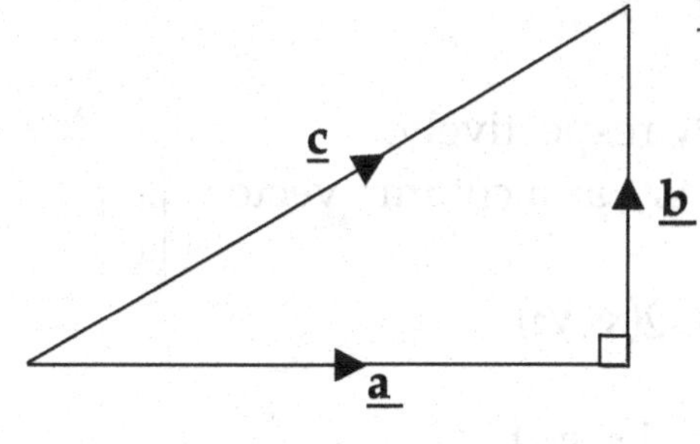

Here, $\underline{\mathbf{a}}$ and $\underline{\mathbf{b}}$ are perpendicular vectors

$$\underline{\mathbf{a}} + \underline{\mathbf{b}} = \underline{\mathbf{c}}$$

$$c^2 = \underline{\mathbf{c}}.\underline{\mathbf{c}}$$

$$= (\underline{\mathbf{a}}+\underline{\mathbf{b}}).(\underline{\mathbf{a}}+\underline{\mathbf{b}})$$

$$= \underline{\mathbf{a}}.\underline{\mathbf{a}}+\underline{\mathbf{b}}.\underline{\mathbf{a}}+\underline{\mathbf{a}}.\underline{\mathbf{b}}+\underline{\mathbf{b}}.\underline{\mathbf{b}}$$

$= a^2 + b^2$ since $\underline{\mathbf{a}}.\underline{\mathbf{b}} = \underline{\mathbf{b}}.\underline{\mathbf{a}} = 0$

$$\therefore a^2 + b^2 = c^2$$

===================================

The Cosine Rule

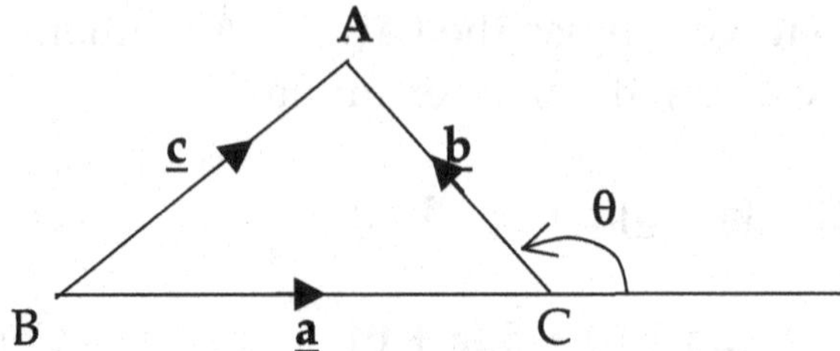

Referring to the above figure:

$$c^2 = \underline{c}.\underline{c} = (\underline{a}+\underline{b}).(\underline{a}+\underline{b})$$

$$= \underline{a}.\underline{a} + \underline{b}.\underline{b} + 2\underline{a}.\underline{b}$$

$$= a^2 + b^2 + 2ab \cos \theta$$

$$= a^2 + b^2 - 2ab \cos C \qquad \text{since } C = 180-\theta$$

==============================

We will denote vectors in **bold** characters and numbers unbolded.

Components in 2D

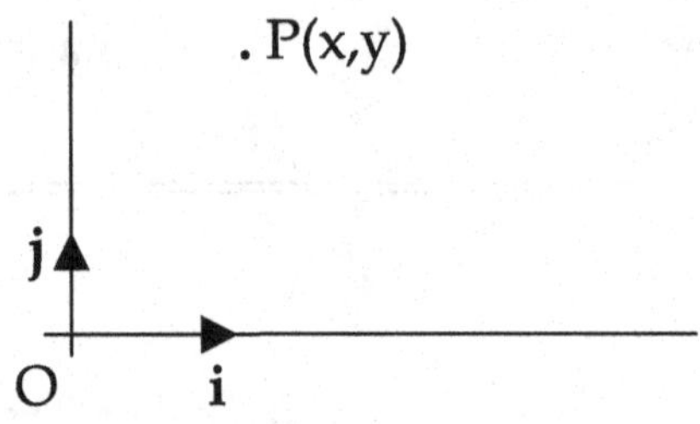

Let **i** and **j** be unit vectors along Ox and Oy respectively.
Let be the point P(x,y) then **OP** = x**i** + y**j** or, as a column vector $\begin{pmatrix} x \\ y \end{pmatrix}$

Suppose that P is the point $P(x_1,y_1)$ and Q is $Q(x_2,y_2)$
Let **OP** = **a** and **OQ** = **b**
Then

$$\mathbf{a.b} = (x_1\mathbf{i} + y_1\mathbf{j}).(x_2\mathbf{i} + y_2\mathbf{j})$$

$$= x_1x_2\, \mathbf{i.i} + x_1y_2\, \mathbf{i.j} + y_1x_2\, \mathbf{j.i} + y_1y_2\, \mathbf{j.j}$$

$$= x_1x_2 + y_1y_2 \qquad \text{since } \mathbf{i.i} = \mathbf{j.j} = 1 \text{ and } \mathbf{i.j} = \mathbf{j.i} = 0$$

Therefore

$$\mathbf{a.b} = x_1x_2 + y_1y_2 \quad ■$$

Components in 3D

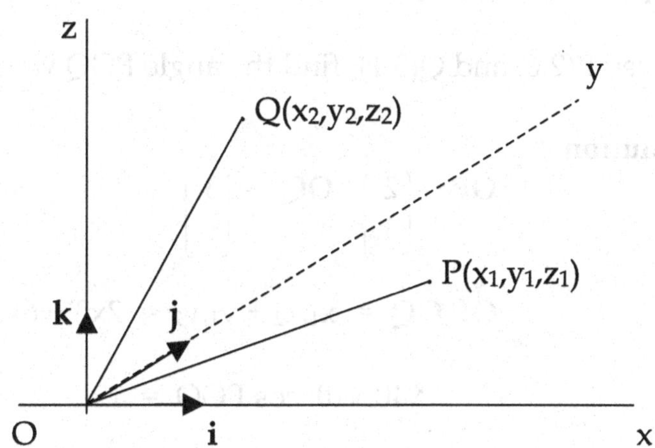

The unit vectors along Ox, Oy and Oz are **i** , **j** and **k**

Let $\quad \mathbf{a} = \mathbf{OP} = x_1\mathbf{i} + y_1\mathbf{j} + z_1\mathbf{k}$

and $\quad \mathbf{b} = \mathbf{OQ} = x_2\mathbf{i} + y_2\,\mathbf{j} + z_2\mathbf{k}$

Since **i** , **j** , and **k** are perpendicular unit vectors we have

$$\mathbf{i}.\mathbf{j} = \mathbf{j}.\mathbf{k} = \mathbf{k}.\mathbf{i} = \mathbf{j}.\mathbf{i} = \mathbf{k}.\mathbf{j} = \mathbf{i}.\mathbf{k} = 0$$

and $\quad \mathbf{i}.\mathbf{i} = \mathbf{j}.\mathbf{j} = \mathbf{k}.\mathbf{k} = 1$

so that $\mathbf{a}.\mathbf{b} = \mathbf{OP}.\mathbf{OQ} = (x_1\mathbf{i} + y_1\mathbf{j} + z_1\mathbf{k}).(x_2\mathbf{i} + y_2\,\mathbf{j} + z_2\mathbf{k})$

and using the distributive law to expand the brackets, with the above results we have

$$\mathbf{a}.\mathbf{b} = x_1.x_2 + y_1.y_2 + z_1.z_2$$

==

Exple 1

Given P(2,6) and Q(3,1), find the angle POQ where O is the origin.

Solution

$$\mathbf{OP} = \begin{pmatrix} 2 \\ 6 \end{pmatrix} \quad \mathbf{OQ} = \begin{pmatrix} 3 \\ 1 \end{pmatrix}$$

$$\mathbf{OP.OQ} = x_1.x_2 + y_1.y_2 = 2x3 + 6x1 = 6+6 = 12$$

$$\sqrt{40}.\sqrt{10} \ \cos POQ = 12$$

$$\cos POQ = \frac{12}{20} = \frac{3}{5}$$

$$\angle POQ = 53.13^{o} \text{ (2d.p.)}$$

======================================

Example 2

Given P(8,4,8), Q(2,4,4) and origin O, find angle POQ

Solution

$$\mathbf{OP.OQ} = 16+16+32 = 64$$

$$\therefore \quad \sqrt{(64+16+64)}.\sqrt{(4+16+16)}.\cos POQ = 64$$

$$\sqrt{(144)}.\sqrt{(36)}.\cos POQ = 64$$

$$72.\cos POQ = 64$$

$$\cos POQ = \frac{8}{9}$$

$$\angle POQ = 27.27^{o}$$

==

Exercise 6

[1] Given P(3,5), Q(6,8) and R(4,5), find ∠P

[2] Given A(2,4,5), B(3,5,-2) and C(6,3,4) find the angles of triangle ABC.

==================================

Compound Angle Formulae

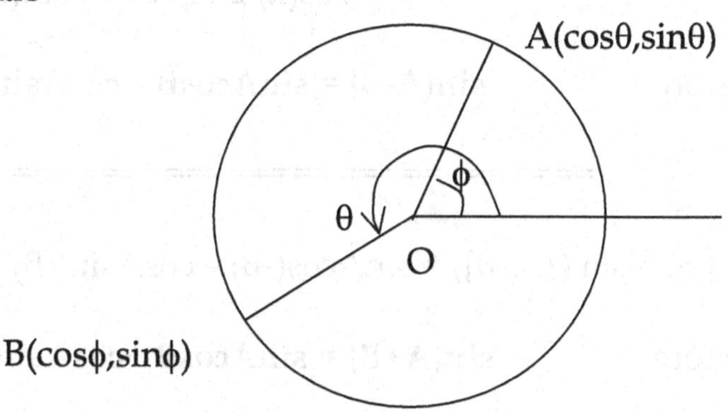

Let A and B be two points on the unit circle at angles θ and ϕ round from the initial line, then their coordinates will be A(cosθ,sinθ) and B(cosϕ,sinϕ).

OA is then the unit vector **OA** = cosθ**i** + sinθ**j**
OB is the unit vector **OB** = cosϕ**i** + sinϕ**j**

Using the dot product, we have

OA.OB = cosθ.cosϕ + sinθ.sinϕ

But **OA.OB** = 1.1.cos(θ-ϕ)

Therefore **cos(θ - ϕ) = cosθ.cosϕ + sinθ.sinϕ**

================================

Thus, in general, for angles of any size, we have

cos(A - B) = cosA.cosB + sinA.sinB

===================================

$\cos(A + B) = \cos(A - (-B)) = \cos A.\cos(-B) + \sin A.\sin(-B)$

giving $\mathbf{\cos(A + B) = \cos A.\cos B - \sin A.\sin B}$

==========================

$\sin(A-B) = \cos(\pi/2 - [A-B]) = \cos([\pi/2-A] + B)$

$= \cos(\pi/2-A)\cos B - \sin(\pi/2-A)\sin B$

Therefore $\mathbf{\sin(A-B) = \sin A\cos B - \cos A\sin B}$

==============================

$\sin(A+B) = \sin(A-[-B]) = \sin A\cos(-B) - \cos A\sin(-B)$

Therefore $\mathbf{\sin(A+B) = \sin A\cos B + \cos A\sin B}$

===============================

These identities are true for angles of any size, positive or negative because the subtraction of the rotations **(θ - ϕ)** does not depend on the angles being in any particular quadrant.

==================================

Answers

Exercise 1 BF = 2$\underline{p}$ DG = 1 ½ $\underline{p}$ EA = -2$\underline{p}$ GB = -2 ½ $\underline{p}$

Exercise 2

1. The mid point of PR and the mid point of QS is $\frac{\underline{a}+\underline{b}+\underline{c}+\underline{d}}{4}$

2. The 3:1 point of division is $\frac{\underline{a}+\underline{b}+\underline{c}+\underline{d}}{4}$ for each of the lines

3. If $\underline{a}' = \frac{\underline{b}+\underline{c}}{2}$, $\underline{b}' = \frac{\underline{c}+\underline{a}}{2}$ and $\underline{c}' = \frac{\underline{a}+\underline{b}}{2}$ then $\frac{\underline{a}'+\underline{b}'+\underline{c}'}{3} = \frac{\underline{a}+\underline{b}+\underline{c}}{3}$

4. $\frac{2\underline{a}-\underline{b}-\underline{c}}{3} + \frac{2\underline{b}-\underline{c}-\underline{a}}{3} + \frac{2\underline{c}-\underline{a}-\underline{b}}{3} = 0$

5. The centroid of each triangle is $\frac{\underline{a}+\underline{b}+\underline{c}+\underline{d}+\underline{e}+\underline{f}}{6}$

6. The centroid of PQR = $\left(\frac{n\underline{a}+m\underline{b}}{m+n} + \frac{n\underline{b}+m\underline{c}}{m+n} + \frac{n\underline{c}+m\underline{a}}{m+n}\right) \Big/ 3 = \frac{\underline{a}+\underline{b}+\underline{c}}{3}$

7. If the tetrahedron is ABCD and the mid points of AB and CD are P and Q then

$$\underline{p} = \frac{\underline{a}+\underline{b}}{2} \text{ and } \underline{q} = \frac{\underline{c}+\underline{d}}{2} \text{ the mid point of PQ is then } \frac{\underline{a}+\underline{b}+\underline{c}+\underline{d}}{4}$$

We will arrive at the same point if we use AC with BD or AD with BC.

Exercise 3 1. OQ = $\begin{pmatrix} 5 \\ 4 \end{pmatrix}$

2. AB = $\begin{pmatrix} 2 \\ 3 \end{pmatrix}$ BC = $\begin{pmatrix} 5 \\ 1 \end{pmatrix}$ CD = $\begin{pmatrix} 3 \\ -2 \end{pmatrix}$ DE = $\begin{pmatrix} -2 \\ -2 \end{pmatrix}$ AE = $\begin{pmatrix} 8 \\ 0 \end{pmatrix}$

Exercise 4

[1] $\mathbf{r} = \begin{pmatrix} 3 \\ 4 \end{pmatrix} + t \begin{pmatrix} 4 \\ 2 \end{pmatrix}$ 2y = x+5; **[2]** (10,13)

Exercise 5

Ans G(11, 8, 14)

Exercise 6

[1] 45

[2] 70.72, 34.90, 74.37 (correct to 2d.p.)

==

CHAPTER 25

Numbers and Complex Numbers

Over the centuries, priests, astronomers and mathematicians have developed number systems designed to provide answers to problems and solutions to equations. Different kinds of number systems were devised to solve different kinds of problems. The simplest of these number systems is the set of natural numbers with two laws of composition, addition and multiplication. The Natural numbers are the counting numbers that answer question like "How many do I have?". The rational numbers or fractions answer questions asking "How much do I have?" for example if you are sharing a cake between a number of people. The negative numbers can answer the question "Where am I?" when we climb up or down a staircase. Number systems have been extended by mathematicians over time, to cope with different kinds of problem, however, the laws of arithmetic that hold for the natural number system have been preserved in all of the extensions that have been made to these different number systems. Indeed, it is the desire to preserve the laws of arithmetic that determines the rules of combination for new numbers that make up an extension to a given number system. In an extension to any number system we like our numbers to obey the same rules of arithmetic that hold in the natural number system. We give here a brief survey of the different number systems that have evolved, leading up to what are called the "complex" numbers and illustrate the laws of arithmetic with simple examples as we go. The complex number system can be regarded as a final point in the journey in the sense that any nth degree polynomial equation with complex number coefficients will have exactly n solutions in the complex number system.

N: The Natural Numbers

The first numbers that we encounter in our school careers are the "counting numbers", designed to answer questions asking "How many?". The set of counting numbers or **Natural numbers** is called **N.** The dictionary defines the set of Natural numbers as the integers **1, 2, 3, 4, ...** but in some areas of study, for example in Formal Methods in computing, the set of Natural numbers includes zero, and you may find, in some books **N = {0, 1, 2, 3, 4, ...}** and the set $\mathbf{N_1}$ defined as $\mathbf{N_1 = \{1,2,3,4,.....\}}$, but we will define our set of

Natural Numbers to be the set of positive whole numbers **N = { 1, 2, 3, 4, 5, ……}**

On the set **N = { 1, 2, 3, 4, 5, ……}** the operations of addition and multiplication are well defined. If **a** and **b** are two numbers in the set of natural numbers **N,** then we can always find **a +b** and **a x b** in the set **N. Addition** tells us how many we have if two piles of objects are combined together. **Multiplication** tells us how many we have if there are a number of piles with the same number of objects in each pile.

The Rules of Arithmetic

Example 1. Carol has **m** marbles and Joe has **n** marbles. They put their marbles into a single pile. How many marbles are there in the pile?

Starting at number **m** in the set of Natural numbers, count **n** places to the right and we have the answer **m+n.**

Addition can be taken as a move to the right on the line of natural numbers. To perform the addition **4 + 5** start at **4** in the line of natural numbers and move **5** places to the right:

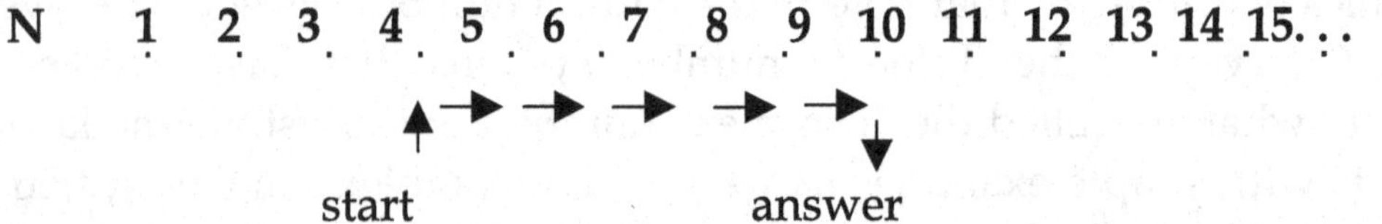

Clearly, there are **m+n** marbles in the pile but if we started at number **n** in the set of natural numbers and counted on **m** places, we would arrive at the same place. It does not matter if Carol puts her marbles in first, or whether Joe puts his marbles in first, we end up with the same total in the pile. This illustrates:

The Commutative rule for addition: m+n = n+m

Example 2. I have a rectangle that is **m** dots long and **n** dots wide. How many dots are there altogether?

8x4 dots, or 4x8 dots

To count the rectangle of dots, start at **1** in the set of natural numbers, count along **4** numbers and do this **8** times. You will arrive at the numbers **4, 8, 12, 16, 20** ….. finishing at the number **32.**

Alternatively you could start at **1** and count along **8** numbers, doing this **4** times and in this case you will go through the numbers **8, 16, 24, 32.**

m columns of **n dots** each, gives **mxn** dots altogether.

Alternatively, **n** rows of **m** dots each gives the answer **nxm.** The order of multiplication does not affect the result. **mxn = nxm.**

This example illustrates:

The Commutative law for multiplication: **mxn = nxm**

Example 3. Carol has **3** marbles, Joe has **4** marbles and Tash has **5** marbles. How many marbles do they have altogether?

Addition has only been discussed for combining **two** numbers, so we have to decide which two piles of marbles to add together first. We could either say

$$3 + (4+5) = 3 + 9 = 12$$

or

$$(3+4) + 5 = 7 + 5 = 12$$

either way, we end up with the same total, so, we could just write **3+4+5** and not bother about the brackets.

This illustrates

The associative Law for Addition; **m + (n+p) = (m+n) + p**

Example 4. I have a **3 by 4 by 5** block of cubes. How many cubes are there altogether?

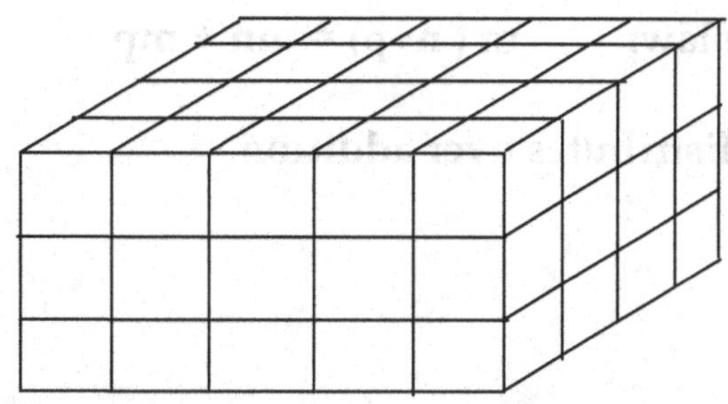

Answer:

I could calculate **3** layers with **(4x5)** blocks in each layer

$$3 \times (4 \times 5) = 3 \times 20 = 60 \text{ blocks}$$

or, I could calculate the number of blocks **(3x4)** wall total **5** times

$$(3 \times 4) \times 5 = 12 \times 5 = 60 \text{ blocks}$$

Thus we could omit the brackets and just write **3x4x5,** since we get the same result, whichever multiplication we do first.

This example illustrates:

The associative law for multiplication: m x (nxp) = (mxn) x p

Example 5. What is the area of this figure?

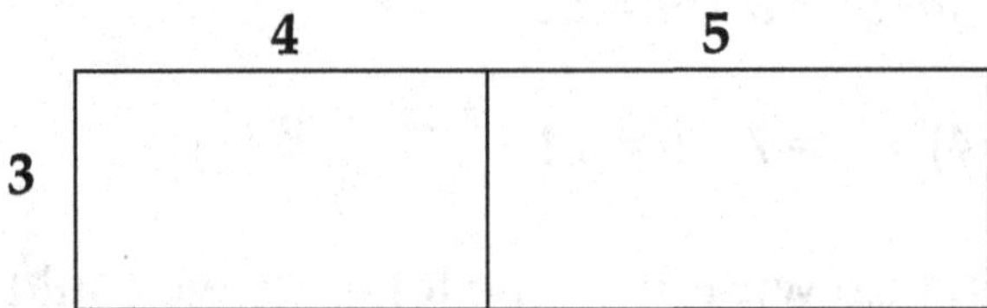

Answer

$$\text{Area} = 3 \times (4 + 5) = 3 \times 9 = 27$$

We also note that the sum of the two parts is **3x4 + 3x5 = 12 + 15 = 27**

$$3 \times (4 + 5) = 3 \times 4 + 3 \times 5$$

This example illustrates:

The distributive law: m (n+p) = mn + mp

Multiplication distributes over addition.

Example 6

Fred has **m** marbles and he gives **n** marbles to Jill. How many marbles does Fred have left?

To answer this question, we invent the operation of subtraction. We start at the number **m** in the line of natural numbers and move **n** places to the left: If Fred has 9 marbles and gives 5 to Jill, then we could illustrate the calculation as follows:

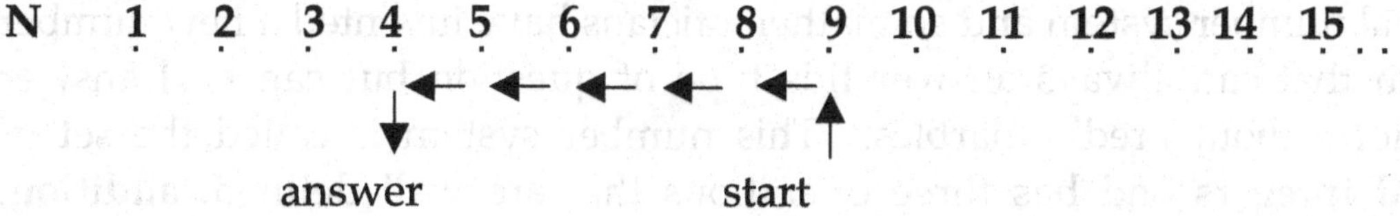

We write the answer, of course, as **m – n = 9 - 5 .**

However, it could be that Fred gives all of his marbles to Jill, in which case, we cannot answer the question using the Natural Numbers **N.** To include this case, we would have to extend our number system to include the number **zero.**

Using the set of numbers **{0, 1, 2, 3, 4, 5, 6, 7, 8, ... }** we can always find an answer to **Example 6** provided that **n** is not greater than **m.**

Zero

The number zero is like a "don't move" operation on the Natural number line so that **m+0** can be read as "start at **m** and move no places". Thus we have the first zero rule:

(i) **m + 0 = 0 + m = m**

Just as for natural numbers, we have, for example, **3x2 = 2+2+2** we would also want **3x0 = 0+0+0 = 0** (note the use of the associative rule)
Thus, we have the second zero rule and define

(ii) **mx0 = 0xm = 0**

Z: The Signed Integers

Example 7
You are in a lift. The lift goes up 3 floors and then goes down 5 floors. Where are you now?

Answer: Any schoolchild would know that going up three and then coming down five is the same as going down 2. So, the answer is, two floors down.

To answer questions like this, we cannot always find an answer in the Natural number system and so mathematicians have invented a new number system that can always answer this type of question but can **still** answer questions about Fred's marbles. This number system is called the set of **signed integers** and has three operations that are well defined: **addition, multiplication and subtraction.** The set of signed integers is called **Z** and we write

$$\mathbf{Z} = \{\ldots\ldots -4, -3, -2, -1, 0, +1, +2, +3, +4, \ldots\ldots\}$$

The **positive integers +1, +2, +3, ...** behave exactly as the natural numbers behave in **N**, but we also have a new type of number here called **negative integers: -1, -2, -3, -4 . . . . .**
We need to define how these negative integers are combined when using the operations of addition, multiplication and subtraction. To do this, we make sure that the **commutative, associative** and **distributive laws** still hold for all of the numbers in **Z.**

Rules of Arithmetic in Z

Addition of signed integers
Using the natural numbers **m +n** was defined by starting at **m** on the line of natural numbers and moving **n** places to the right.
Using the signed integers, we can still move right or left. If **n** is negative, then we find **m +n** by moving to the left. If **n** is positive then we move to the right. This is best illustrated using examples:

(+3) + (+4) start at +3 and move 4 places to the right: answer +7

(+3) + (-4) start at +3 and move 4 places to the left: answer –1

(-3) + (+4) start at –3 and move 4 places to the right: answer +1

(-3) + (-4) start at –3 and move 4 places to the left: answer –7

Zero is dealt with as follows:

0 + (+4) start at 0 and move 4 places right **0+(+4) = +4**

0 + (-4) start at 0 and move 4 places left **0+(-4) = –4**

(+4) + 0 start at +4 and don't move: **(+4)+0 = +4**

(-4) + 0 start at –4 and don't move: **(-4)+0 = -4**

Rules for Zero

In general for any signed integer **n** we have

Rule 1 for zero **n + 0 = 0 + n = n**

In the natural number system, we have, for example, 3x2 = 2+2+2, so that we write, for example, 3x0 = 0+0+0 = 0.
In general, for any signed integer **n,** we have

Rule 2 for zero **nx0 = 0xn = 0**

It is not difficult, though a little tedious, to show that the commutative and associative laws are obeyed for these definitions of addition in the system of signed integers.

The negative of a number

(+2) + (-2) can be interpreted as, start at (+2) and move 2 places left, arriving at 0 so that (+2)+(-2) = 0. Similarly, starting at (-2) on the number line and moving 2 places right also brings us to 0, so that (-2)+(+2) = 0.
In general for any natural number **m,** we have in the system of signed integers,

(+m) + (-m) = (-m) + (+m) = 0

When two numbers add up to zero, we say that one is **the negative** of the other, thus (+2) + (-2) = 0

hence, (-2) is the negative of (+2) and (+2) is the negative of (-2).
In general, since **(+m) + (-m) = (-m) + (+m) = 0**
(-m) is the negative of **(+m)**

and **(+m)** is the negative of **(-m)**

Subtraction

Subtraction is defined as "adding the negative"

Thus (+2) – (+3) = (+2) + (-3) = -1

(-2) – (+3) = (-2) + (-3) = -5

(+2) – (-3) = (+2) + (+3) = +5

(-2) – (-3) = (-2) + (+3) = +1

subtraction does not obey the commutative rule:

(-4) – (-3) = (-4) + (+3) = -1

(-3) – (-4) = (-3) + (+4) = +1

Multiplication

The product of two positive numbers is defined to match the corresponding product for the two corresponding natural numbers, thus, for example, since 3x2 = 6, we have
(+3) x (+2) = +6.

The products for other cases are defined so that the distributive law will hold. We illustrate with numerical examples:

Using the rules for negatives and zero we have, for example:

(+3) x { (+2) + (-2) } = (+3) x 0 = 0

Applying the distributive law, this gives

$$(+3) \times (+2) + (+3) \times (-2) = 0$$

hence $+6 + (+3) \times (-2) = 0$

therefore **(+3) x (-2)** must be the negative of 6, i.e. **(+3) x (-2) = -6**

requiring the commutative law to hold now gives

$$\mathbf{(-2) \times (+3) = -6}$$

Again, for example, $(-2) \times ((+3) + (-3)) = (-2)\times 0 = 0$

So that $(-2)\times(+3) + (-2)\times(-3) = 0$

$$-6 + (-2)\times(-3) = 0$$

hence **(-2) x (-3)** must be the negative of –6 i.e. **(-2) x (-3) = +6**

We can summarize these results as:

$$\mathbf{(+m) \times (+n) = +(mn)}$$

$$\mathbf{(-m) \times (+n) = -(mn)}$$

$$\mathbf{(+m) \times (-n) = -(mn)}$$

$$\mathbf{(-m) \times (-n) = +(mn)}$$

where **m** and **n** are symbols for natural numbers (or zero) and **mn** is the product of these two natural numbers (or zero).

The unit

m x n gives the total of " **m** rows of **n** dots " hence, **1 x n** gives the total of **1** row of **n** dots. **1** is a "unit" for multiplication in that **nx1 = 1xn = n.**

The Laws of arithmetic

We could now check that the following laws hold for the system of signed integers; if **a,b** and c are any signed integers then

Commutative rules $\quad \mathbf{a + b = b + a}$

$\mathbf{a \times b = b \times a}$

associative rules $\quad \mathbf{a + (b + c) = (a + b) + c}$

$\mathbf{a \times (b \times c) = (a \times b) \times c}$

distributive rule $\quad \mathbf{a \times (b + c) = a \times b + a \times c}$

rules for zero $\quad \mathbf{a + 0 = 0 + a = 0, \quad a \times 0 = 0 \times a = 0}$

the unit $\quad \mathbf{a \times 1 = 1 \times a = a}$

negatives exist $\quad \mathbf{a + (-a) = 0}$

subtraction is defined: **a – b** means **a + (-b)**

The Natural Numbers and Z

We note that the set of natural numbers **N={1, 2, 3, 4, . . . }** with the operations of addition and multiplication behaves exactly as the set of positive integers **Z^+={+1, +2, +3, +4, . . . }** under **+** and **x.**
We can say that **(N, +, x)** is isomorphic to **(Z^+, +, x).**

Some books state that the Natural numbers are "embedded" in the system of signed integers; some writers say that the natural number system is "extended" to the system of signed integers. However, the two numbers systems are distinct and strictly, we should not regard a natural number as a special kind of signed integer. The problem solver should adopt the number system that is appropriate to the problem. Thus, when Fred gives his marbles away, we should use **N,** although if Fred owes Jill two marbles I'm sure that many teachers would be happy to say that Fred now has minus two marbles. If Jill jumps into a lift, use **Z.**

Whichever view is chosen the fact is that we all use the symbol 2 when we should use +2. A further complication is provided by the ambiguous use of the + and – signs.
The simple statement **-3 + 5 = +2**
can be read as "the sum of the integers –3 and +5 is +2"
or as "going down 3 then up 5 is the same as going up 2"
The **+ and –** signs have a double meaning: they can either refer to an **action,** i.e. add or subtract or move left or right, or they could refer to a **position** on the number line.
If we go back to the lift, we could invent a different number system that avoids the ambiguity:

Let + mean the action go up
Let – mean the action go down
Let **U** indicate positions above ground floor
Let **D** indicate positions below ground floor
Ground floor is position **0**

Then we could write **+2 – 5 = – 3** meaning "going up 2 and then down 5 is the same as going down 3"

Whereas **U2 – 5 = D3** means, "start at **U2,** go down 5 to **D3"**

For example, **D2 + 3 – 5**

Could be evaluated as **(D2 + 3) – 5 = U1 – 5 = D4**

or **D2 + (3 – 5) = D2 – 2 = D4**

This number system is sufficient to answer problems about going up and down in the lift, but note, does it make any sense to state that **U3 x D2 = D6 ?** Clearly not. The justification for statements like **+3 x –2 = -6** comes from the laws that we want our numbers to obey, that is, the commutative, associative and distributive laws that hold for the natural numbers.

+, - and =

In our system of signed integers, the + and – signs have two ambiguous meanings. With reference to our line of signed numbers, the + or - signs can either indicate a position on the number line, or they could indicate a move to the left (-) or a move to the right (+). Some books distinguish between these two meanings by the use of a superscript, so that;

$^{+}3$ and $^{-}2$ are positions

while $+3$ and -2 are actions

The right hand side of the an equation could mean, either an action or a position, so that we could either interpret = to mean "is the same as" or to mean "lands you up at".

Thus, for example,

$$^{+}5 - 3 = 2 \text{ could mean}$$

"start at 5 floors up and go down 3 floors and you end up at the second floor"

whereas

$$+5 - 3 = 2 \text{ could mean}$$

"going up 5 floors and then coming down 3 floors is the same as going up 2 floors"

Going up and down in the lift is a problem area that can be solved using arithmetic in Z, however, we should not let the problem define the arithmetic. $+5 - (-3) = +8$ because of the laws of arithmetic in Z **not** because "going down minus 3 floors is the same as going up 3 floors".

Q: The Rational Numbers

Sharing Numbers

The rational numbers are concerned with sharing or dividing something into equal parts. We write

$\frac{\mathbf{a}}{\mathbf{b}}$ ← number to be shared

← number of parts into which **a** is divided

Example 1:
I have one cake and share it equally between Fred, Jill and Tash. How much cake does each person have?

Answer:
Of course, we would say that each person has one third of the cake but we need a new kind of sharing number to describe this amount of cake.
If one cake is shared equally between 3 persons, we represent each share as

$\underline{1}$ .... one cake
3.......... shared between three people

Which we could represent diagrammatically:

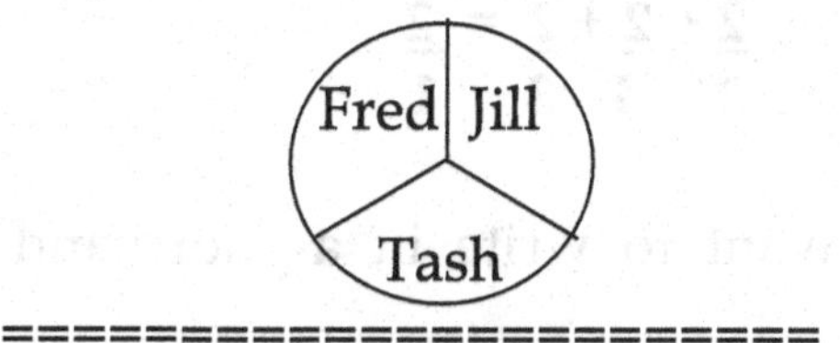

========================

The Rules of Arithmetic in Q

Example 2:
2 cakes are to be shared equally between three friends Fred, Jill and Tash. How much cake does Fred get?

Answer:
The equal shares are represented by

$\underline{2}$ ⟵ number of cakes
3 ⟵ number of shares

thus Fred, Jill and Tash each get a $\frac{2}{3}$ amount of cake.

This gives Fred twice the amount of cake that he had in example 1.

Thus, we would like our "sharing numbers" to obey a rule something like

$$\frac{1}{3}+\frac{1}{3} = 2\times\frac{1}{3} = \frac{2}{3}$$

or, combining the three shares together, we would naturally want

$$\frac{2}{3}+\frac{2}{3}+\frac{2}{3} = 2$$

but, our use of the number 2 here is not strictly correct; we should say, "the same share as two cakes shared by one person" i.e.

$$\frac{2}{3}+\frac{2}{3}+\frac{2}{3} = \frac{2}{1}$$

We would naturally want to write in a shorthand to correspond to the natural numbers:

$$3\times\frac{2}{3} = \frac{2}{1}$$

or, again, more strictly

$$\frac{3}{1}\times\frac{2}{3} = \frac{2}{1}$$

Example 3: Three cakes shared by Fred alone gives Fred a share $\frac{3}{1}$

Four cakes shared by Fred alone gives Fred a share $\frac{4}{1}$

How much cake does Fred now have?

Answer: Clearly, Fred has seven cakes (all shared by Fred alone). Thus, we want our sharing numbers to obey

$$\frac{3}{1}+\frac{4}{1}=\frac{7}{1}$$

If we compare this to **3 + 4 = 7** in the system of natural numbers, we see that the system of Natural numbers **N** is "embedded" in **Q**, being represented by the rational numbers $\frac{1}{1}\ \frac{2}{1}\ \frac{3}{1}\ \frac{4}{1}\ \frac{5}{1}$ ......

Example 5: Zero cakes are shared between **n** people.

Whatever the value of **n**, each person gets no cake! We just write **0** for a share of no cake, thus we want

$$\frac{0}{n} = 0 \quad \text{for any value of } n.$$

Example 6: (simple additions)
2 cakes are shared by Fred, Jill and Tash.

Fred has a **2/3** share

Fred has a 2/3 share but later on they are given four more cakes to share out:

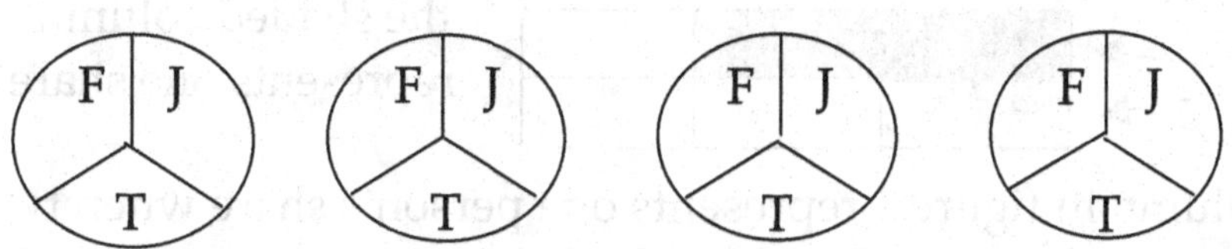

Fred now has a further **4/3** share

Answer: Fred now has the share of six cakes when divided between three friends, thus we want our sharing numbers to obey a rule like:

$$\frac{2}{3} + \frac{4}{3} = \frac{6}{3}$$

In the next few examples, we change the shape of our cakes...purely for illustrative purposes. Our cakes are now rectangular so the illustration of example 6 would now look like this:

		Fred	Jill	Tash
two	cake 1 →	■		
cakes	cake 2 →	■		

Fred's share is $\frac{2}{3}$

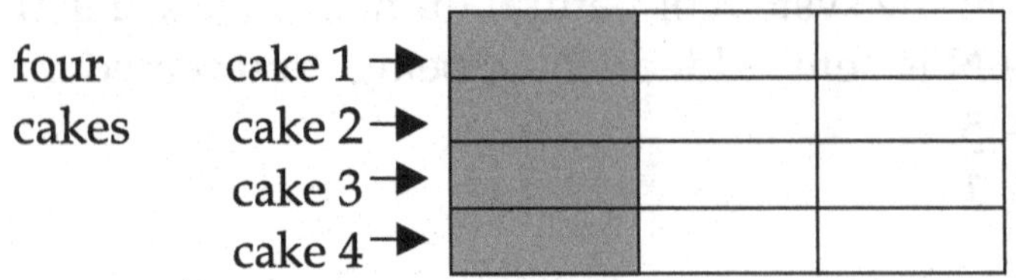

Fred's share of the four cakes is $\frac{4}{3}$

Fred's share of all six cakes is **6/3** which means, that our rational numbers should obey a rule that gives $\frac{2}{3} + \frac{4}{3} = \frac{6}{3}$

Thus we define Addition Rule 1 $\frac{a}{n} + \frac{b}{n} = \frac{a+b}{n}$

The cancelling rule: We consider: 2 cakes shared between 3 persons
4 cakes shared between 6 persons
6 cakes shared between 9 persons

[1] Two cakes shared between three persons

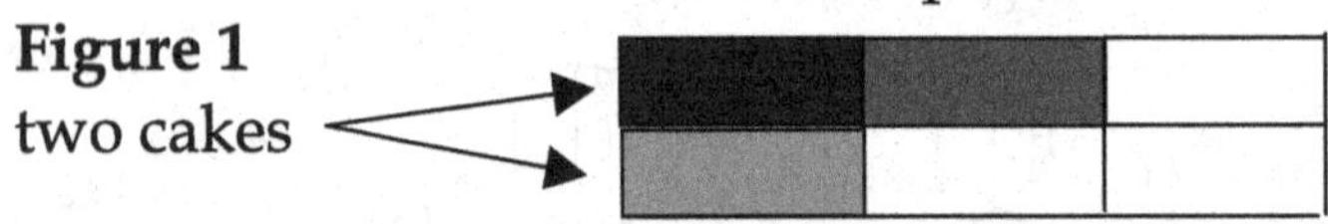

Figure 1

the shaded column represents one share of $\frac{2}{3}$

The shaded column in figure 1 represents one person's share when two cakes are shared between three persons. The shaded row represents two shares of one cake that has been shared between three persons. We would say that either of these shaded areas represents two thirds of a cake.

[2] Four cakes shared between six persons

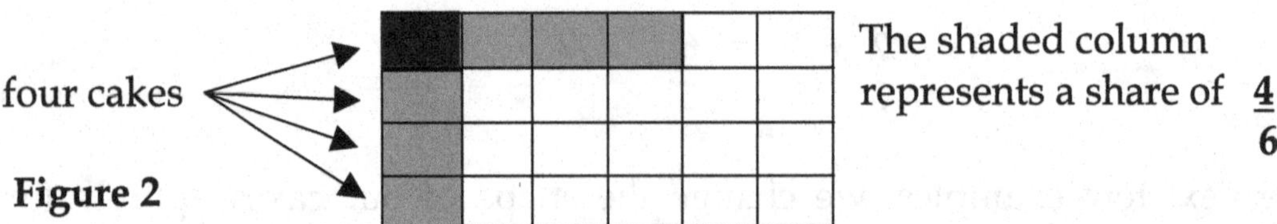

Figure 2

The shaded column represents a share of $\frac{4}{6}$

The shaded column in figure 2 represents one person's share when four cakes are shared between six persons. The shaded row represents four shares of one cake that has been divided equally between six persons. We would say that either of these shaded areas represents four sixths of a cake. We also see, when we compare fig. 1 with fig. 2, that the two rows, in fig.1 and fig.2 represent the same share of cake.

Thus, $\frac{4}{6} = \frac{2}{3}$

[3] Six cakes shared between nine persons

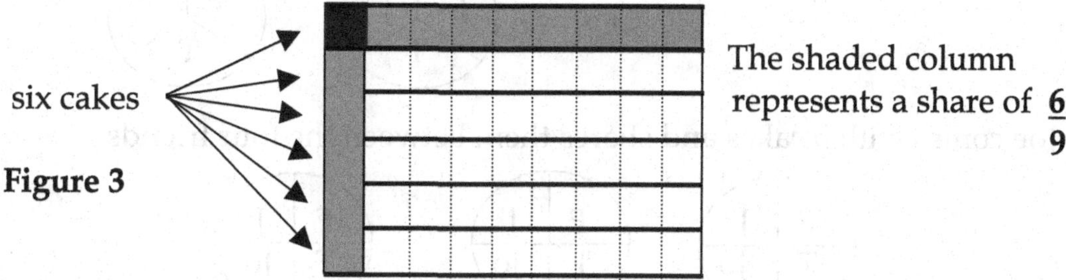
six cakes
The shaded column represents a share of 6/9
Figure 3

The shaded column in figure 3 represents one persons share when six cakes are shared between nine persons. The shaded row represents six portions of the share of one person when a single cake is divided between nine persons. We would say that either of these shaded areas represents six ninths of a cake.
In each of these figure, the shaded column is equal to the shaded row but the shaded rows in each of the figures are equal. Therefore, the shaded columns in each figure are equal. Thus, if two cakes are divided between 3 people, four cakes are divided between six people, or six cakes are divided between 9 people, in each case, the share of one person is the same:

$$\frac{2}{3} = \frac{4}{6} = \frac{6}{9}$$

More generally, if 2 cakes are shared between 3 persons, and we multiply the numbers of cakes by **n** but also multiply the number of persons by **n** then each person will still get the same share of cake.

$$\frac{2n}{3n} = \frac{2}{3}$$

this illustrates:

The cancelling Rule

$$\frac{a.n}{b.n} = \frac{a}{b}$$

Addition of rational numbers

Example 7

Fred, Jill and Tash share two cakes

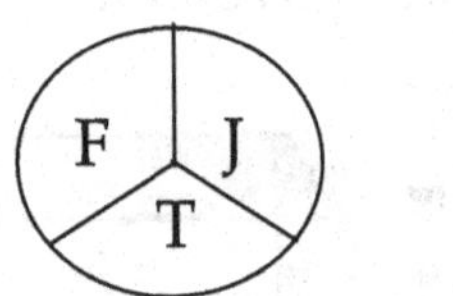

Now Joe comes with 3 cakes and shares them between the four friends

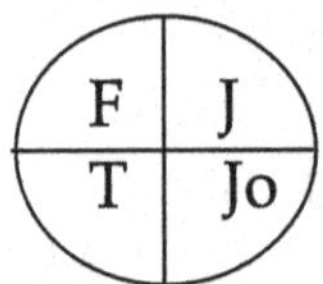

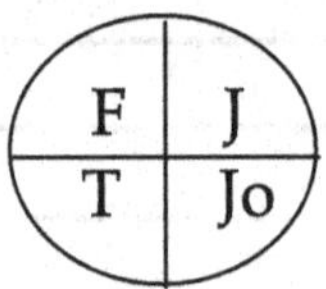

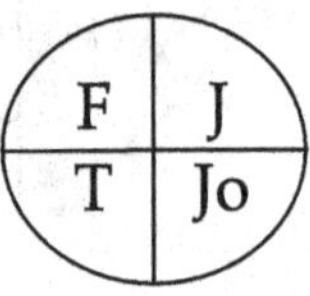

How much cake does Fred now have? i.e. what is $\frac{2}{3} + \frac{3}{4}$

Answer Fortunately, the cancelling rule provides us with a quick solution:

For, $\frac{2}{3} = \frac{2x4}{3x4}$ 2 cakes shared between 3 people gives the same share as 8 cakes shared between 12 people

and $\frac{3}{4} = \frac{3x3}{4x3}$ 3 cakes shared between 4 people is the same share as 9 cakes shared between 12 people

Hence $\frac{2}{3} + \frac{3}{4} = \frac{2x4}{3x4} + \frac{3x3}{4x3} = \frac{2x4 + 3x3}{3x4}$ **(by addition rule 1)**

Fred has $\frac{17}{12}$ share of cake

In general we have $\frac{a}{b} + \frac{c}{d} = \frac{ad}{bd} + \frac{cb}{db} = \frac{ad + bc}{bd}$

The Addition Rule

$$\frac{a}{b} + \frac{c}{d} = \frac{ad + bc}{bd}$$

Multiplication

So far, we have emphasized that **2/3** is the share that each person gets when two cakes are shared between three persons. Thus, if Fred, Jill and Tash share two cakes equally between themselves, then Fred's share will be represented by the rational number **2/3.**
If, however, one cake is shared equally between Fred, Jill and Tash and then Fred pinches Jill's piece, then Fred will have **2** portions of **1/3** of the cake:

Case 1: Fred Jill and Tash share two cakes

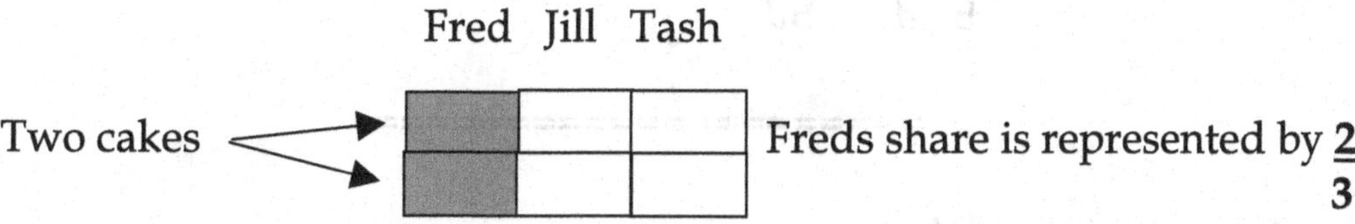

Case 2: Fred, Jill and Tash share one cake but Fred pinches Jill's share

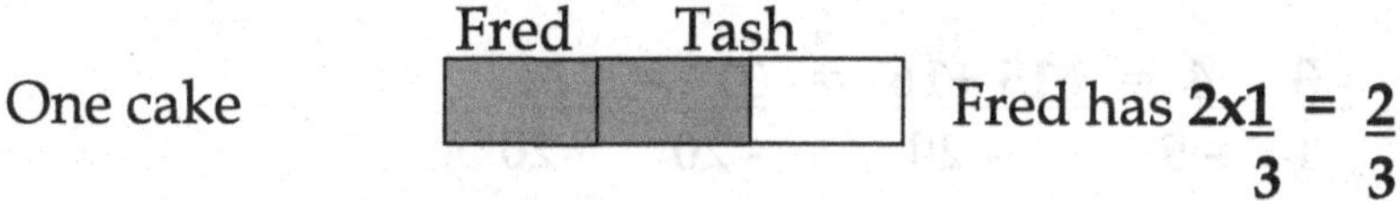

In case 2, we interpret **2/3** to mean "divide into three parts and take two of those parts ".
This is how we can give a meaning to expressions like $\frac{2}{3} \times \frac{4}{5}$

We divide the **4/5** share into **3** parts and take **2** of those parts. This is illustrated in the following diagram:

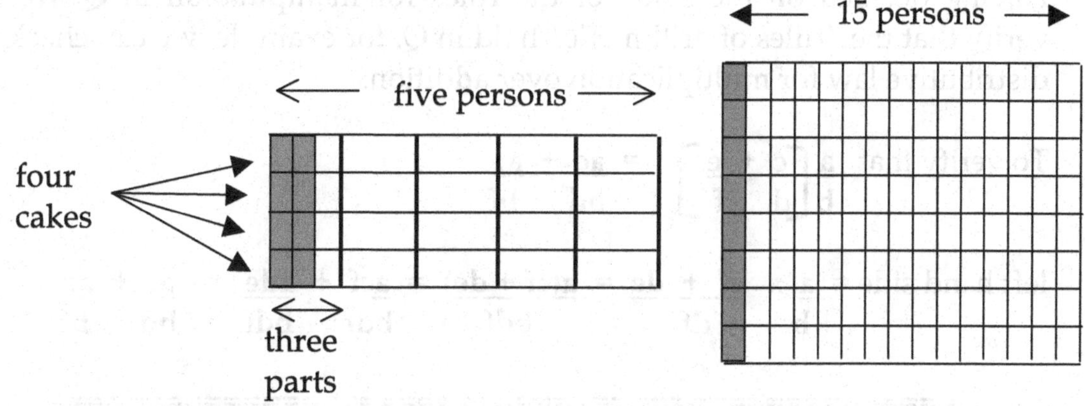

The **4/5** share is divided into three parts and we take two of those parts. This is the same as sharing 2x4 = 8 cakes between 3x5 = 15 persons, thus we want

$$\frac{2}{3}\frac{x4}{5} = \frac{2x4}{3x5} = \frac{8}{15}$$

thus we have:

The Multiplication rule

$$\frac{a}{b} \times \frac{c}{d} = \frac{ac}{bd}$$

====================

Negative rational numbers

For the negative rationals, we carry over the rules for signed integers so that

$$\frac{-3}{4} + \frac{4}{-5} = \frac{+15 + 16}{-20} = \frac{31}{-20} = \frac{-31}{20}$$

$$\frac{-3}{4} \times \frac{4}{-5} = \frac{-3 \times 4}{4 \times -5} = \frac{-3}{-5} = \frac{3}{5}$$

Note: $\frac{31}{-20} = \frac{31 \times (-1)}{-20 \times (-1)} = \frac{-31}{+20}$ and $\frac{-3}{-5} = \frac{3 \times (-1)}{5 \times (-1)} = \frac{3}{5}$ using the cancelling rule with (-1)

Having decided on the some of the rules for manipulation in Q, we can verify that the "rules of arithmetic" hold in Q, for example, we can check the distributive law for multiplication over addition:

To verify that $\frac{a}{b}\left[\frac{c}{d} + \frac{e}{f}\right] = \frac{ac}{bd} + \frac{ae}{bf}$

left hand side = $\frac{a}{b} \times \frac{cf + de}{df} = \frac{a(cf + de)}{bdf} = \frac{acf}{bdf} + \frac{ade}{bdf} = \frac{ac}{bd} + \frac{ae}{bf}$

==

The number systems N, Z and Q

We have briefly looked at the following number systems:

N .. the natural numbers .. **{1, 2, 3, …}** with addition and multiplication.
In **N**, we cannot solve a simple linear equation such as $x + b = c$ unless $c>b$

Z .. the signed integers .. { . . -2, -1, 0, +1, +2, . . } with addition, multiplication and subtraction. In **Z**, we an always solve $x + b = c$.

The solution of $x+b = c$ in **Z** is $x = c-b$,
because $(c-b)+b = c+(-b+b) = c+0 = c$

Q .. the rational numbers .. are numbers represented by **a/b** where **a** and **b** are signed integers (but **b** cannot be zero, because we want **a/b** x **b** to give **a** and also **a/b x 0** to give **0**). In **Q** we have addition, subtraction, multiplication and division (not by zero).

In **Q** we can solve $ax + b = c$. $(a \neq 0)$

The solution of $ax+b = c$ in **Q** is $x = \frac{(c-b)}{a}$

because $a.\frac{(c-b)}{a} + b = (c-b) + b = c +(-b + b) = c + 0 = c$

using the **cancelling rule, associative rule, properties of the negative and zero.**

================================

Every generalization of a number system has at first presented itself as needed for the description of a simple problem, however, these extensions of number systems are not created merely by the need to solve problems. They are created by their definition.

A more formal definition of the rational number system regards a rational number to be an ordered pair of numbers from **Z**:

A Formal Construction of Q

Definition
A rational number is an ordered pair of signed integers written **[a,b]** (**b**≠0)

The rules for arithmetic in **Q** are

Equality **[a,b] = [x,y]**

means **x=a.n** and **y=b.n** or **a=x.n** and **b=y.n**
for some signed integer **n.**

Addition rule
[a,b] + [c,d] = [ad+bc, bd]

Multiplication rule
[a,b] x [c,d] = [ac, bd]

Cancellation rule
[an,bn] = [a,b]

From these rules, we can verify that the laws of arithmetic hold in **Q**, for example, the distributive rule can be verified as follows:

[a,b]{ [c,d] + [e,f] } = [a,b] { [cf + de, df] }

= [a(cf+de) , bdf]

whereas

[a,b] x [c,d] + [a,b] x [e,f] = [ac,bd] + [ae,bf]

= [acbf + bdae , bdbf]

= [acf + dae, dbf] cancelling **b**

= [a(cf+de), bdf] as required

The zero element in **Q** is **[0 , N]** where **N** is any signed integer.

The unit element in **Q** is **[1, 1]**

since **[a,b] x [1,1] = [1,1] x [a,b] = [a,b]**

The inverse, or reciprocal of **[a,b]** is then **[b,a]**

since **[a,b] x [b,a] = [ab.ab] = [1,1]** providing **a≠0** and **b≠0**

Division is then defined as multiplying by the reciprocal

[a,b] ÷ [c,d] = [a,b] x [d,c] = [ad,bc] **(b≠0, c≠0, d≠0)**

==============================

Embedding Z in Q

We can easily find an algebraic structure in **Q** that is isomorphic to **Z.**
We pick out all those elements of **Q** that have a denominator 1 (we assume that we have used the cancelling rule if possible).
Then addition, subtraction and multiplication on the set

{.... -3/1, -2/1, -1/1 , 0 +1/1, +2/1, +3/1 .... }

has the same structure as **Z.**

==============================

R: The Real Numbers

Irrational numbers

One of the most central problems of algebra is the solution of equations such as **ax + b = c,** (linear equations) and **ax² + bx + c = 0,** (quadratic equations) and equations of higher degree.
Linear equations can always be solved in **Q,** but there are many quadratic equations that do not have solutions in **Q.**
In around 300BC, Euclid showed that **x² = 2** has no solution in the rational number system:

Proof Suppose that $\left(\frac{a}{b}\right)^2 = 2$ where **a** and **b** have no common factor.

then

$$a^2 = 2b^2$$

therefore a^2 is even, so that **a** must then be even because the square of an odd number is odd.
Since **a** is even, we can write $a = 2k$ and then we have $4.k^2 = 2b^2$ so that $b^2 = 2k^2$. Which means that **b** would have to be even too. Thus **a** and **b** have a common factor 2 which contradicts the original supposition.

In fact, a similar argument can show that **if n is an integer that is not a perfect square, then √n is not a rational number.**
Such numbers belong to the **real number system** and are called **irrational numbers.**

√2, √3, √5, ³√4, ⁴√7, π, are all example of irrational numbers.

The Real Number Line

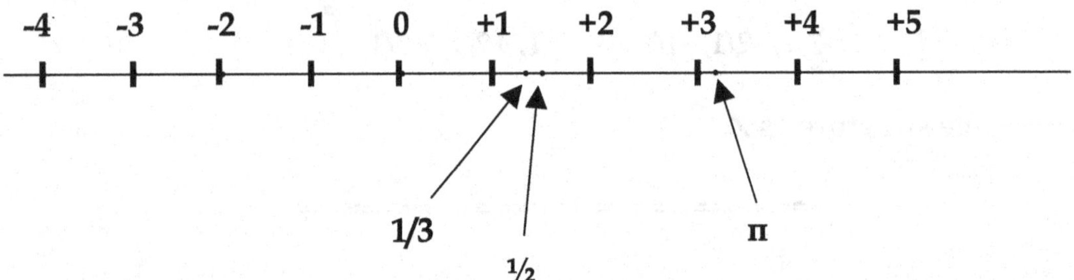

The diagram shows the signed integers laid out along a number line. The rational numbers correspond to points on the number line, and in fact, there are infinitely many rationals between any two rational numbers on the line.
We can easily find any number of rational numbers between any two rational numbers we choose. To find a rational number between $\frac{3}{4}$ and $\frac{4}{5}$ all we need do is to add the top and add the bottom numbers to get $\frac{7}{9}$

Similarly, to get a rational number between $\frac{3}{4}$ and $\frac{7}{9}$ do the same to get $\frac{10}{13}$

We say, the rational numbers are dense on the number line. However, there are some points on the number line that do not correspond to rational

numbers. The numbers $\sqrt{2}$ and π, for example, have their positions on the number line, but $\sqrt{2}$ and π are not a rational numbers. They cannot be written as a fractions of the form **a/b** where **a** and **b** are integers.
These numbers that have their place on the number line but do not correspond to rational number fractions are called **irrational numbers.** The rationals together with the irrationals complete the number line. The rationals together with the irrationals and the laws of arithmetic complete the real number system. Every position on the number line corresponds to a real number.

==================================

C: The Complex Numbers

Every schoolchild know that there are equations such as $x^2 = -4$ that have no solutions in the real number system because if **x** is any real number then x^2 will be positive.

Every schoolchild also knows that the solutions to the quadratic equation

$$ax^2 + bx + c = 0$$

are given by the formula $x = \dfrac{-b \pm \sqrt{(b^2-4ac)}}{2a}$

Proof $$ax^2 + bx + c = 0$$

transpose **c** to the other side and divide by **a** $$x^2 + \frac{bx}{a} = -\frac{c}{a}$$

complete the square $$x^2 + \frac{bx}{a} + \left(\frac{b}{2a}\right)^2 = -\frac{c}{a} + \left(\frac{b}{2a}\right)^2$$

giving $$\left(x + \frac{b}{2a}\right)^2 = \frac{b^2 - 4ac}{4a^2}$$

$$x + \frac{b}{2a} = \frac{\pm\sqrt{(b^2 - 4ac)}}{2a}$$

$$x = \frac{-b \pm \sqrt{(b^2 - 4ac)}}{2a}$$

Example 1 Solve $x^2 - 5x + 6 = 0$

Solution

$$x = \frac{5 \pm \sqrt{(25 - 4(1)(6))}}{2}$$

$$= \frac{5 \pm 1}{2}$$

$$x = 3 \text{ or } 2$$

graphically, we are finding the points where the curve $y = x^2 - 5x + 6$ cuts the x axis:

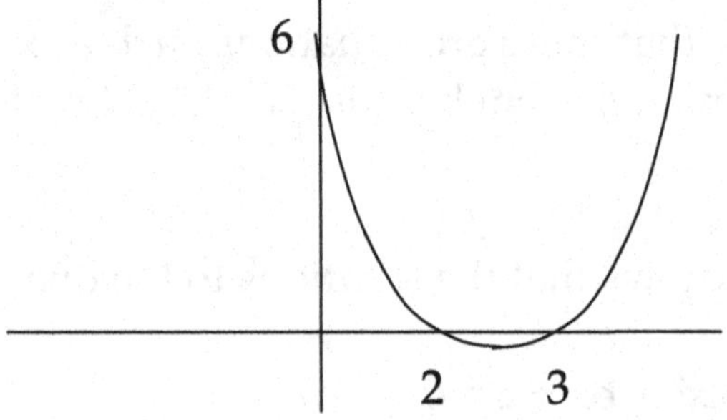

Exercise 1 Solve

[1] $x^2 + 2x - 15 = 0$

[2] $x^2 - 5x + 4 = 0$

[3] $x - 2\sqrt{x} - 3 = 0$

[4] $x^4 - 5x^2 + 4 = 0$

[5] $(2x + 1)^2 - 8(2x + 1) + 15 = 0$

[6] $(x^2 + 2x)^2 - 11(x^2 + 2x) + 24 = 0$

[7] Find the Cartesian equation for this curve.

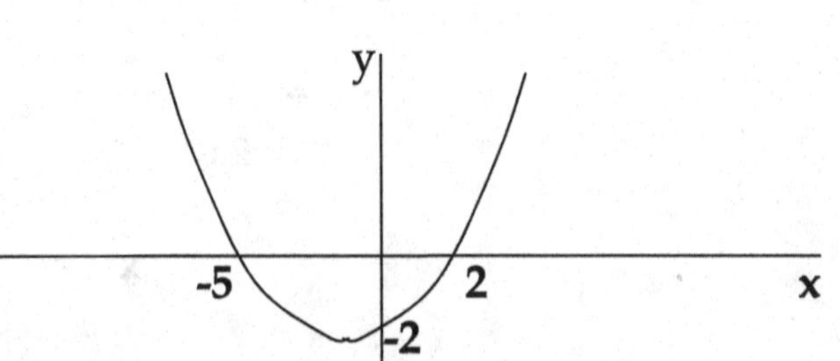

Example 2

Solve (i) $x^2 - 4x + 3 = 0$

(ii) $x^2 - 4x + 4 = 0$

(iii) $x^2 - 4x + 5 = 0$

Solutions

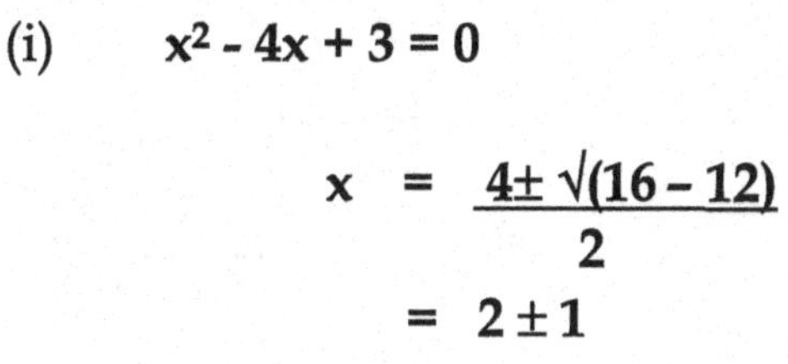

(i) $x^2 - 4x + 3 = 0$

$$x = \frac{4 \pm \sqrt{(16 - 12)}}{2}$$

$$= 2 \pm 1$$

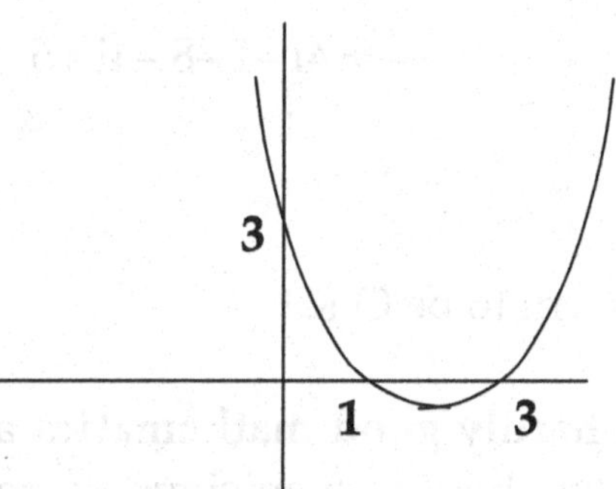

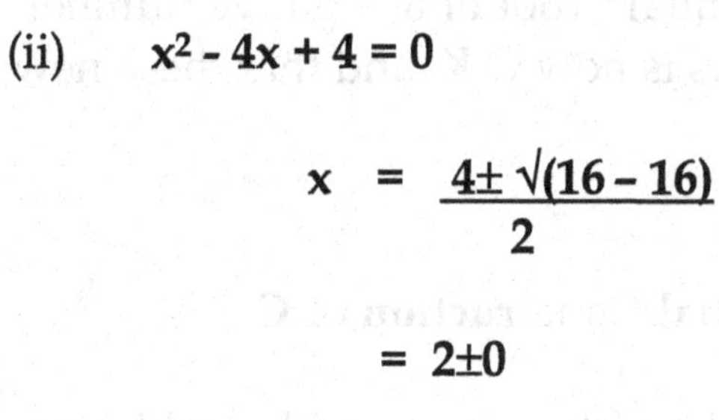

(ii) $x^2 - 4x + 4 = 0$

$$x = \frac{4 \pm \sqrt{(16 - 16)}}{2}$$

$$= 2 \pm 0$$

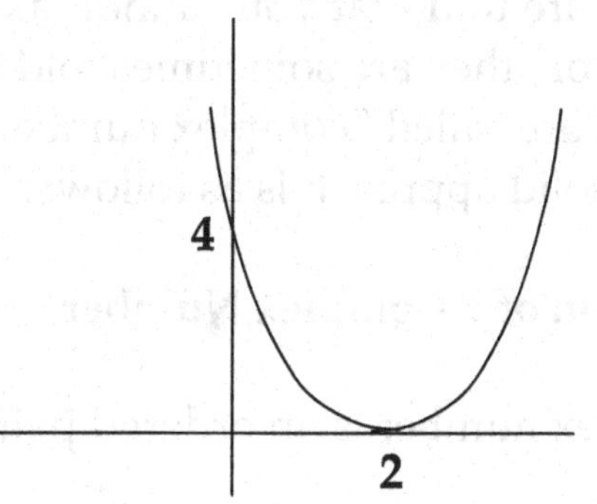

(iii) $x^2 - 4x + 5 = 0$

$$x = \frac{4 \pm \sqrt{(16 - 20)}}{2}$$

$$= 2 \pm \tfrac{1}{2}\sqrt{(-4)}$$

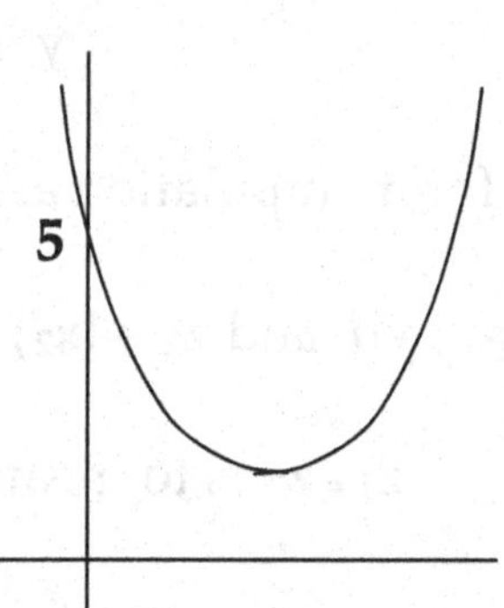

In ex(i), the curve cuts the real number line at two points, x=1 and x=3.
In ex(ii), the curve is lifted up by 1 and as a result touches the line at x=2.
In ex(iii), the curve is again lifted up by 1 and so does not cut the real number line. There are no real solutions to the equation of ex(iii).

Students are often told such things as
"suppose that $\sqrt{(-1)} = i$, so that $\sqrt{(-4)} = \sqrt{4} \times \sqrt{(-1)} = 2i$"
Then the quadratic formula would give the two solutions $x=2+i$ and $x=2-i$ for ex(iii).
Try these solutions in the equation for example:

$$(2+i)^2 - 4(2+i) + 5 = 4 + 4i + i^2 - 8 - 4i + 5$$

$$= 4 + 4i - 1 - 8 - 4i + 5$$

$$= 0$$

so, things seem to be O.K.!

but this is hardly good mathematics and any good student of mathematics would be left highly suspicious as to what is going on. Up to O level, students are told that you cannot take the square root of a negative number, but later on, they are sometimes told that this is now O.K. and that these new numbers are called "complex numbers".
A more solid approach is as follows:

Definition of a Complex Number : A Formal Construction of C

A complex number is an ordered pair of real numbers, say x and y, which we write as

$[x, y]$ x is called the real part
y is called the complex part

The rules for manipulation are as follows:

Let $z_1 = [x_1, y_1]$ and $z_2 = [x_2, y_2]$

equality $z_1 = z_2$ is to mean $x_1 = x_2$ **and** $y_1 = y_2$

addition

$$z_1 + z_2 = [x_1+x_2, y_1+y_2]$$

multiplication

$$z_1 \times z_2 = [x_1x_2 - y_1y_2, x_1y_2 + x_2y_1]$$

zero

The zero complex number is **[0,0]**, and we write **[0,0] = 0**

It is easy to show that **z + 0 = 0 + z = 0** for any complex number **z.**

The negative of a complex number

Since **[x,y] + [-x,-y] = 0** we call **[-x,-y]** the negative of **[x,y]**

Subtraction

Subtraction is defined as "adding the negative"

The Unit Complex number

The unit complex number is **[1,0]** and this is the only complex number **u** say, for which **z x u = z** for any complex number **z.**

For, suppose that **[x,y] x [a,b] = [x,y]**

Then
$$x = ax - by \quad \dots (1)$$
$$y = bx + ay \quad \dots (2)$$

(1).x gives $x^2 = ax^2 - bxy$

(2).y $y^2 = bxy + ay^2$

add

$$(x^2+y^2) = a(x^2+y^2)$$

assuming that **x** and **y** are not both zero cancel (x^2+y^2) to get **a=1**

(1).y gives $xy = axy - by^2$

(2).x gives $xy = bx^2 + axy$

subtract

$$0 = b(x^2+y^2)$$

assuming that **x** and **y** are not both zero cancel (x^2+y^2) to get **b=0**

Thus the only complex number **[a,b]** for which

[x,y] x [a,b] = [x,y]

is **[a,b] = [1,0]** (unless **[x,y]=0**)

The reciprocal of a Complex number

Definition

The reciprocal of [x,y] is the complex number [a,b] such that

$$[x,y] \times [a,b] = [1,0]$$

i.e.

$$ax - by = 1 \quad \ldots\ldots (1)$$
$$bx + ay = 0 \quad \ldots\ldots (2)$$

(1).x gives $ax^2 - bxy = x$

(2).y gives $bxy + ay^2 = 0$

add

$$a(x^2+y^2) = x \qquad a = \frac{x}{x^2+y^2}$$

thus

$$a = \frac{x}{x^2+y^2} \qquad b = \frac{-y}{x^2+y^2}$$

Thus reciprocal of [x,y] is $\left[\frac{x}{x^2+y^2}, \frac{-y}{x^2+y^2} \right]$

Division

Division is defined as "multiplying by the reciprocal".

The Laws of Arithmetic in C

Using these rules, we can show that the **laws of arithmetic** hold for these complex numbers, so that, if $z_1 = [x_1, y_1]$, $z_2 = [x_2, y_2]$ and $z_3 = [x_3, y_3]$ then

$$z_1 + z_2 = z_2 + z_1$$

$$z_1.z_2 = z_2.z_1$$

$$z_1 + (z_2+z_3) = (z_1+z_2) + z_3$$

$$z_1(z_2.z_3) = (z_1.z_2)z_3$$

$$z_1(z_2 + z_3) = z_1.z_2 + z_1.z_3$$

In **C**, the system of complex numbers, we have the zero **[0,0]**, the unit **[1,0]**, subtraction and division (not by zero).
In short, we can do arithmetic in the complex number system.

The real numbers in C
If we consider the subset of the complex numbers that have a zero imaginary part, we see that this subsystem of the complex numbers system behaves just like (i.e. is isomorphic to) the real number system.
e.g.

$$[x_1, 0] + [x_2, 0] = [x_1+x_2, 0]$$

$$[x_1, 0].[x_2, 0] = [x_1x_2, 0]$$

and some would say that, in a sense, the real number system is embedded in the complex number system. Since the real number system and the real numbers system obey the same laws of arithmetic, it is easy to use real numbers as a shorthand for this subset of the complex number system, thus, for example, we can write

$$2 + 3 = 5$$

when, strictly speaking, we should write

$$[2,0] + [3,0] = [5,0]$$

Using this shorthand, instead of **[x,0]** we write just **x**.

The Complex Number i

The complex number **[0,1]** is called **i.**

We have $i^2 = [0,1]\text{x}[0,1] = [-1,0] = -1$

and so, we naturally write **i = √(-1)** but with reservations, because there are no positives or negatives in the complex number system. It can be shown that we cannot say if one complex number is bigger than another without runnimg into problems. There is no "order system" in C, so that we cannot always say for example, **z1>z2.** (Of course, this is O.K. if **z1** and **z2** happen to be "**real**".

Note also, that

$$(-i)^2 = [0\,,-1]\ \text{x}\ [0\,,-1] = [-1\,,0] = -1\,.$$

Theorem $x + iy = [x, y]$

Proof

If **x** and **y** are the real numbers, our notation gives

$$x + iy = [x, 0] + [0, 1] \times [y, 0] = [x, 0] + [0, y] = [x, y]$$

Order amongst the Complex Numbers

We are all familiar with the statement $a<b$, meaning that the number a is less than the number b. On the number line it means that number a is to the left of number b:

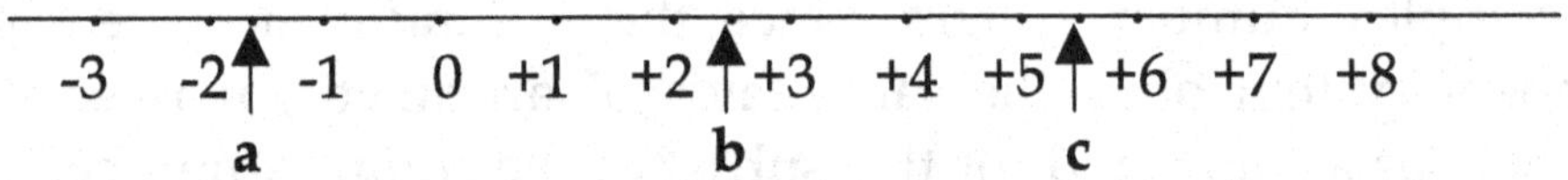

Referring to this number line, we can say, $a<b$, $b>a$, $c>a$, $b<c$ etc.

However, if we try to do such a thing with the complex numbers, we run into trouble.

There is no order relation in the complex number system

If z and w are any two complex numbers, for an order relation $<$ to hold we need the following three rules:

(A) either $z<w$ or $z=w$ or $w<z$

(B) $z<w$ implies that $z+c<w+c$ for any complex number c

(C) $0<z$ and $0<w \Rightarrow 0<zw$

Now suppose that $0<i$

Then by (C) $0<i.i$ therefore $0<-1$

By (C) again $0<-1.i$ so that $0<-i$

Now add i to both sides, then by (B) we have $i<0$ which contradicts (A)

In short, **we cannot say one complex numbers is larger than another** without running into trouble.

Just as we solved the general quadratic in the real numbers system to give the quadratic formula, since we have the same algebraic rules in C, we can solve equations with complex coefficients in the same way, for example:

Example 8

Solve $(2+3i)z^2 + (3-4i)z + 7 - i = 0$

Solution

$$z = \frac{-3+4i \pm\sqrt{(3-4i)^2 - 4(2+3i)(7-i)}}{2(2+3i)}$$

$$= \frac{-3+4i \pm\sqrt{-7-24i - 4(17+19i)}}{4+6i}$$

$$= \frac{-3+4i \pm\sqrt{-75-100i}}{4+6i}$$

$$= \frac{-3+4i \pm 5i\sqrt{3+4i}}{4+6i}$$

$$= \frac{-3+4i \pm 5i(2+i)}{4+6i} \qquad \textbf{(see next section on square roots)}$$

$$= \frac{-8+14i}{4+6i} \quad \text{or} \quad \frac{2-6i}{4+6i}$$

$$= \frac{(-8+14i)(4-6i)}{(4+6i)(4-6i)} \quad \text{or} \quad \frac{(2-6i)(4-6i)}{(4+6i)(4-6i)} \qquad \textbf{(see note)}$$

$$= \frac{13+26i}{13} \quad \text{or} \quad \frac{(1-3i)(2-3i)}{13}$$

$$= 1+2i \quad \text{or} \quad \frac{7}{13} - \frac{9}{13}i$$

There are two steps in the calculation that require some explanation: (1) how we find that $\sqrt{3+4i} = 2+i$ and (2) how we do the division
We tackle the easier one first:

The Sum of Two Squares

The difference of two squares is well known: $x^2-y^2 = (x-y)(x+y)$

Now, we can factorize the sum of two squares: $x^2+y^2 = (x-iy)(x+iy)$

Thus $4+9 = (2-3i)(2+3i)$, $(5-2i)(5+2i) = 25+4 = 29$

We use this method to simplify division:

Example 9

Perform the division $\dfrac{15+20i}{2+i}$

Solution $\dfrac{15+20i}{2+i} = \dfrac{(15+20i)(2-i)}{(2+i)(2-i)} = \dfrac{50+25i}{4+1} = 10+5i$

Exercise Show that $\dfrac{5+10i}{4+3i} = 2+i$

Finding the Square Root

In the solution of the complex quadratic equation above, we have stated

$$\sqrt{3+4i} = 2+i$$

but how do we find the square root of a complex number?

One method is to try $\sqrt{3+4i} = x+iy$

Then $$3+4i = x^2 - y^2 + 2ixy$$

Equate parts, $x^2 - y^2 = 3$ and $2xy = 4$

$$x^2 - (2/x)^2 = 3$$

$$x^4 - 3x^2 - 4 = 0$$

$$(x^2 - 4)(x^2 + 1) = 0$$

Now **x** must be real, therefore $x = \pm 2,\ y = \pm 1$ (signs matching)

Hence we have the result $\sqrt{3+4i} = 2 + i \text{ or } -2 - i$

==================

In general, we can always find square roots for any complex number **a+ib**

put $\sqrt{a+ib} = x + iy$

Then $a+ib = x^2 - y^2 + 2ixy$

Equate parts, $x^2 - y^2 = a$ and $2xy = b$

$$x^2 - (b/2x)^2 = a$$

$$4x^4 - 4ax^2 - b^2 = 0$$

$$4x^4 - 4ax^2 + a^2 = a^2 + b^2$$

$$(2x^2 - a)^2 = (a^2 + b^2)$$

$$2x^2 = \pm\sqrt{(a^2 + b^2)} + a$$

x must be real therefore $x^2 = \frac{1}{2}(\sqrt{(a^2 + b^2)} + a)$

and this gives $y^2 = \frac{1}{2}(\sqrt{(a^2 + b^2)} - a)$

Hence, a the square root of **a + ib** is

$$\sqrt{\frac{\sqrt{(a^2 + b^2)} + a}{2}} + i\sqrt{\frac{\sqrt{(a^2 + b^2)} - a}{2}}$$

and the other root is the negative of this.

The Argand Diagram

Argand's diagram is a representation of the complex number **$x + iy$** on x-y axes. The complex number is represented by the point **$P(x,y)$:**

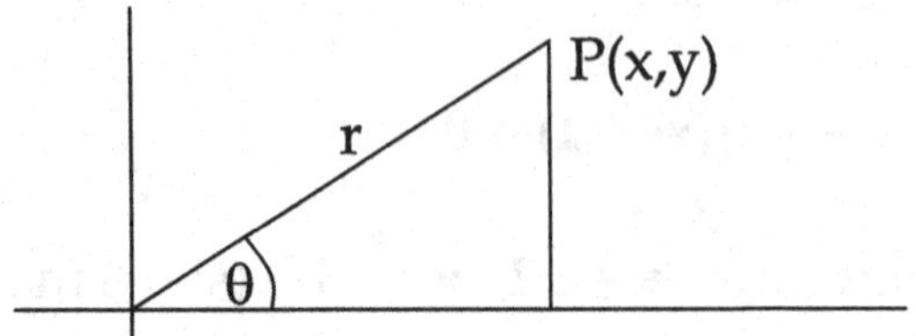

Since $x = r\cos\theta$ $\;y = r\sin\theta$ we have $z = x + iy$ so that **$z = r\cos\theta + i\, r\sin\theta$** for any value of **$\theta$.**
Addition of two complex numbers, $z_1 = x_1 + i\, y_1$, $z_2 = x_2 + i\, y_2$ is represented on the Argand diagram by the parallelogram law:

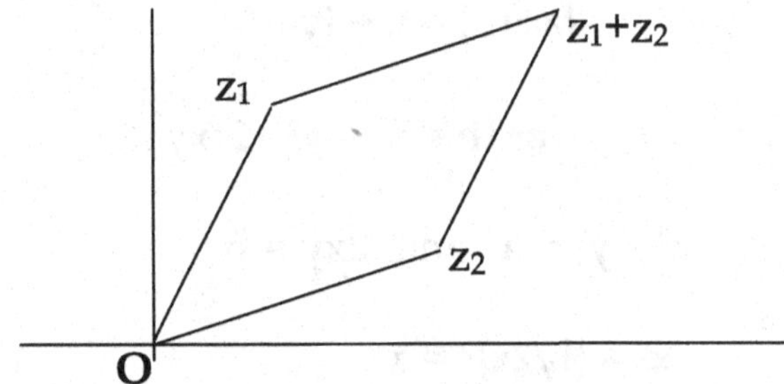

Which corresponds to addition of 2 dimensional vectors:

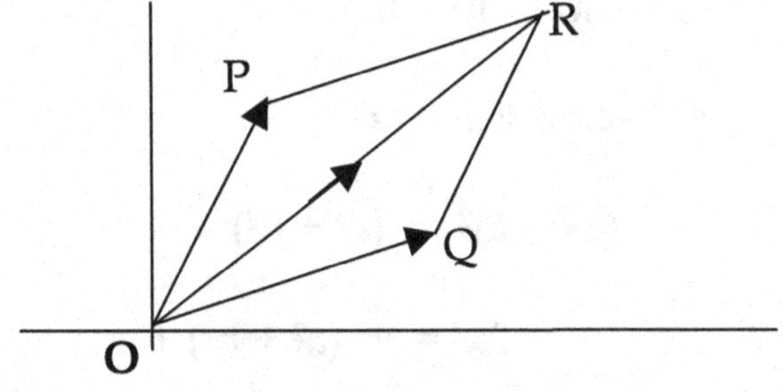

Thus $\quad$ **$(x_1 + i\, y_1) + (x_2 + i\, y_2) = (x_1 + x_2) + i\,(y_1 + y_2)$**

$$[x_1, y_1] + [x_2, y_2] = [x_1+x_2, y_1+y_2]$$

and $$\begin{bmatrix} x_1 \\ y_1 \end{bmatrix} + \begin{bmatrix} x_2 \\ y_2 \end{bmatrix} = \begin{bmatrix} x_1+x_2 \\ y_1+y_2 \end{bmatrix}$$

All describe the vector equation

$$\mathbf{OP} + \mathbf{OQ} = \mathbf{OR}$$

Modulus and argument

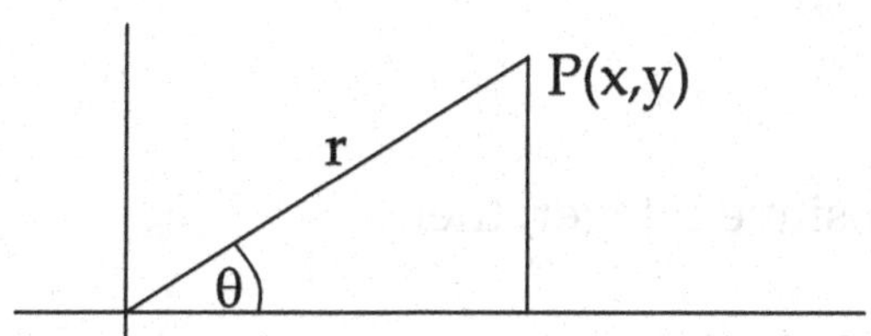

Here, we have $z = r(\cos\theta + i \sin\theta)$

r is called the modulus of z and we write $|z| = r = \sqrt{(x^2+y^2)}$

θ is called an argument of z and we write $\arg(z) = \theta$

The argument is not unique for any value $\theta \pm 2n\pi$ will function as an argument for the same z.

The **principle argument** is determined by $-\pi < \arg(z) \leq \pi$

==============================

Multiplying on the Argand Diagram

We shall use the shorthand **cisθ** for **cosθ + isinθ.**

Let $z_1 = r_1.\text{cis}\theta_1$ and $z_2 = r_2.\text{cis}\theta_2$

Then $z_1.z_2 = r_1.r_2(\cos\theta_1+i\sin\theta_1)(\cos\theta_2+i\sin\theta_2)$

$= r_1.r_2(\cos\theta_1.\cos\theta_2 - \sin\theta_1.\sin\theta_2 + i(\sin\theta_1.\cos\theta_2 + \cos\theta_1.\sin\theta_2)$

$= r_1.r_2(\cos(\theta_1+\theta_2) + i \sin(\theta_1+\theta_2))$

$= r_1r_2\text{cis}(\theta_1+\theta_2)$

This result can be readily extended thus $z_1.z_2.z_3 = r_1.r_2.r_3 \text{ cis}(\theta_1+\theta_2+\theta_3)$ etc

and putting $\theta_1=\theta_2=\theta_3$ ….. we have $z^2 = r^2\text{cis}(2\theta)$, $z^3 = r^3\text{cis}(3\theta)$ etc

DeMoivres Theorem

Let $z = \cos\theta + i\sin\theta$

(i) If **n** is any positive integer, then

$$z^n = (\cos\theta + i\sin\theta)^n$$

$$= \cos(\theta+\theta+\theta+\ldots n \text{ terms}) + i.\sin(\theta+\theta+\theta+\ldots n \text{ terms})$$

$$= \cos(n\theta) + i\sin(n\theta)$$

Thus we can write $\text{cis}^n\theta = \text{cis}(n\theta)$ **De Moivre's theorem** for +ve integral **n**

(ii) Let **n** be a negative integer, **n = -m** where **m>0**

Then $z^n = (\cos\theta + i\sin\theta)^{-m}$

$$= \frac{1}{(\cos\theta + i\sin\theta)^m}$$

$$= \frac{1}{\cos m\theta + i\sin m\theta}$$

$$= \frac{1}{\cos m\theta + i\sin m\theta} \times \frac{\cos m\theta - i\sin m\theta}{\cos m\theta - i\sin m\theta}$$

$$= \frac{\cos m\theta - i\sin m\theta}{\cos^2 m\theta + \sin^2 m\theta}$$

$$= \cos m\theta - i\sin m\theta$$ since $\cos^2 + \sin^2 = 1$

$$= \cos(-m\theta) + i\sin(-m\theta)$$

Thus $\text{cis}^n\theta = \text{cis}(n\theta)$ for any negative integer **n**.

(iii) Let q be any signed integer and suppose that $z = \cos\theta + i\sin\theta$

We note that $\left[\cos\frac{2n\pi+\theta}{q} + i\sin\frac{2n\pi+\theta}{q}\right]^q$

$= \cos(2n\pi+\theta) + i\sin(2n\pi+\theta)$ using (i) and (ii)

$= \cos\theta + i\sin\theta$

So that $z^{1/q}$ is a **q** valued expression with values

$\text{cis}\,\frac{\theta}{q}$, $\text{cis}\,\frac{2\pi+\theta}{q}$, $\text{cis}\,\frac{-2\pi+\theta}{q}$, $\text{cis}\,\frac{4\pi+\theta}{q}$, $\text{cis}\,\frac{-4\pi+\theta}{q}$, $\text{cis}\,\frac{6\pi+\theta}{q}$ . . . .

================================

De Moivres theorem allows us to Find Roots of any complex number.

Example 1 Find $\sqrt{(2+2i)}$

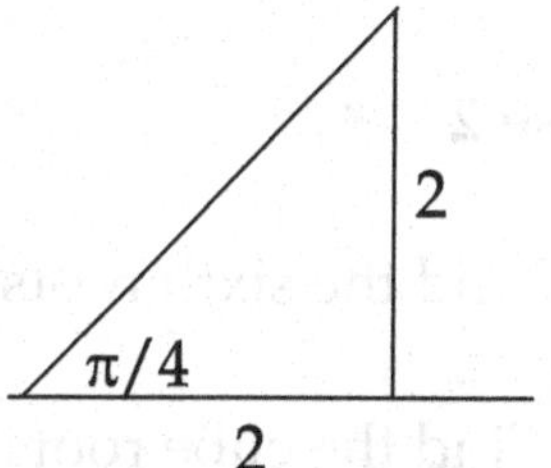

Solution

$2+2i = 2\sqrt{2}\,\text{cis}(\pi/4)$

$= 2\sqrt{2}\,\text{cis}(2n\pi + \pi/4)$

$\sqrt{(2+2i}$ $= (2\sqrt{2})^{1/2}\,\text{cis}(n\pi + \pi/8)$ using De Moivre

$= 2^{3/4}\,\text{cis}(\pi/8)$ or $2^{3/4}\,\text{cis}(\pi + \pi/8)$

================================

Example 2 Find the fifth roots of $1 + i\sqrt{3}$

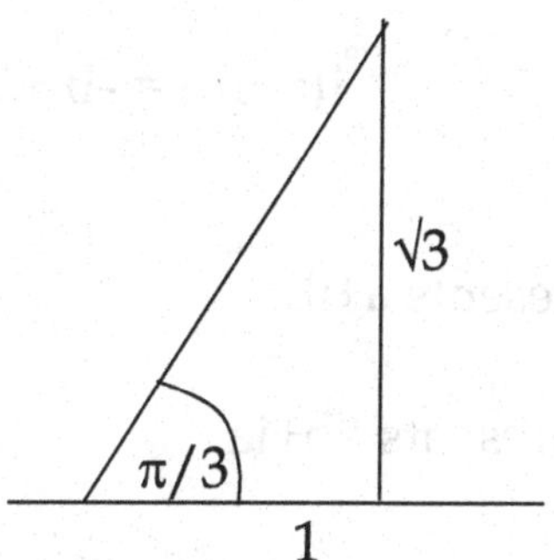

Solution

$z = 2\,\text{cis}(\pi/3)$

$= 2\,\text{cis}(2n\pi + \pi/3)$

$z^{1/5} = 2^{1/5}\,\text{cis}(2n\pi/5 + \pi/15)$

The five values of $z^{1/5}$ are

$2^{1/5}\text{cis}(\pi/15)$, $2^{1/5}\text{cis}(7\pi/15)$, $2^{1/5}\text{cis}(13\pi/15)$, $2^{1/5}\text{cis}(19\pi/15)$, $2^{1/5}\text{cis}(25\pi/15)$

The fifth roots of $1 + i\sqrt{3}$ are represented on the Argand diagram as five equally spaced points on a circle of radius $2^{1/5}$

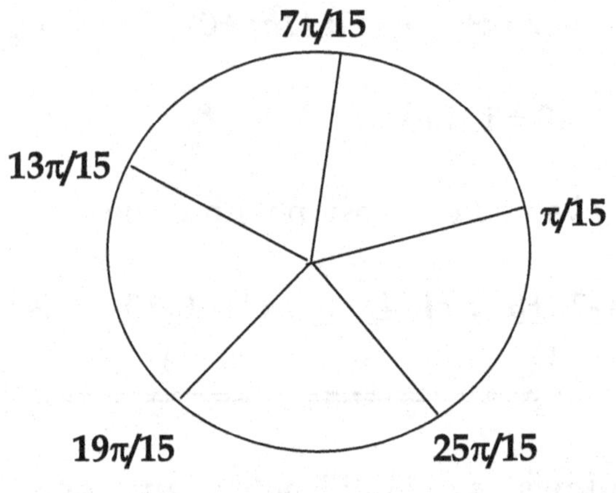

==

Exercise 2

[1] Find the sixth roots of 8, by writing $8 = 8\text{cis}(2n\pi)$

[2] Find the cube roots of –1 by writing $-1 = \text{cis}(2n\pi + \pi)$

==

Multiplication by i

Consider any complex number **a+ib** then we have

$$i(a+ib) = -b + ia$$

Let OP represents a+ib

and OQ represents -b+ia

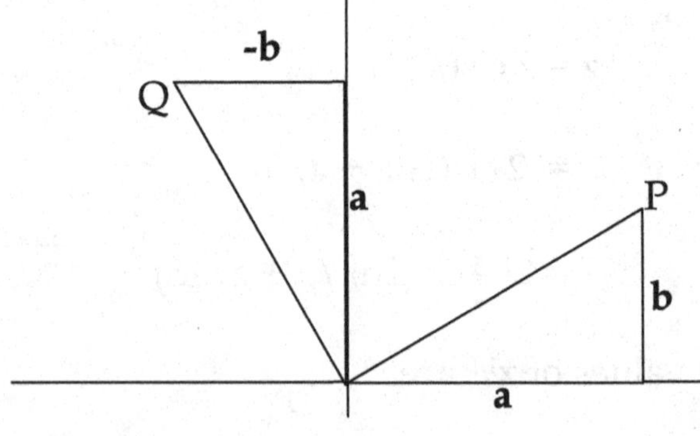

then the angle between OP and OQ will be 90°.

We may note that the dot product $\begin{pmatrix} a \\ b \end{pmatrix} \cdot \begin{pmatrix} -b \\ a \end{pmatrix} = -ab + ba = 0$

or we may note that the transformation $(x,y) \longrightarrow (-y,x)$ is a clockwise rotation of 90° (see chapter 19 or chapter 12).
thus

Theorem

Multiplying by i has the effect of a clockwise rotation of 90° in the Argand diagram.

Vectors and Complex numbers

In the x-y plane, each point P(x,y) is associated with its position vector $\begin{pmatrix} x \\ y \end{pmatrix}$

and P is also associated with the complex number x+iy on the Argand diagram.

If we compare vector addition with addition of complex numbers, we see the same structure:

Vectors	Complex numbers
$\begin{pmatrix} a \\ b \end{pmatrix} + \begin{pmatrix} c \\ d \end{pmatrix} = \begin{pmatrix} a+c \\ b+d \end{pmatrix}$	[a,b] + [c,d] = [a+c, b+d]

Two points $A(x_1,y_1)$ and $B(x_2,y_2)$ determine the vector

$$AB = \begin{pmatrix} x_2-x_1 \\ y_2-y_1 \end{pmatrix}$$

and also, the complex number $[x_2-x_1, y_2-y_1]$

We phrase these results in the form of a theorem that we will use for a pair of geometrical proofs:

Theorem

If **a** and **b** are the complex numbers associated with the points **A** and **B** on the Argand diagram, then the line **AB** represents the complex number **b-a**

Proof

Let **a=x+iy, b=p+iq**

so that we have **A**(x,y) and **B**(p,q)

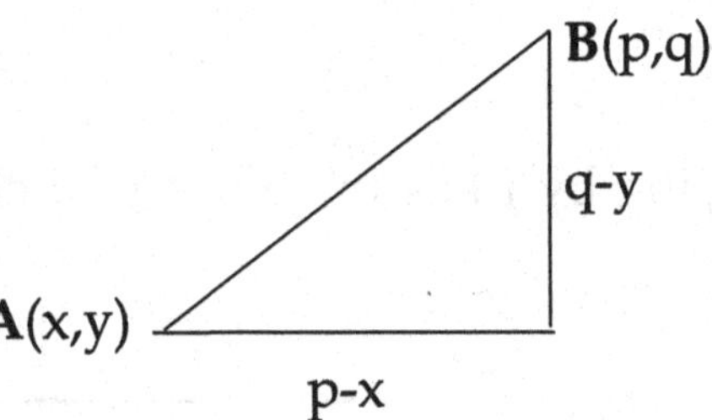

The vector **AB** = $\begin{pmatrix} p-x \\ q-y \end{pmatrix}$

The complex number **b-a = (p-x) + i(q-y)**

Hence **AB** is represented by the complex number **b-a**

================

Thus

If AB is rotated through 90^0 to AC, then AC is represented by **i(b-a)**

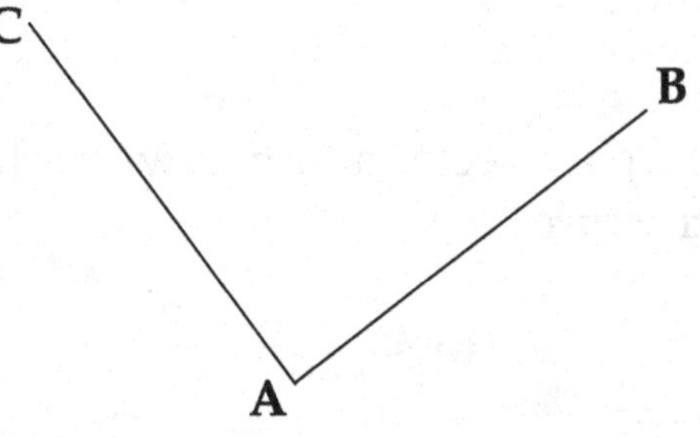

We can write

AC = iAB

for **c-a = i(b-a)**

====================

Mid Points

The mid point of **AB** is represented by the complex number **m** where

m = ½(a+b)

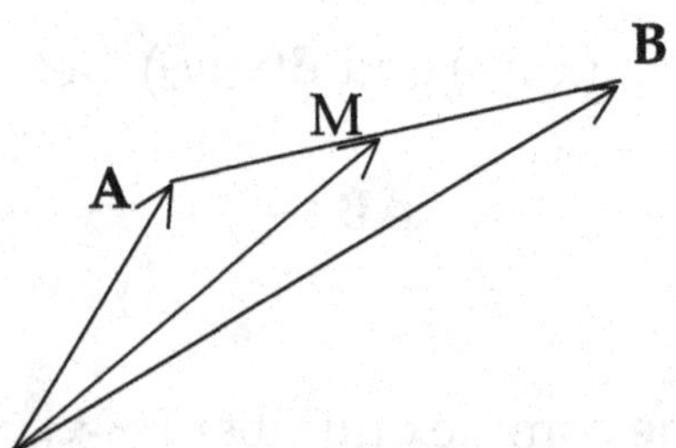

O

We now give two geometrical results that can be proved using complex numbers in the complex plane:

Example 1 (see chapter 4 problem 1)

Squares are drawn on the sides of a triangle ABC. J,K and L are the centres of the three squares as shown in the diagram.

Prove that **JK = LB** and **JK ⊥ LB.**

Proof

We use lower case letters to represent the complex number of the corresponding point.

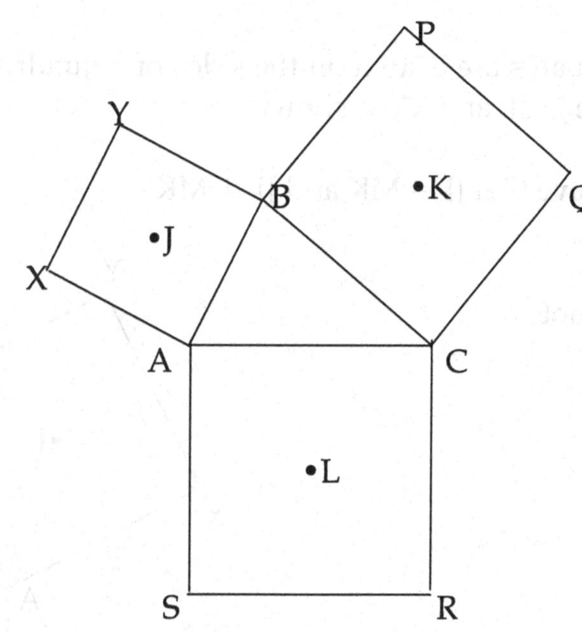

AB rotates to AX
BC rotates to BP
CA rotates to CR

Hence:

$$AX = iAB$$
$$BP = iBC$$
$$CR = iCA$$

Therefore

$$\mathbf{x - a = i(b - a)} \quad \text{giving} \quad \mathbf{x = a + i(b - a)}$$
$$\mathbf{p - b = i(c - b)} \quad \text{giving} \quad \mathbf{p = b + i(c - b)}$$
$$\mathbf{r - c = i(a - c)} \quad \text{giving} \quad \mathbf{r = c + i(a - c)}$$

also, using the mid point formula $\mathbf{j = \frac{1}{2}(b+x),\ k = \frac{1}{2}(c+p),\ l = \frac{1}{2}(a+r)}$

therefore $\mathbf{BL = l - b = \frac{1}{2}(a+r) - b = \frac{1}{2}(a+c+i(a-c)) - b}$

now $\mathbf{iKJ = I(j-k) = i(\frac{1}{2}(b+x) - \frac{1}{2}(c+p))}$

$$=\tfrac{1}{2}i(b+a+i(b-a) - c - b - i(c-b))$$

$$=\tfrac{1}{2}i(a - c - i(a+c) + 2ib)$$

$$= \tfrac{1}{2}(i(a-c) + a+c - 2b)$$

$$= \tfrac{1}{2}(a+c+i(a-c)) - b = \mathbf{BL}$$

hence **iKJ = BL** so that **KJ** rotates to **BL** as required.

======================

Example 2 (see chapter 4 problem 3)

Squares are drawn on the sides of a quadrilateral. The centres of the squares are J,K,L and M, as shown.

Prove that JL = MK and JL ⊥ MK

Proof

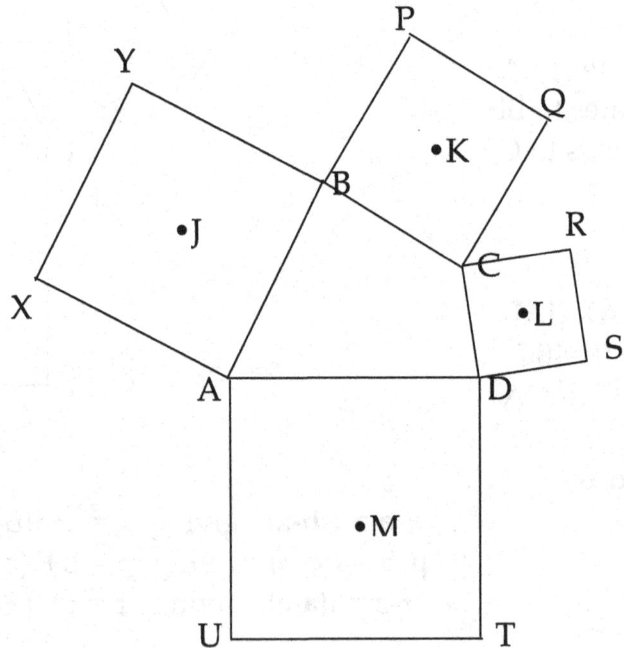

Following the method of example 1, we have

$$\begin{array}{lll} iAB = AX, & i(b-a) = x-a, & x = i(b-a)+a \\ iBC = BP, & i(c-b) = p-b, & p = i(c-b)+b \\ iCD = CR, & i(d-c) = r-c, & r = i(d-c)+c \end{array}$$

$$iDA = DT, \quad i(a-d) = t-d, \quad t = i(a-d)+d$$

using the mid point formula

$$j = \tfrac{1}{2}(x+b), \; k = \tfrac{1}{2}(p+c), \; l = \tfrac{1}{2}(r+d), \; m = \tfrac{1}{2}(a+t)$$

now $\quad LJ = j-l = \tfrac{1}{2}(x+b - r-d) = \tfrac{1}{2}(i(b-a)+a + b - i(d-c)-c - d)$

$$= \tfrac{1}{2}(a(1-i) + b(1+i) + c(-1+i) + d(-1-i))$$

also $\quad iMK = i(k-m) = \tfrac{1}{2}i(p+c - a-t) = \tfrac{1}{2}i(i(c-b)+b + c - a -i(a-d)-d)$

$$= \tfrac{1}{2}i(a(-1-i) + b(1-i) + c(1+i) + d(-1+i))$$

$$= \tfrac{1}{2}(a(-i+1) + b(i+1) + c(i-1) + d(-i-1))$$

$$= LJ$$

thus **iMK = LJ** so that MK rotates through 90° to LJ giving the required result

===========================

Solving Equations

Solving equations is one of the central activities in algebra. We have seen that in **Z**, the system of signed integers, we can solve equations of the form $x+b=c$ but **Z** had to be extended to **Q**, the system of rational numbers, where we have division, in order to solve the general linear equation, of the form $ax+b=c$.

Example 1

2x-3 = 0 cannot be solved in **Z**
but it can be solved in **Q**

The rational numbers were extended to **R**, the system of real numbers, before we could solve equations such as $x^2=3$.

Example 2

$x^2-3 = 0$ cannot be solved in **Q**
but it can be solved in **R**

In order to solve quadratic equations, we needed to extend **R** to the system of complex numbers, **C.**

Example 3

$x^2+3 = 0$ cannot be solved in **R**
but it can be solved in **C**

In **C,** we can always solve the general quadratic equation $ax^2+bx+c=0$. We can always use the formula for the roots of the quadratic and find two (possibly coincident) solutions in **C.**
The question arises, do we need another extension of the number system in order to solve a cubic equation such as

$$x^3+2x^2+3x+4 = 0$$

Oddly enough, no! We do not need any further extensions of the number system in order to find solutions to other polynomial equations.
In 1799, **Carl Friedrich Gauss** proved that every polynomial equation has at least one root in **C.** This is often called the Fundamental theorem of algebra and leads to the surprising result that
every polynomial equation of degree n has exactly n roots in C.
(some of the roots, of course, may be repeated roots).
Gauss's theorem means that there in no need to extend the number system any further in order to solve equations of higher degrees.

Another question arises, "has anyone found a formula for solving a cubic?"
The answer is **yes!**
The general cubic equation $x^3+ax^2+bx+c = 0$

Can be solved by substituting $x = t - a/3$

To reduce it to the form $t^3+pt+q = 0$

Now putting $u^3 - v^3 = q$ and $uv = p/3$, we solve for **u** and **v**

Solving for u^3 we arrive at $27u^6 - 27u^3q - p^3 = 0$

Which is a quadratic in u^3
Giving

$$u^3 = \frac{q}{2} \pm \sqrt{\left(\frac{q}{2}\right)^2 + \left(\frac{p}{3}\right)^3}$$

solving for **v** we find

$$v^3 = -\frac{q}{2} \pm \sqrt{\left(\frac{q}{2}\right)^2 + \left(\frac{p}{3}\right)^3}$$

the values of **t** are then given by **v - u** and hence then $x = t - a/3$.

====================================

Quadratic equations can be solved using the quadratic formula, with its square roots.
Cubic equations can also be solved using the above method, leading to a formula involving cube roots and square roots.
Quartic equations can also be solved using methods that lead to formulae involving square roots and cube roots.
However, in 1824, **Neils Abel** proved a remarkable theorem that shows that there is no general formula for a Quintic (5th degree) equation.
For polynomial equations of degree 5 and higher, we have to resort to iterative methods which "home in" on root from an educated guess.

To find a real root of a real equation we can draw a graph of the expression and try to find where the graph cuts the x axis.

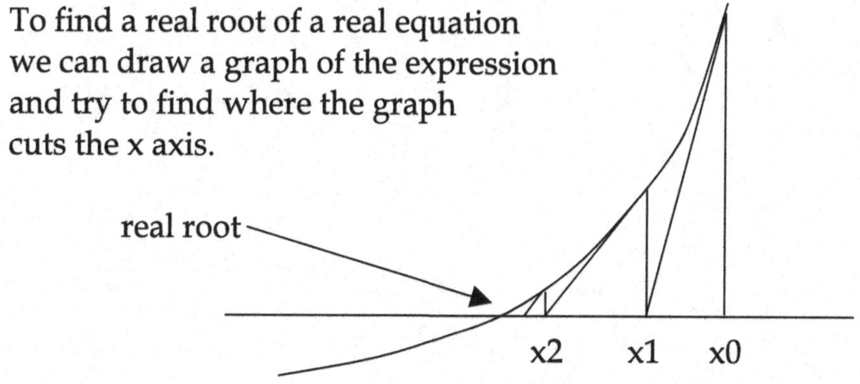

Make a guess x0; draw the line up to the curve; draw the tangent to the curve and calculate x1 where the tangent cuts the x axis; x1 should be closer to the root.
Repeat the process using x1 as the starting point and the calculation should bring you closer and closer to the root of the equation.
But you will never quite get there!
Don't be disheartened however as most calculations do not give us exact answers. We cannot get an exact answer to $\sqrt{2}$. In fact we cannot get any square root exactly unless it happens to be a perfect square!

End of Numbers

====================

Answers

Exercise 1

[1] $x = 3$ or -5
[2] $x = 1$ or 4

[3] $x = 9$ ($\sqrt{x}$ is the positive square root, $x = 1$ is the solution to $x+2\sqrt{x} - 3 = 0$)

[4] $x = -2, -1, +1$ or $+2$

[5] $x = 1$ or $x = 2$

[6] $x = 1, 2, -3$ or -4

[7] $5y = (x-2)(x+5)$

Exercise 3

[1] **$\sqrt{2}$, $\sqrt{2}\text{cis}(\pi/3)$, $\sqrt{2}\text{cis}(2\pi/3)$, $\sqrt{2}\text{cis}(\pi)$, $\sqrt{2}\text{cis}(4\pi/3)$, $\sqrt{2}\text{cis}(5\pi/3)$**

$\sqrt{2}$, $\sqrt{2}(\frac{1}{2} + i\sqrt{3}/2)$, $\sqrt{2}(- \frac{1}{2} + i\sqrt{3}/2)$, $-\sqrt{2}$, $\sqrt{2}(-\frac{1}{2} - i\sqrt{3}/2)$, $\sqrt{2}(\frac{1}{2} - i\sqrt{3}/2)$

[2] **$\text{cis}(\pi/3)$, $\text{cis}(\pi)$, $\text{cis}(5\pi/3)$**

$\frac{1}{2} + i\sqrt{3}/2$, -1, $-\frac{1}{2} - i\sqrt{3}/2)$

============================

CHAPTER 26

Calculus

Dr Fred and his friend Tash were walking along the road when a man in a red sports car roared up from behind and quickly disappeared into the distance. It was exactly 12 o'clock when it passed them and Tash snapped the car on her digital camera just as it went by. The car accelerated into the distance and 10 seconds later it was a quarter of a mile away.
"Wow", said Tash, " He was going at a fair lick how fast was that?"
"Quite simple ", Dr Fred replied, "To find the speed of the car you divide the distance covered by the time taken. That quarter of a mile was covered in 10 seconds, which is 10÷60÷60 hours. Therefore, the speed was

$$\frac{1}{4} \div \frac{10}{60x60} = \frac{1 \times 60 \times 60}{4 \times 10} = 90 \text{ mph.}$$

"Yes, but how fast was he going when it passed us?", said Tash
"Ah well that's a much more interesting question", said Dr Fred, "if you look carefully at the photo you took you see a still picture and you might say, 'no distance covered, no time taken speed zero over zero', but of course that would be wrong. The time for your picture to be taken might have been one millisecond and in that time, the car might have moved half an inch or so. I suspect that there might be a faint blur to your picture if you look closely, but of course that doesn't answer the question 'What was its speed **at 12 o'clock?'**. We need to do some further analysis.

Let us suppose that the formula for the distance that the car covers after it has passed us is given by the formula

$$s = 0.02\,t + 0.005\,t^2$$

where t is the time in seconds and s it how far it has gone passed us in miles. To find out how far it is ahead of us after time t seconds, we substitute the value of t in the formula. Then we can find the average speed it was going at by dividing the value we get for s by t. The trick to finding the speed **at 12 o'clock** is to make the value of t smaller and smaller. We could proceed like this, but we will see later that, thanks to Isaac Newton and his calculus of

fluxions, that there is an easy way to get to the answer. Look at the values we get when we do the arithmetic:
At time t=0, the sports car is just passing us, so here let t be the number seconds after **12 o'clock**. We calculate the average speed between **t = 0** and **t** seconds after **12 o'clock** by dividing **s**, the number of miles covered, by **t**, the time taken in seconds. This gives us the average speed in miles per second.

t = 0 to 10 seconds
s= 0.2 + 0.5 = 0.7 miles speed = 0.7/10 = 0.07 miles per second

t = 0 to 5 seconds
s= 0.1 + 0.125 = 0.225 miles speed = 0.225/5 = 0.045 mps

t = 0 to 2 seconds
s= 0.04 + 0.02 = 0.06 miles speed = 0.06/2 = 0.03 mps

t = 0 to 1 seconds
s= 0.02 + 0.005 = 0.025 miles speed = 0.025/1 = 0.025 mps

t = 0 to 0.1 seconds
s= 0.002+0.00005 =0.00205 miles speed = 0.00205/0.1 = 0.0205 mps

t = 0 to 0.01 seconds
s=0.0002+0.0000005=0.0002005 m speed = 0.0002005/0.01 = 0.02005 mps

t = 0 to 0.001 seconds
s=0.00002+0.000000005 = 0.000020005 m
speed= 0.000020005/0.001 = 0.020005 mps

If the time is continually divided by ten, we see a pattern in the values we calculate for the average speed:

Time after passing	average speed
t=1	speed = 0.025
t=0.1	speed = 0.0205
t=0.01	speed = 0.02005
t=0.001	speed = 0.020005
t=0.0001	speed = 0.0200005

The value that the average speed approaches is called the speed **at** 12 o'clock. The speed at 12 o'clock appears to be 0.02 miles per second and that works out at 72 miles per hour."

"I thought it was about that.", said Tash.

==============================

The Speed at time t

The point P moves along the x axis Ox, so that its x coordinate OP at time t seconds after we start observations is given by

$$x = t^2 \text{ metres}$$

When we start observations, t=0 and P is at the origin O.
At time t=1, P has moved to the point A. At time t=2, P is at B and after 3 seconds P is passing point C.

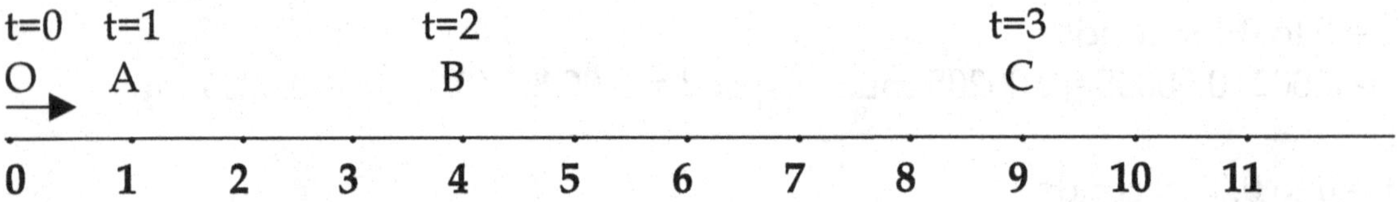

t=0	t=1			t=2					t=3		
O →	A			B					C		
0	1	2	3	4	5	6	7	8	9	10	11

P moves with increasing speed, accelerating further and faster away from O. The average speed in any time interval can be worked out from the formula

$$\text{Average speed} = \frac{\text{distance covered}}{\text{time taken}}$$

For example, at time t=1 P is at point A and at time t=3 P is at C. P covers 9-1 = 8 metres in 2 seconds so its average speed in moving from A to C is 8/2 = 4 metres per second.

We pose the question: **"What is the speed of P when t=2?"**

If we imagine an instant snap shot of P when time t=2, it appears as if P is frozen in time and covers no distance, so the formula $\frac{\text{distance}}{\text{time}}$ becomes $\frac{0}{0}$

and does not work.

Exercise 1

Calculate the average speed for the point P

(i) between t=2 and t=2.5

(ii) between t=2 and t=2.1

(iii) between t=2 and t=2.01

(iv) between t=2 and t=2.001

(v) between t=2 and t=2.0001

Suggest the speed at time t=2

==========================

A General Treatment

To find the speed at time t=2, we calculate the average speed in the time interval t=2 to t=2+h and let the value of h tend to zero. The speed of P **at time t=2** seconds is defined to be the limiting value of the average speed in the time interval t to t+h, as h tends to zero.

The distance from O at time t+h is $(2+h)^2$ and the distance from O at time t=2 is 2^2. The time taken is h seconds and therefore

$$\text{average speed} = \frac{(2+h)^2 - 2^2}{h} = \frac{4+4h+h^2 - 4}{h} = \frac{4h + h^2}{h} = 4+h \ \text{ms}^{-1}.$$

As h → 0 then 4+h → 4 thus we say, the speed at time t=2 is 4 ms^{-1}.

==========================

Exercise 2

[1] Use this method to find the speed when t = 3.

[2] Find the speed when t = T.

The Distance Time Graph

The figure shows the graph of $s=t^2$. In this case, a parabola through the origin.
B represents the position at time t=2, when P is 2 metres from O and C represents the position at time t=3, when P is 9 metres from O.

The average speed between time t=2 and time t=3 is represented by

$$\frac{\text{distance covered}}{\text{time taken}} = \frac{CN}{BN}$$

which is the gradient of the chord BC.

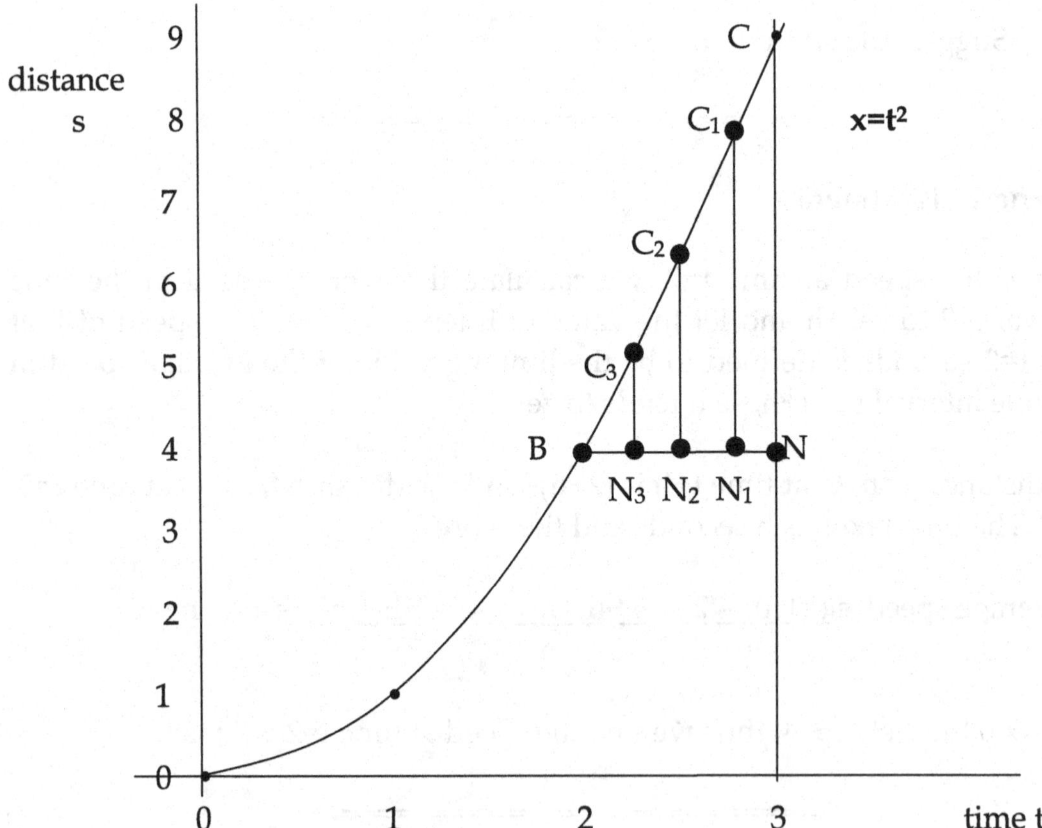

BN represents the time taken between t=2 and t=3 and CN represents the distance covered in this time.
BN_1 represents a smaller time interval of length h seconds from t=2 to t=2+h. As $h \rightarrow 0$ the length BN_1 becomes BN_2, BN_3, BN_4 .... representing shorter and shorter time intervals.

The corresponding distances covered in these shorter time intervals are C_1N_1, C_2N_2, C_3N_3.... and the average speeds in these intervals are C_1N_1/BN_1, C_2N_2/BN_2, C_3N_3/BN_3...

The average speed in any time interval is represented by the gradient of the chord on the distance time graph and as $h \rightarrow 0$, the chord gets closer and closer to the tangent at B and therefore, as $h \rightarrow 0$ the gradient of the chord becomes closer and closer to the gradient of the tangent at B.

We conclude that the limiting value of the average speed is given by the slope of the tangent at B on the distance time graph.

Thus the slope of the tangent at B on the curve $x=t^2$ is 4.

The speed of P at time t=2 seconds is 4 metres per second.

===================================

Differentiation

This limiting process is called differentiation and is a purely algebraic process but it shows that the speed of a moving point at any time is given by the slope of its distance time graph.

Example. A particle moves along the a straight line so that its distance from the origin O on the line, at time t sec is given by $s = t^2 + 2t$. Find its speed at time t=2 seconds.

Solution

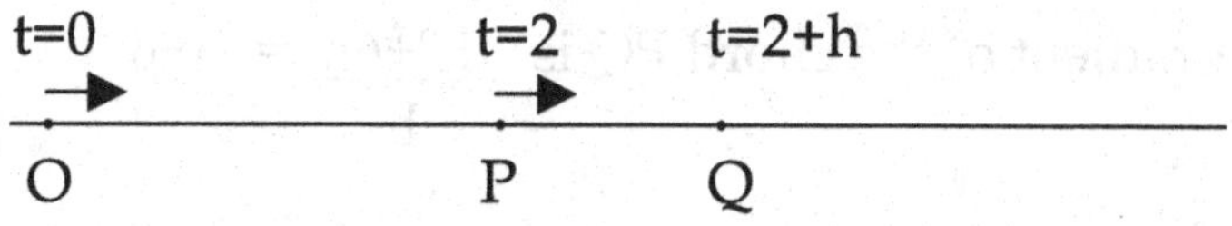

t=0 t=2 t=2+h

O P Q

To find the speed at time t=2, we find the average speed during a small time interval t=2 to 2+h, and then let $h \rightarrow 0$.

The following diagram illustrates this on a distance time graph:

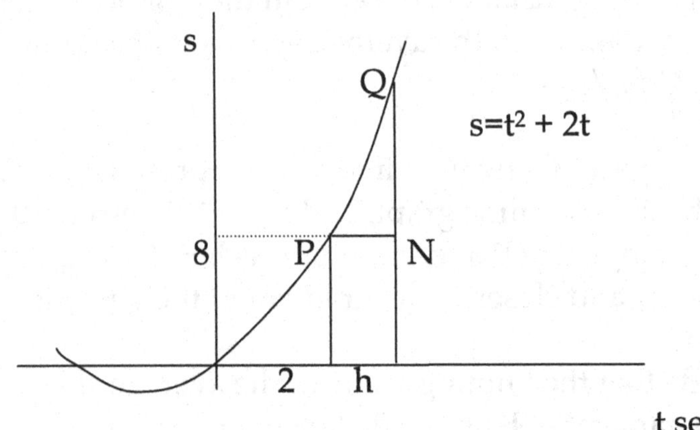

At time t=2 sec the particle is a distance 8 metres from O, represented by the point P(2,8) on the graph $s = t^2 + 2t$.

Let Q be the point on the curve with time coordinate 2+h seconds.

The average speed between time t=2 and t=2+h is represented by the distance covered, QN, divided by the time taken PN.

The average speed from P to Q is therefore the gradient of the chord PQ on the distance time graph.

The s coordinate of Q will be $(2+h)^2 + 2(2+h) = h^2 + 6h + 8$

Subtracting the s coordinate of p from the s coordinate of q we get

$$QN = (h^2 + 6h + 8) - 8 = h^2 + 6h$$

and $\quad PN = h$

Therefore, the gradient of the chord PQ is $\dfrac{h^2+6h}{h} = h+6$

This represents the average speed of the particle between time t=2 and t=2+h seconds.
As $h \rightarrow 0$ the point Q approaches the point P and the gradient of the chord approaches the gradient of the tangent at p.

Now grad(PQ) = h+6

Therefore grad(PQ) → 6 as h → 0

But grad(PQ) → the gradient of the tangent at p.

Therefore the average speed → the gradient of the tangent on the distance time graph. This is what we define to be the speed **at time t=2.**

We conclude that the speed at time t=2 seconds is therefore 6 m/sec.

==

Delta notation

An increase in the value of a variable will be denoted using the delta prefix, thus δx represents an increase (usually intended to be small) in the value of x and δy represents an increase in the value of y.
In our examples on the speed of a moving particle, the average speed can be represented by $\delta s/\delta t$, which is the distance travelled divided by the time taken As $\delta t \to 0$, $\delta s/\delta t \to$ the speed at time t.

The gradient of a curve

Let P(x,y) be a point on the curve y=f(x).

Q is a nearby point also on the curve. The increase in the x coordinate from P to Q is called δx and the corresponding increase in the y coordinate is δy. Thus Q is the point $Q(x+\delta x, y+\delta y)$.

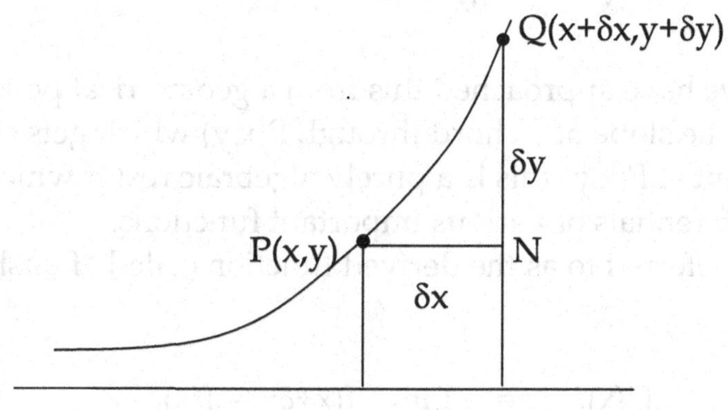

The gradient of the chord PQ is $\frac{QN}{PN} = \frac{\delta y}{\delta x}$

Therefore, the gradient of the tangent at P is the limit of $\frac{\delta y}{\delta x}$ as $\delta x \to 0$

this is written $\lim\limits_{\delta x \to 0} \frac{\delta y}{\delta x}$

If P(x,y) is any point on the curve y = f(x), and Q is the point $(x+\delta x, y+\delta y)$ then, since Q is a point on the curve, we have

$$y+\delta y = f(x+\delta x)$$

therefore

$$\delta y = f(x+\delta x) - f(x)$$

The formula for the gradient at any point P(x,y) on the curve y=f(x) is therefore

$$\lim_{\delta x \to 0} \frac{\delta y}{\delta x} = \lim_{\delta x \to 0} \frac{f(x+\delta x) - f(x)}{\delta x}$$

The symbol for the gradient of the tangent is $\frac{dy}{dx}$ and this is called the

differential of f(x). Thus we now have the formula for the differential of the function f(x) in the form

$$\frac{dy}{dx} = \lim_{\delta x \to 0} \frac{f(x+\delta x) - f(x)}{\delta x}$$

Although we have approached this from a geometrical point of view, being the limit of the slope of a chord through P(x,y) which gets closer and closer to the tangent at P(x,y) this is a purely algebraic result which is used us to find the differentials of various important functions.
This is also referred to as the derived function called "f dash of x" and we often write

$$f'(x) = \lim_{\delta x \to 0} \frac{f(x+\delta x) - f(x)}{\delta x}$$

Notation

Newton and Leibnitz

In a manuscript dated 20 May 1665, by Isaac Newton, aged 23, he shows that he had developed the idea of this kind of limiting process sufficiently to be able to find the equations of tangents to smooth curves and the calculate rates of change. He called these rates of change "fluxions". The rate of change of the distance covered by a moving particle with time is now recognized as the speed of the particle. We still use Newton's dot notation for fluxions when we write "x dot" for dx/dt, i.e. $\dot{x} = \frac{dx}{dt}$

Ten or eleven years later, Gottfried Leibnitz had also, independently, invented the calculus but used the d(x) and δx notation that has now been adopted worldwide.

Theorem 1

If $y=x^n$ then $dy/dx = nx^{(n-1)}$ when n is a positive integer:

Proof

$$\frac{dy}{dx} = \lim_{\delta x \to 0} \frac{f(x+\delta x) - f(x)}{\delta x}$$

$$= \lim_{\delta x \to 0} \frac{(x+\delta x)^n - x^n}{\delta x}$$

$$= \lim_{\delta x \to 0} \frac{x^n+nx^{n-1}\delta x+\text{terms in }\delta x^2 - x^n}{\delta x}$$

(using the binomial expansion)

$$= \lim_{\delta x \to 0} nx^{n-1}+\text{terms in }\delta x$$

$$= nx^{n-1}$$

==

A slight adjustment to the proof allows for negative or rational values of n.

Theorem 2
If $y=x^n$ then $dy/dx = nx^{(n-1)}$ when n is rational or a negative integer:

Proof

$$\frac{dy}{dx} = \lim_{\delta x \to 0} \frac{f(x+\delta x) - f(x)}{\delta x}$$

$$= \lim_{\delta x \to 0} \frac{(x+\delta x)^n - x^n}{\delta x}$$

$$= \lim_{\delta x \to 0} \frac{x^n(1+\delta x/x)^n - x^n}{\delta x}$$

$$= \lim_{\delta x \to 0} \frac{x^n+nx^{n-1}\delta x+\text{terms in }\delta x^2 - x^n}{\delta x}$$

(using Newtons Binomial for rational or negative index)

$$= \lim_{\delta x \to 0} (nx^{n-1}+\text{terms in }\delta x)$$

$$= nx^{n-1}$$

==================================

Example 1 What is the differential of y=4

Solution The graph y=4 gives a straight line parallel to the x axis.
The gradient of the line is zero
Therefore

$$\frac{dy}{dx} = 0$$

===========================

Example 2 What is the differential of y=3x+4

Solution The graph of y=3x+4 is a straight line of gradient m=4
Therefore

$$\frac{dy}{dx} = 4$$

==========================

Example 3 Find the equation of the tangent to $y=x^3$ at the point P(2,8).

Solution $\frac{dy}{dx} = 3x^2$

Therefore, the gradient at P(2,8) is $3.\,2^2 = 12$

Using $y=mx+c$ we have the equation $y = 12x + c$

substitute (2,8) to get $8 = 24 + c$ giving $c = -16$

the equation of the tangent is therefore $y=12x-16$

========================

Exercise 3

[1] Find the equation of the tangent to $y=x^2$ at (1,1)

[2] Find the equation of the tangent to $y=x^3$ at (1,1)

[3] Find the equation of the tangent to $y=x^4$ at (1,1)

[4] Find the equation of the tangent to $y=x^5$ at (1,1)

==

Theorem 3

If $y=u+v$ where $u(x)$ and $v(x)$ are functions of x, then $\frac{dy}{dx} = \frac{du}{dx} + \frac{dv}{dx}$

Proof

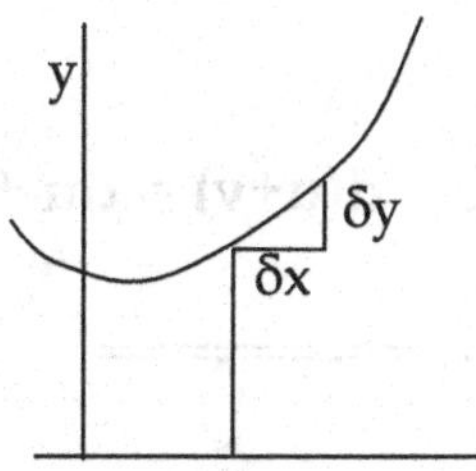

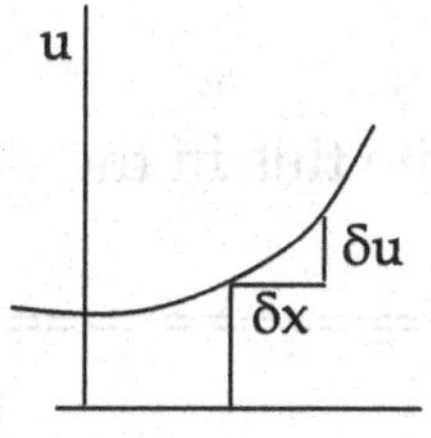

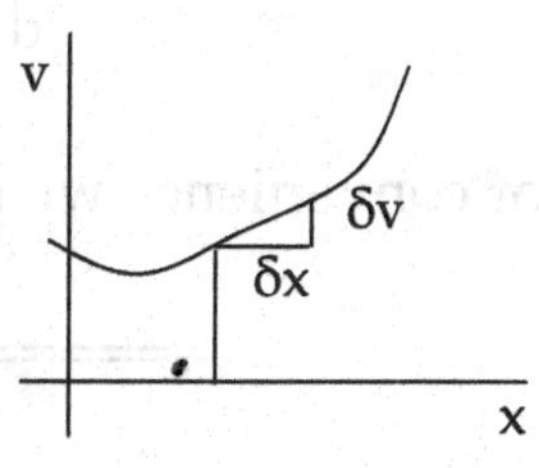

The three figures illustrate the graphs of y, u and v plotted against x but we should realize that the theorem is a purely algebraic result.

Suppose that an increase δx in the value of x produces increases δy in y, δu in u and δv in v.
Then we can therefore write

$$y = u+v$$

$$y+\delta y = u+\delta u + v+\delta v$$

subtracting gives $\delta y = \delta u + \delta v$

divide by δx to get $\frac{\delta y}{\delta x} = \frac{\delta u}{\delta x} + \frac{\delta v}{\delta x}$

and take the limit of both sides to get $\mathbf{\frac{dy}{dx} = \frac{du}{dx} + \frac{dv}{dx}}$

==================================

alternatively, we could express the same argument as

$$\frac{dy}{dx} = \lim_{\delta x \to 0} \frac{u(x+\delta x) + v(x+\delta x) - u(x) - v(x)}{\delta x}$$

$$= \lim_{\delta x \to 0} \frac{u(x+\delta x) - u(x)}{\delta x} + \frac{v(x+\delta x) - v(x)}{\delta x}$$

$$= \lim_{\delta x \to 0} \frac{u(x+\delta x) - u(x)}{\delta x} + \lim_{\delta x \to 0} \frac{v(x+\delta x) - v(x)}{\delta x}$$

$$= \frac{du}{dx} + \frac{dv}{dx}$$

for convenience we may write this in the shorthand **d(u+v) = du + dv**

==

Theorem 4 If y=k.f(x) where k is a constant then $\frac{dy}{dx} = k.f'(x)$

Proof

$$\frac{dy}{dx} = \lim_{\delta x \to 0} \frac{k.f(x+\delta x) - k.f(x)}{\delta x}$$

$$= k. \lim_{\delta x \to 0} \frac{f(x+\delta x) - f(x)}{\delta x}$$

$$= k.f'(x)$$

d(k.u) = k.du

====================================

Example Find the equation of the tangent to $y = x^3+x^4$ at the point (1,2)

Solution $\frac{dy}{dx} = 3x^2 + 4x^3$

the gradient of the tangent is therefore m = 3 + 4 = 7

Use $y-y1 = m(x-x1)$

To get $y - 2 = 7(x - 1)$

The tangent at (1,2) is therefore **y = 7x – 5**

============================

Exercise 4

[1] Use dy/dx and the equation $y-y_1 = m(x-x_1)$ to find the equation of the tangent to $y = 4x^2+4$ at the point P(2,20).

[2] Find the equation of the tangent to $y = x^2 + 2x + 1$ at the point P(1,4) showing that this tangent passes through the origin.

[3] Find the equation of a tangent to the curve $y = x^2+2x+4$, that passes through the point (-1 , -1)

============================

Maximum and Minimum

A farmer wishes to make a rectangular chicken run from 100 feet of chicken wire, using an existing wall for one of the sides.

He could make a long thin run

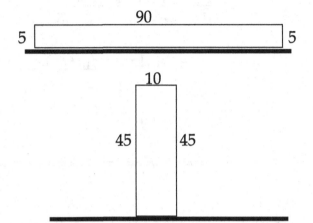

90

5 5

or a narrow wide chicken run

10

45 45

Somewhere between these two extremes, the farmer figures there must be a chicken run that gives the maximum area for his chickens:

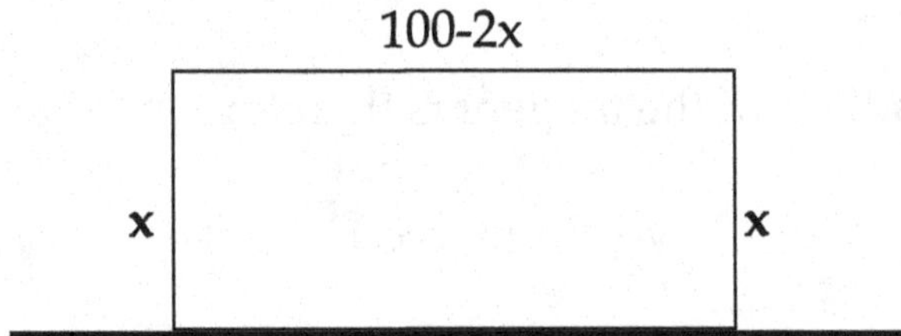

Suppose that the dimensions of the ideal chicken run are x ft by 100-2x ft.

Then $\quad$ Area = $100x - 2x^2$

If we graph the area against x then we get this parabola shape:

area

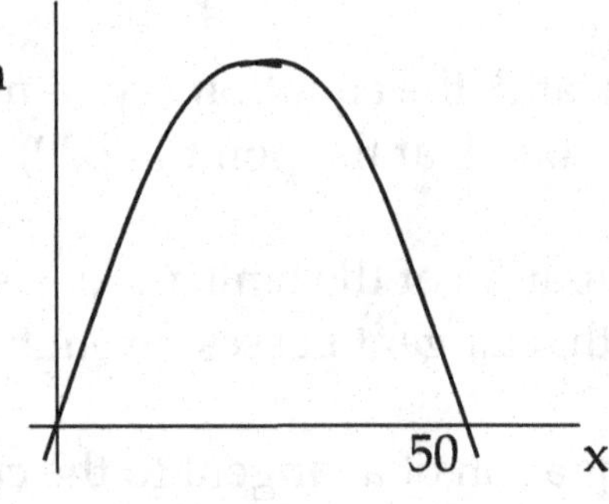

The parabola cuts the x axis at x=0 and x=50.

The farmer knows that the parabola is symmetrical, so, he deduces that the peak occurs when x=25.

The farmer therefore constructs a 25 by 50 chicken run to give his chickens the maximum area.

Problem

Suppose that we did not know that the parabola was symmetrical? How would we know what value of x gives the maximum area?

The solution is to use the gradient of the graph. At the maximum, the gradient of the graph is zero.

Solution: $A = 100x - 2x^2$

$$\frac{dA}{dx} = 100 - 2x$$

$$\frac{dA}{dx} = 0 \text{ when } 2x = 100$$

so the maximum occurs when x = 50

=====================================

Example 1 Find the maximum value of $6x - 4 - 2x^2$

Solution

$y = 8x - 4 - 2x^2$

gives $dy/dx = 8 - 4x$

dy/dx is zero when x = 2 which gives the maximum $y_{max} = 16 - 4 - 8 = 4$

=====================================

Example 2 Find the minimum value of x^3-x^2 for positive values of x.

Solution

Let $y=x^3-x^2$ then the sketch of the curve has a maximum at the origin where the curve touches the x axis and a minimum between x=0 and x=1.

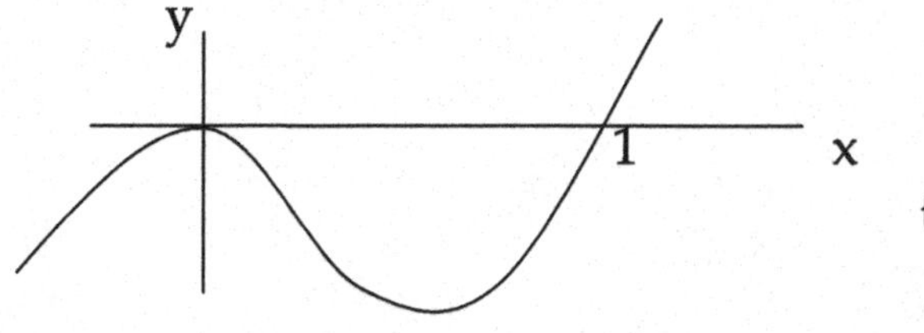

the graph $y=x^3-x^2$

Now $$y = x^3 - x^2$$

gives $$\frac{dy}{dx} = 3x^2 - 2x$$

$\frac{dy}{dx} = 0$ when x = 0 at the origin or where x = 2/3, at the minimum.

Substituting x = 2/3 we have $y_{min} = (2/3)^3 - (2/3)^2$

$$= (4/9)[-1/3] = -4/27$$

=====================================

Exercise 5

[1] Find the maximum value of $2 + 3x^2 - 2x^3$ for positive x values.

[2] Find the maximum and minimum values of the curve

$$y = 8x^3 - x^4 - 22x^2 + 24x \quad \text{for values of x between 0 and 4}$$

[3]

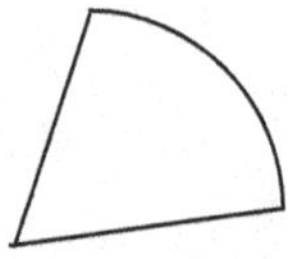

The perimeter of this sector is 40 cm. Given that the radius is r cm, show that the area of the sector is $A = 20r - r^2$ and hence find the maximum possible area of the sector.

[4] An open box is to be made by cutting squares from the corners of a 100cm x 100cm square of cardboard and then folding up the flaps. What is the maximum possible volume of the box?

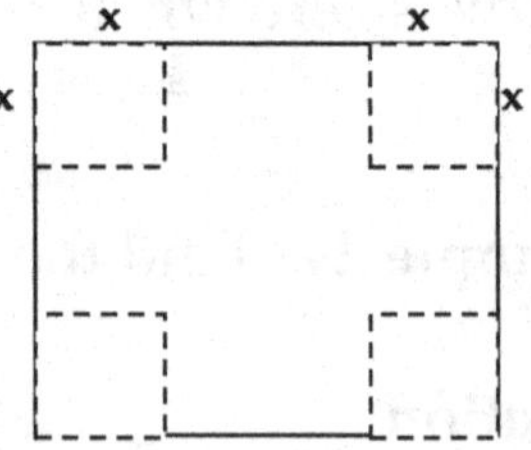

========================

Turning Points, Stationary Points

Example A particle moves along the x axis so that its distance from the origin O at time t seconds is given by

$$s = 4t - t^2$$

When is the particle at rest?

Solution

The velocity is given by

$$\frac{ds}{dt} = 4 - 2t$$

The particle is at rest when $4 - 2t = 0$.

Thus, when t=2, the velocity is instantaneously zero.

If we plot the distance time graph we see a peak in the graph at t=2, giving a local maximum value.

At t=0 the particle is at the origin.
It moves away from O for 2 seconds.
At t=2, the particle turns and moves back towards the origin.

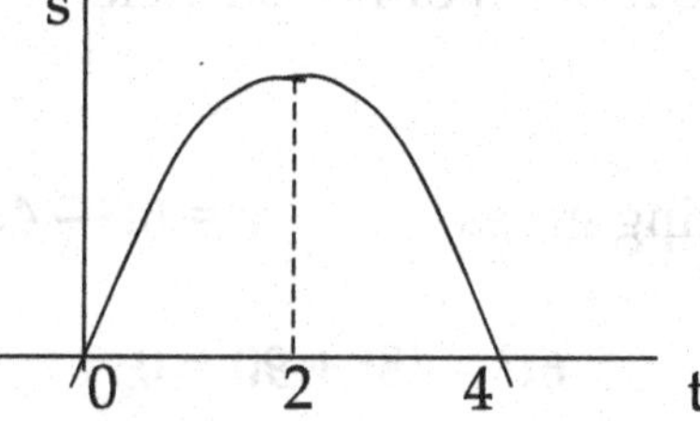

When t=2, the particle is instantaneously stationary.

The point t=2, s=4 is called a **stationary point,** or a **turning point** of the graph because the particle turns when t=2 and is instantaneously stationary.

(Note: in such problems there is nothing wrong with using negative values for time. A value of t = -2 in the above problem would refer to the instant of time 2 seconds before we started our stop watch. If time zero is now, then time minus 2 days is the day before yesterday.

==================================

Kinematics

If we are given the position **s** of a particle at time **t** then we know that its velocity is given by $\mathbf{v} = \frac{\mathbf{ds}}{\mathbf{dt}}$

This gives the velocity of the particle as a function of **t.**

In a similar way, the acceleration of the particle is given by the slope of the velocity-time graph so that the acceleration **a** is given by

$$\mathbf{a} = \frac{\mathbf{dv}}{\mathbf{dt}}$$

Example. A particle moves in a straight line so that its position at time t relative to an origin O on the line is given by

$$\mathbf{s = 2t^3 - 24t^2 + 90t}$$

Describe the motion of the particle.

Solution
Differentiating gives $\mathbf{v = 6t^2 - 48t + 90}$

v=0 when $\mathbf{6t^2 - 48t + 90 = 0}$

$$\mathbf{t^2 - 8t + 15 = 0}$$

$$\mathbf{(t\text{-}3)(t\text{-}5) = 0}$$

there are therefore turning points at **t=3** and **t=5** where the motion changes direction.
Differentiating **v** gives
$\mathbf{a = 12t - 48}$ which is zero at time **t=4.**

The particle leaves the origin with speed **90.** At **t=3** the particle changes direction and moves back toward the origin until **t=5** when it again changes direction and speeds away from O.
The motion is best described by sketches of the distance-time graph, the velocity-time graph and the acceleration-time graph.

The Distance time graph

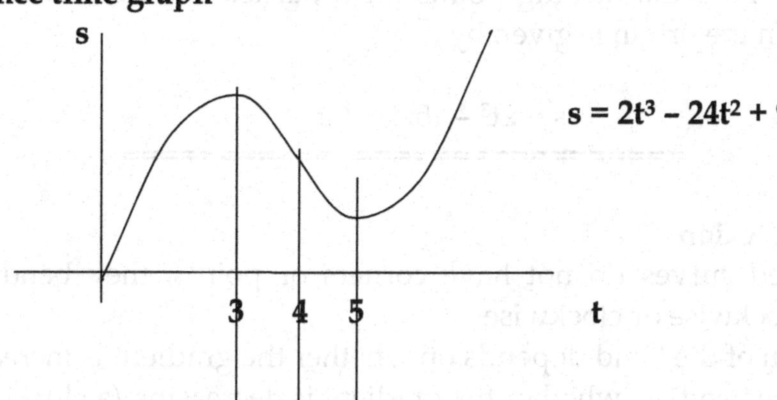

The Velocity time graph

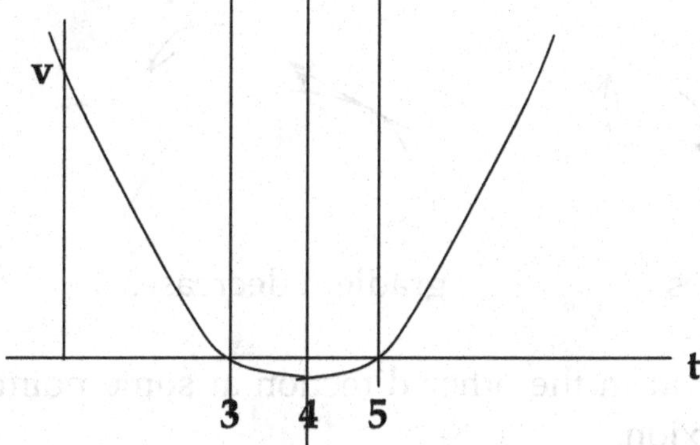

The Acceleration time graph

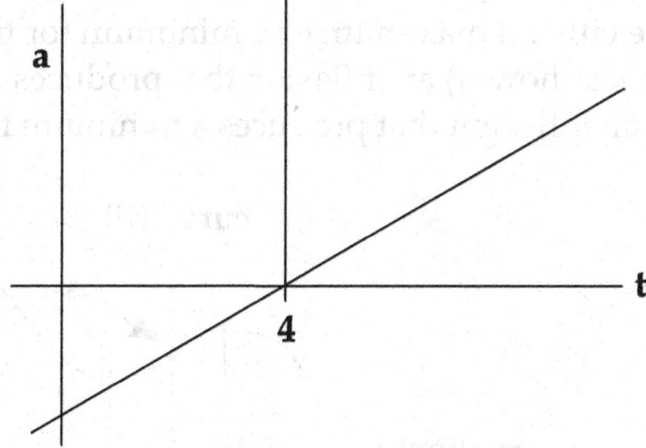

The turning points, where **v=0,** are local maxima and minima on the **s-t** graph. The acceleration is zero at any maximum or minimum on the **v-t** graph.

===============================

Exercise 6. Find the turning points for a particle P that moves so that its distance from the origin is given by

$$s = 2t^3 - 15t^2 + 36t$$

================================

Points of Inflexion

Well behaved curves do not have corners or points, they bend smoothly either anticlockwise or clockwise.

The direction of the bend depends on whether the gradient is increasing (an anticlockwise bend) or whether the gradient is decreasing (a clockwise bend).

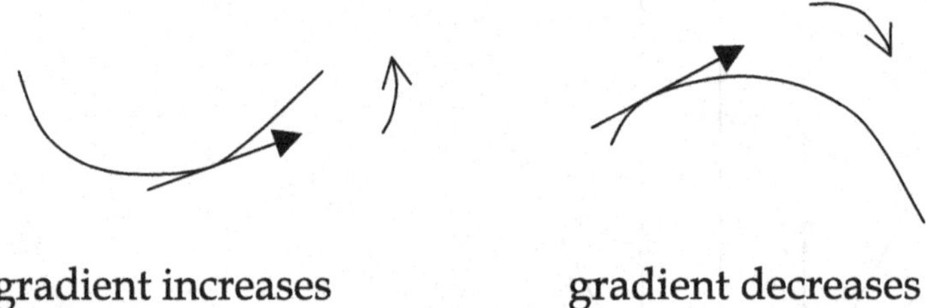

gradient increases gradient decreases

If the curve starts to bend in the other direction at some point, then we say we have a point of inflexion.

On one side of the inflexion the gradient increases and on the other side the gradient decreases.

This means that we have either a maximum or a minimum for the gradient:

The following two sketches show (i) an inflexion that produces a maximum for the gradient and (ii) an inflexion that produces a minimum for the gradient.

curve (i) **curve (ii)**

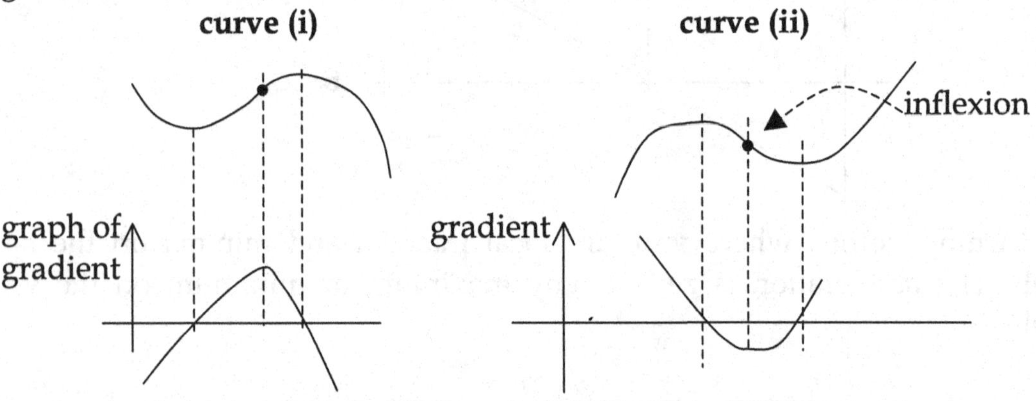

The gradient is zero at the two turning points and has a maximum at the inflexion.

Here, the gradient is again zero at the two turning points but has a minimum at the infexion.

The second differential, d^2y/dx^2

If we differentiate **dy/dx** we obtain the second differential of y with respect to x and we write

$$\frac{d(dy/dx)}{dx} = \frac{d^2y}{dx^2}$$ pronounced D 2 Y by D X squared

Since the gradient has a maximum (or minimum) at points of inflexion on the curve, the gradient of the gradient is zero. Thus, at these points of inflexion we have $d^2y/dx^2 = 0$
but:

Vertical inflexions

Note that this last result cannot apply to a vertical inflexion, for example, the graph $y=\sqrt[3]{x}$, the cube root of x, has an inflexion at the origin but d^2y/dx^2 is not zero.

$y=\sqrt[3]{x}$ has a vertical inflexion at x=0

The Shapes of Curves

An example of a polynomial curve is $y=ax^5+bx^4+cx^3+dx^2+ex+f$. The degree of the polynomial is the highest power of **x** and in this example the degree of the curve is **5.** A polynomial of degree 5 has at most 5 real zeros and so this curve can cut the x axis in at most five points. In general, a polynomial curve of degree **n** can cut the x axis in at most **n** points. If a polynomial of degree n is differentiated then it becomes a polynomial of degree (n-1) and therefore an nth degree polynomial curve can have at most (n-1) maximum or minimum points. Further, there can be at most (n-2) infexions.

Linear equations: $y=ax+b$
The graph of a linear equation is, of course, a straight line.
It has no turning points and slopes upwards or downwards according to the sign of the term in x;
The linear curve has at most one zero. (None if it is parallel to the x axis)

Example sketches

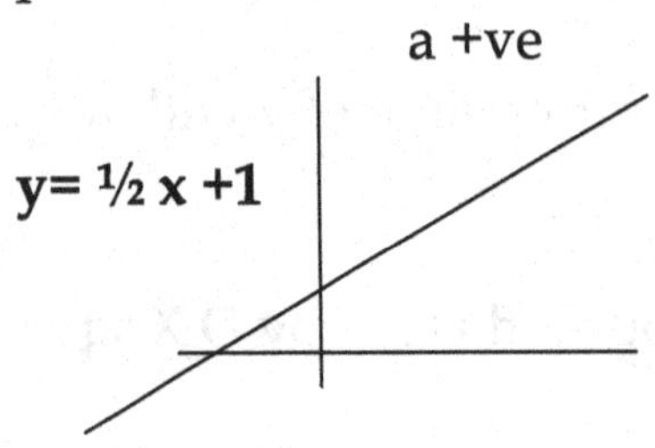

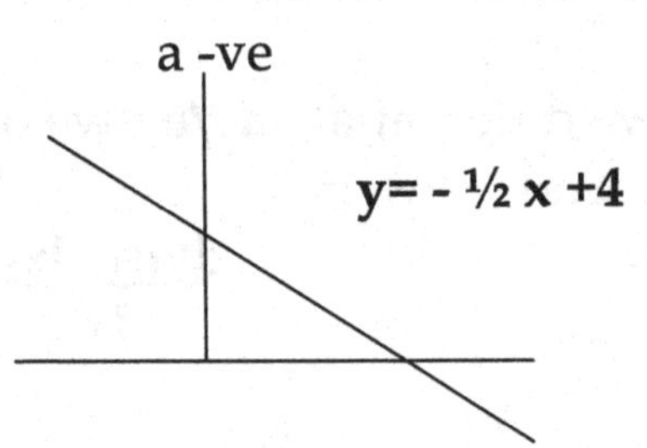

Quadratic equations: $y=ax^2 + bx + c$

A quadratic has one turning point and at most 2 real zeros.
The graph of a quadratic is symmetrical about the vertical axis through the turning point.

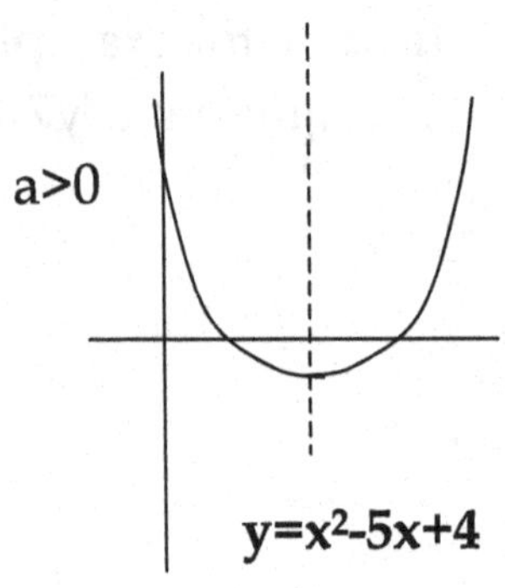

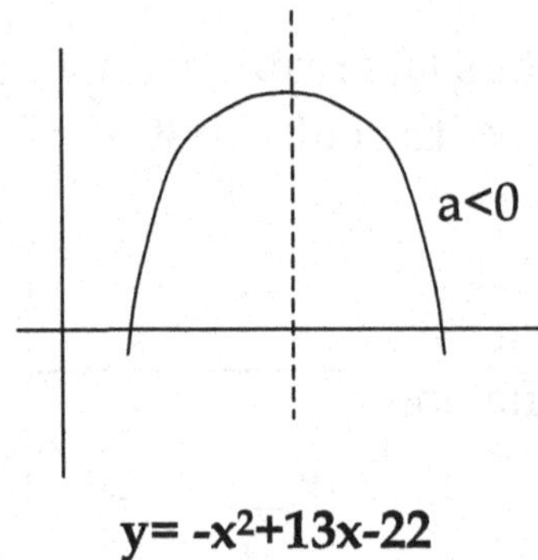

Cubic equations: $y=ax^3 + bx^2 + cx + d$

A cubic graph has at most 3 real zeros, and at most 2 real turning points.
A cubic graph has one point of inflexion where the curve starts to bend in the opposite direction.
A cubic graph has rotational symmetry about the point of inflexion.

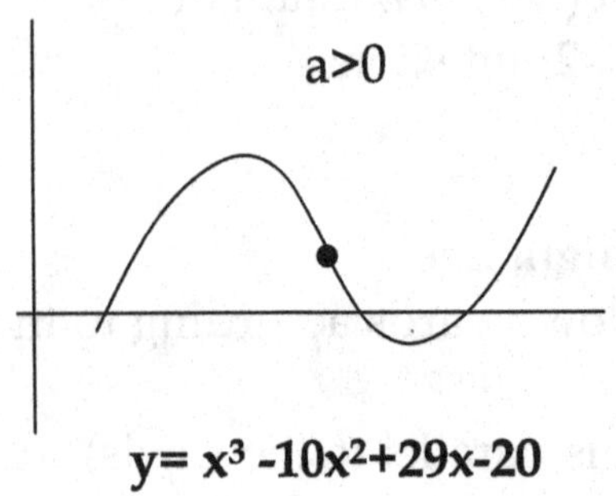

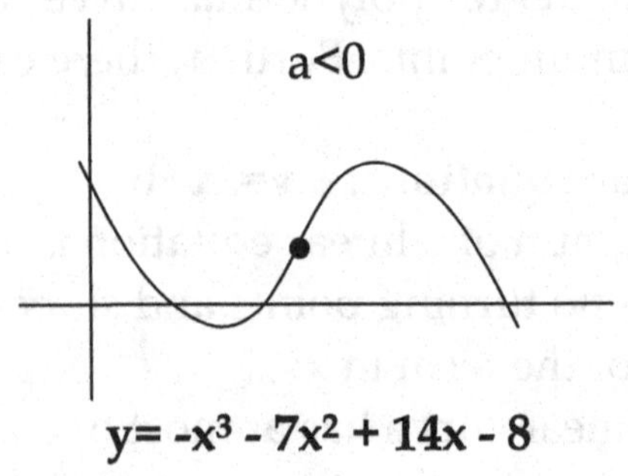

Quartic equations: $y=ax^4 + bx^3 + cx^2 + d$

A fourth degree graph has at most 4 real zeros, 3 turning points and 2 inflexions.
There is in general no symmetry.

The shapes of Quartic Curves

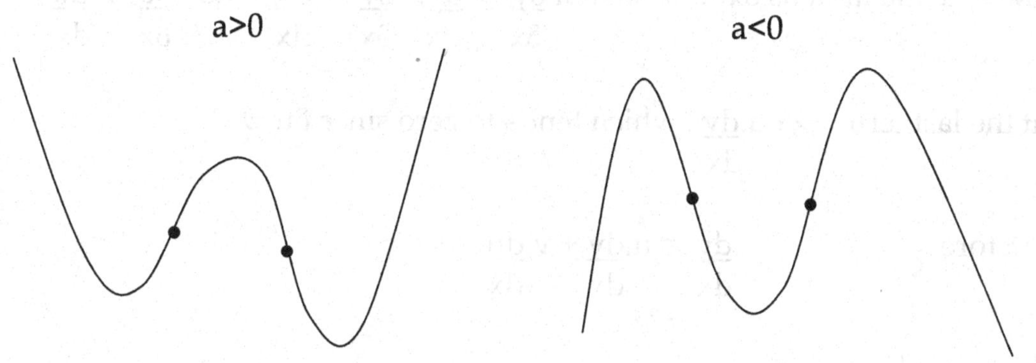

Exercise 7

[1] Find the turning points and the point of inflexion for $y=x^3-6x^2+9x+2$
Use this information to draw a sketch of the curve.

============================

Theorem 5 The product rule. If $y = u.v$ then $\frac{dy}{dx} = u.\frac{dv}{dx} + v.\frac{du}{dx}$

Proof u(x) and v(x) are functions of x.

Let δu, δv and δy be the increases in u, v and y produced by an increase δx in x.

Then $$\delta y = (u+\delta u)(v+\delta v) - uv$$

$$= uv+u\delta v + \delta uv + \delta u\delta v - uv$$

$$= u.\delta v + \delta u.v + \delta u.\delta v$$

divide by δx to get

$$\frac{\delta y}{\delta x} = u.\frac{\delta v}{\delta x} + \frac{\delta u}{\delta x}.v + \delta u.\frac{\delta v}{\delta x}$$

Now take the limit as $\delta x \rightarrow 0$ so that $\frac{\delta y}{\delta x} \rightarrow \frac{dy}{dx}$, $\frac{\delta v}{\delta x} \rightarrow \frac{dv}{dx}$ and $\frac{\delta u}{\delta x} \rightarrow \frac{du}{dx}$

but the last term $\rightarrow \delta u.\frac{dv}{dx}$ which tends to zero since $\delta u \rightarrow 0$.

therefore $$\frac{dy}{dx} = u.\frac{dv}{dx} + v.\frac{du}{dx}$$

which we abbreviate to

$$\mathbf{d(uv) = u.dv + v.du}$$

=====================================

A Graphical illustration

Consider the curve $y = x^5$ which we can write as the product $y = x^2.x^3$
The graphs of these three functions are:

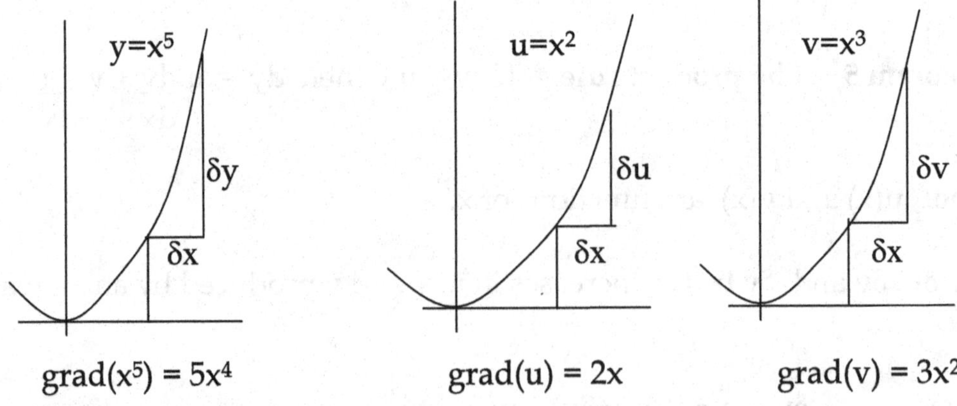

Suppose that a small increase δx, in the value of x, produces increases δy in y, δu in u(x) and δv in v(x).

The product rule gives

$$\text{grad}(x^2.x^3) = x^2.\text{grad}(v) + x^3.\text{grad}(u)$$

$$= x^2.\ 3x^2 + x^3.2x$$

$$= 3x^4 + 2x^4$$

$$= 5x^4$$

However, in practice, none of this detail is necessary as differentiation is essentially a purely algebraic exercise.

===

Theorem 6 The Quotient rule. If $y = \frac{u}{v}$ then $\frac{dy}{dx} = \frac{v.du - u.dv}{v^2}$

Proof

Suppose that $y = \frac{u}{v}$

and that an increase δx in x produces increases δy in y, δu in u and δv in v.

then $$y + \delta y = \frac{u + \delta u}{v + \delta v}$$

so that $$\delta y = \frac{u + \delta u}{v + \delta v} - \frac{u}{v}$$

$$= \frac{uv + v.\delta u - uv - u.\delta v}{v.(v + \delta v)}$$

$$= \frac{v.\delta u - u.\delta v}{v.(v + \delta v)}$$

divide by δx and we have

$$\frac{\delta y}{\delta x} = \frac{v.\delta u/\delta x - u.\delta v/\delta x}{v.(v + \delta v)}$$

Le t $\delta x \to 0$ then $\frac{\delta y}{\delta x} \to \frac{dy}{dx}$, $\frac{\delta u}{\delta x} \to \frac{du}{dx}$, $\frac{\delta v}{\delta x} \to \frac{dv}{dx}$ and $\delta v \to 0$

Then we have $$\frac{dy}{dx} = \frac{v.du/dx - u.dv/dx}{v^2}$$

which we usually abbreviate to

$$\mathbf{d(u/v) = \frac{v.du - u.dv}{v^2}}$$

===

Exercise 8

[1] Differentiate $y=(x^2+1)^3(x^3+1)^2$

[2] Differentiate $y=\frac{x^3+1}{x^2+1}$

================================

Implicit Functions

$y=x^2$ is a simple curve with the gradient function $dy/dx = 2x$.

Now if $y=x^2$ it is easy to show that $y^2 + x^2y = 2x^4$

so that we have $$y^2 + x^2y - 2x^4 = 0$$

This is an example of an **implicit function.**
There is a certain set of points (x,y) that satisfy this equation and describe a curve in the x-y plane.
It is implied that we should be able to find y as some function of x, but although we expect that $y=x^2$ should appear as we will soon see, we cannot always take it for granted that we should always get y as a function of x.

In this case, we have $y^2 + x^2y - 2x^4 = 0$

giving
$$y = \frac{-x^2 \pm \sqrt{(x^4 - 4(1)(-2x^4)}}{2}$$
$$= \frac{-x^2 \pm 3x^2}{2}$$

so that we have two functions, $y=x^2$ and $y=-2x^2$
let us now differentiate the original equation term by term:

$$y^2 + x^2y - 2x^4 = 0$$

gives $2y.dy + x^2.dy + y.2xdx - 8x^3 dx = 0$

so that
$$\frac{dy}{dx} = \frac{2xy - 8x^3}{2y + x^2}$$

If we choose $y= x^2$ we get
$$\frac{dy}{dx} = \frac{2x^3 - 8x^3}{2x^2 + x^2} = \frac{6x^3}{3x^2} = 2x$$

If we choose $y= -2x^2$ we get
$$\frac{dy}{dx} = \frac{-4x^3 - 8x^3}{-4x^2 + x^2} = \frac{12x^3}{-3x^2} = -4x$$

Both of these results are correct for the chosen curve.

The question arises: "Can we always do this?". Well not all "implicit" equations give smooth curves. Some "implicit" equations that do produce smooth curves often have points where the above method does not work, for example, where the curve intersects itself or has a vertical gradient.

It depends on the "Implicit Function Theorem" which needs the use of partial derivatives.
If F(x,y) is a function of two variables x and y, we can differentiate F(x,y) assuming that y is a constant. This gives the partial derivative $\partial F/\partial x$.
If F(x,y) is differentiated with respect to y, assuming that x is a constant then we have the partial derivative with respect to y, $\partial F/\partial y$.

Example Given $F(x,y) = y^2 + x^2y - 2x^4$

We have $\partial F/\partial x = 2xy - 8x^3$ and $\partial F/\partial y = 2y + x^2$

The Implicit Function Theorem states that if $F(x,y)$, $\partial F/\partial x$ and $\partial F/\partial y$ are continuous at $P(x,y)$ then there is a unique function $y=f(x)$ near $P(x,y)$ and also, that

$$\frac{dy}{dx} = -\frac{\partial F/\partial x}{\partial F/\partial y}$$ provided that $\partial F/\partial y$ is not zero

In our example, if we choose the point P(1,1) on $y= x^2$ implicit differentiation gives the gradient of $y= x^2$ at the point (1,1).

If we choose the point P(1,-2) on $y= -2x^2$ then implicit differentiation gives the gradient of $y= -2x^2$ at P(1,-2).

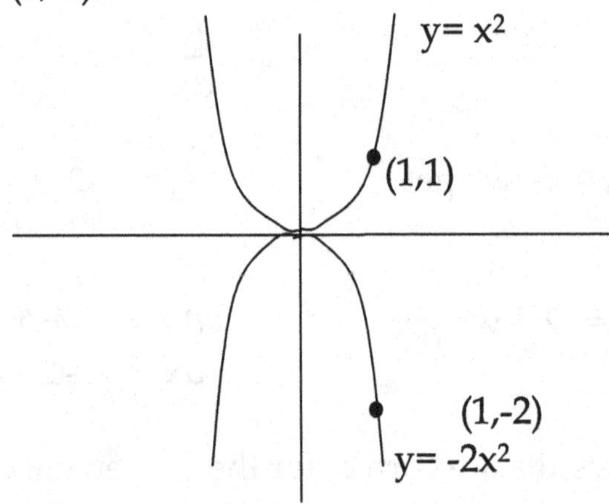

however, if we choose the point (0,0), the result does not work.

Example

Find the equation of the tangent to the ellipse $4x^2 + 9y^2 = 25$ at P(2,1).

Solution

$$\text{Let } F(x,y) = 4x^2 + 9y^2 - 25 = 0$$

Then

$$\partial F/\partial x = 8x \text{ and } \partial F/\partial y = 18y$$

Therefore, by the implicit function theorem

$$\frac{dy}{dx} = -\frac{8x}{18y}$$

the gradient at P(2,1) is therefore -16/18

Using $y-y_1 = m(x-x_1)$, the equation of the tangent is therefore

$$y - 1 = -16/18\ (x - 2)$$

$$18y - 18 = -16x + 32$$

$$18y + 16x - 50 = 0$$

Exercise 9

[1] Use implicit differentiation to find the gradient of the circle $x^2 + y^2 - 6x - 8y = 0$ at the point P(6,8)
Hence show that the equation of the tangent at P is $4y + 3x = 48$
Why does this method not work for the point (8,4)?

[2] Use implicit differentiation to find the gradient of the curve $3x^2 - 4y^2 + 2xy - 4 = 0$ at the point (2,2). Hence find the equation of the tangent to the curve at that point.

==

Of course not all functions F(x,y) can trusted to produce "well behaved " curves in the x-y plane, for example:

[1] $x^2 + y^2 = 1$ is the locus of a circle, radius 1

[2] $x^2 + y^2 = 0$ is only true for the origin

[3] $x^2 + y^2 = -1$ is does not hold for any points in the x-y plane,

however, implicit differentiation gives

$$\partial F/\partial x = 2x \quad \text{and} \quad \partial F/\partial y = 2y \qquad \text{so that} \quad \frac{dy}{dx} = -\frac{x}{y}$$

for each of the equations but the gradient function is only valid for the first equation.

==============================

Some equations have very strange graphs:

Given $$\sin x + \frac{1}{\cos y} = 0$$

we note that sin x is never greater than 1 but $\frac{1}{\cos y}$ is never less than 1.

We can only get the required zero if one term is +1 and the other –1.
Either we have sin x = -1 and cos y = 1
thus x= ......-5π/2, -π/2, 3π/2, 7π/2...... and y= ... -4π, -2π, 0, 2π, 4π.....

or we have sin x = +1 and cos y = -1.
so that x= .....-7π/2, -3π/2, π/2, 5π/2...... and y= ....-5π, -3π, -π, π, 3π ......

Giving an infinite set of isolated points:

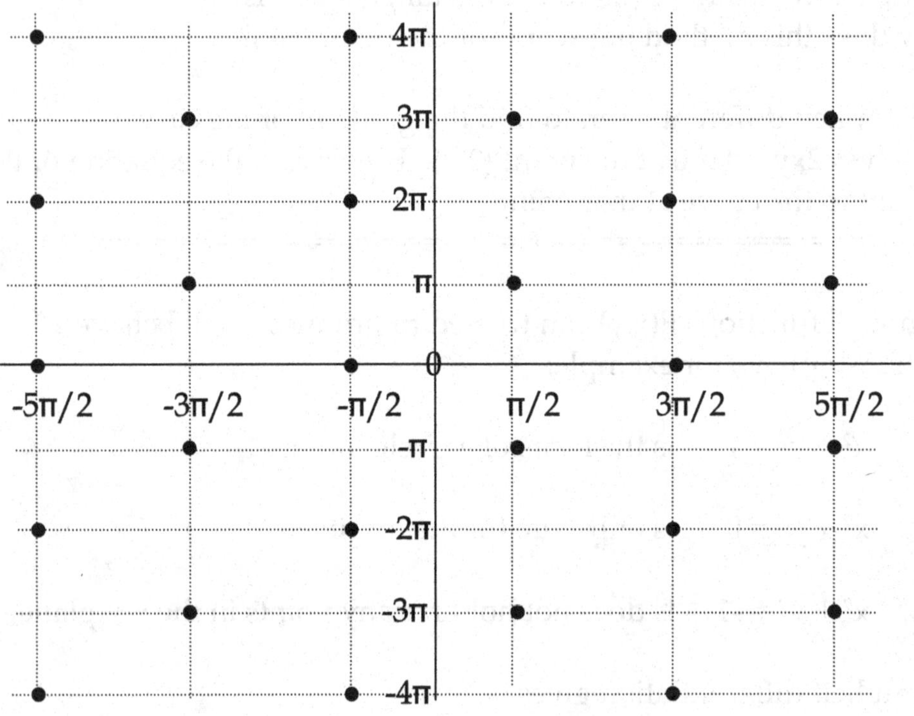

The "Graph" of $$\sin x + \frac{1}{\cos y} = 0$$

Trig functions

Theorem 7 $\frac{d}{dx}(\sin x) = \cos x$

Proof

$$\frac{d}{dx}(\sin x) = \lim_{\delta x \to 0} \frac{\sin(x+\delta x) - \sin x}{\delta x}$$

$$= \lim_{\delta x \to 0} \frac{2\cos(x+\delta x/2).\sin(\delta x/2)}{\delta x}$$

$$= \lim_{\delta x \to 0} \frac{\cos(x+\delta x/2).\sin(\delta x/2)}{(\delta x/2)}$$

Now $\cos(x+\delta x/2) \to \cos x$ and $\frac{\sin(\delta x/2)}{(\delta x/2)} \to 1$ (see chapter 20)

hence $\mathbf{\frac{d}{dx}(\sin x) = \cos x}$

==

Theorem 8 $\frac{d}{dx}(\cos x) = -\sin x$

Proof

$$\frac{d}{dx}(\sin x) = \lim_{\delta x \to 0} \frac{\cos(x+\delta x) - \cos x}{\delta x}$$

$$= \lim_{\delta x \to 0} \frac{-2\sin(x+\delta x/2).\sin(\delta x/2)}{\delta x}$$

$$= \lim_{\delta x \to 0} \frac{-\sin(x+\delta x/2).\sin(\delta x/2)}{(\delta x/2)}$$

Now $\sin(x+\delta x/2) \to \sin x$ and $\frac{\sin(\delta x/2)}{(\delta x/2)} \to 1$ (see chapter 20)

hence $\mathbf{\frac{d}{dx}(\cos x) = -\sin x}$ ■

We use the results of Theorems 8 and 9 along with the quotient rule to find the differential of **tan x**

Theorem 9 $\frac{d}{dx}(\tan x) = \sec^2 x$

Proof

Using the quotient rule, $\frac{v.du - u.dv}{v^2}$

we have

$$\frac{d}{dx}\left(\frac{\sin x}{\cos x}\right) = \frac{\cos x \,.\, \cos x \;-\; \sin x \,.(-\sin x)}{\cos^2 x}$$

$$= \frac{\cos^2 x + \sin^2 x}{\cos^2 x}$$

$$= \frac{\cos^2 x + \sin^2 x}{\cos^2 x}$$

$$\mathbf{\frac{d}{dx}(\tan x) = \sec^2 x}$$

===================================

Exercise 10

[1] Use $\frac{d}{dx}\cot x = \frac{d}{dx}\left(\frac{\cos x}{\sin x}\right)$ to prove that $\mathbf{\frac{d}{dx}\cot x = -\operatorname{cosec}^2 x}$

===================================

Sec and Cosec

Example We note that $\sec^2 x = 1 + \tan^2 x$

Differentiating both sides we have (using D for d/dx)

$$2\sec x.\, D(\sec x) = 2\tan x. D(\tan x)$$

thus $2\sec x. D(\sec x) = 2\tan x.\sec^2 x$

hence $D(\sec x) = \tan x.\sec x$

or, more usually $\mathbf{\frac{d}{dx}\sec x = \sec x.\tan x}$

Exercise 11

Use $\text{cosec}^2 x = 1 + \cot^2 x$ to prove that $\frac{d}{dx}\text{cosec}\, x = -\text{cosec}\, x.\cot x$

====================================

The Indefinite Integral

Integration is the reverse of differentiation thus, since the differential of x^4 is $4x^3$ we say, the integral of $4x^3$ is x^4.
However, since the differential of a constant is zero, we do not know if there were any constants that differentiated out hence the name indefinite integral.
We write:

$$\int 3x^2\,dx = x^4 + c$$ where c is called the arbitrary constant

The integral sign $\int$ is simply an elongated S for sum. It was invented by Leibnitz towards the end of the 17th century.

The general formula for integrating powers of x is

$$\int x^n\,dx = \frac{x^{n+1}}{n+1} + c.$$

Exercise 12

Write down

[1] $\int (x^2 + x^3 + x^4 + x^5 + x^6)\,dx$

[2] $\int (1 + 2x + 3x^2 + 4x^3)\,dx$

[3] $\int (x^{-1/2} + x^{-2} + x^{-3} + x^{-4} + x^{-5})\,dx$

================================

We now use the indefinite integral to derive (i) the area of a circle, (ii) the volume of a sphere, (iii) the surface area of a sphere and (iv) the volume of a cone.

Examples using the Indefinite Integral

Example 1: The area of a Circle

Suppose that the area of a circle of radius r is given by A.
Now we know that the only variable that the area A depends on is the radius r so that A is some function of r that is A = A(r).

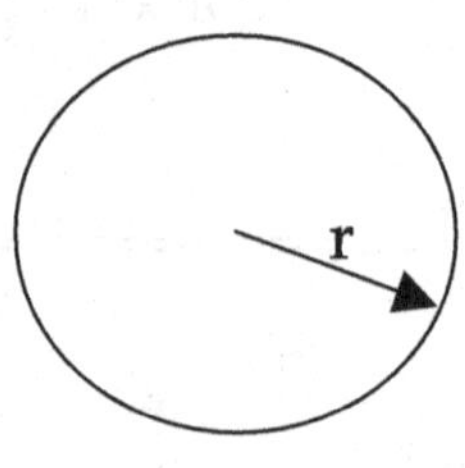

Since all circles are similar, the ratio of the circumference to the diameter of any circle is the same value, that is π.
Thus

$$C = \pi d$$

or

$$C = 2\pi r$$

Now suppose that the radius of the circle is increased by δr producing an increase in the area of δA.

The area δA will be a ring of width δr and length approximately C.
Thus we have

$$\delta A = C \times \delta r$$

thus $$\frac{\delta A}{\delta r} = 2\pi r \qquad \text{(approximately)}$$

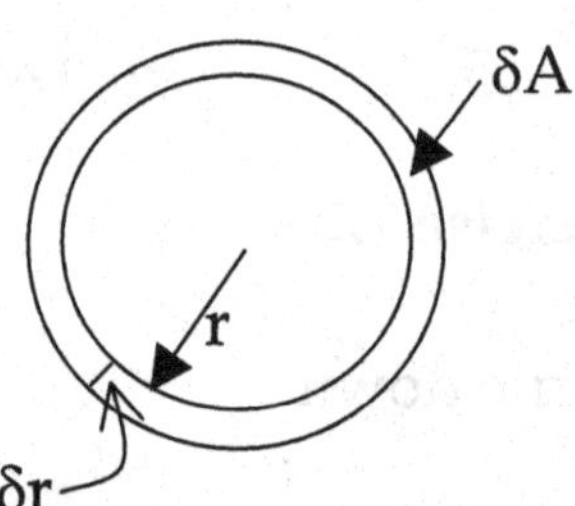

and we conclude that

$$\frac{dA}{dr} = 2\pi r$$

integrating gives $A = \pi r^2 + c$ and clearly A=0 if r=0 so therefore c=0

Thus

$$\mathbf{A = \pi r^2}$$

==============================

The Volume of a Sphere

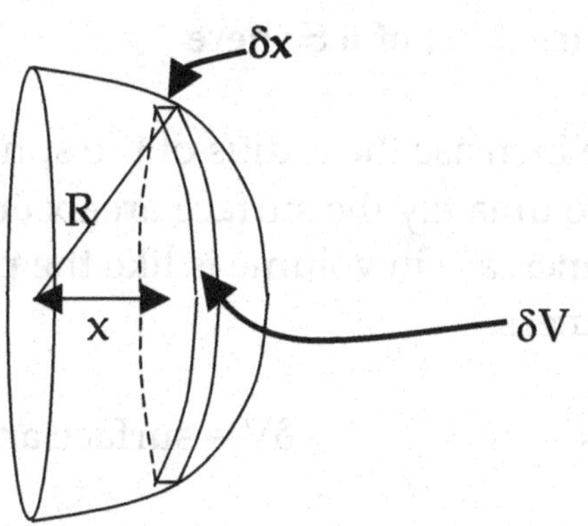

The diagram represents a hemi sphere of radius R.
V(x) represents the volume that part of the sphere from the circular base, to a distance x horizontally along the x axis. Thus V(x) depends on the constant value R and the variable x.
Let δx represent a small increase in the value of x.
Let δV be the corresponding increase in V.
δV will be a circular slice of thickness δx and radius y where $y^2 = R^2 - x^2$
We therefore have the approximate relation

$$\delta V = \pi y^2 . \delta x$$

$$= \pi(R^2 - x^2) . \delta x$$

we deduce that

$$\frac{d}{dx}V(x) = \pi(R^2 - x^2)$$

integrate

$$V(x) = \pi R^2 .x - \pi \frac{x^3}{3} + c$$ (V=0 when x=0 hence c=0)

putting x=R and c=0 we have the volume of the hemisphere as

$$\text{hemi sphere} = \pi R^3 - \pi \frac{R^3}{3} = \frac{2\pi R^3}{3}$$

The volume of a sphere, radius R is therefore

$$\mathbf{V = \frac{4\pi R^3}{3}}$$

■

Surface Area of a Sphere

If we increase the radius of the sphere by δr, then the volume increases by approximately the surface area x δr.
The increase in volume is like the peel of an orange with volume = area x thickness.

Thus $\delta V = \text{surface area} \times \delta r$

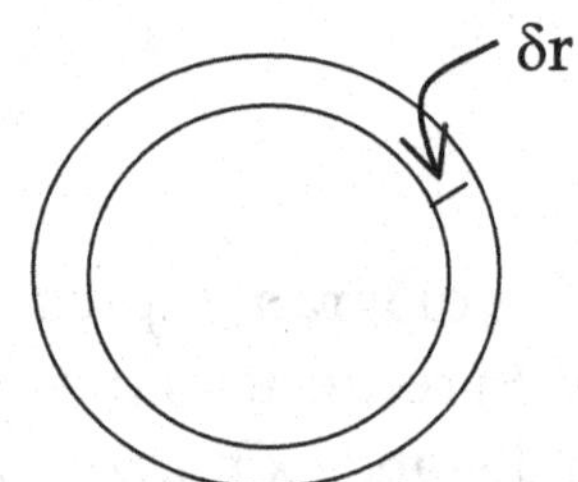

From which we conclude that

$$\text{Surface area} = \frac{dv}{dr} = \frac{d}{dr}\left(\frac{4\pi R^3}{3}\right) = 4\pi R^2$$

The surface area of the sphere is therefore $4\pi R^2$

==================================

Volume of a Cone

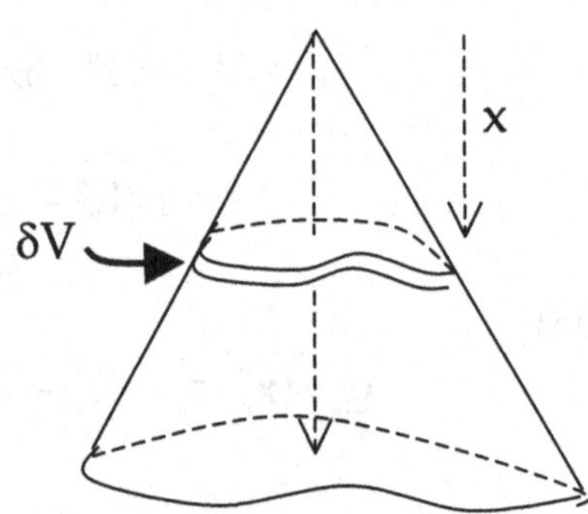

The area of the base of the cone shaped solid is **A.**
Thus the cone shaped solid could be a circular cone, a square pyramid, a tetrahedron...etc.
The perpendicular height of the solid is **h.**
For a given shaped base, the volume will be a function of the distance down from the vertex.
Let **V(x)** be the volume of the solid down a distance **x** from the vertex and let the area of the base of this solid be **a**. Increasing **x** by **δx** increases the volume by **δV** where

$$\delta V = \text{area} \times \text{thickness}$$

$$= a \,.\, \delta x$$

from which we deduce that $\frac{dV}{dx} = a$

Now because the smaller cone shape and the total cone shape are similar figures we have

$$\frac{a}{A} = \frac{x^2}{h^2}$$

and we conclude that

$$\frac{d}{dx}V(x) = \frac{Ax^2}{h^2}$$

Integrating gives

$$V(x) = \frac{Ax^3}{3h^2}$$

(the constant of integration is clearly 0)

Putting **x=h**, we have the general formula for the volume of a cone shaped solid as

$$V = \frac{1}{3} \text{ area of base x perpendicular height}$$

====================================

The Area under a Curve

Problem: Find the area under the curve $y=x^2$ between the values x=2 and x=3 and the x axis:

Solution

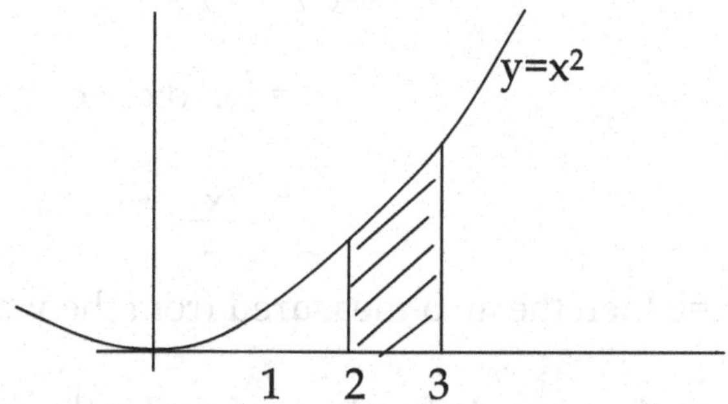

The area under the curve from x=0 to x is some function of x, say A(x) that depends on the equation of the curve.
Given the value of x. we hope to be able to find the value of A(x).
Then the answer to the problem will be A(3) – A(2).

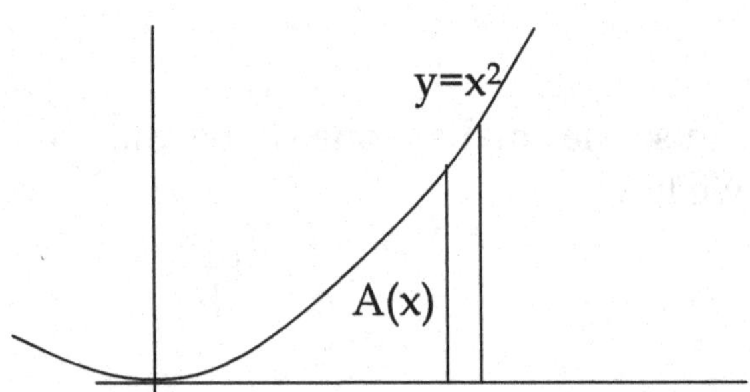

If we increase x by a small amount δx, then we get an increase in the area which we can call δA.

The area δA is approximately $y.\delta x + \frac{1}{2}.\delta x.\delta y$ (a more formal derivation is given below)

This being made up of a narrow rectangle plus a small triangle on top of it. The height of the rectangle is the value of y which is, in this case, x^2. Dividing by δx we have

$$\frac{\delta A}{\delta x} = y + \tfrac{1}{2}\delta y$$

Letting $\delta x \rightarrow 0$ we have

$$\frac{dA}{dx} = y$$

The formula for the area under the curve is therefore given by

$$A(x) = \int y \, dx + c \qquad \text{(where c is some constant)}$$

$$= \int x^2 \, dx + c$$

$$= \frac{x^3}{3} + c$$

now if x=0 then the area measured from the y axis will be zero, therefore c=0.

Therefore, the formula for the area under the curve from the y axis up to x is

$$A(x) = \frac{x^3}{3}$$

If we want the area from x=2 to x=3, we have to subtract A(3) – A(2). Using a special square bracket notation for this we write:

$$A(3) - A(2) = \left[A(x)\right]_2^3$$

The solution to our problem then is

$$A_2^3 = \int_2^3 x^2\,dx = \left[x^3/3\right]_2^3 = 27/3 - 8/3 = 19/3$$

==

Note: A more formal derivation of the formula $\frac{dA}{dx} = y$ proceeds as follows:

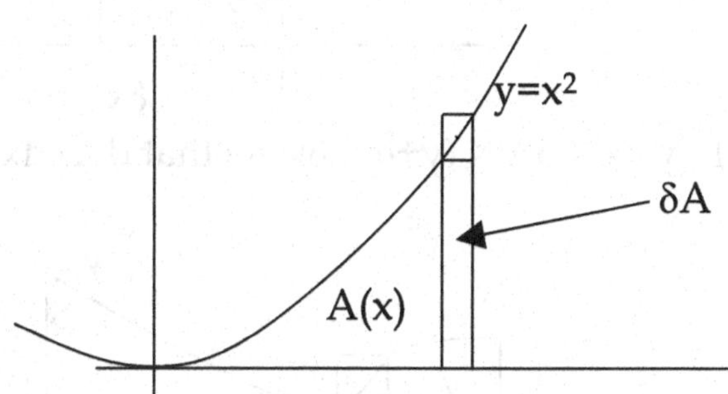

The area δA in this diagram, is greater than the lower rectangle y.δx and less than the upper rectangle (y+δy).δx
Therefore we have

$$y.\delta x < \delta A < (y+\delta y).\delta x$$

divide by δx to get

$$y < \frac{\delta A}{\delta x} < y+\delta y$$

and let $\delta x \to 0$ and since $\frac{\delta A}{\delta x}$ is sandwiched between y and y+δy
if the curve is continuous then when $\delta x \to 0$ we will have $\delta y \to 0$ so we conclude that the limit of $\frac{\delta A}{\delta x}$ must be y

thus $$\frac{dA}{dx} = y$$ ■

Clearly, there are other cases to be examined before we can quote a general formula for finding the area under a graph. These cases depend on whether the slope of the graph is up or down and which quadrant the curve lies in. If the slope is up then δy is positive but if the slope is down then δy is negative. The other factor that needs to be considered is which quadrant the area is in. If the area is above the x axis then y is positive but if the area is below the x axis then y will be negative.

One other adjustment is needed. Instead of measuring the area from the y axis, we take some arbitrary position to the left, so that **δx will always be measured to the right** and will therefore always be positive.

Case 1

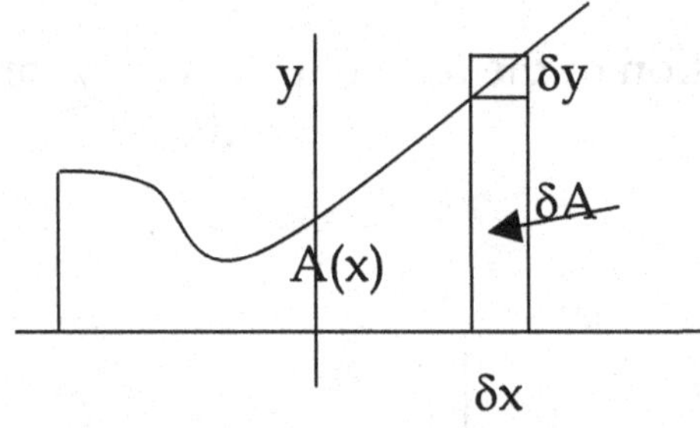

In case 1 $y.\delta x < \delta A < (y+\delta y).\delta x$ so that $dA/dx = y$

Case 2

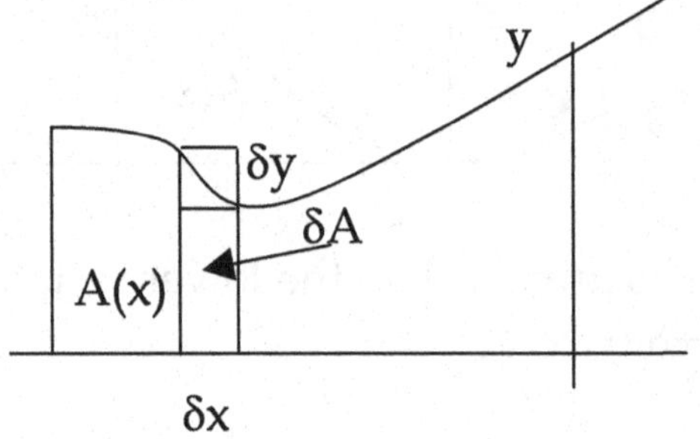

In case 2, δy is negative, $y.\delta x > \delta A > (y+\delta y).dx$ so that $dA/dx = y$

case 3

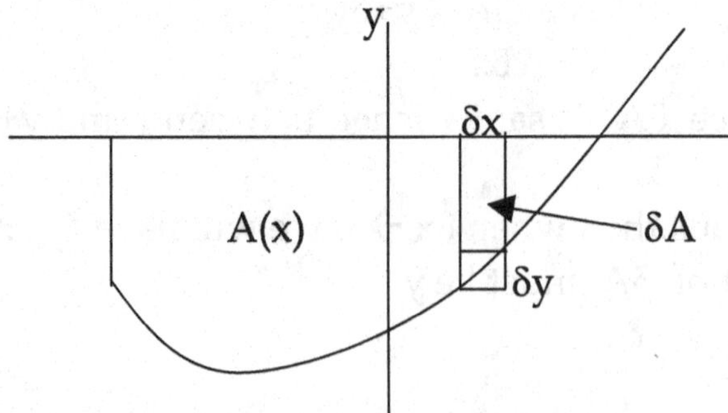

in case 3, δy is positive but y is negative and we have

$$- y.\delta x > \delta A > (-y+\delta y).dx \qquad \text{so that } dA/dx = -y$$

case 4

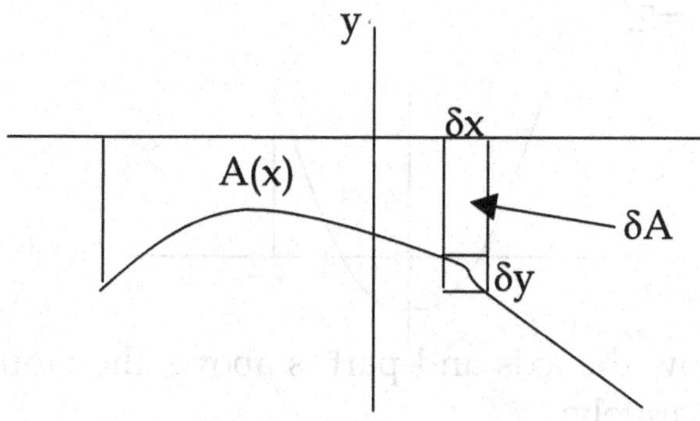

In case 4, δy is negative and y is negative so we have

$$-y.\delta x < \delta A < (-y-\delta y).dx \qquad \text{so that } dA/dx = -y$$

In summary, we have

$A(x) = \int y\,dx + c$ whenever the curve is above the x axis

$A(x) = -\int y\,dx + c$ whenever the curve is below the x axis

and we measure areas from left to right.

The Definite Integral

Using the square bracket notation, for an area above the x axis between the values **a** and **b**, where **a<b,** the area under the curve y=f(x) is given by

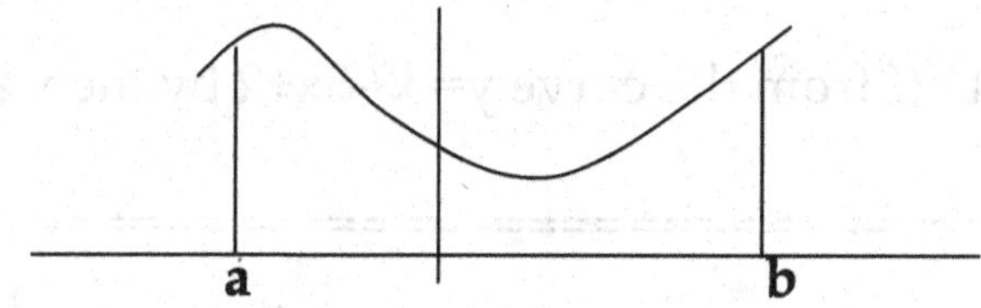

$$\textbf{Area} = \int_a^b y\,dx = \int_a^b f(x)\,dx = [\,A(x)\,]_a^b = A(b) - A(a)$$

For areas below the x axis the answer has a negative sign.
If **b<a** then the answer will also have a negative sign.

Example Find the area between the curve $y=x^2-1$ and the x axis between $x=0$ and $x=2$.

Solution

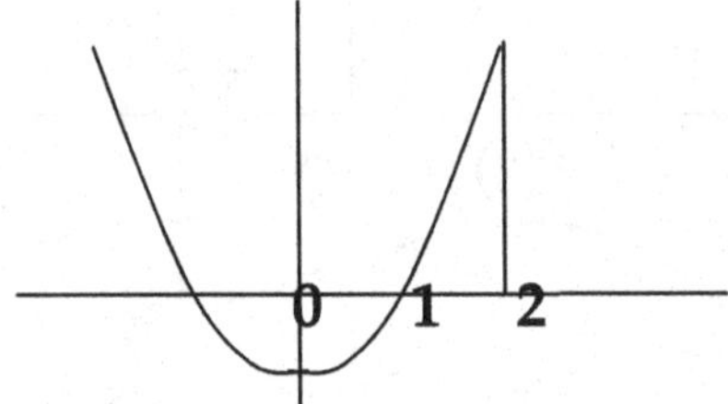

Part of the area is below the axis and part is above, therefore the two parts must be calculated separately:

Area below axis = $\int_0^1 (x^2-1)dx = \left[x^3/3 - x \right]_0^1 = (1/3 - 1) - 0 = -2/3$

Area above axis = $\int_1^2 (x^2-1)dx = \left[x^3/3 - x \right]_1^2 = (8/3 - 2) - (1/3 - 1) = 4/3$

The total area is therefore **4/3 + 2/3 = 2**

==

Exercise 13

[1] Find the area under the curve $y= x^3$ from $x=2$ to $x=4$.

[2] Find the area under the curve $y= x^2+4$ from $x=3$ to $x=6$

[3] Find the area cut off from the curve $y= x^2-3x+2$ by the x axis.

==============================

Evaluation of integrals has always been a major source of questions on examination papers for higher level maths. The two favourite methods tested are called **"integration by substitution"** and **"integration by parts"**.
The use of **reduction formulae** is usually reserved for more advanced students.

Integration by Substitution (or 'change of variable')

The method is best illustrated by examples;

Example 1

Calculate

$$\int_0^2 x\sqrt{(2x^2+1)}\,dx$$

Solution let $2x^2+1 = t$

Then $4x.dx = dt$

When x=0, t=1 and when x=2, t=9

$$\int_0^2 x\sqrt{(2x^2+1)}\,dx = \int_{t=1}^{t=9} \sqrt{t}.\frac{dt}{4}$$

$$= \int_{t=1}^{t=9} t^{1/2}\,\frac{dt}{4}$$

$$= \left[\frac{2\,t^{3/2}}{3.4}\right]_1^9$$

$$= \left(\frac{1}{6}.27\right) - \left(\frac{1}{6}\right)$$

$$= \frac{26}{6} = 4^1/_3$$

================================

Example 2

Calculate

$$\int_0^{\pi/2} \cos x.\sin^4 x.dx$$

Solution Let $\sin x = t$

Then $\cos x.dx = dt$

When $x=0, t=0$ and when $x=\pi/2, t=1$

$$\int_0^{\pi/2} \cos x.\sin^4 x.dx = \int_{t=0}^{t=1} t^4\, dt$$

$$= \left[\frac{t^5}{5}\right]_{t=0}^{t=1}$$

$$= 1/5$$

=====================================

Exercise 14

Evaluate

[1] $$\int_0^{\pi/2} 4\cos x.\sin^3 x.dx$$

[2] $$\int_0^1 6x.(x^2+1)^2\, dx$$

[3] $$\int_0^2 9x^2 \sqrt{x^3+1}.dx$$

[4] $$\int_1^2 15(x+1)\sqrt{x-1}\, dx$$

=============================

Integration by Parts

The product formula for differentiation leads to the formula for integrating a product by the method called "integration by parts"

$$d(uv) = u\,dv + v\,du$$

after transposing the vdu term and then integrating gives

$$\int u\,dv = uv - \int v\,du$$

Example 1

$$\int x \cos x\,dx = \int x\,d(\sin x)$$

$$= x.(\sin x) - \int \sin x\,dx$$

$$= x \sin x + \cos x + \text{const.}$$

There is nothing really new here, we still have to use our intuition (or guess) that x cos x comes from differentiating x sin x.

We could avoid "integration by parts" altogether and proceed as follows:

$$d(x \sin x) = x(\cos x\,dx) + \sin x.dx$$

$$\therefore \quad x \sin x = \int x \cos x\,dx - \cos x + \text{const.}$$

$$\therefore \quad \int x \cos x\,dx = x \sin x + \cos x + \text{const.}$$

which uses the differential of a product as in the first equation of this section.

==============================

Exercise 15

[1] Find $\int x \cos x\,dx$

[2] Find $\int x^2 \sin x\,dx$

Reduction formulae

It may seem rather daunting to to be asked to evaluate $\int_0^{\pi/2} \cos^{10}\theta.d\theta$

but the fact that the limits are 0 and $\pi/2$ allows the value to be written down, almost at sight, after we have done a little investigation. The method uses integration by parts:

Consider

$$I_n = \int_0^{\pi/2} \cos^n\theta.d\theta$$

$$= \int_0^{\pi/2} \cos^{n-1}\theta.d(\sin\theta)$$

$$= \Big[\sin\theta.\cos^{n-1}\theta\Big]_0^{\pi/2} - \int_0^{\pi/2} \sin\theta.(n-1)\cos^{n-2}\theta.\ d(\cos\theta)$$

$$= - \int_0^{\pi/2} - \sin^2\theta.(n-1)\cos^{n-2}\theta\ d\theta$$

$$= (n-1)\int_0^{\pi/2} (1-\cos^2\theta).\cos^{n-2}\theta\ d\theta$$

$$= (n-1)\{I_{n-2} - I_n\}$$

Transposing gives $I_n = \dfrac{(n-1)}{n} I_{n-2}$

Therefore, we can write

$$\int_0^{\pi/2} \cos^{10}\theta.d\theta = \frac{9}{10}.I_8 = \frac{9.7}{10.8} I_6 = \frac{9.7.5}{10.8.6} I_4 \ \ldots\ldots\ \frac{9.7.5.3.1}{10.8.6.4.2} I_0$$

(that is, after careful inspection of what happens at I_0)

The same method yields

$$\int_0^{\pi/2} \cos^9\theta.d\theta = \frac{8.6.4.2}{9\,7\,5\,3}.\,I_1$$

Exercise 16

Show that $I_0 = \pi/2$ and that $I_1 = 1$

Thus "odd powers end in 1 and even powers end in $\pi/2$"

The same method also reveals that powers of sin θ obey exactly the same formulae, for example,

$$\int_0^{\pi/2} \sin^8\theta.d\theta = \frac{7.5.3.1}{8\,6\,4\,2}.\,I_0 = \frac{7.5.3.1}{8\,6\,4\,2}.\pi/2$$

Exercise 17
Write down

[1] $\int_0^{\pi/2} \cos^2\theta.d\theta$ $\int_0^{\pi/2} \cos^3\theta.d\theta$ $\int_0^{\pi/2} \cos^4\theta.d\theta$ $\int_0^{\pi/2} \cos^5\theta.d\theta$

[2] $\int_0^{\pi/2} \sin^2\theta.d\theta$ $\int_0^{\pi/2} \sin^3\theta.d\theta$ $\int_0^{\pi/2} \sin^4\theta.d\theta$ $\int_0^{\pi/2} \sin^5\theta.d\theta$

The results for $\int_0^{\pi/2} \cos^n\theta.d\theta$ and $\int_0^{\pi/2} \sin^n\theta.d\theta$

are special cases of a more general formula that is useful for anyone studying engineering:

$$\int_0^{\pi/2} \sin^m\theta.\cos^n\theta.d\theta = \frac{(m-1)(m-3)\ldots\ (n-1)(n-3)\ldots}{(m+n)(m+n-2)\ldots}\ \Theta$$

where $\Theta = \pi/2$ when both m and n are even but 1 otherwise.

Exercise 18 Evaluate

[1] $\int_0^{\pi/2} \sin^3\theta.\cos^5\theta.d\theta$ [2] $\int_0^{\pi/2} \sin^4\theta.\cos^2\theta.d\theta$ [3] $\int_0^{\pi/2} \sin\theta.\cos\theta.d\theta$

Answers

Exercise 1

(i) 4.5 mps (ii) 4.1 mps (iii) 4.01 mps (iv) 4.001 mps (v) 4.0001 mps

The speed at t=2 is 4 metres per second.

Exercise 2 [1] $6+h \rightarrow 6$ [2] $2T + h \rightarrow 2T$

Exercise 3

$y=2x-1$, $y=3x-2$, $y=4x-3$, $y=5x-4$

Exercise 4

[1] $y=16x - 12$
[2] $y=4x$
[3] $y=4x+3$ or $y=-4x-5$

Exercise 5

[1] maximum value is 3
[2] max at x=1 is 9, min at x=2 is 8, max at x=3 is 9
[3] maximum area is 100
[4] Vmax = 50 x 50 x 25 cm^3 when x = 25 cm

Exercise 6

Turning points are t=2, s=28 and t=3, s=27

Exercise 7

Maximum at (1,6), minimum at (3,2), inflexion at (2,4)

Exercise 8

[1] $6x(x^3+1)(x^2+1)^2(2x^3+x+1)$

[2] $\dfrac{x^4+3x^2-2x}{x^4+2x^2+1}$

Exercise 9

[2] $m=4/3$, $3y - 4x + 2 = 0$

Exercise 12

[1] $\frac{x^3}{3} + \frac{x^4}{4} + \frac{x^5}{5} + \frac{x^6}{6} + \frac{x^7}{7}$ + constant

[2] $x + x^2 + x^3 + x^4$ + const

[3] $2x^{½} - x^{-1} - \frac{x^{-2}}{2} - \frac{x^{-3}}{3} - \frac{x^{-4}}{4}$ + const

Exercise 13

[1] 60, [2] 75, [3] -1/6

Exercise 14

[1] 1 [2] 7 [3] 52 [4] 26

Exercise 15

[1] $x.\sin x + \cos x + c$

[2] $-x^2\cos x + 2x.\sin x + 2\cos x + c$

Exercise 17

[1] $\pi/4$ 3/2 $3\pi/16$ 8/15

[2] $\pi/4$ 3/2 $3\pi/16$ 8/15

Exercise 18

[1] 1/24 [2] $\pi/32$ [3] ½

================================

CHAPTER 27

Logs and Exponentials

When the Ark came to rest on mount Ararat, Noah said to the animals " Go forth and multiply!". But the two snakes replied, "Mr Noah we are adders we don't know how to multiply!". Noah said "Do not worry, there are some logs over there in the forest. The logs will help you to multiply."

Indices

In this chapter we will see how the simple idea of a positive integral power, for example 2^3 which is just a shorthand for 2x2x2, can be extended to negative numbers and fractions. We see how meanings can be given to symbols such as 2^{-3} and $2^{2/3}$ based on the rules for positive integral indices. The chapter ends by showing how meaning is given to complex powers such as 2^{3+2i} and i^i

How Adders Multiply

Logarithm: a figure representing the power to which a base must be raised, used to simplify calculations since addition and subtraction of logarithms is equivalent to multiplication and division [Greek: logos = calculate; arithmos = number] .... from the Concise Oxford Dictionary.

This is a table of the powers of 2:

	2^0	2^1	2^2	2^3	2^4	2^5	2^6	2^7	2^8	2^9	2^{10}
power	0	1	2	3	4	5	6	7	8	9	10
value	1	2	4	8	16	32	64	128	256	512	1024

In order to multiply two of these values, look up the powers, add the powers and then look up the answer using the table backwards:

Example Multiply 16 by 64

Solution

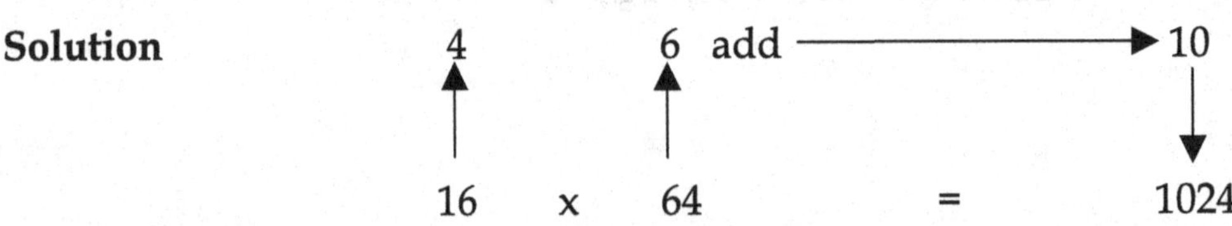

Exercise 1 Use the table of powers to find 8x32.

Laws of Indices

These rules are called **Laws** because, although they can be proved for integer indices, it is by demanding that they also hold for rational, negative and other types of indices, that definitions are given to powers that have these kinds of index.
Index is another word for power. When we say " 2 raised to the power 4 is 16" then the power 4 is the index.

2^4 is a shorthand for 2x2x2x2

2^6 is a shorthand for 2x2x2x2x2x2

$2^4 \times 2^6$ written out in full is has 4+6 = 10 twos

$2^4 \times 2^6 = 2^{10}$

The First Law of Indices

When multiplying powers, add the indices

$$a^x \times a^y = a^{x+y}$$

=======================================

If we divide powers find that we subtract the indices;

$2^6 \div 2^4$ written out in full gives $\dfrac{2x2x2x2x2x2}{2x2x2x2}$

four of the twos cancel leaving 6 – 4 twos

The Second Law of Indices

When dividing powers, subtract the indices

$$a^x \div a^y = a^{x-y}$$

Consider $(2^4)^3$ written out in full:

We have $(2^4)^3 = (2^4) \times (2^4) \times (2^4) = (2x2x2x2) \times (2x2x2x2) \times (2x2x2x2)$

Giving altogether, 4x3 = 12 twos multiplied together

The Third Law of Indices

When raising a power to another power, multiply the indices

$$(a^x)^y = a^{xy}$$

=======================================

Consider $(ab)^3 = ab \times ab \times ab = axaxa \times bxbxb = a^3 b^3$

This illustrates a useful rule that not all books give as a law of indices, however

The Fourth Law of Indices

$$(ab)^x = a^x.b^x$$

==================================

These are laws that we know hold for positive integer indices. We now extend the definitions of powers so that the same laws hold for indices that may be negative, zero or fractional.
At a higher level the definitions for powers are extended to all real numbers so that we can use expressions such as 2^π.
At a further level we extend the definitions to include complex powers of complex numbers.

Definitions

Consider $a^3 \div a^2 = a^{3-2} = a^1$ (by the second law)

but $a^3 \div a^2 = a$ which leads us to

Definition 1 $a^1 = a$

============================

Consider $a^n \div a^n = a^{n-n} = a^0$ (by the second law)

but $a^n \div a^n = 1$ which leads us to

Definition 2 $a^0 = 1$

============================

By the first law we have

$$a^n \times a^{-n} = a^{n-n} = a^0 = 1$$

divide both sides by a^n and we have

Definition 3

$$a^{-n} = \frac{1}{a^n}$$

===================================

By law 3

$(a^{1/n})^n = a^{1/n \times n} = a^1 = a$ which leads to

Definition 4

$$a^{1/n} = \sqrt[n]{a}$$

==================================

By law 3

$(a^{1/n})^m = a^{m/n}$ which leads to

Definition 5

$$a^{m/n} = \left(\sqrt[n]{a}\right)^m$$

====================================

Thus for a fractional index, we have

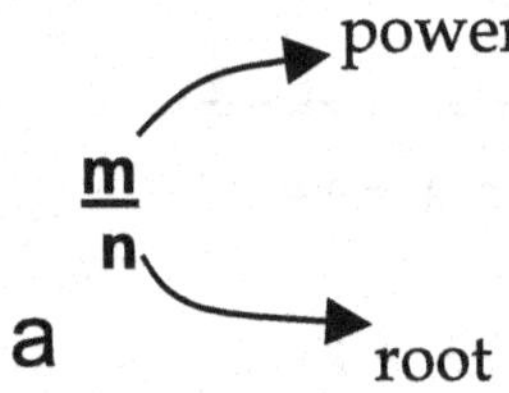

and it does not matter whether we take the power first or the root first, for example;

power first $$8^{2/3} = \sqrt[3]{64} = 4$$

root first $$8^{2/3} = (\sqrt[3]{8})^2 = 2^2 = 4$$

Fractional indices can, of course, be negative. The negative index means that somewhere in the calculation we have a "one over" to perform.

Example 1

$$27^{-2/3} = \frac{1}{27^{2/3}} = \frac{1}{(\sqrt[3]{27})^2} = \frac{1}{3^2} = \frac{1}{9}$$

Example 2

$$\left(\frac{4}{9}\right)^{-3/2} = \left(\frac{9}{4}\right)^{3/2} = \left(\sqrt{\frac{9}{4}}\right)^3 = \left(\frac{\sqrt{9}}{\sqrt{4}}\right)^3 = \left(\frac{3}{2}\right)^3 = \frac{27}{8}$$

Example 3

$$2^{1/2} \times 8^{1/2} = 16^{1/2} = 4 \quad \text{(using law 4)}$$

Example 4

$$2^{1/3} \times (13.5)^{1/3} = 27^{1/3} = 3$$

Exercise 2

Evaluate

[1] $8^{2/3} \times 8^{4/3} \div 8^{2}$

[2] $8^{2/3} \div 64^{1/6}$

[3] $27^{2/3} \div 81^{1/4}$

[4] $64^{2/3} \times 2^{-2}$

[5] $125^{2/3} \times 25^{-1/2}$

[6] $6^{3/4} \times 35^{-1/4} \times 210^{1/4}$

[7] $14^{2/3} \times 21^{1/3} \div 12^{1/3}$

[8] $80^{3/4} \times 25^{-3/8}$

[9] $189^{2/3} \times 49^{-1/3}$

[10] $(0.001)^{-1/3}$

Logarithms

In concise terms, the logarithm is the index, for example

$$2^4 = 16 \text{ is rewritten } \log_2 16 = 4$$

We say, the logarithm of 16 to base 2 is 4. The logarithm is the power:

$$2^5 = 32 \text{ can be written } \log_2 32 = 5$$

Now $16 \times 32 = 2^4 \times 2^5 = 2^9$

Thus $\log_2 (16 \times 32) = 9$

If we go back to our table of powers we have:

N	1	2	4	8	16	32	64	128	256	512	1024
$\log_2$ N	0	1	2	3	4	5	6	7	8	9	10

To find 16 x 32 using this table of logs we look up log 16 = 4, and log 32 = 5, add the logs to get 9 and then look up the anti-log of 9 to get 512.

Before the invention of the calculator, log tables were printed using base 10. Your calculator has a log button which gives you the power of ten for the number so that if you key in

log 100 you get the answer 2

because $100 = 10^2$

Exercise 3

Use the calculator button to find

log 0.1 log √10 log (10^8)

log 10^3 log(log(10^10))

We do not really need the log button any more because we can do any calculations that we want using the calculator directly, however, mathematicians have found another base for logarithms that is extremely useful. This number is called **e** (for exponential) and logs to base **e** are so important that they have been given a special button marked **ln.**
ln stands for **natural logarithm.**
Thus

$\ln 50 = \log_e 50 = 3.912$ and means that $e^{3.912} = 50$

where **e = 2.718281828**

====================

The Laws of Logs

Let $P = a^x$ then we can write $\log_a P = x$

The laws of logs hold for any base and so we do not bother here, to specify any particular base unless it is necessary to do so.

The first law of logs

$$\log PQ = \log P + \log Q$$

Proof

Let $P = a^x$ and $Q = a^y$ so that $x = \log P$ and $y = \log Q$

Then $PQ = a^x . a^y = a^{x+y}$

Therefore $\log PQ = x+y = \log P + \log Q$

======================

The second law of logs

$$\log P/Q = \log P - \log Q$$

Proof

Let $P = a^x$ and $Q = a^y$ so that $x = \log P$ and $y = \log Q$

Then $P/Q = a^x / a^y = a^{x-y}$

Therefore $\log P/Q = x\text{-}y = \log P - \log Q$

======================

The third Law of logs

$$\log(P^n) = n.\log P$$

Proof

Let $P = a^x$ so that $x = \log P$

Then $P^n = (a^x)^n = a^{nx}$

Therefore $\log P^n = nx = n.\log P$

====================

Since $a^0 = 1$ for any non zero number,

$\log_a 1 = 0$ for any base a

==================================

Exercise 4

Evaluate

[1] $\log_4 8 \times \log_8 4$

[2] $\log_4 8 + \log_4 2$

[3] $\log_2 1 - \log_{½} 8$

[4] $\log_4 16 \times \log_2 4$

[5] $\log_6 32 + \log_6 243$

[6] $\log_6 (3^5) + \log_6 32$

[7] $\log_4 16 + \log_{16} 4$

[8] $\log_{12} 1 + \log_{12} 2 + \log_{12} 3 + \log_{12} 4 + \log_{12} 12$

============================

Change of Base

$$\log_a N = \frac{\log_b N}{\log_b a}$$

Proof

Let $\log_a N = x$

then $N = a^x$

Take logs of both sides to base b, then we have

$$\log_b N = x.\log_b a$$

therefore $x = \dfrac{\log_b N}{\log_b a}$

hence $\mathbf{\log_a N = \dfrac{\log_b N}{\log_b a}}$

==

Example 1

Solve $7^x = 5$

Solution

Take logs of both sides to base 10

$$x \log_{10} 7 = \log_{10} 5 \quad \text{(by the third law of logs)}$$

$$x = \frac{\log_{10} 7}{\log_{10} 5} = 1.20906 \text{ (5 d.p.)}$$

Example 2

Solve $x^7 = 5$

Solution

$$x = 5^{1/7}$$

$$= 1.25850 \text{ (5 d.p.)}$$

===========================

Exercise 5

Solve

[1] $3^x = 4$

[2] $4^y = 5$

[3] $3^z = 5$

[4] Show that xy = z

===============================

Exponentials

We know that $2^2 = 2x2 = 4$ but if we want the value of $2^{2.1}$ we would have to work out

$$2^{21/10}$$

which is the tenth root of 2 raised to the power 21. Our calculator is happy to work out this value for us:

$2^{2.1} = 4.287$ using 2 ^ (2.1) = 4.28709385

We give here, values for other powers of 2, between $2^2 = 4$ and $2^3 = 8$

We have

$2^2 = 4$
$2^{2.1} = 4.287$
$2^{2.2} = 4.595$
$2^{2.3} = 4.925$
$2^{2.4} = 5.278$
$2^{2.5} = 5.657$
$2^{2.6} = 6.063$
$2^{2.7} = 6.498$
$2^{2.8} = 6.964$
$2^{2.9} = 7.464$
$2^3 = 8$

Other values $2^0 = 1$; $2^1 = 2$; $2^{1/2} = \sqrt{2} = 1.414$; $2^{-1} = ½ = 0.5$; $2^{-2} = ¼ = 0.25$ etc..

Now if we plot these values on an x-y graph, something magical happens: all the points lie on a beautifully smooth curve.

Question: But what would be the curve for the equation $y= (-2)^x$?

This is an example of an **exponential curve,** in this case, $y=2^x$.

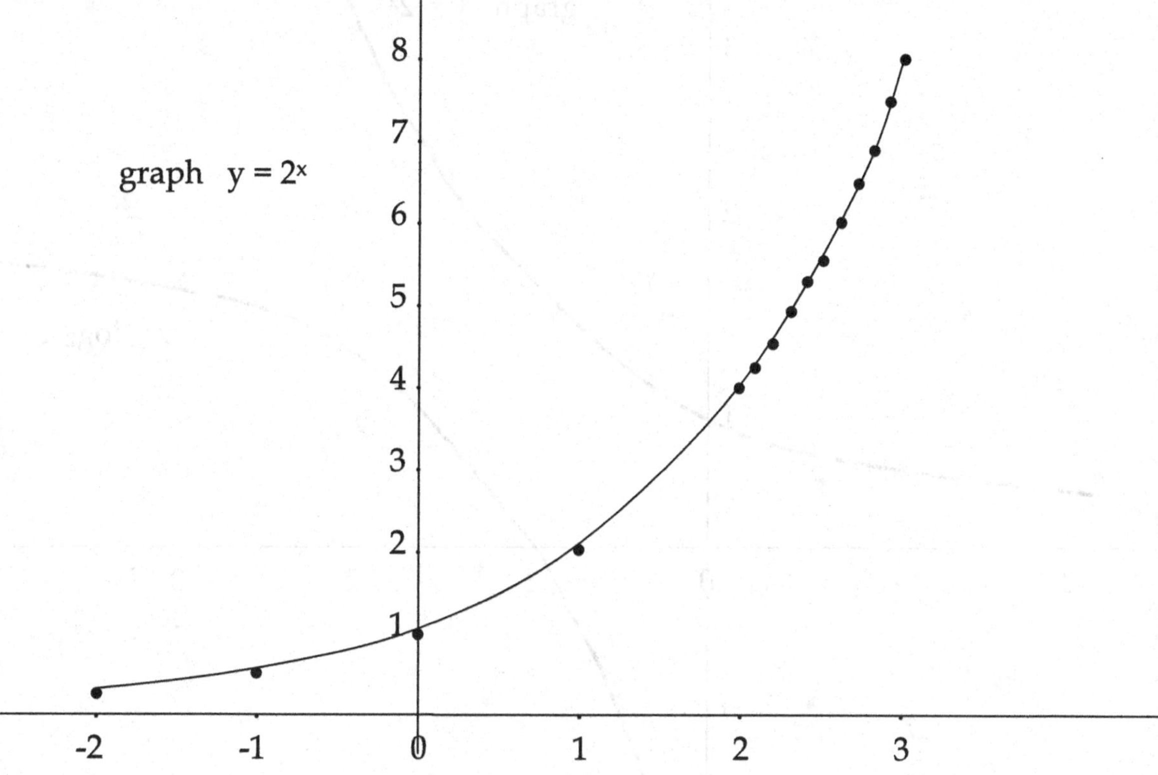

These values are plotted on the graph above.

We find that the graph $y = 2^x$ is a smooth curve.

Given $y = 2^x$

we have $x = \log_2 y$

Showing that the function $y = \log_2 x$ is the inverse of the function $y = 2^x$.

Each function "undoes" the other.

$$x = \log_2 y = \log_2 (2^x) = x$$

$$y = 2^x = 2^{\log y} = y$$

==================================

The graphs of two inverse functions are reflections of each other in the line y=x thus we have:

graph $y = 2^x$

$y = \log_2 x$

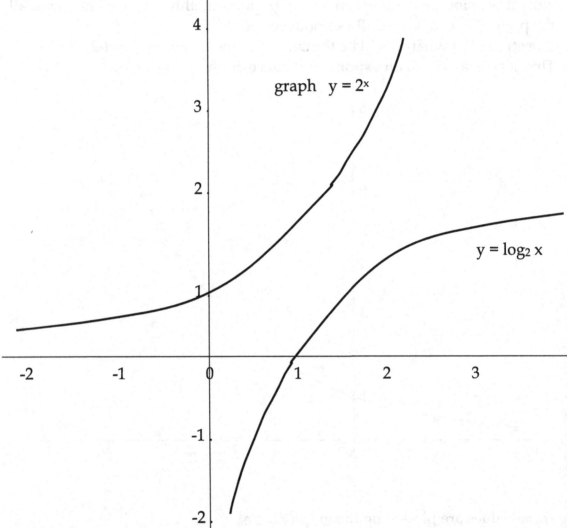

An exponential function and its corresponding log function are inverses of each other, thus we have, in general:

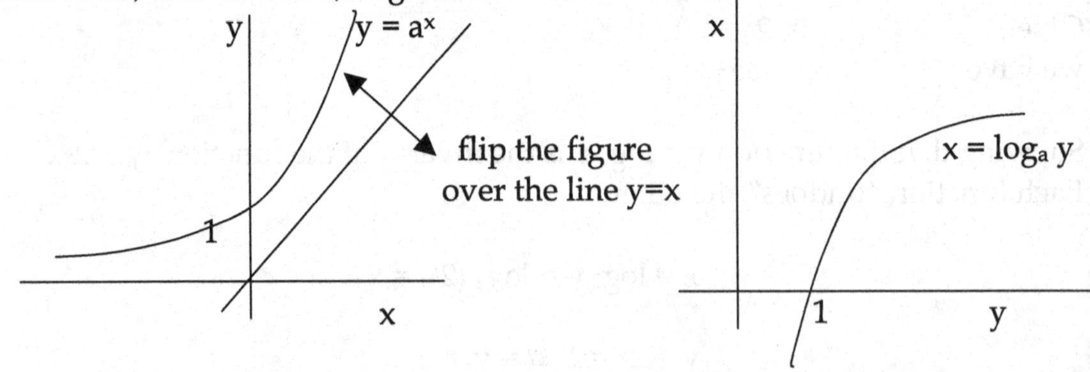

These two graphs are the same shape. The second graph is the first graph reflected in the line y=x. If we superpose the two graphs on the same pair of axes then we have the exponential curve $y = a^x$ and its inverse $y = \log_a x$.

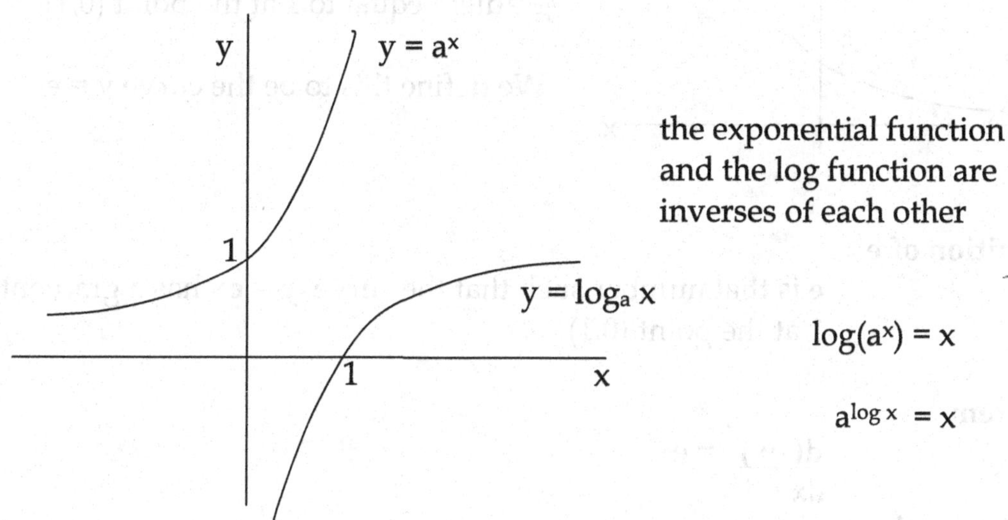

The Gradient of an Exponential Curve

Exponential curves for various values of **a** are sketched below:

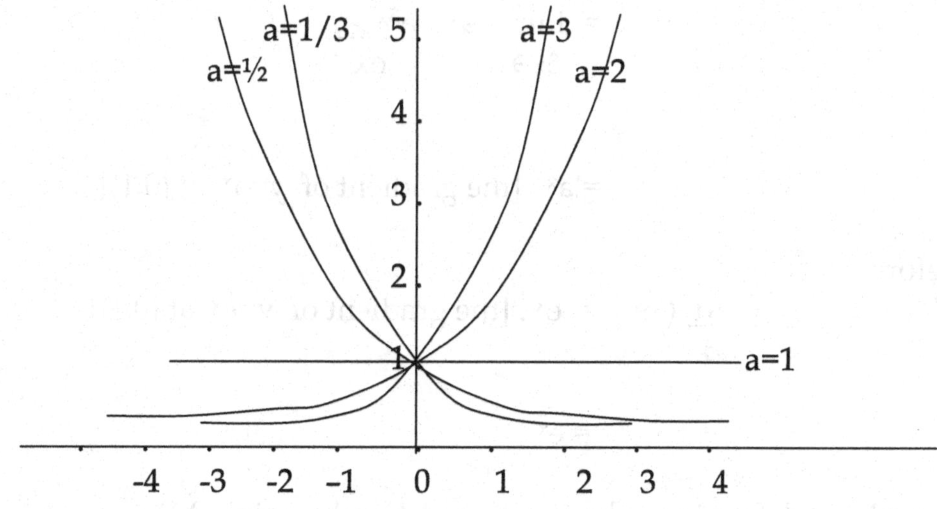

Consider the graph of $y = a^x$ where a is some positive constant greater than 1.

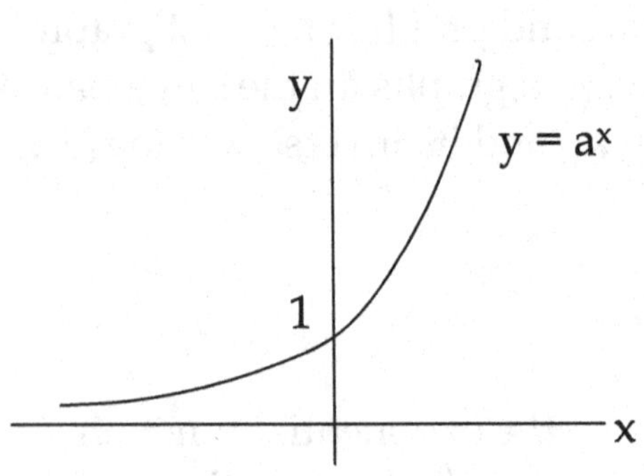

Whatever the value of a, the graph passes through the point (0,1) on the y axis.
One of these curves will have a gradient equal to 1 at the point (0,1).

We define this to be the curve $y = e^x$.

Definition of e

e is that number such that the curve $\mathbf{y = e^x}$ has a gradient 1 at the point (0,1).

Theorem

$$\frac{d(e^x)}{dx} = e^x$$

Proof

The gradient of the curve $y = a^x$ is given by the limit

$$\frac{d}{dx}(a^x) = \lim_{\delta x \to 0} \frac{a^{x+\delta x} - a^x}{\delta x}$$

$$= \lim_{\delta x \to 0} a^x \cdot \frac{a^{\delta x} - 1}{\delta x}$$

$= a^x$. [the gradient of $y=a^x$ at (0,1)]

Therefore

$\frac{d}{dx}(e^x) = e^x$. [the gradient of $y=e^x$ at (0,1)]

$= e^x$

since we have defined e^x to have gradient 1 at the point (0,1) .

■

Natural Logarithms

Logs to base e are so important that we use a special notation for them.

If $y = e^x$ then we write $\mathbf{x = \ln y}$ for $x = \log_e y$.

ln y is called the natural logarithm of y.

==================================

Theorem

$$\int \frac{1}{x}\,dx = \ln x + c$$

Proof

If $y = e^x$ then $x = \ln y$

and $\frac{dy}{dx} = e^x = y \qquad (y>0)$

$\therefore \quad \frac{dx}{dy} = \frac{1}{y}$

hence $\frac{d(\ln y)}{dy} = \frac{1}{y}$

integrate both sides
thus $\ln y = \int \frac{1}{y} + \text{const}$

and since y here is now a dummy variable, we can change y into x :

$$\int \frac{1}{x}\,dx = \ln x + c$$

=======================================

The Definite Integral

We wish to determine the definite integral

$$\int_a^b 1/x \ dx$$

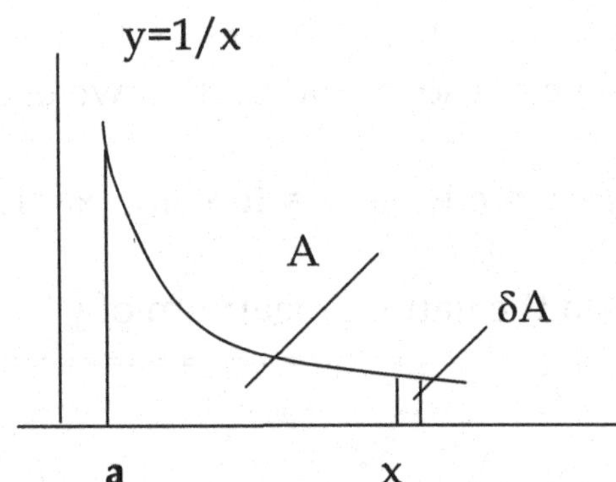

for a range $0 < a < b$ in other words, the area between x=a , x=b and the x axis under the curve y=1/x.
The area under the curve y = 1/x from **a** up to x on the x axis is some function of x that we will call A(x) so that A(**a**) = 0
If we increase the value of x by δx then the area increases by δA.
and we have, approximately

$$\delta A = y.\delta x = 1/x \, . \, \delta x$$

thus $\quad \dfrac{\delta A}{\delta x} = \dfrac{1}{x}$ which becomes, in the limit $\dfrac{dA}{dx} = \dfrac{1}{x}$

Integrate both sides to get

$$A(x) = \ln x + \text{const}$$

But A(**a**) = 0 therefore $\quad 0 = \ln \mathbf{a} + \text{const}$

Therefore the value of the constant is -ln **a**

Therefore $\quad A(x) = \ln x - \ln \mathbf{a}$

This is the formula for the area under the curve y=1/x from x = **a** to x.

Therefore $\quad A\,(b) = \ln a - \ln b$

Giving $\quad \displaystyle\int_a^b 1/x \ dx = [\ \ln x]_a^b$ **for 0<a<x<b or 0<b<x<a**

(the second range simply changes signs) <u>call this result A</u>

Example 1

Calculate the area under the graph y=1/x , between x=1 and x=4.

Solution

$$A = \int_1^4 \frac{1}{x}\,dx$$

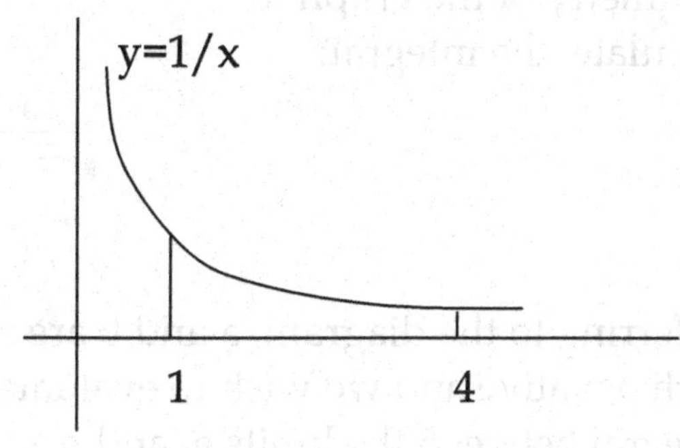

$$= [\ \ln x\ ]_1^4$$

$$= \ln 4 - \ln 1 = \ln 4$$

Example 2

Calculate the area under the graph y=1/x , between x= ½ and x=4.

Solution

$$A = \int_{1/2}^4 \frac{1}{x}\,dx$$

y=1/x

½ 4

$$= [\ \ln x\ ]_{1/2}^4$$

$$= \ln 4 - \ln \tfrac{1}{2} = \ln 4 + \ln 2 \qquad \text{(third law)}$$

$$= \ln 8 \qquad \text{(first law)}$$

Example 3 The area under y=1/x from 1 to a^n = n x (area from 1 to a)

Proof $A_1^{a^n} = \int_1^{a^n} 1/x\,dx = [\ \ln x\ ]_1^{a^n} = \ln a^n - \ln 1 = n \ln a - 0 = n\,.A_1^a$

Negative Limits

If the limits for the integral are negative, then we use the symmetry of the graph to calculate the integral:

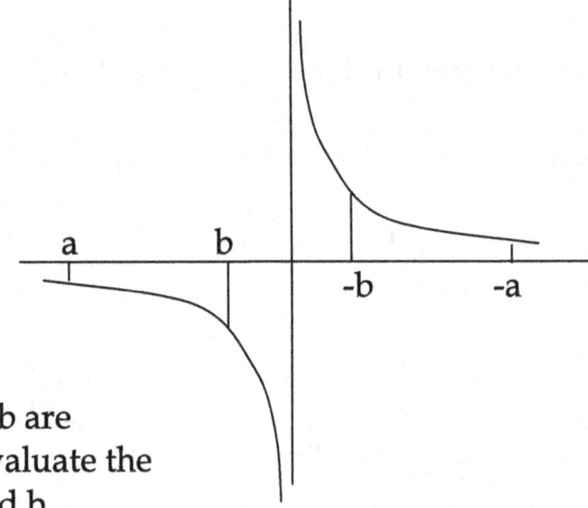

Referring to the diagram, a and b are both negative and we wish to evaluate the integral between the limits a and b.

Due to the symmetry of the graph, we know that

A_a^b is negative but has the same magnitude as A_{-b}^{-a} .

Remember, we cannot at this stage, give a meaning to the log of a negative number, so we cannot write ln(a) or ln(b) if a and b are negative.

but, $A_a^b = -A_{-b}^{-a} = -(\ln(-a) - \ln(-b)) = \ln(-b) - \ln(-a) = [\ln x]_{-a}^{-b}$

Giving

$$\int_a^b 1/x \; dx = [\ln x]_{-a}^{-b} \qquad \text{for } a<x<b<0 \text{ or } b<x<a<0$$

(again, the second range reverses the signs on both sides...call this result B)

==================================

It is not often wise to ignore positive or negative signs in mathematics, but the two results result A (above) and result B can be combined into one formula which is:

$$\int_a^b 1/x \; dx = [\ \ln|x|\ ]_a^b$$

Note that the range [a,b] must lie either along the positive x axis or along the negative x axis. We cannot skip over x=0 (where $1/x = \infty$)

Example 1

(a) $$\int_2^6 1/x \, dx = [\ln|x|]_2^6 = \ln 6 - \ln 2 = \ln 3$$

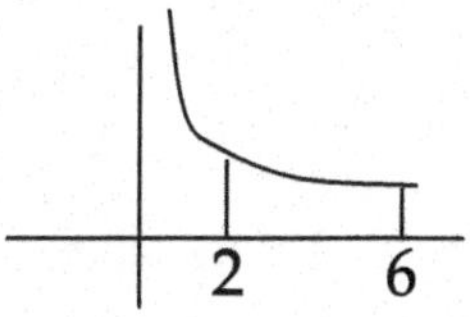

(b) $$\int_{-6}^{-2} 1/x \, dx = [\ln|x|]_{-6}^{-2}$$

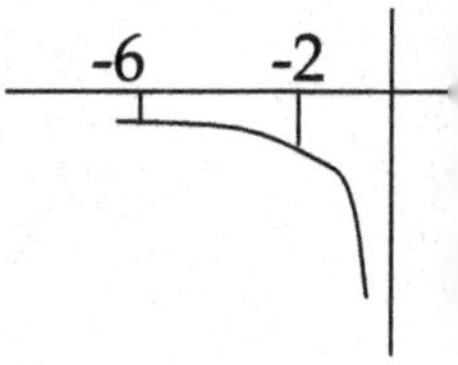

$= \ln 2 - \ln 6$ (wiping out the sign)

$= -\ln 3$ (area below x axis)

==================================

Books often quote the general result as;

$$\int_p^q \frac{dx}{ax+b} = \left[\frac{1}{a}\ln|ax+b|\right]_p^q$$

Example 2

$$\int_4^6 \frac{dx}{3-5x} = \left[\frac{1}{-5}\ln|3-5x|\right]_4^6$$

$$= \frac{-1}{5}\left(\ln|-27| - \ln|-17|\right)$$

$$= \frac{-1}{5}\ln\left(\frac{27}{17}\right)$$

==============================

alternatively, you may prefer a substitution:

$$\int_4^6 \frac{dx}{3 - 5x} \qquad \text{let } t = 3 - 5x$$

$$dt = -5\,dx$$

$$\int_{t=-17}^{t=-27} \frac{-1}{5}\frac{dt}{t} = \left[\frac{1}{-5}\ln|t|\right]_{-17}^{-27}$$

$$= \frac{-1}{5}\left(\ln|-27| - \ln|-17|\right)$$

$$= \frac{-1}{5}\ln\left(\frac{27}{17}\right)$$

========================

Exercise 6

Evaluate

[1]

$$\int_1^{e^2} \frac{dx}{e + x}$$

[2]

$$\int_{18}^{13} \frac{dx}{x - 23}$$

[3]

$$\int_1^{21} \frac{dx}{3 + 2x}$$

[4]

$$\int_1^{85} \frac{dx}{1 + 3x}$$

============================

Functions of a Complex variable

Expressions such as z^2, $\frac{2z+1}{2z-1}$ or **sin(2z+1)** where **z=x+iy,** are functions of the complex variable **z.** It might seem that they are simply functions of the two real variables **x** and **y**, for example we have $z^2 = x^2 - y^2 + 2ixy$ however, to qualify for the title of **function of a complex variable** we insist that the **x** and **y** always appear as a package wrapped in the form **(x+iy)** so that expressions such as **x+3iy** or **sin(2x+3iy)** do not qualify for the title. We remark here, that we have not yet defined what is to be the meaning of expressions such as **sin(2z+1).**

Differentiating Functions of z

The differential of the real function f(x) has been defined as

$$\frac{d\,(f(x)}{dx} = \lim_{\delta x \to 0} \frac{f(x+\delta x) - f(x)}{\delta x}$$

and that this was found to represent the gradient of the x-y graph y=f(x).

The differential of **f(z)**, a function of the complex variable **z=x+iy,** is similarly defined as

$$\frac{d\,(f(z)}{dz} = \lim_{\delta z \to 0} \frac{f(z+\delta z) - f(z)}{\delta z}$$

but this has no graphical interpretation and we need to be aware that **δz** can go to zero (**→ 0**) in an infinite numbers of different ways. For example, if **z** represents a point on the Argand diagram, then the point **z+δz** could approach **z** parallel to the x axis, parallel to the y axis, parallel to y=x or in any wiggly fashion it chose. (if the result is always the same, regardless of the path then **f(z)** is said to be differentiable and if **f(z)** also a one valued function then **f(z)** would be called a regular function.

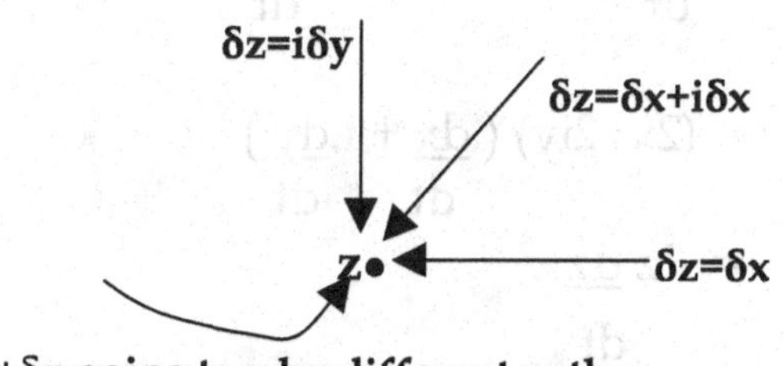

z+δz **going to z by different paths**

Example 1

Differentiate $\mathbf{f(z) = z^2}$

Solution

$$\frac{d z^2}{dz} = \lim_{\delta z \to 0} \frac{(z+\delta z)^2 - z^2}{\delta z}$$

$$= \lim_{\delta z \to 0} \frac{z^2 + 2z.\delta z+\delta z^2 - z^2}{\delta z}$$

$$= \lim_{\delta z \to 0} \frac{2z.\delta z+\delta z^2}{\delta z}$$

$$= \lim_{\delta z \to 0} 2z+\delta z \qquad = 2z$$

================================

Naturally, we want the rules for differentiating real functions to carry over into the complex domain.
If x and y are both functions of the real variable t, then we can differentiate the function **f(z)** with respect to t. We illustrate using $\mathbf{f(z) = z^2}$

Example 2 **x** and **y** are functions of the real variable t, differentiate $\mathbf{f(z) = z^2}$ with respect to t.

We will verify that $\mathbf{\frac{d z^2}{dt} = \frac{d z^2}{dz} . \frac{dz}{dt} = 2z \frac{dz}{dt}}$

Solution

$$z^2 = x^2 - y^2 + 2ixy$$

$$\frac{d z^2}{dt} = 2x \frac{dx}{dt} - 2y \frac{dy}{dt} + 2ix \frac{dy}{dt} + 2iy \frac{dx}{dt}$$

$$= \frac{dx}{dt} (2x + 2iy) + i.\frac{dy}{dt} (2iy + 2x)$$

$$= (2x+2iy) \left(\frac{dx}{dt} + i.\frac{dy}{dt} \right)$$

$$= 2z \frac{dz}{dt}$$

■

Example 3 In particular, if **z = x + iy** then $\frac{dz}{dt} = \frac{dx}{dt} + i.\frac{dy}{dt}$

==================================

Complex Powers

We began this chapter with a discussion of the basic idea of a positive integral index and the laws that these indices obey. The laws were then used to help define other kinds of index, namely, negative indices and fractional indices. The idea of a power or exponential function is then extended to all real numbers.
We now see how the idea of a power can be further extended to include complex powers of real numbers and complex power of complex numbers.

Let z be the complex function of θ given by the equation

$$z = \cos\theta + i\sin\theta$$

then, as in example 3 above, we can differentiate with respect to θ to get

$$\frac{dz}{d\theta} = -\sin\theta + i\cos\theta$$

$$= i(\cos\theta + i\sin\theta)$$

$$= i\,z$$

thus $$\frac{dz}{d\theta} = i\,z$$

separate the variables to get

$$\frac{dz}{z} = i\,d\theta$$

We have not yet defined the log of a complex number, so we now

define $$\ln z = \int dz/z$$

now integrate both sides between the limits z = 1 to z and θ = 0 to θ then

$$[\ln z]_1^z = i\,[\theta]_0^\theta$$

so that $$\ln z = i\theta$$

but in real variables ln x = θ gives $x = e^\theta$ so we therefore define the complex power $e^{i\theta}$ by a similar rule. We define

$$e^{i\theta} = z$$

or

$$e^{i\theta} = \cos\theta + i\sin\theta$$

=====================================

note that $$e^{i\theta} \times e^{i\varphi} = (\cos\theta + i\sin\theta)\times(\cos\varphi + i\sin\varphi)$$

$$= \cos(\theta + \varphi) + i\sin(\theta + \varphi)$$

$$= e^{i(\theta + \varphi)}$$

leading to $(e^{i\theta})^n = e^{i\theta n}$ etc.

We can show that the laws of indices are satisfied by $e^{i\theta}$.

Real Powers of Complex Numbers

The function $f(x) = x^2$ always gives a single result. It is a single valued function. If x = 2 the f(x) = 4, if $x = \sqrt{2}$ then f(x) = 2.

The inverse of this function is, say, $g(x) = x^{1/2}$. In the set of natural numbers, we do not always get a result for g(x). In N there is no $2^{1/2}$ but we do have a single value for $16^{1/2}$ in N, which is 4.
In the field of complex numbers, however, we always have two answers for g(x).

In C, $16^{1/2} = 4$ or -4,
$2^{1/2} = +1.414\ldots$ or $-1.414\ldots$

Using De Moivres Theorem, we can always get two square roots for any complex number.
In the field of Complex numbers, $z^{1/2}$ is always a two valued function (except when z=0) and in general if p/q is a rational number, then $z^{p/q}$ will always be a q valued function of z.

Example 1 find $(1+i)^{1/2}$

Solution

$$1+i = \sqrt{2}(\cos \pi/4 + i \sin \pi/4)$$

$$= \sqrt{2} \text{ cis } (\pi/4 + 2n\pi)$$

$$\therefore \quad (1+i)^{1/2} = \left\{ 2^{1/2} \text{ cis } (\pi/4 + 2n\pi) \right\}^{1/2}$$

$$= 2^{1/4} \text{ cis } (\pi/8 + n\pi)$$

Substituting integer values for n we have

$$(1+i)^{1/2} = 2^{1/4} \text{ cis } (\pi/8) \quad \text{or} \quad 2^{1/4} \text{ cis } (9\pi/8)$$

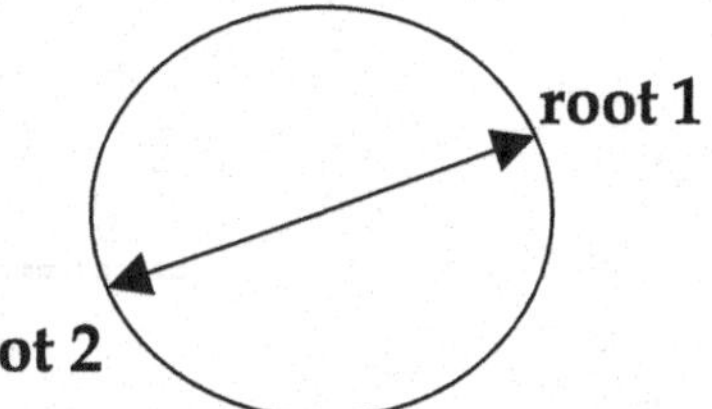

Example 2 Find the five values of $(\sqrt{3} + i)^{1/5}$

Solution We first write the complex number in modulus argument form:

$$(\sqrt{3} + i) = 2(\cos \pi/6 + i\sin \pi/6)$$

$$\therefore \quad (\sqrt{3} + i)^{1/5} = \left\{ 2\text{cis } (\pi/6 + 2n\pi) \right\}^{1/5}$$

$$= 2^{1/5} \text{cis}(\pi/30 + 2n\pi/5)$$

2 1 π/6 √3

Substituting n= 0,1,2,3,4 we get the five values for the fifth root;

$2^{1/5}\text{cis}(\pi/30)$, $2^{1/5}\text{cis}(\pi/30 + 2\pi/5)$, $2^{1/5}\text{cis}(\pi/30 + 4\pi/5)$,
$2^{1/5}\text{cis}(\pi/30 + 6\pi/5)$ and $2^{1/5}\text{cis}(\pi/30 + 8\pi/5)$

On the Argand diagram, these five roots are represented by five equally spaced points on a circle of radius $2^{1/5}$

$$(\sqrt{3} + i)^{1/5} =$$

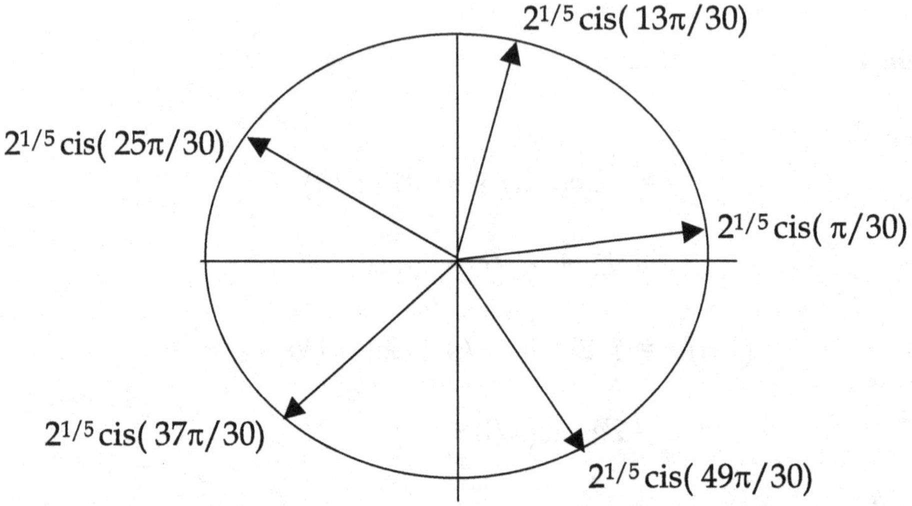

$2^{1/5}\text{cis}(13\pi/30)$

$2^{1/5}\text{cis}(25\pi/30)$

$2^{1/5}\text{cis}(\pi/30)$

$2^{1/5}\text{cis}(37\pi/30)$

$2^{1/5}\text{cis}(49\pi/30)$

==========================

Infinitely Many Valued Functions

$\sin\theta$ is a single value function. Given the value of θ then there is only one value for $\sin\theta$, (and this is true whether θ is real or complex).

Assuming that θ is real, the graph of $\sin\theta$ is the familiar sine wave:

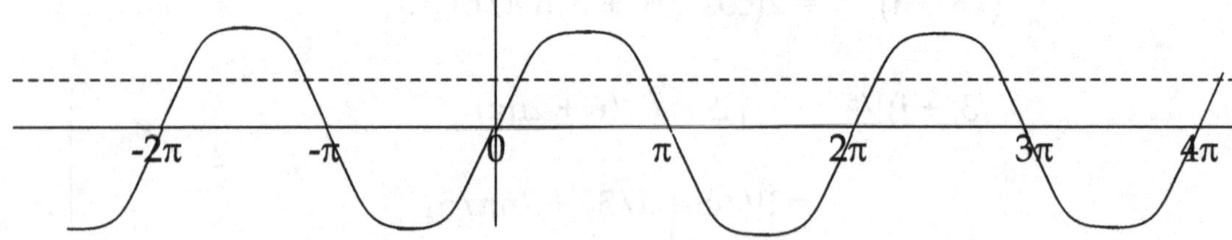

If we put $\sin\theta = ½$ (the dotted line), then we can find an infinite number of values for θ: ..... $-2\pi+\pi/3$, $-\pi-\pi/3$, $\pi/3$, $\pi-\pi/3$, $2\pi+\pi/3$, $3\pi-\pi/3$, $4\pi+\pi/3$.....

The inverse function $\sin^{-1}\theta$ is an infinitely many valued function.

Complex powers of complex numbers and complex logs are also infinitely many valued functions.

The Log of a Complex Number

Let $z = r(\cos\theta + i\sin\theta)$

then θ can be increased by any integer multiple of 2π without changing the value of z, thus

$$z = r\text{cis}(\theta+2n\pi)$$

$$= re^{i(\theta+2n\pi)}$$

this suggests an infinitely many valued Log z which we define as follows:

$$\text{Log } z = \ln r + i(\theta+2n\pi) \qquad \text{where n is any integer}$$

If $z=re^{i\theta}$ with $(-\pi<\theta\leq\pi)$ then we write $\log z = \ln r + i\theta$ so that log z is a single valued function called the **principal value of Log z.**

Note that log zw is not necessarily equal to log z + log w when we use the principal values, as the next example illustrates:

Example Let $z=w=r\ \text{cis}(2\pi/3)$

then $zw = r^2\,\text{cis}(4\pi/3) = r^2\,\text{cis}(-2\pi/3)$

so that $\log zw = \ln r^2 - i2\pi/3$

since the log function uses the principal value of the argument, in the range $-\pi< \theta \leq \pi$.

now $\log z + \log w = \ln r + i2\pi/3 + \ln r + i2\pi/3$

$$= \ln r^2 + i4\pi/3$$

Therefore in this case $\log z + \log w \neq \log zw$

==============================

Equations for Many valued Functions

We can, however, write

$$\text{Log } zw = \text{Log } z + \text{Log } w$$

(But since these are many valued functions, we have to modify our understanding of the = sign.)

Proof let $z = r \text{ cis } \theta$ and $w = s \text{ cis } \varphi$

Then $z.w = rs \text{ cis}(\theta+\varphi + 2n\pi)$

$$\text{Log } zw = \ln rs + i(\theta+\varphi + 2n\pi)$$

$$\text{Log } z + \text{Log } w = \ln r + i(\theta+2k\pi) + \ln s + i(\varphi + 2m\pi)$$

$$= \ln rs + i(\theta+\varphi + 2(k+m)\pi)$$

clearly, both $2n\pi$ and $2(k+m)\pi$ cover all integer multiples of 2π

Here, every value of the Right hand side is also a value of the Left hand side and also, every value of the Left hand side is a value of the Right hand side.

The truth sets for each side are identical and we say that the equation is **completely true.**

The equation $\text{Log } z^m = m \text{ Log } z$

however is **not completely true.**

Proof Let $z = r \text{ cis } (\theta+2n\pi)$

Then $z^m = r^m \text{ cis } m(\theta+2n\pi) + 2N\pi$

$$\text{Log } z^m = \ln r^m + i[m\theta+2mn\pi) + 2N\pi]$$

but $m \text{ Log } z = m \ln r + m[i(\theta+2n\pi)]$

$= \ln r^m + i[\, m\theta+2mn\pi)\,]$ here, we only have to add multiples of $2m\pi$

but because of the $2N\pi$ term in Log z^m, we have all possible values of 2π to add on in the right hand side. The set of values for the Right hand side (Log z^m), has all values of m Log z, but the set of values of m Log z misses some of the values of the of the right hand side.

The relation is not completely true

==================================

Example 1

Consider Log z^4 and 4 Log z

Let $z = r\,\text{cis}(\theta+2n\pi)$ so that $z^4 = r^4\,\text{cis}(4\theta+2k\pi)$

$$\text{Log } z^4 = \ln r^4 + i(4\theta+2k\pi)$$

$$4\text{Log } z = 4[\, \ln r + i(\theta+2n\pi)]$$

$$= \ln r^4 + i(4\theta+8n\pi)$$

we see that Log z^4 adds on all multiples of 2π
whereas 4 Log z adds on only multiples of 8π
for example, $\ln r^4 + i(4\theta+2\pi)$ is a value of Log z^4 but not a value of 4Log z.

Example 2

Log (z^{-1}) = -Log z is completely true

Proof

Let $z=r\,\text{cis}(\theta + 2n\pi)$

then $z^{-1} = r^{-1}\,\text{cis}(-\theta + 2k\pi)$

$\therefore$ $\text{Log }(z^{-1}) = \ln r^{-1} + i(-\theta + 2k\pi)$

But $-\text{Log } z = -[\, \ln r + i(\theta + 2n\pi)\,]$

$$= \ln r^{-1} + i(-\theta - 2n\pi)$$

Both formulae cover all integral multiples of 2π hence the equation is completely true ■

Example 3 Find Log i and state the principal value of Log i.

Solution

$$i = \cos \pi/2 + i \sin \pi/2 = e^{i\pi/2} = e^{i(\pi/2 + 2n\pi)}$$

where n is any integer

$$\therefore \quad \text{Log } i = i(\pi/2 + 2n\pi)$$

The principal value of Log i is therefore $\log i = i\pi/2$

==================================

Exercise 7

As in example 3 above, find the following multi-valued Logs and state each principal value.

[1] Log 2i [2] Log (i+1) [3] Log 3

The Exponential Function e^z

If Log w = z then we write $e^z = w$

Thus e^z is the inverse function of Log z

Just as $\sin^{-1}x$, the inverse function of sinx, is an infinitely many valued function although sin x is single valued, so also Log z is infinitely many valued and its inverse, e^z is single valued.

Theorem 1 **e^z is single valued**

Proof Let $e^z = r \operatorname{cis} \theta$ and let $e^z = s \operatorname{cis} \varphi$

then $z = \text{Log}(r \operatorname{cis} \theta)$ and $z = \text{Log}(s \operatorname{cis} \varphi)$

thus $z = \ln r + i(\theta + 2n\pi)$

and $z = \ln s + i(\varphi + 2m\pi)$

Equate real parts to get $\ln r = \ln s$ $\quad \therefore r = s$

Equate complex parts: $\theta + 2n\pi = \varphi + 2m\pi$

$$\theta = \varphi + 2(m-n)\pi$$

therefore $\quad \text{cis}\,\theta = \text{cis}\,\varphi$

thus $r\,\text{cis}\,\theta = s\,\text{cis}\,\varphi$ so that e^z is single valued.

======================================

Theorem 2 $\quad e^{x+iy} = e^x.e^{iy}$

Proof Let $z=x+iy$ and suppose that $e^z = r\,\text{cis}(\theta + 2n\pi)$

then $\quad z = \text{Log}\, e^z = \ln r + i(\theta + 2n\pi)$

$\therefore \quad x+iy = \ln r + i(\theta + 2n\pi)$

$\therefore \quad x = \ln r$ so that $r=e^x$

and $\quad y = \theta + 2n\pi$

therefore $\quad e^z = r\,\text{cis}(\theta + 2n\pi) = e^x\,\text{cis}\,y = e^x.e^{iy}$

hence $\quad e^{x+iy} = e^x.e^{iy}$ ■

======================================

Theorem 3 $\quad e^z.e^w = e^{z+w}$

Proof Let $z=x+iy$, $w=p+iq$, $z+w = x+p + i(y+q)$

Then, by Theorem 2 $\quad e^z.e^w = (e^{x+iy}).(e^{p+iq}) = e^x.e^{iy}.e^p.e^{iq} = e^{x+p}.e^{i(p+q)}$

$= e^{x+p+i(y+q)}$ By Theorem 2

$= e^{z+w}$ ■

The general Definition of a^z

Let $a = r \text{ cis } \theta = r\, e^{i(\theta + 2n\pi)}$

Then we define $\mathbf{a^z} = \mathbf{e^{z.\text{Log } a}} = \mathbf{e^{z(\ln r + i(\theta + 2n\pi))}}$

which is infinitely many valued, due to the $2n\pi$ term

unless z is real, in which case a^z is single valued when z is an integer but q valued if z is a rational number p/q in its lowest terms.

==================================

Example 1 $(1+i)^i$

Solution $1+i = \sqrt{2} \text{ cis } \pi/4$

$$(1+i)^i = e^{i\text{Log}(\sqrt{2} \text{ cis } \pi/4)}$$

$$= e^{i(\ln \sqrt{2} + i(\pi/4 + 2n\pi))}$$

$$= e^{-(\pi/4 + 2n\pi)}.e^{i(\ln \sqrt{2})}$$

$$= e^{-(\pi/4 + 2n\pi)} \text{ cis } (\ln \sqrt{2})$$

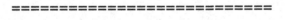

Note of caution:

Avoid $2^{3+2i} = 2^3.2^{2i}$!!!

If a is a real number, are we justified in writing $a^{x+iy} = a^x . a^{iy}$???

For example, suppose that **a** is real so that we have, in modulus argument form

$$a = a \text{ cis}(2n\,\pi)$$

Then $a^{x+iy} = e^{(x+iy)\text{Log } a}$

$= e^{(x+iy)(\ln a + 2n\pi i)}$

$= e^{(x\ln a - 2n\pi y)} e^{i(y\ln a + 2n\pi x)}$

$= a^x.e^{-2n\pi y}.\text{cis}(y\ln a + 2n\pi x)$

$= a^x.e^{-2n\pi y}.\text{cis}(\ln a^y + 2n\pi x)$

Now $a^x . a^{iy} = a^x . e^{iy \text{ Log } a}$

$= a^x . e^{iy(\ln a + 2n\pi i)}$

$= a^x . e^{-2n\pi y}. e^{i(y\ln a)}$

$= a^x.e^{-2n\pi y}.\text{cis}(\ln a^y)$

Thus we see that the two are only the same when **x is an integer.** If x is an integer, then the $2n\pi x$ term in the first expression has no effect.

=====================================

Examples on Complex Powers

In these exercises, n is any integer, positive, zero or negative.

Example 1 $-2^{1/2}$

Now $-2 = 2\text{cis}(\pi + 2n\pi)$

Therefore $-2^{1/2} = e^{1/2 \text{Log}(-2)}$

$= e^{1/2(\ln 2 + i(\pi + 2n\pi))}$

$= e^{1/2 \ln 2}.e^{i(\pi/2 + n\pi))}$

$= e^{\ln\sqrt{2}}\text{cis}(\pi/2 + n\pi)$

$$= \sqrt{2}\,\text{cis}(\pi/2) \text{ or } \sqrt{2}\,\text{cis}(\pi/2 + \pi)$$

$$= \sqrt{2}\,i \text{ or } -\sqrt{2}\,i$$

Note: This exercise can be done directly using DeMoivre's Theorem:

$$-2^{½} = \left[2\text{cis}(\pi+2n\pi)\right]^{½} = \sqrt{2}\ \text{cis}(\pi/2 + n\pi)$$

$$= \sqrt{2}\,\text{cis}(\pi/2) \text{ or } \sqrt{2}\,\text{cis}(\pi/2 + \pi)$$

$$= \sqrt{2}\,i \text{ or } -\sqrt{2}\,i$$

Example 2 $(-2)^{2i}$

$$(-2)^{2i} = e^{2i\text{Log}(-2)}$$

$$= e^{2i(\ln 2 + i(\pi+2n\pi))}$$

$$= e^{-(2\pi+4n\pi)}.e^{2i\ln 2}$$

$$= e^{-2\pi(1+2n)}.\text{cis}(\ln 4)$$

Example 3 $(-2)^{½+2i}$

$$(-2)^{½+2i} = e^{(½+2i)\text{Log}(-2)}$$

$$= e^{(½+2i)(\ln 2 + i(\pi+2n\pi))}$$

$$= e^{(½\ln 2 - (2\pi+4n\pi))}.e^{i(2\ln 2 + \pi/2+n\pi)}$$

$$= e^{(\ln\sqrt{2} - 2\pi(1+2n))}.e^{i(\ln 4 + \pi/2+n\pi)}$$

$$= \sqrt{2}\,e^{-2\pi(1+2n)}\,\text{cis}(\ln 4 +\pi/2+n\pi)$$

Now $\text{cis}(\ln 4 +\pi/2+n\pi) = \text{cis}(\ln 4) \times \text{cis}(\pi/2+n\pi)$

$$= \text{cis}(\ln 4) \times (+i) \text{ or } \text{cis}(\ln 4) \times (-i)$$

giving two sets of values for example 3, $\pm i\sqrt{2}\, e^{-2\pi(1+2n)}.\text{cis}(\ln 4)$

so that we cannot rely on **ex 1** x **ex 2** being equal to **ex 3**

$$-2^{1/2} \times (-2)^{2i} = (-2)^{1/2+2i}$$ **is not completely true**

Example 4 $(2i)^{1/2}$

Here we can use De Moivre's Theorem for a rational index, thus

$$2i = 2\,\text{cis}(\pi/2 + 2n\pi)$$

$$(2i)^{1/2} = \sqrt{2}\,\text{cis}\left(\frac{\pi/2 + 2n\pi}{2}\right) = \sqrt{2}\,\text{cis}(\pi/4 + n\pi)$$

$$= \sqrt{2}\,\text{cis}(\pi/4) \text{ or } \sqrt{2}\,\text{cis}(\pi/4+\pi)$$

$$= \sqrt{2}\left(\frac{1}{\sqrt{2}} + i.\frac{1}{\sqrt{2}}\right) \text{ or } \sqrt{2}\left(\frac{-1}{\sqrt{2}} + i.\frac{-1}{\sqrt{2}}\right)$$

$$= 1 + i \text{ or } -(1 + i)$$

Alternatively, and slightly longer, we could work from the definition for a complex power:

$$2i = 2\,\text{cis}(\pi/2 + 2n\pi)$$

$$(2i)^{1/2} = e^{1/2\text{Log}2i}$$

$$= e^{1/2(\ln 2 + i(\pi/2+2n\pi))}$$

$$= e^{(1/2\ln 2)}.e^{i(\pi/4+n\pi)}$$

$$= \sqrt{2}\,\text{cis}(\pi/4+n\pi)$$

$$= \sqrt{2}\,\text{cis}(\pi/4) \text{ or } \sqrt{2}\,\text{cis}(\pi/4+\pi)$$

$$= \sqrt{2}\left(\frac{1}{\sqrt{2}} + i.\frac{1}{\sqrt{2}}\right) \text{ or } \sqrt{2}\left(\frac{-1}{\sqrt{2}} + i.\frac{-1}{\sqrt{2}}\right)$$

$$= 1 + i \text{ or } -(1 + i)$$

Example 5 $(2i)^{2i}$

$$(2i)^{2i} = e^{2i\,\text{Log}2i}$$

$$= e^{2i\,(\ln 2 + i(\pi/2+2n\pi))}$$

$$= e^{-(\pi+4n\pi)}\,.e^{i\,(2\ln 2)}$$

$$= e^{-\pi\,(1+4n)}\,.\text{cis}(\ln 4)$$

Example 6 $(2i)^{½+2i}$

$$(2i)^{½+2i} = e^{(½+2i)\text{Log}2i}$$

$$= e^{(½+2i)(\ln 2 + i\,(\pi/2+2n\pi))}$$

$$= e^{½\ln 2}\,.e^{-\,(\pi+4n\pi)}\,.e^{i\,(2\ln 2 + \pi/4+n\pi)}$$

$$= e^{½\ln 2}\,.e^{-\,(\pi+4n\pi)}.\ \text{cis}(\ln 4)\ .\ \text{cis}(\ \pi/4+n\pi)$$

$$= \pm(1+i).e^{½\ln 2}\,.e^{-\,(\pi+4n\pi)}.\ \text{cis}(\ln 4)$$

(compare this answer with **ex 4 x ex 5**)

Example 7 $(-2+2i)^{½}$

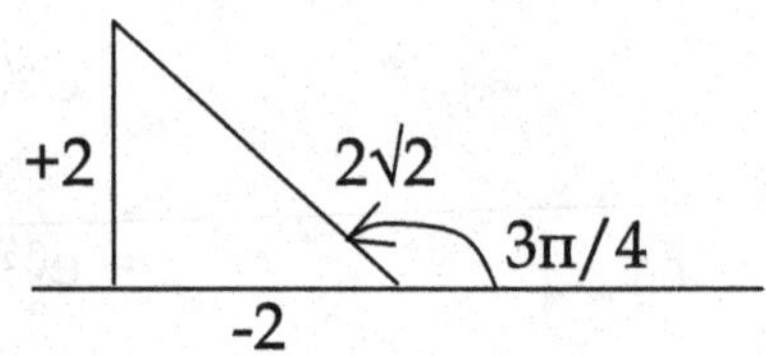

Here we need $-2+2i = 2\sqrt{2}\ \text{cis}(3\pi/4 + 2n\pi)$

$$(-2+2i)^{½} = e^{½\,\text{Log}(-2+2i)}$$

$$= e^{½\,(\ln 2\sqrt{2} + i\,(3\pi/4+2n\pi))}$$

$$= e^{\frac{1}{2}\ln 2\sqrt{2}}.e^{i(3\pi/8+n\pi)}$$

$$= \sqrt{2\sqrt{2}}.\text{cis}(3\pi/8 + n\pi)$$

$$= \sqrt[4]{8}.\text{cis}(3\pi/8) \text{ or } \sqrt[4]{8}.\text{cis}(11\pi/8)$$

============================

Or, using De Moivre's Theorem $(-2+2i)^{\frac{1}{2}} = \left[\sqrt{8}\,\text{cis}(3\pi/4 + 2n\pi)\right]^{\frac{1}{2}}$ etc.

Example 8 $(-2+2i)^{2i}$

$$(-2+2i)^{2i} = e^{2i\text{Log}(-2+2i)}$$

$$= e^{2i(\ln 2\sqrt{2} + i(3\pi/4+2n\pi))}$$

$$= e^{-(3\pi/2+4n\pi)}.e^{2i\ln 2\sqrt{2}}$$

$$= e^{-(3\pi/2+4n\pi)}.\text{cis}(\ln 8)$$

Example 9 $(-2+2i)^{\frac{1}{2}+2i}$

$$(-2+2i)^{\frac{1}{2}+2i} = e^{(\frac{1}{2}+2i)\text{Log}(-2+2i)}$$

$$= e^{(\frac{1}{2}+2i)(\ln 2\sqrt{2} + i(3\pi/4+2n\pi))}$$

$$= e^{\frac{1}{2}\ln 2\sqrt{2} - (3\pi/2+4n\pi)}.e^{i(2\ln 2\sqrt{2}+3\pi/8+n\pi)}$$

$$= \sqrt[4]{8}.e^{-(3\pi/2+4n\pi)}.\text{cis}(\ln 8 + 3\pi/8 + n\pi)$$

$$= \sqrt[4]{8}.e^{-(3\pi/2+4n\pi)}.\text{cis}(\ln 8).\text{cis}(3\pi/8 + n\pi)$$

$$= \sqrt[4]{8}.e^{-(3\pi/2+4n\pi)}.\text{cis}(\ln 8).\text{cis}(3\pi/8)$$

$$\text{or} \quad \sqrt[4]{8}.e^{-(3\pi/2+4n\pi)}.\text{cis}(\ln 8).\text{cis}(11\pi/8)$$

====================================

Exercise 8

[1] Find i^i

[2] Find $(-i)^i$

[3] Find i^{-i}

[4] Find $(-i)^{-i}$

================================

Answers

Exercise 1 8x32 = 256

Exercise 2 1, 2, 3, 4, 5, 6, 7, 8, 9, 10

Exercise 3 -1, 0.5, 8, 3, 1

Exercise 4 1, 2, 3, 4, 5

Exercise 5
[1] 1.26186, [2] 1.16096, [3] 1.46497

Exercise 6
[1] 1 [2] ln2 [3] ln3 [4] ln4

Exercise 7

[1] $\text{Log } 2i = \ln 2 + i\,\pi/2 + 2n\pi i$

principal value $\ln 2 + i\,\pi/2$

[2] $\text{Log}(1+i) = \ln\sqrt{2} + i\,\pi/2 + 2n\pi i$

principal value $\ln\sqrt{2} + i\,\pi/2$

[3] $\text{Log } 3 = \ln 3 + 2n\pi i$

principal value $\ln 3$

Exercise 8

[1] $e^{-(\pi/2+2n\pi)}$ $= e^{2k\pi - \pi/2}$

[2] $e^{-(3\pi/2+2n\pi)}$ $= e^{2k\pi - 3\pi/2}$

[3] $e^{\pi/2+2n\pi}$ $= e^{2n\pi + \pi/2}$

[4] $e^{3\pi/2+2n\pi}$ $= e^{2n\pi + 3\pi/2}$

end

CHAPTER 28

Conic Sections and Hyperbolic Functions

Isaac Newton was a farmer living at the Manor House at Woolsthorpe in Lincolnshire but all was not well and towards the end of the year 1642, the year of Galileo's death, he died, just three months before his son was born. His son, also called Isaac, was born on the 4th January 1643.

But if you were to examine the record of the birth at the village of Woolsthorpe-by-Colsterworth, not far from Grantham, you would find that the birth date is recorded as the 25th of December 1642.

The new Gregorian calendar, instituted by Pope Gregory X111 in 1582, was not adopted by Britain until 1752 and by then, the old style calendar was 11 days behind the new Gregorian calendar and the British were left wondering who had stolen their eleven days. Russia did not adopt the new calendar until after the October Revolution, of 1917, which is why the October Revolution of 25th October is celebrated on November the 7th.

Isaac Newton went to The King's School, Grantham and his talents as a scholar and inventor were quickly recognized. He became head boy at the school and later, his uncle, reverend William Ayscough, persuaded Newton's mother to send him to Cambridge University and in 1661, he was admitted to Trinity College.

In 1663, Henry Lucas, member of parliament for the University, founded the chair of Lucasian Professor of Mathematics at Cambridge. Over the years it has become the most famous academic position in the world. It is now (2008) occupied by Stephen Hawking who was born in Oxford in 1942 exactly 300 years after the death of Galileo. He was elected Lucasian Professor of Mathematics in 1980. Newton's tutor at Trinity, Dr Isaac Barrow was elected the first Lucasian Professor of Maths in 1664.

As a student at Trinity College he started experimental investigations into the nature of light, showing how white light can be split into the colours of the spectrum and explaining the diffraction of light by thin films. Newton's rings are the diffraction patterns caused by the interference of reflected rays formed when a convex lens is placed on a flat plate.

In 1665 when the great plague swept through England, Cambridge University was closed and Newton was sent home to work at Woolsthorpe. In this period, from 1665 to 1666, Newton discovered his binomial theorem, the expansion of $(a+x)^n$ for negative and fractional values of n, started work on his theory of fluxions that led to the differential and integral calculus and formulated the basis of his theory of mechanics that was to lead to the book that we know as "Principia Mathematica". In 1667 Newton was elected a fellow of Trinity College and started working on the laws of mechanics and on a new theory of orbits and planetary motion.

In 1669 Barrow resigned from the post in favour of his student Isaac Newton. From 1670 to 1672, Newton lectured on optics and invented the reflecting telescope. In 1671 he was elected a Fellow of the Royal Society and demonstrated his telescope to fellow members. In 1684 he proved that an inverse square law of force would hold a planet or satellite in an elliptical orbit.

His most famous work: Philosophiae Naturalis Principia Mathematica was published (in Latin) in 1687.

Woolsthorpe Manor is now a historic building open to the public.

Principia Mathematica and the Laws of Motion

Principia consists of three books but in Book 1 Newton defines the laws of motion that form the axioms for the study of dynamics:

Law 1: Corpus omne perseverare in statu suo quiescendi vel movendi uniformiter in directum, nisi quatenus a viribus impressi cogitur statum illum mutare.

(Every body remains in a state of rest or uniform motion in a straight line unless it is made to change that state by impressed forces).

At the time, this law appeared to contradict common sense for our common experience shows that any moving object comes to rest sooner or later.

Law 2: Mutationem motus proportionalem esse vi motrici impressae et fieri secundum lineam rectam qua vis illa imprimatur.

(Change in motion is proportional to the impressed force and acts along the same straight line).

We know this as F=ma, force equals mass times acceleration, but Newton uses the phrase "change in motion". His measure of the amount of motion is what we call momentum, being equal to mass times velocity, so we should be writing $F=\frac{d(mv)}{dt}$ which is the same as F=ma provided that the mass does not change.

Law 3: Actioni contrariam semper et aequalem esse reactionem: sive corporum duorum actiones in se mutuo semper esse aequales et in partes contrarias dirigi.

(Action is always opposite and equal to reaction)

Book 2 applies the calculus to solution of problems on resisted motion for example where the resistance is given by $R=k.v$ or $R=k.v^2$, v being the speed.
The universal law of gravitation $F = \frac{Gm_1.m_2}{d^2}$ is found in Book 3.

G is a universal gravitational constant and $\mathbf{m_1}$ and $\mathbf{m_2}$ are the masses of two particles a distance **d** apart.
Theorem V111 proves that the gravity due to a sphere varies as the distance from the centre of the sphere and Theorem X111 shows that the planets move around the sun in elliptical orbits that have one focus at the sun.
Theorem VIII shows the the force of gravity on a mass **m** near the surface of the Earth is $\mathbf{F = G\ Me.m/R^2}$ where **G** is the universal gravitational constant and **R** is the radius of the Earth. This gives the force of gravity , or **weight,** on the mass **m** equal to **mg** where $\mathbf{g = G.Me/ R^2}$.
Then, using **F = ma,** we get **mg = ma** showing that the acceleration due to gravity is **g** which is a constant
Newton left Cambridge for London in 1696 for a post as Warden of the Royal Mint and in 1705 became Master of the Royal Mint.
He was elected President of the Royal Society in 1703 and in 1705 was knighted by Queen Anne, but not for services to science, rather for services to the Royal mint. Newton died in 1727 and is buried in Westminster Abbey.

Conic Sections

When a plane intersects an extended cone, the outline of the curve produced is of one of three types: ellipse, parabola or hyperbola.
The extended cone in this context is illustrated in the following diagrams:

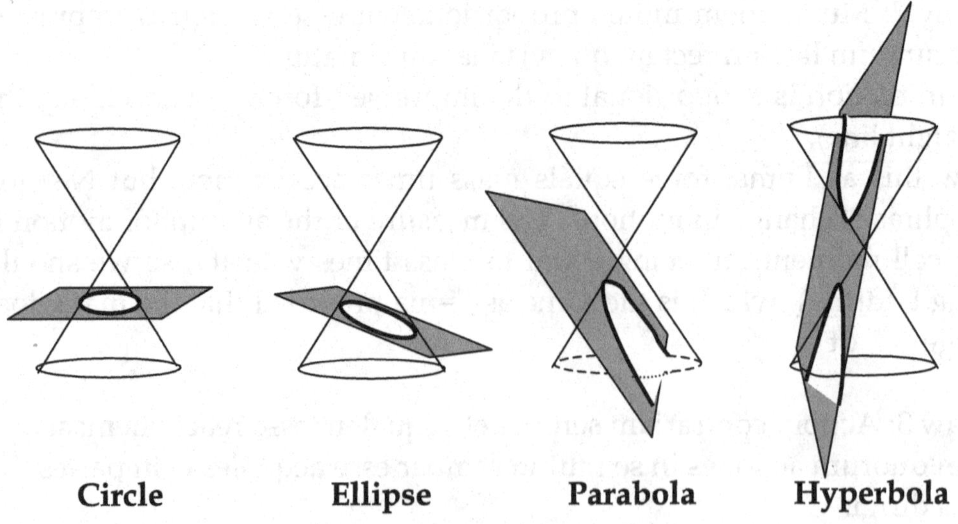

A slice perpendicular to the axis of the cone produces a circle, which is a special kind of ellipse.

A slice parallel a generator of the cone (a straight line of the cone) produces a parabola.

A plane that cuts both parts of the extended cone produces a hyperbola.

The Ellipse

Any gardener will tell you that an elliptical flower bed can be marked out by tying a length of string to two posts and tracing the outline by keeping the string taut:

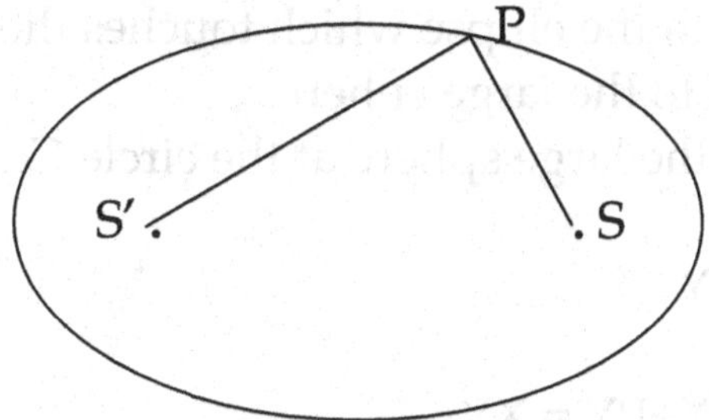

If S and S′ are the two posts, and SPS′ is the length of string, then if P is on the ellipse, SP + PS′ will always be the same length, so your gardener can mark out the shape of the ellipse by sliding P round while keeping the string taut. It is not too difficult to see why this is so:

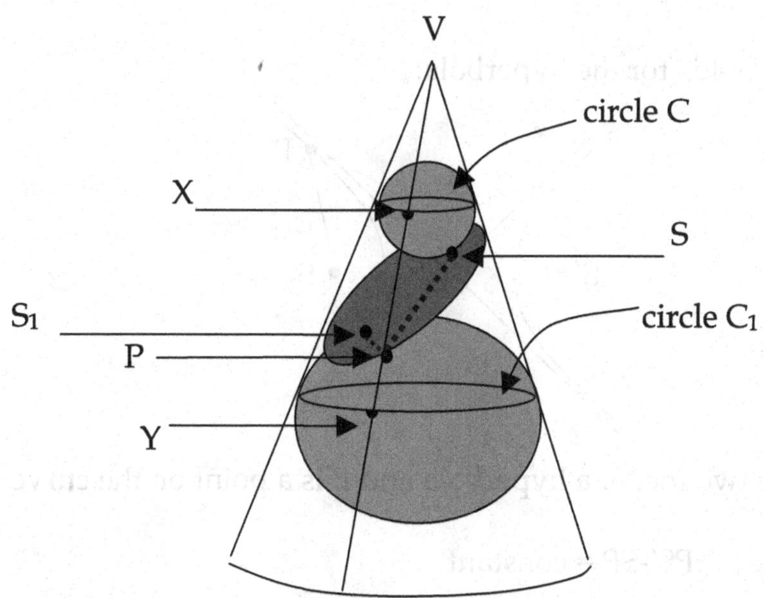

The straight lines represent generators of the cone and the dark region represents the ellipse where a plane intersects the cone.
P is a point on the ellipse.
The small gray area is a sphere that touches the cone in the circle C and touches the ellipse at the point S.
The large gray area is a sphere that touches the cone in the circle C_1 and touches the ellipse at the point S_1.
The line PS (dotted), lies in the ellipse which touches the small sphere and therefore PS is a tangent to the small sphere.
The line PX also touches the small sphere at the circle C.
Therefore

$$PS = PX \qquad \text{(both are tangents to the small sphere)}$$

The line PS_1 (dotted) lies in the ellipse which touches the large sphere and therefore PS_1 is a tangent to the large sphere.
The line PY also touches the large sphere at the circle C_1.
Therefore

$$PS_1 = PY$$

Therefore $\quad PS+PS_1 = PX+PY = XY$

But XY is the fixed distance between the two circles C and C_1.
Therefore

$PS+PS_1$ is constant wherever P lies on the ellipse.

=====================================

The Hyperbola
A similar result holds for the hyperbola:

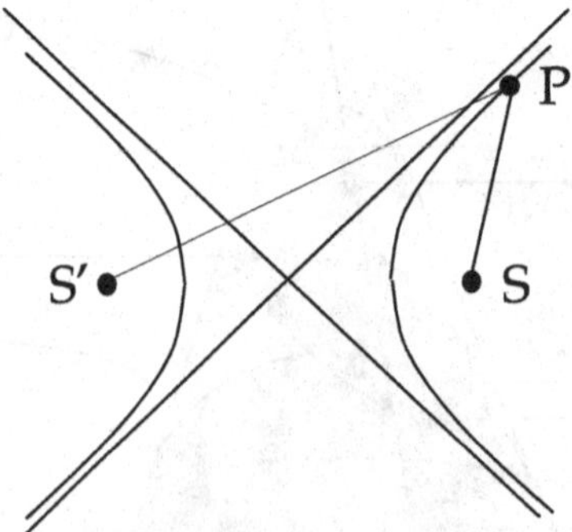

S and S′ are the two foci of a hyperbola and P is a point on the curve.
Here we have

$$PS'-SP = \text{constant}$$

The proof follows:

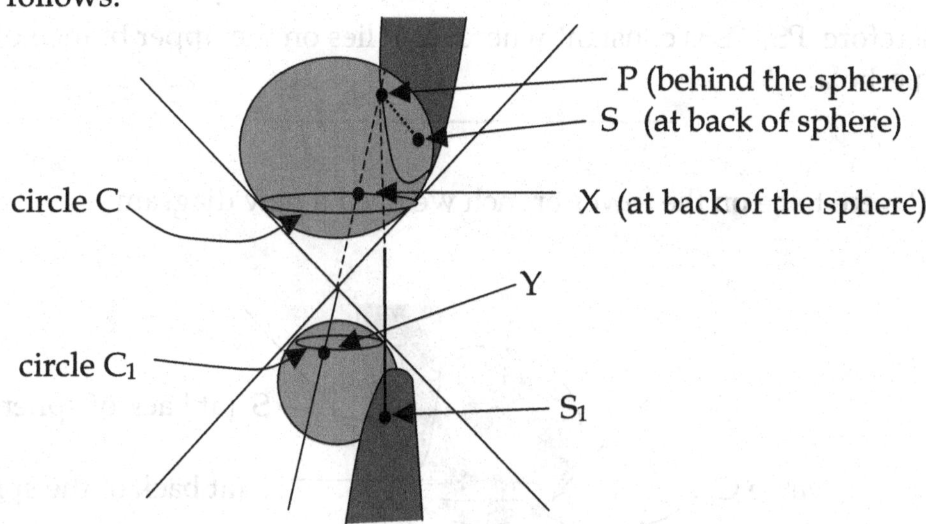

The straight lines represent generators of the cone and the dark region represents the intersection of the plane of the hyperbola with the cone.
P is a point on the hyperbola.
Note that the generator PXY goes to the back surface of the cone where it is dotted.
The large gray area is a sphere that touches the cone in the circle C and touches the plane of the hyperbola at the point S at its back surface.
The small gray area is a sphere that touches the cone in the circle C_1 and touches the hyperbola at the point S1 on its front surface.
The line PS (dotted), lies in the plane of the hyperbola which touches the large sphere at S and therefore PS is a tangent to the large sphere.
The line PX touches the large sphere at the circle C and is therefore also a tangent to the sphere.
Therefore

$PS = PX$ (both are tangents to the large sphere)

The line PS_1 lies in the plane of the hyperbola which touches the small sphere at S_1 and therefore PS_1 is a tangent to the small sphere.
The line PY touches the small sphere at the circle C_1 and is therefore also a tangent to the small sphere.
Therefore

$PS_1 = PY$ (both are tangents to the small sphere)

Therefore $PS_1 - PS = PY - PX = XY$

But XY is the fixed distance between the two circles C and C1.

Therefore PS_1-PS is constant wherever P lies on the upper branch of the hyperbola.

If P were to be on the lower branch we need a new diagram:

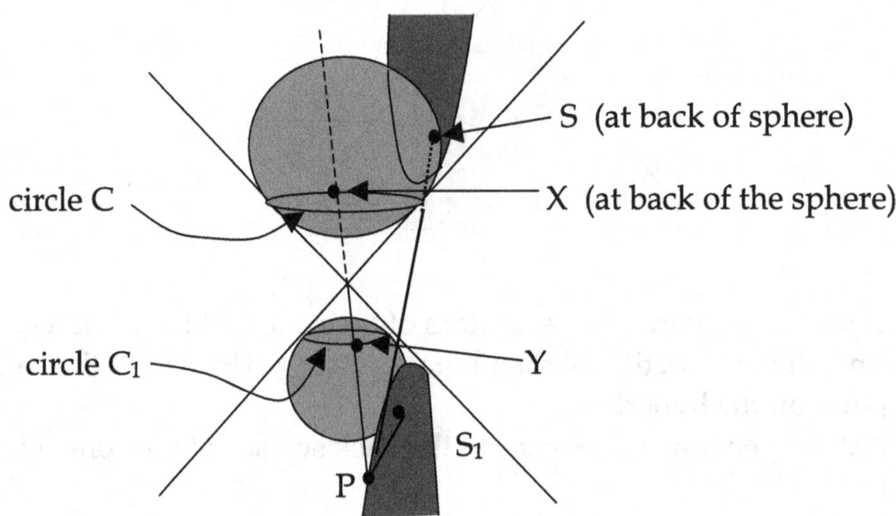

The straight lines represent generators of the cone and the dark region represents the intersection of the plane of the hyperbola and the cone. Note that the generator PYX goes to the back surface of the cone when it passes the vertex of the cone, where dotted.
The large gray area is a sphere that touches the cone in the circle C and touches the hyperbola at the point S at its back surface.
The small gray area is a sphere that touches the cone in the circle C_1 and touches the hyperbola at the point S_1 on its front surface.
The line PS lies in the plane of the hyperbola which touches the large sphere at S so that PS is a tangent to the large sphere.
The line PX touches the large sphere at the circle C and is therefore also a tangent to the sphere.
Therefore

PS = PX (both are tangents to the large sphere)

The line PS_1 lies in the plane of the hyperbola which touches the small sphere at S_1 so that PS_1 is a tangent to the small sphere.

The line PY touches the small sphere at the circle C_1 and is therefore also a tangent to the small sphere.

Therefore $PS_1 = PY$ (both are tangents to the small sphere)

Therefore $PS - PS_1 = PX - PY = XY$

But XY is the fixed distance between the two circles C and C_1.

Therefore $PS - PS_1$ is constant wherever P lies on the lower branch of the hyperbola.

Loci

The parabola can also be described as the locus of a point that moves so that its distance from a fixed point, its focus, is equal to its distance from a fixed line, called its directrix.

The Equation for a Parabola

The simplest equation for the parabola is derived by taking the y axis mid way between the focus and the directrix:

Let the focus be the point S(a,0) and the directrix be the line x = -a.

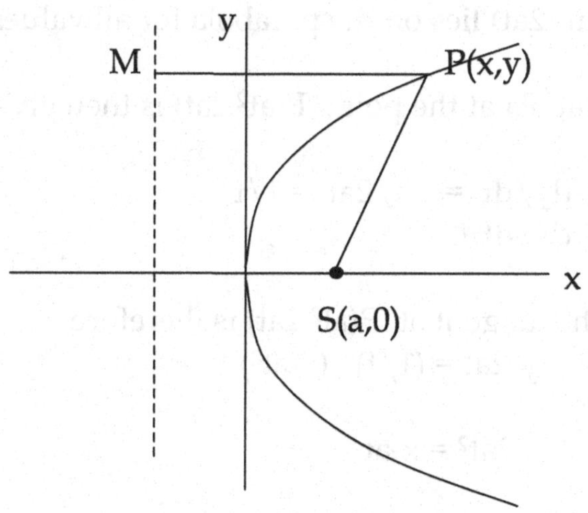

The point P(x,y) moves so that its distance for the focus S(a,0) is equal to its distance from the directrix x = -a.

$$PS^2 = PM^2$$

Gives $$(x-a)^2 + y^2 = (x+a)^2$$

$$x^2-2ax+a^2 + y^2 = x^2+2ax+a^2$$

which reduces to

$$y^2 = 4ax$$

If a bulb is placed at the focus of a parabolic reflector, then the reflected light rays will form a parallel beam. Parabolic reflectors are used for car headlights.

Conversely if you have installed satellite TV then the signals you receive will be beamed from a satellite that is about 25000 miles out in space, to a parabolic dish. The low noise block that shields your signal from any unwanted noise and receives the TV signal is at the focus of the parabolic dish.

To prove that the parabolic reflector gives a parallel beam we need to find a convenient form for the equation of a tangent to the parabola.

The Equation of a tangent

The most convenient coordinates for points on the parabola $y^2=4ax$ are $x=at^2$, $y=2at$ so that $P(at^2,2at)$ lies on the parabola for all values of the parameter t.

The gradient of the parabola at the point $P(at^2,2at)$ is then given by

$$dy/dx = \frac{dy/dt}{dx/dt} = 2a/2at = 1/t$$

using $y-y_1 = m(x-x_1)$, the tangent at $P(at^2,2at)$ is therefore

$$y-2at = (1/t) . (x-at^2)$$

or $$ty-2at^2 = x-at^2$$

which reduces to $$\mathbf{ty = x + at^2}$$ ■

The Parabolic Reflector

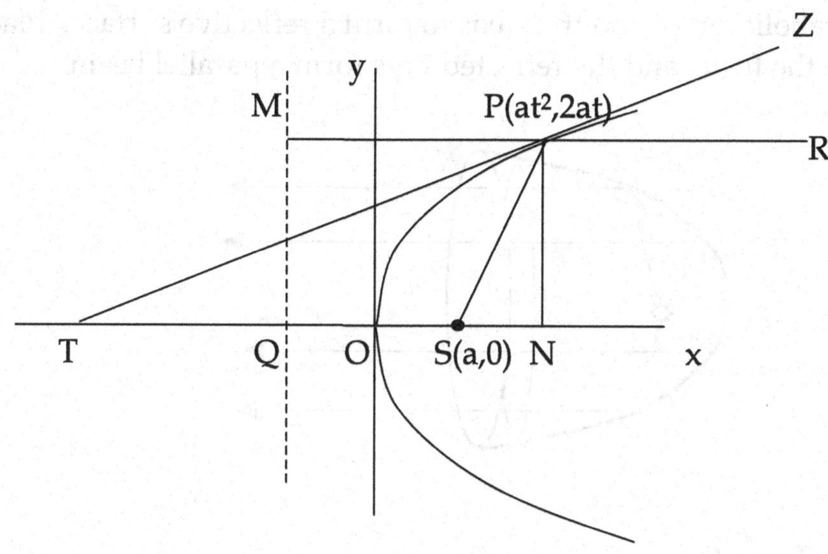

Let the tangent at P meet the x axis at T.
PN is the perpendicular from P to the x axis.

Then, using $ty = x + at^2$ with $y=0$, T will be the point $(-at^2, 0)$

Therefore $$OT = ON$$

$$OT + a = ON + a$$

$$TS = QN$$

$$TS = MP$$

But MP=PS, therefore

$$TS = PS$$

Therefore, parallelogram MPST is a rhombus. The angles of a rhombus are bisected by its diagonals and therefore, $\angle SPT = \angle MPT$

But $\angle MPT = \angle ZPR$ (see figure)

Therefore $\angle SPT = \angle ZPR$ which shows that a ray emitted from S to P will be reflected along PR, which is parallel to the x axis.

Therefore any rays of light emitted from the focus S will be reflected by the parabola in a beam of rays parallel to the x axis.
Rotate the parabolic curve about its axis to form a reflective surface. Place a light source at the focus and the reflected rays form a parallel beam

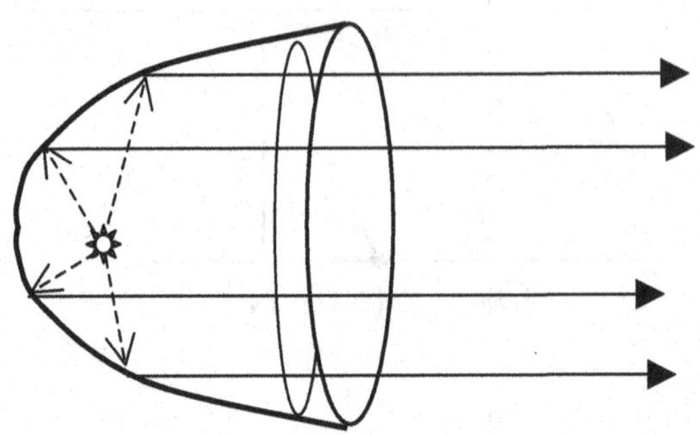

The Parabolic Trajectory

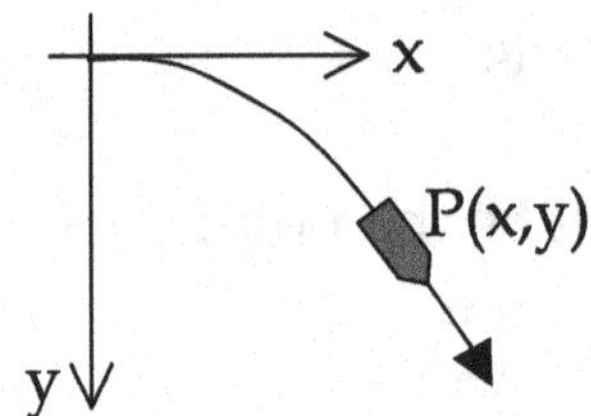

A projectile is fired horizontally with a speed u.
After a time t seconds, the projectile has coordinates P(x,y) referred to axes through the point of projection.

Horizontal motion proceeds with constant speed u, so that $x = ut$

Vertical motion proceeds with constant acceleration g so that $y = \frac{1}{2} gt^2$

The path of the projectile is therefore given by $y = \dfrac{g x^2}{2u^2}$

which represents a parabola with a vertical axis.

The motion of a projectile is symmetrical about the vertical through the highest point, therefore the path of a projectile moving under gravity, (assuming the Earth is flat), is a parabola.

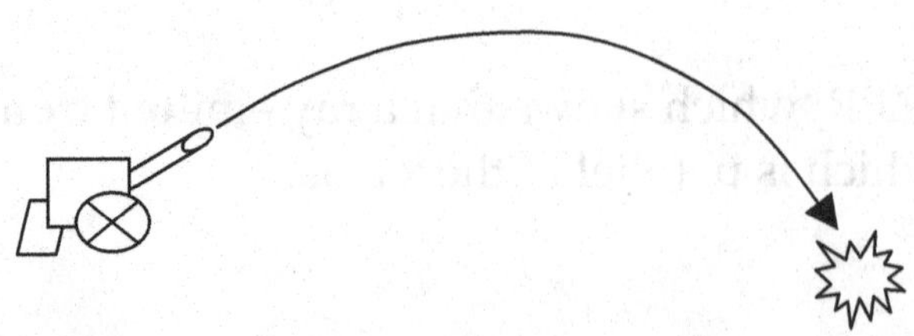

The Equation for an Ellipse

A point P moves so that its distance from a fixed point S, the focus, is always e times its distance from a fixed line called the directrix, where e is a constant $e<1$. The constant e is called the eccentricity.

Let ZK be the directrix and S be the focus.
Take the x axis to be ZS perpendicular to ZK.
There will be two points A and A′ on the x axis that are also on the curve. Let AA′=2a. We call a the semi major axis. Let the y axis bisect AA′.

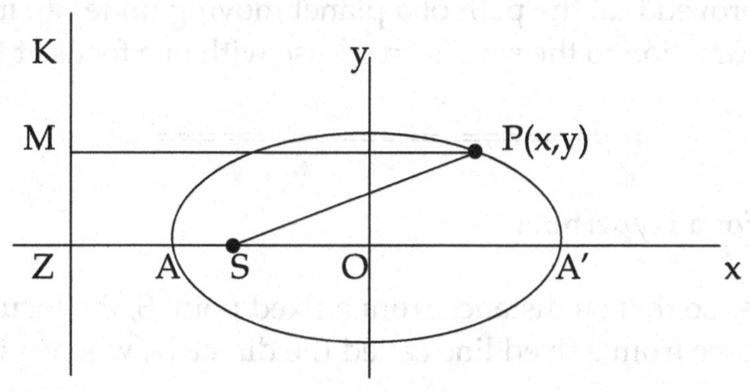

Then

$$AS = e.AZ \quad \text{and} \quad A'S = e.A'Z$$

add $\quad AS+A'S = e(OZ-a) + e(OZ+a)$

so that $\quad 2a = 2e.OZ \quad$ or $\mathbf{OZ = a/e}$

thus $\quad AZ = OZ - a = a/e - a$

but $\quad AS = e.AZ \quad$ therefore $\quad \mathbf{AS = a - ae}$

hence $\quad OS = a - AS = ae \quad$ therefore $\quad$ S is the point $(-ae,0)$

(there is another focus with coordinates $S'(ae,0)$

Now $\quad PM = x+OZ = x+a/e \;$ so $\; PS = e.PM = ex+a$

Use $\quad PS^2 = (ex+a)^2$ to get $\quad (x+ae)^2 + y^2 = (ex+a)^2$

$$x^2+2aex+a^2e^2 + y^2 = e^2x^2+2aex+a^2$$

or $$x^2(1-e^2)+ y^2 = a^2(1-e^2)$$

let $b^2=a^2(1-e^2)$ and we have $\frac{x^2}{a^2} + \frac{y^2}{b^2} = 1$

This is the simplest form for the equation of an ellipse and we find that b is the height of the ellipse from the origin, called the semi minor axis.

If a=b, then we get a rather special ellipse $x^2 + y^2 = a^2$, a circle radius a.

Isaac Newton proved that the path of a planet moving under an inverse square law of attraction to the sun, is an ellipse with one focus at the sun.

========================

The Equation for a Hyperbola

A point P moves so that its distance from a fixed point S, the focus, is always e times its distance from a fixed line called the directrix, where e is a constant e>1.
The constant e is called the eccentricity.

Let ZK be the directrix and S be the focus.
Take the x axis to be ZS perpendicular to ZK.
There will be two points A and A′ on the x axis that are also on the curve. Let AA′=2a. We call a the semi major axis. Let the y axis bisect AA′.

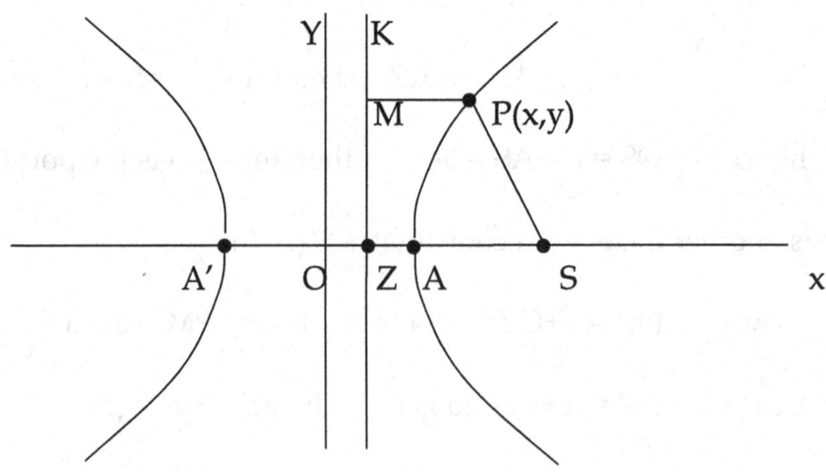

Then

$$A'S = e.A'Z \quad \text{and} \quad AS = e.AZ$$

subtract $A'S - AS = e(a+OZ) - e(a-OZ)$

so that $2a = 2e.OZ$ or $\mathbf{OZ = a/e}$

thus $AZ = a - OZ = a - a/e$

but $AS = e.AZ$ therefore $AS = ae - a$

hence $OS = a + AS = ae$ therefore **S is the point (ae,0)**

(there is another focus with coordinates S′(-ae,0)

Now $PM = x - OZ = x - a/e$

so $PS = e.PM = ex - a$

Use $PS^2 = (ex-a)^2$

to get $(x-ae)^2 + y^2 = (ex-a)^2$

$$x^2 - 2aex + a^2e^2 + y^2 = e^2x^2 - 2aex + a^2$$

or $x^2(1-e^2) + y^2 = a^2(1-e^2)$

now $e>1$ so we rewrite this equation as

$$x^2(e^2-1) - y^2 = a^2(e^2-1)$$

let $b^2 = a^2(e^2-1)$ and we have

$$\frac{x^2}{a^2} - \frac{y^2}{b^2} = 1$$

This is the simplest form for the equation of a hyperbola.

================================

Some Conic Sections and their Equations

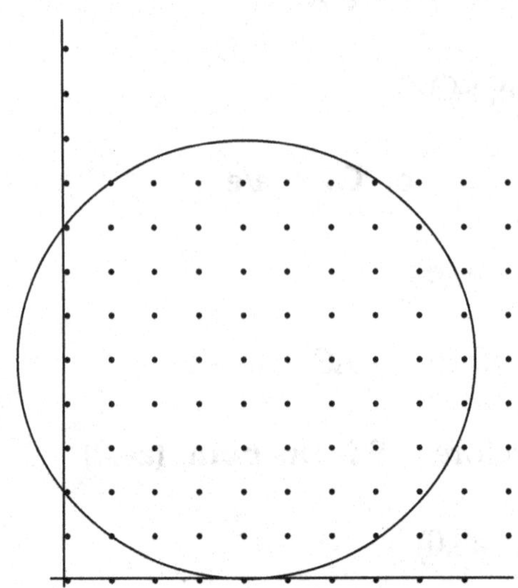

$x^2+y^2-8x-10y+16 = 0$

is a circle, centre (4,5) radius 5

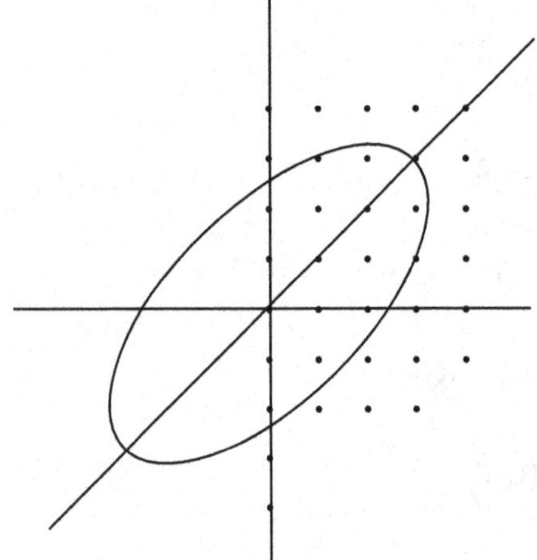

$5x^2+5y^2-6xy-32 = 0$

is a central ellipse with major axis 4 and minor axis 2

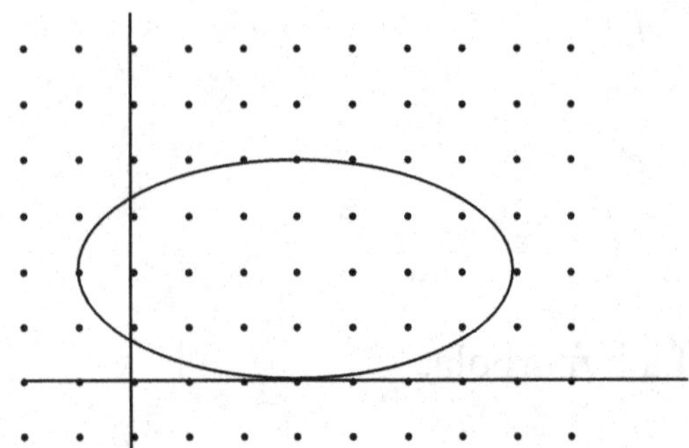

$x^2+4y^2-6x-16y+9 = 0$

is an ellipse with centre (3,2), major axis 4 and minor axis 2

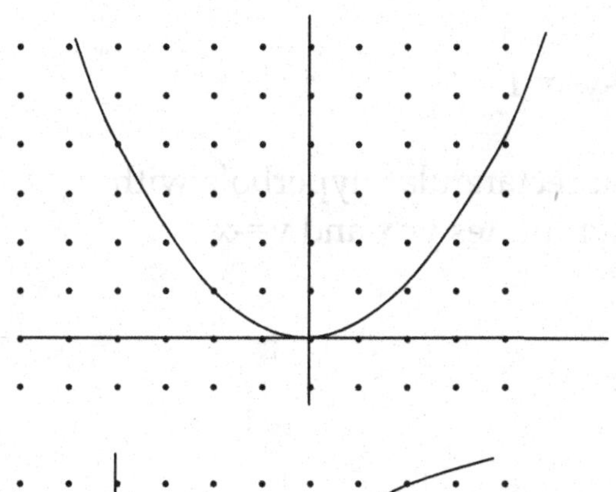

$x^2-4y = 0$

is a parabola, symmetrical about the y axis

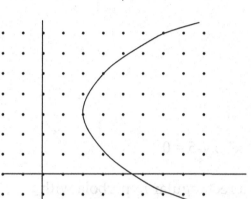

$y^2-4x-6y+17 = 0$

is a parabola with vertex at (2,3) and axis y=3

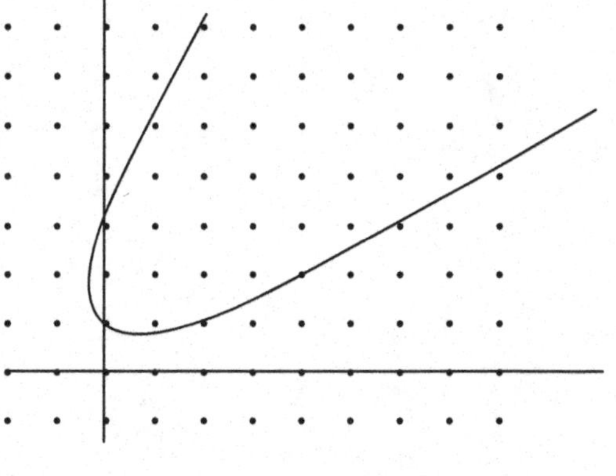

$x^2+y^2-2xy-4y+3 = 0$

is a parabola with vertex at (0,1) and axis y=x+1

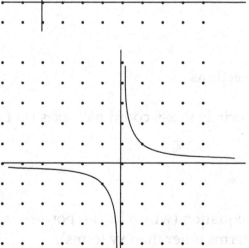

$xy = 1$

is a rectangular hyperbola with asymptotes x=0 and y=0

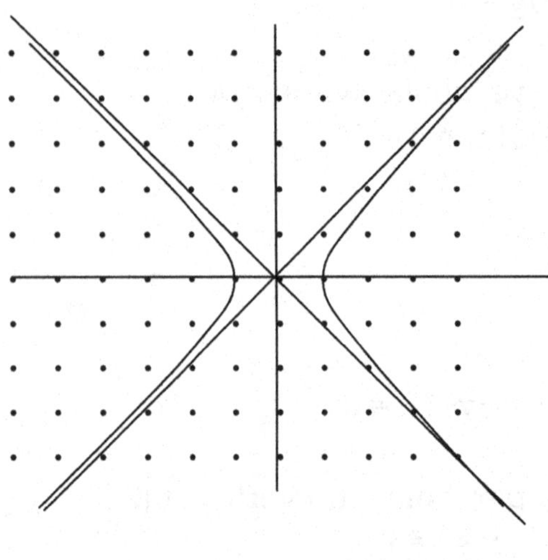

$x^2-y^2 = 1$

is a rectangular hyperbola with asymptotes y=x and y=-x

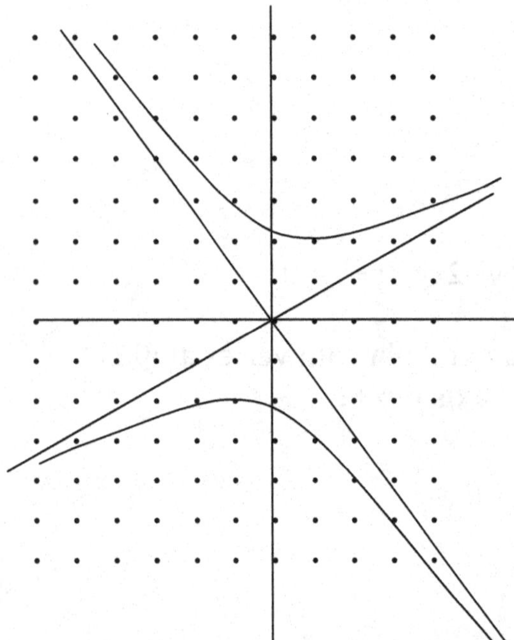

$y^2-x^2+xy-5 = 0$

is a rectangular hyperbola with asymptotes $2y+x(1\pm\sqrt{5}) = 0$

Second degree equations represent conic sections

In chapter 11 we saw that the equation for a circle in x-y coordinate axes is of the form

$$x^2+y^2+2gx+2fy+c = 0$$

This is a particular case of a second degree equation (which has no power terms other than x^2 and y^2 and no product terms other than **xy** terms).

The most general equation of the second degree is of the form

$$ax^2+by^2+2hxy+2gx+2fy+c = 0$$

(the three twos that appear in the equation are there only to make the algebra more satisfying)

Any equation of the second degree that can be plotted on a pair of axes always represents one of the conic sections, an ellipse, parabola or hyperbola or in degenerate cases a pair of lines but given a second degree equation, how do we know which kind of conic section it represents?

What kind of Conic?

If we have the Cartesian equation of a second degree curve, we can decide what kind of conic it represents by considering whether the curve can escape from the axes and move off to infinity.
We have to see if we can get to infinity by travelling along the curve. We can never get to infinity by travelling round an ellipse. The parabola certainly does go to infinity, but both branches of the parabola point in the same direction. The two branches of the parabola get more and more parallel to each other as the curve gets further away from the origin.
The hyperbola however, goes off to infinity in two different directions.

How to get to infinity

To see if we can get to infinity along a curve, we can try to make the **x** and/or **y** coordinates very large and see what happens to the equation.
If x and y are very large, in the equation

$$ax^2+by^2+2hxy+2gx+2fy+c = 0$$

then we should be able to ignore the constant c and the first degree terms 2gx and 2fy.
To make this argument more concrete, divide the whole equation by x^2 and put $y/x = m$. Remember, that y/x is the gradient of the line joining the point (x,y) to the origin.

The equation becomes

$$\frac{ax^2}{x^2} + \frac{by^2}{x^2} + \frac{2hxy}{x^2} + \frac{2gx}{x^2} + \frac{2fy}{x^2} + \frac{c}{x^2} = 0$$

i.e.

$$a+b(m)^2+2h(m)+2g(1/x)+2f(m/y)+c(1/x^2) = 0$$

Now if x and y are very large then the terms $2g(1/x)$, $2f(m/y)$, $c(1/x^2)$ can be ignored because they have finite numerators but denominators that are "going to infinity". In the limit, therefore, as x and y go to infinity, the equation becomes

$$a+bm^2+2hm = 0$$

or

$$bm^2 + 2hm + a = 0$$

Now this is a quadratic equation for the gradient m of the line joining a point(x,y) on the curve, to the origin, assuming that $x\to\infty$ and $y\to\infty$.

If this equation has two real roots ($h^2>ab$) then we have two distinct directions for which the curve can go to infinity and the curve must be a hyperbola.

If the equation has equal roots (**$h^2=ab$**) then there is only one direction for the curve to go off to infinity and the curve is a parabola.

If the equation has no real roots (**$h^2<ab$**) then we cannot find a direction for the curve to go off to infinity and the curve is an ellipse.

Example 1

What kind of conic section is represented by the equation

$$2x^2 - 3xy - 4y^2 + x + y + 4 = 0$$

Solution

For this equation, $h^2 - ab = 9/4 - (2)(-4) > 0$

Two real solutions for the quadratic → two directions → hyperbola

$2x^2 - 3xy - 4y^2 + x + y + 4 = 0$ is therefore the equation of a hyperbola ■

Example 2

Sketch the curve $x^2 - xy + 3y + 6 = 0$

Solution

The second degree terms give $x^2 - xy = 0$

$$x(x - y) = 0$$

giving two directions for which the curve can go to infinity; $x = 0$, and $y = x$

The curve is therefore a hyperbola with asymptotes parallel to $x=0$ and $y= x$

The equation can be rearranged to give

$$y(3-x) + x^2 + 6 = 0$$

or

$$y = \frac{x^2 + 6}{x - 3}$$

one asymptote is therefore the line $x = 3$ since, as $x \to 3$ we have $y \to \infty$

To find the other asymptote rewrite the numerator so that we can divide by the denominator (x-3)

$$y = \frac{x(x-3) + 3x + 6}{(x-3)}$$

$$y = \frac{x(x-3) + 3(x-3) + 9 + 6}{(x-3)}$$

now, dividing by (x-3) we have

$$y = x + 3 + \frac{15}{(x-3)}$$

If we let $x \to \infty$ and we can see that the curve approaches the line $y = x+3$
The other asymptote is therefore the line $y = x+3$.

The two asymptotes for the hyperbola are $x=3$ and $y = x+3$

If we draw the axes with the asymptotes, then we can fit the hyperbola onto the diagram:

Sketch graph of the hyperbola

$x^2 - xy + 3y + 6 = 0$

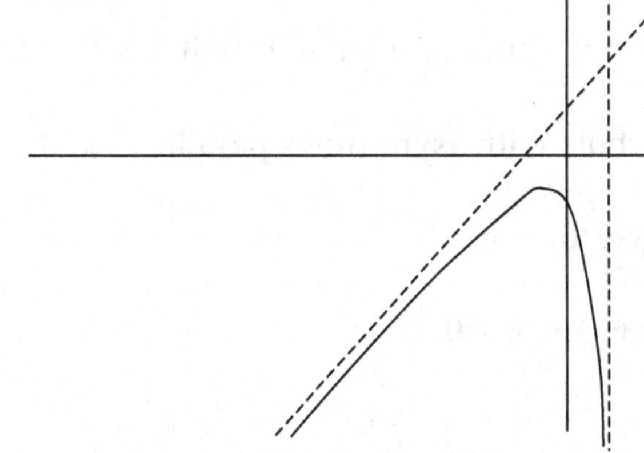

We can confirm the shape further by finding the maximum and minimum points of the graph. To do this, we can differentiate the implicit function:

$$2x.dx - x.dy - y.dx + 3.dy = 0$$

gives $dy/dx = (-2x + y)/(-x + 3)$

so that $dy/dx = 0$ when $y = 2x$

substitute in the equation for the curve to get

$$x^2 - x.2x + 6x + 6 = 0$$

giving $x^2 - 6x + 9 = 15$ or $(x-3)^2 = 15$

so that $x = 3 \pm \sqrt{15} = 3 \pm 3.9$ (1 d.p)

the maximum is therefore (-0.9, -1.8) and the minimum is (6.9, 13.8) to 1 dp and the curve cuts the y axis at (0, -2) ■

Exercise 1

Determine the kind of conic section represented by the following equations, sketch the curves and find any asymptotes and maximum or minimum points:

[1] $2x^2 - xy - 4x + 2y + 2 = 0$

[2] $2x^2 + y^2 - 4x + 2y + 2 = 0$

[3] $2x^2 - y^2 - 4x + 2y + 2 = 0$

[4] $x^2 - 2xy + y^2 - 2x + y + 1 = 0$

[5] $x^2 + y^2 - 2x + y + 1 = 0$

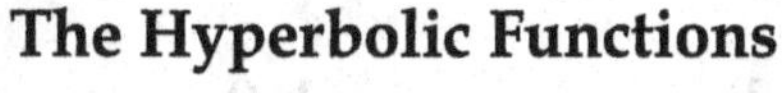

The Hyperbolic Functions

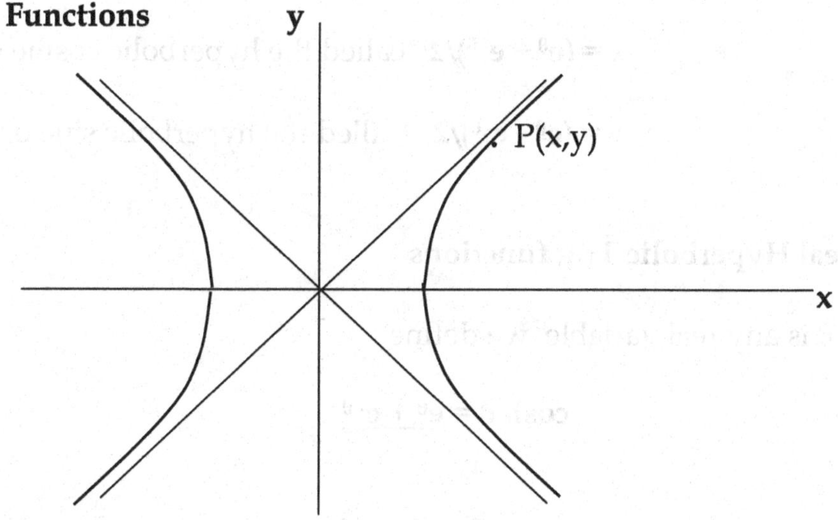

Let P(x,y) be a point on the hyperbola $\mathbf{x^2 - y^2 = 1}$

Suppose we put $x=(e^{\theta} + e^{-\theta})/2$ and $y = (e^{\theta} - e^{-\theta})/2$

and evaluate $x^2 - y^2 = [(e^{\theta} + e^{-\theta})/2]^2 - [(e^{\theta} - e^{-\theta})/2]^2$

$$= \frac{(e^{2\theta} + 2 + e^{-2\theta})}{4} - \frac{(e^{2\theta} - 2 - e^{-2\theta})}{4}$$

$$= 2/4 + 2/4 = 1$$

Thus we find that $x=(e^{\theta}+e^{-\theta})/2$ and $y=(e^{\theta}-e^{-\theta})/2$ are parametric equations for the hyperbola $x^2-y^2=1$.

Compare with $x=(e^{i\theta}+e^{-i\theta})/2$ and $y=(e^{i\theta}-e^{-i\theta})/2$ as parametric equation for the circle $x^2+y^2=1$

Parametric equations for the circle $x^2+y^2=1$ can be taken as

$$x=\cos\theta=(e^{i\theta}+e^{-i\theta})/2$$

$$y=\sin\theta=(e^{i\theta}-e^{-i\theta})/2$$

so **sin θ** and **cos θ,** understandably, are called the circular functions.

Similarly, we can take parametric equations for the hyperbola $x^2-y^2=1$ to be

$x=(e^{\theta}+e^{-\theta})/2$ called the hyperbolic cosine or **cosh θ**

$y=(e^{\theta}-e^{-\theta})/2$ called the hyperbolic sine or **sinh θ**

Real Hyperbolic Trig functions

If θ is any real variable, we define

$$\cosh\theta=\frac{e^{\theta}+e^{-\theta}}{2}$$

$$\sinh\theta=\frac{e^{\theta}-e^{-\theta}}{2}$$

$$\tanh\theta=\frac{\sinh\theta}{\cosh\theta}$$

$$\coth\theta=\frac{\cosh\theta}{\sinh\theta},\quad \text{sech}\,\theta=\frac{1}{\cosh\theta},\quad \text{cosech}\,\theta=\frac{1}{\sinh\theta}$$

now we know that $\cos\theta+i\sin\theta=e^{i\theta}$ (see chapter 27) and $\cos\theta - i\sin\theta = e^{-i\theta}$

add to get $\cos\theta = \dfrac{e^{i\theta} + e^{-i\theta}}{2}$

subtract to get $\sin\theta = \dfrac{e^{i\theta} - e^{-i\theta}}{2i}$

Comparing the formulae for **cos** and **cosh, sin** and **sinh** we see that

$$\cos\theta = \cosh i\theta \qquad (1)$$

$$i\sin\theta = \sinh i\theta \qquad (2)$$

and further, by changing θ to $i\theta$ in the above relations we have

$$\cos i\theta = \cosh(i.i\theta) = \cosh(-\theta) = \cosh\theta \qquad (3)$$

$$i\sin i\theta = \sinh(i.i\theta) = \sinh(-\theta) = -\sinh\theta$$

dividing by i this gives $\sin i\theta = i\sinh\theta$ (4)

dividing equation (4) by (3) we have $\tan i\theta = i\tanh\theta$ (5)

dividing equation (2) by (1) we have $\tanh i\theta = i\tan\theta$ (6)

In Summary:

$$\cos\theta = \cosh i\theta \qquad (1)$$

$$i\sin\theta = \sinh i\theta \qquad (2)$$

$$\cos i\theta = \cosh\theta \qquad (3)$$

$$\sin i\theta = i\sinh\theta \qquad (4)$$

$$\tan i\theta = i\tanh\theta \qquad (5)$$

$$\tanh\theta = i\tan\theta \qquad (6)$$

Using these results, any formula involving **sin θ, cos θ, tan θ** we can be transformed into a corresponding formula connecting **sinh θ, cosh θ, tanh θ**.

Example 1: The compound angle formula for cosine is

$$\cos(\theta + \Phi) = \cos\theta.\cos\Phi - \sin\theta.\sin\Phi$$

giving $$\cos i(\theta + \Phi) = \cos i\theta.\cos i\Phi - \sin i\theta.\sin i\Phi$$

hence $$\cosh(\theta + \Phi) = \cosh\theta.\cosh\Phi - i.\sinh\theta.i.\sinh\Phi$$

thus $$\cosh(\theta + \Phi) = \cosh\theta.\cosh\Phi + \sinh\theta.\sinh\Phi$$

Example 2: To convince ourselves we could check by substituting from the original definitions, for example for the last equation we have

$$\text{L.H.S} = \cosh(\theta+\Phi) = \frac{e^{\theta+\Phi} + e^{-(\theta+\Phi)}}{2}$$

$$\text{R.H.S} = \cosh\theta.\cosh\Phi + \sinh\theta.\sinh\Phi = \frac{e^{\theta} + e^{-\theta}}{2}.\frac{e^{\Phi} + e^{-\Phi}}{2} + \frac{e^{\theta} - e^{-\theta}}{2}.\frac{e^{\Phi} - e^{-\Phi}}{2}$$

$$= \frac{e^{\theta+\Phi} + e^{\theta-\Phi} + e^{-\theta+\Phi} + e^{-\theta-\Phi} + e^{\theta+\Phi} - e^{\theta-\Phi} - e^{-\theta+\Phi} + e^{-\theta-\Phi}}{4}$$

$$= \frac{2e^{\theta+\Phi} + 2e^{-\theta-\Phi}}{4} = \text{L.H.S}$$

■

Example 2 $$\tan 2\theta = \frac{2\tan\theta}{1 - \tan^2\theta}$$

gives $$\tan i.2\theta = \frac{2\tan i\theta}{1 - \tan^2 i\theta}$$

hence $$i\tanh 2\theta = \frac{2.i.\tanh\theta}{1 - i.\tanh\theta\ .\ i.\tanh\theta}$$

$$\tanh 2\theta = \frac{2\tanh\theta}{1 + \tanh^2\theta}$$

■

Rule: To change a trig formula into a hyperbolic trig formula:
Replace any **$\sin^2$ (or $\tan^2$)** term by **$-\sin^2$ (or $-\tan^2$)**

It is intriguing to find this link between the trig. functions that were defined in terms of coordinates on the unit circle (chapter 19):

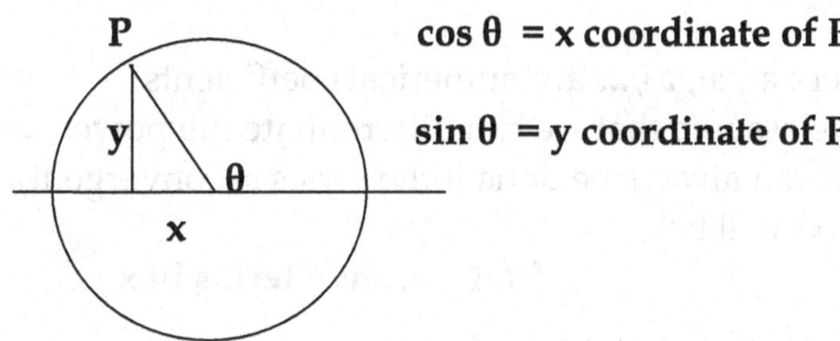

and the exponential function $\mathbf{e^x}$ where **e** is the point at which the area under the curve $\mathbf{y = \frac{1}{x}}$ from **x = 1**, is equal to 1:

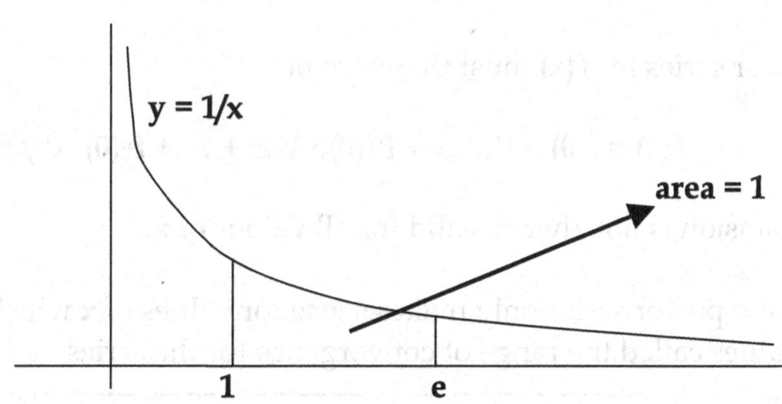

A completely different approach to trig. functions, the exponential function, hyperbolic and log functions uses **series expansions** in their definition.

Colin Maclaurin (1698-1746) was a Scottish mathematician who became professor of mathematics at Edinburgh University and these series expansions are named after him.

The Maclaurin Series

Suppose that **y=f(x)** is a function that can be written out as a series of ascending powers of **x** so that

$$f(x) = a_0 + a_1.x + a_2.x^2 + ... + a_n.x^n + ...$$

where $a_0, a_1, a_2, ...$ are numerical coefficients.
If we assume that we can differentiate the power series any number of times (this can always be done if the series is convergent), then, the nth differential of f(x) will be

$$f^n(x) = a_n.\ n! + \text{terms in } x$$

Now put **x=0** and we have

$$f^n(0) = a_n.\ n!$$

So that

$$a_n = f^n(0)/\ n!$$

The power series for **f(x)** must therefore be

$$f(x) = f(0) + f'(0).x + f''(0).x^2/\ 2! + ...+ f^n(0).\ x^n\ /\ n! + ...$$

The expansion is not always valid for all values of **x**.

An infinite power series only makes sense for values of **x** which lie a certain set of values called the range of convergence for the series.

Example 1 Newton's Binomial theorem

$$f(x) = (1 + x)^n \qquad f(0) = 1$$

$$f'(x) = n(1 + x)^{n-1} \qquad f'(0) = n$$

$$f^2(x) = n(n-1)(1 + x)^{n-2} \qquad f^2(0) = n(n-1)$$

$$f^3(x) = n(n-1)(n-2)(1 + x)^{n-3} \qquad f^3(0) = n(n-1)(n-2)$$

Hence $$(1+x)^n = 1 + nx + \frac{n(n-1)}{2!}x^2 + \frac{n(n-1)(n-2)}{3!}x^3 + \ldots.$$

If **n** is a positive integer then the series terminates at the term in x^n which is

$$f^n(0)\,x^n = \frac{n(n-1)(n-2)\ldots\ldots3.2.1}{1.2.3\ldots(n-2)(n-1)n}\,x^n = x^n$$

otherwise, the range of convergence is $-1 < x < +1$

Example 2 **Expand $f(x) = (1+x)^{-1}$**

Solution

$f(x) = (1+x)^{-1}$ $\qquad f(0) = 1$

$f'(x) = -(1+x)^{-2}$ $\qquad f'(0) = -1$

$f^2(x) = 1.2(1+x)^{-3}$ $\qquad f^2(0) = 1.2$

$f^3(x) = -1.2.3(1+x)^{-4}$ $\qquad f^3(0) = -1.2.3$

$$f(x) = 1 - x + \frac{1.2}{2!}x^2 - \frac{1.2.3}{3!}x^3 + \ldots..$$

$$= 1 - x + x^2 - x^3 + \ldots.$$

We recognize the sum to infinity of a GP with range of convergence $-1 < x < 1$

Example 3 Find the series expansion for e^x

Solution $f(x) = e^x$ $\qquad f(0) = 1$

$f'(x) = e^x$ $\qquad f'(0) = 1$

$f^2(x) = e^x$ $\qquad f^2(0) = 1$

Therefore

$$e^x = 1 + x + \frac{x^2}{2!} + \frac{x^3}{3!} + \frac{x^4}{4!} + \ldots$$ (valid for all values of x)

==============================

Example 4 expansions for **sin x** and **cos x**

$$e^{ix} = 1 + ix + \frac{(ix)^2}{2!} + \frac{(ix)^3}{3!} + \frac{(ix)^4}{4!} + \ldots$$

$$e^{-ix} = 1 - ix + \frac{(ix)^2}{2!} - \frac{(ix)^3}{3!} + \frac{(ix)^4}{4!} + \ldots$$

Add the two series, (this is always possible for convergent series), thus

$$e^{ix} + e^{-ix} = 2(1 - \frac{x^2}{2!} + \frac{x^4}{4!} - \frac{x^6}{6!} +$$

hence $$\frac{e^{ix} + e^{-ix}}{2} = 1 - \frac{x^2}{2!} + \frac{x^4}{4!} - \frac{x^6}{6!} +$$

Thus $$\cos x = 1 - \frac{x^2}{2!} + \frac{x^4}{4!} - \frac{x^6}{6!} + \qquad \text{(valid for all x)}$$

==

Subtracting the two expansions

$$e^{ix} - e^{-ix} = 2(ix - \frac{(ix)^3}{3!} + \frac{(ix)^5}{5!} - \frac{(ix)^7}{7!} +$$

hence $$\frac{e^{ix} - e^{-ix}}{2i} = x - \frac{x^3}{3!} + \frac{x^5}{5!} - \frac{x^7}{7!} +$$

Thus $$\sin x = x - \frac{x^3}{3!} + \frac{x^5}{5!} - \frac{x^7}{7!} + \qquad \text{(valid for all x)}$$

==

Exercise 2

[1] Write down the expansion for $e^{i\theta}$.

Show that the real part is the expansion for cosθ and that the imaginary part is the expansion for i.sinθ.

[2] Differentiate the sin x series term by term and show that the result is the series for cos x.

[3] Differentiate the cos x series term by term and show that the results is the series for –sin x

Example 5 Expansions for **cosh x** and **sinh x**

By definition we have $\cosh x = \frac{e^{x} + e^{-x}}{2}$ $\sinh x = \frac{e^{x} - e^{-x}}{2}$

Now

$$e^{x} = 1 + x + \frac{x^2}{2!} + \frac{x^3}{3!} + \frac{x^4}{4!} + \ldots$$

and

$$e^{-x} = 1 - x + \frac{x^2}{2!} - \frac{x^3}{3!} + \frac{x^4}{4!} - \ldots$$

Adding and dividing by 2 we get

$$\cosh x = 1 - \frac{x^2}{2!} + \frac{x^4}{4!} - \frac{x^6}{6!} + \ldots$$

===

Subtracting and dividing by 2 we get

$$\sinh x = x - \frac{x^3}{3!} + \frac{x^5}{5!} - \frac{x^7}{7!} + \ldots$$

(these two expansions are also valid for all values of x)

===

Exercise 3 Using these expansions:

[1] Show that d/dx (sinh x) = cosh x

[2] Show that d/dx (e^x) = e^x

[3] Write down the expansion for $(1+x)^{-1}$ (valid for $-1<x<+1$)

Integrate your expansion term by term to show that

$$\ln(1+x) = x - \frac{x^2}{2} + \frac{x^3}{3} - \frac{x^5}{5} + \ldots \qquad \text{(valid for } -1<x \leq 1)$$

Notice that we can get the log of 2 from this expansion:

$$\ln 2 = 1 - 1/2 + 1/3 - 1/4 + 1/5 - 1/6 + \ldots\ldots$$

Definitions for e

So far, we have had three definitions for the number

$$e = 2.718281828\ldots$$

Def 1: e is the coordinate for which the area under **y=1/x from x=1 to x=e** is 1

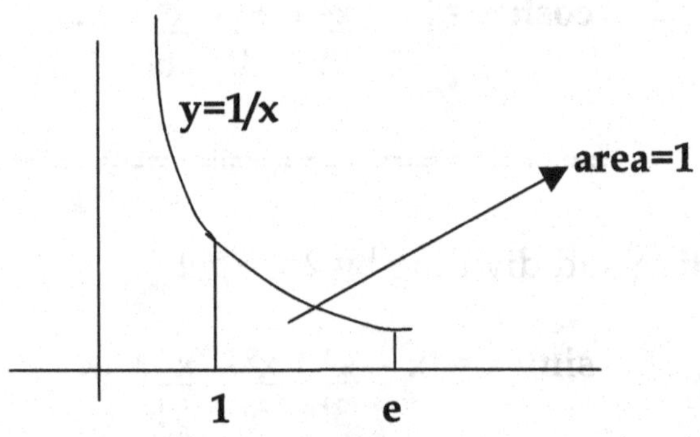

Def 2: e is the value of **a** for which the exponential curve **y=a^x** has a gradient **1** where it crosses the **y** axis

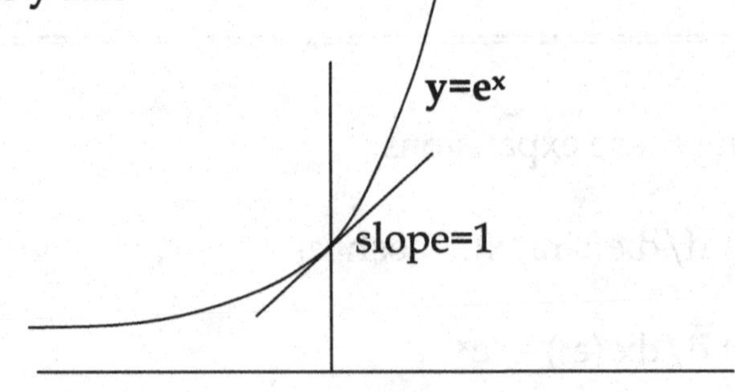

Def 3:

$$e = 1 + 1 + \frac{1}{2!} + \frac{1}{3!} + \frac{1}{4!} + \ldots$$

A fourth definition uses the limit of a binomial expansion.

Def 4:

Consider the binomial series expansion:

$$(1 + x/n)^n = 1 + n(x/n) + \frac{n(n-1)(x/n)^2}{1.2} + \frac{n(n-1)(n-2)(x/n)^3}{1.2.3} + \ldots$$

$$= 1 + x + \frac{(1 - 1/n)x^2}{1.2} + \frac{(1 - 1/n)(1 - 2/n)x^3}{1.2.3} + \ldots$$

Now let $n \rightarrow \infty$ so that 1/n, 2/n, 3/n … all → 0

Then the right hand side becomes $1 + x + \frac{x^2}{2!} + \frac{x^3}{3!} + \ldots$

Which is the series expansion for e^x

Thus we have $\lim_{n \rightarrow \infty} (1 + x/n)^n = e^x$

Putting **x=1** we have our fourth definition for e:

$$e = \lim_{n \rightarrow \infty} (1 + 1/n)^n$$

==============================

Exercise 4

Using the expansions for sin x and cos x we know that

$$\sin 2x = 2x - \frac{(2x)^3}{3!} + \frac{(2x)^5}{5!} - \frac{(2x)^7}{7!} + \ldots$$

and that

$$2\sin x.\cos x = 2\left(x - \frac{x^3}{3!} + \frac{x^5}{5!} - \frac{x^7}{7!} + \ldots.\right)\left(1 - \frac{x^2}{2!} + \frac{x^4}{4!} - \frac{x^6}{6!} + \ldots.\right)$$

Verify that this product gives the first four terms of the series expansion for **sin2x.**

==========================

The Graph of $y = \cosh x$

We start with the graph of $y = e^x$

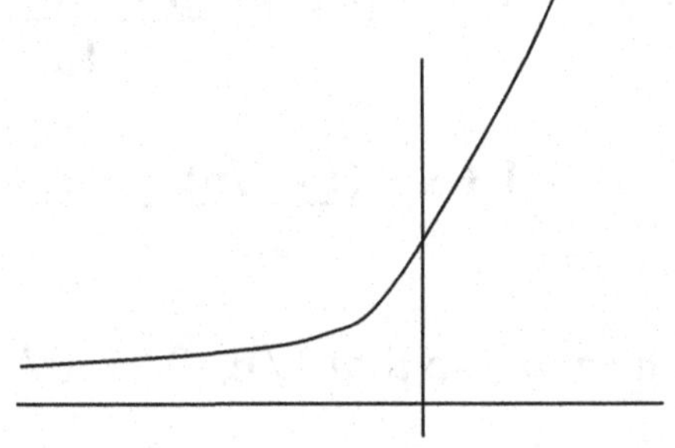

$y = e^{-x}$ is then a reflection in the y axis

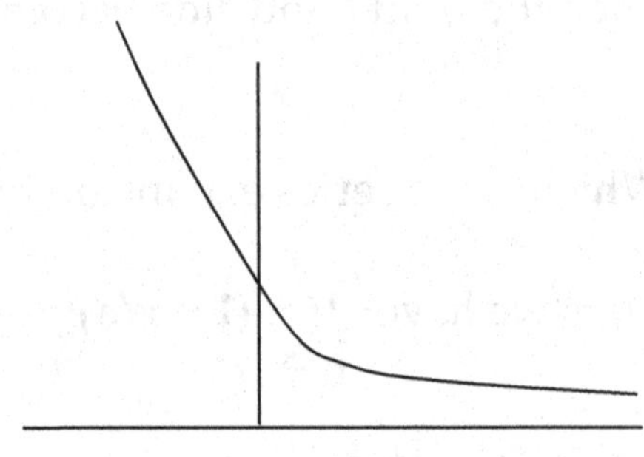

Add these two graphs and we get this dotted graph:

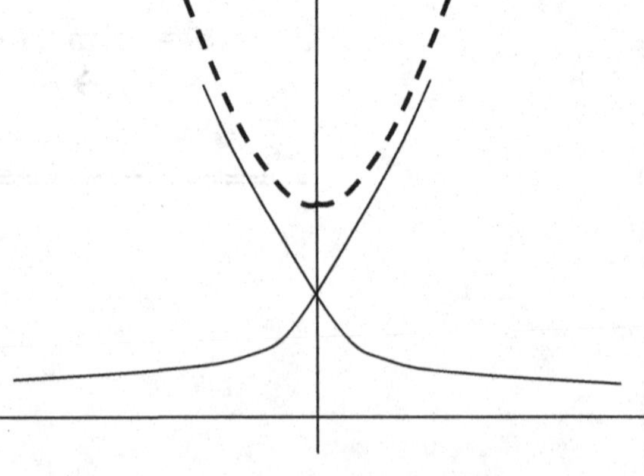

Now divide by 2 to get the graph of **y = coshx,** a **U** shaped graph passing through the point **(0,1)**

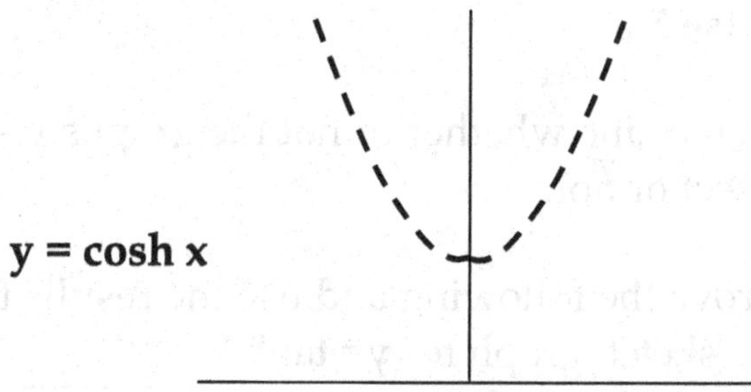

The Graph of y = sinhx

Now put **$y=e^x$** and **$y=e^{-x}$** on the same graph and subtract.

We then get the dotted curve here:

This curve has the equation **$y = e^x - e^{-x}$**

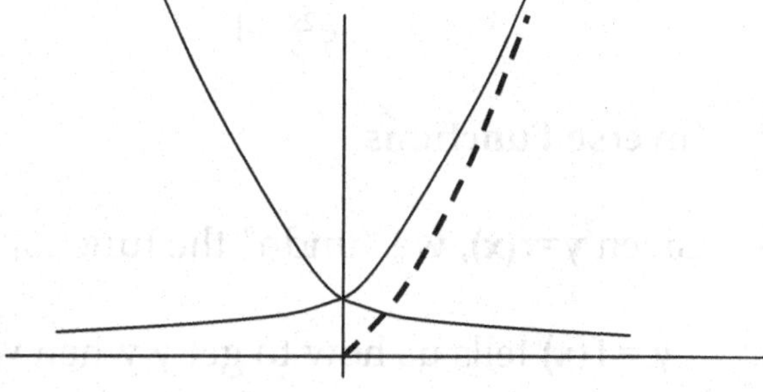

It is easy to show that this curve has rotational symmetry, for if **(a,b)** is a point on the curve, then $b = e^a - e^{-a}$, so that $-b = e^{-a} - e^a$ and this shows that **(-a, -b)** is also on the curve.
Thus, if **(a,b)** is a point on the curve, then so is **(-a,-b)** and so the curve has 180° rotational symmetry.

We now halve the curve and add the piece in the third quadrant, to give the rotational symmetry, and we then have the graph of $y = \dfrac{e^x - e^{-x}}{2}$

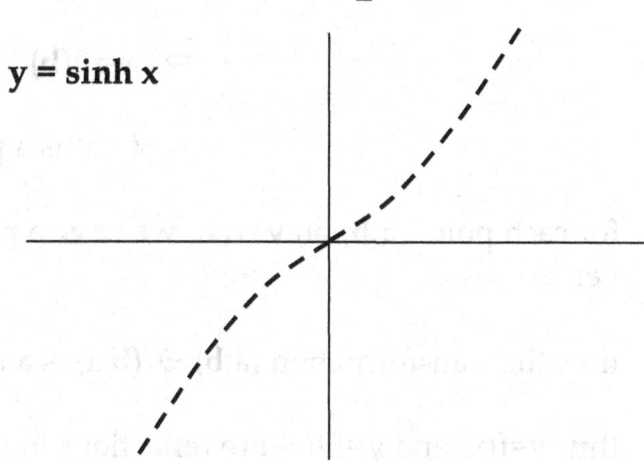

Exercise 5

[1] Determine whether or not the graphs y = sinh x and y=cosh x intersect or not.

[2] Prove the following and use the results to verify the sketch graph for y= tanh x

$$\tanh x = \frac{e^{2x} - 1}{e^{2x} + 1}$$

$$= 1 - \frac{2}{e^{2x} + 1}$$

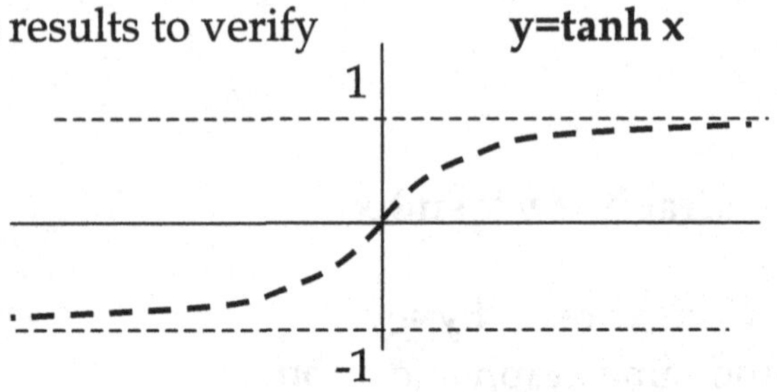

y=tanh x

1

-1

Inverse Functions

Given **y=f(x),** we "undo" the function **f(x)** using the notation **$x=f^{-1}(y)$.**

y = **f(x)** tells us how to get **y** when we know **x.**

x = **$f^{-1}(y)$** tells us how to get back **x** when we know **y.**

Thus, we actually have two functions defined by the expression for **f.**

y = **f(x)** is our function, **y** = **$f^{-1}(x)$** is the inverse function.

The two graphs, **y** = **f(x)** and **y** = **$f^{-1}(x)$** are reflections of each other in the line **y=x** for

If **(a,b)** is a point on **y=f(x)** then **b=f(a)**

⇨ **$a=f^{-1}(b)$**

⇨ **(b,a)** is a point on **$y=f^{-1}(x)$**

for each point **(a,b)** on **y=f(x),** we have a point **(b,a)** on **$y=f^{-1}(x)$** (and vice versa)

now the transformation **(a,b)** → **(b,a)** is a reflection in the line **y=x,**

thus y=f(x) and $y=f^{-1}(x)$ are reflections in y=x.

The simplest way of finding the inverse function is to rewrite the equation $y = f(x)$ as say $x = g(y)$, then switch the x and y to get $y = g(x)$ then this will be the inverse of $y = f(x)$.

Example 1

Find the inverse of the function $y = 2x + 4$

Solution

$$y = 2x + 4 \text{ gives } x = \tfrac{1}{2}(y-4)$$

Now switch x and y to get the inverse function $y=\tfrac{1}{2}x - 2$

Note: the inverse of the inverse gets you back to where your were, thus

$$\mathbf{f(f^{-1}(x)) = x} \quad \text{and} \quad \mathbf{f^{-1}(f(x)) = x}$$

that is, the inverse of the inverse is the identity.

We can often check this for example, in example 1 here,

$$2(\tfrac{1}{2}x - 2) + 4 = x - 4 + 4 = x$$

and

$$\tfrac{1}{2}(2x + 4) - 2 = x + 2 - 2 = x$$

Example 2

If $y=e^x$ then we have $x=\ln(y)$, thus $y=e^x$ and $y=\ln(x)$ are inverses of each other;

Graphically:

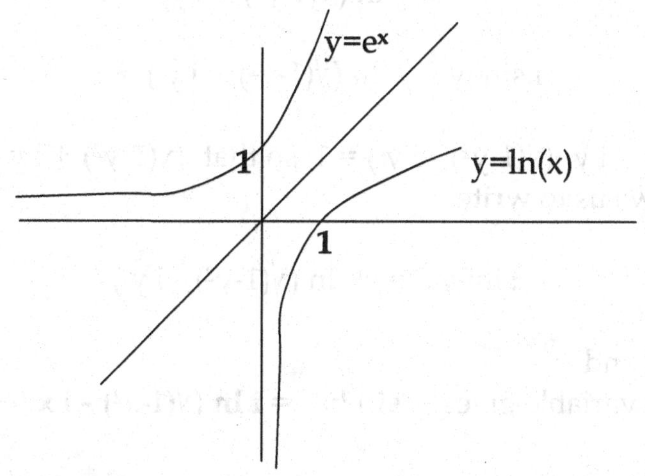

The Inverse Sine Function

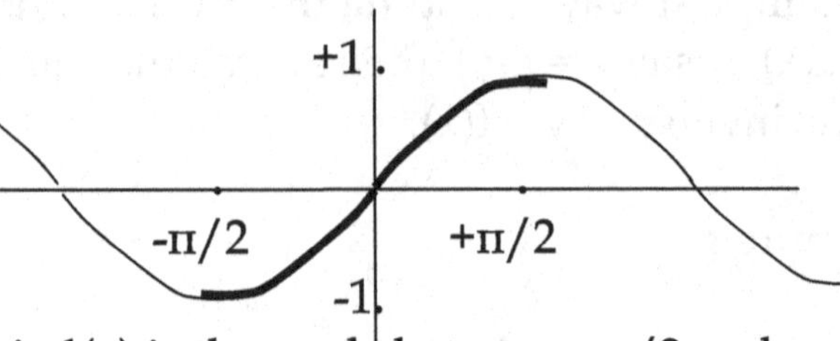

If the graph of y=sin(x) is restricted to the principal range $-\pi/2$ to $+\pi/2$ then $\sin^{-1}(x)$ will be a single valued function. The principal value of $\sin^{-1}(x)$ is the angle between $-\pi/2$ and $+\pi/2$ whose sine is equal to x.

The graph of $\mathbf{y=\sin^{-1}x}$

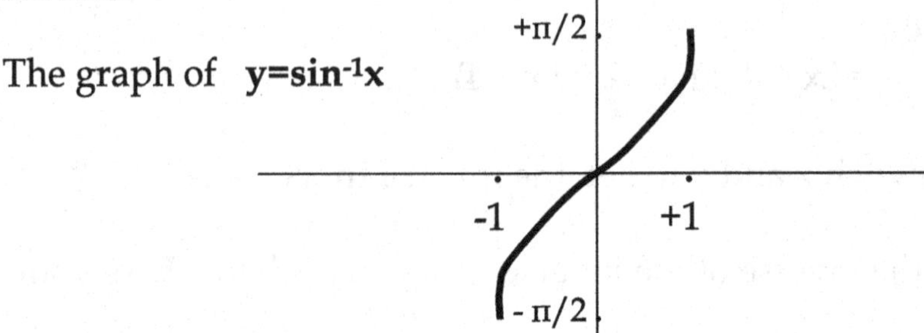

The log Formula for $\sin^{-1}x$

Let P(x,y) be the point P(cos θ, sin θ) on the unit circle.

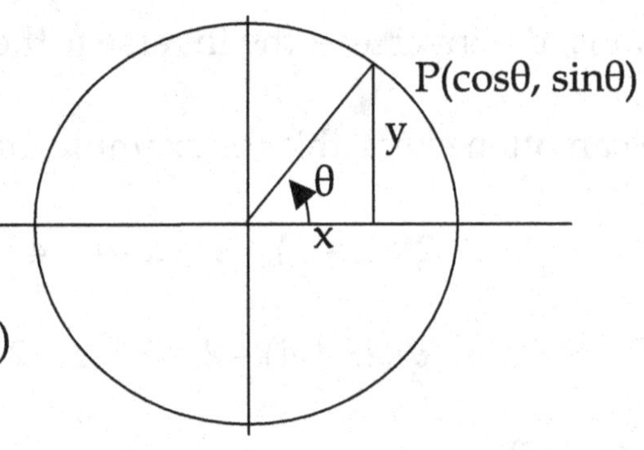

Then $\theta = \sin^{-1}y$ where θ is in the principal range $-\pi/2 \leq \theta \leq \pi/2$

By Pythagoras, $\cos\theta = x = +\sqrt{(1-y^2)}$

and using $\cos\theta + i\sin\theta = e^{i\theta}$

we have $e^{i\theta} = \sqrt{(1-y^2)} + i\,y$

$$i\theta = \ln(\sqrt{(1-y^2)} + i\,y)$$

$$i.\sin^{-1}y = \ln(\sqrt{(1-y^2)} + i\,y)$$

now $(\sqrt{(1-y^2)} + i\,y)(\sqrt{(1-y^2)} - i\,y) = 1$ so that $(\sqrt{(1-y^2)} + i\,y) = (\sqrt{(1-y^2)} - i\,y)^{-1}$ and this allows us to write

$$i.\sin^{-1}y = -1.\ln(\sqrt{(1-y^2)} - i\,y)$$

dividing by **i** and changing the variable gives $\mathbf{\sin^{-1}x = i\ln(\sqrt{(1-x^2)} - i\,x)}$

Exercise 6

[1] Check that the log formula gives π/6 for $\sin^{-1} ½$

The Inverse Cosine

If the graph of y=cos(x) is restricted to the principal range 0 to +π then cos-1(x) will be a single valued function.

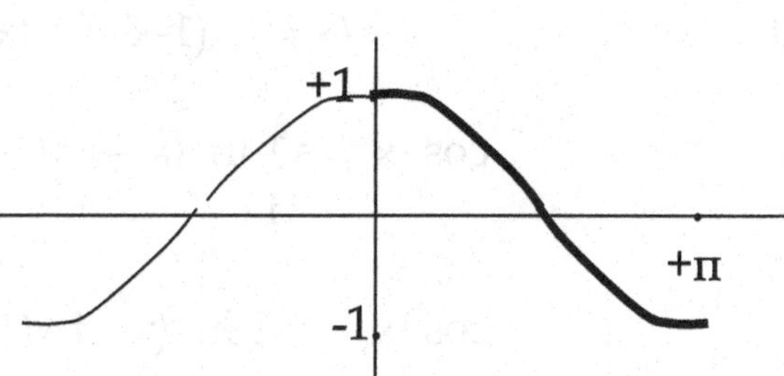

The principal value of $\cos^{-1}(x)$ is the angle between 0 and +π whose cosine is equal to x.

This gives us the following graph

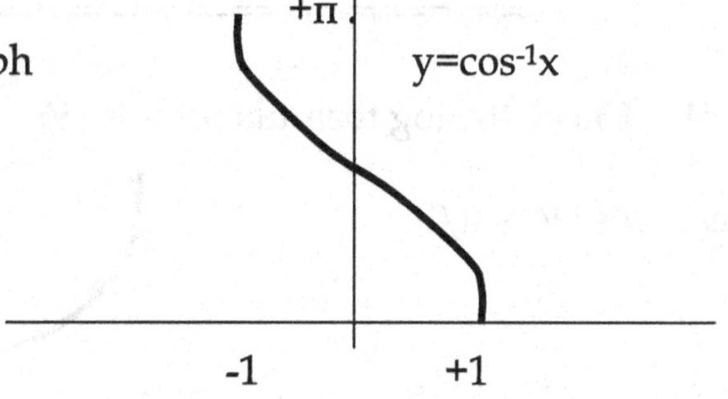

The log Formula for $\cos^{-1}x$

Let P(x,y) be the point P(cos θ, sin θ) on the unit circle.

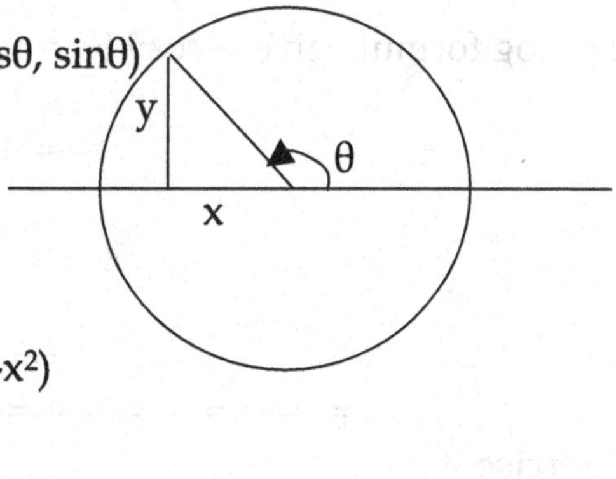

P(cosθ, sinθ)

Then $\theta = \cos^{-1}x$

y θ x

where θ is in the principal range $0 \leq \theta \leq \pi$

also $\sin\theta = y = +\sqrt{(1-x^2)}$

using $\cos\theta + i\sin\theta = e^{i\theta}$

we have $e^{i\theta} = x + i\sqrt{(1-x^2)}$

$$i\theta = \ln\left(x + i\sqrt{(1-x^2)}\right)$$

$$\cos^{-1}x = \frac{1}{i} \ln\left(x + i\sqrt{(1-x^2)}\right)$$

we note that $(x + i\sqrt{(1-x^2)})(x - i\sqrt{(1-x^2)}) = 1$

so that $$(x + i\sqrt{(1-x^2)}) = (x - i\sqrt{(1-x^2)})^{-1}$$

i.e. $$\cos^{-1}x = \frac{1}{i} \ln\left(x - i\sqrt{(1-x^2)}\right)^{-1}$$

$$\cos^{-1}x = -\frac{1}{i} \ln\left(x - i\sqrt{(1-x^2)}\right)$$

hence $$\mathbf{\cos^{-1}x = i \ln\left(x - i\sqrt{(1-x^2)}\right)}$$

==================================

Example: Check the log formula for $\cos^{-1} ½$

We test $\cos^{-1} ½ = \pi/3$

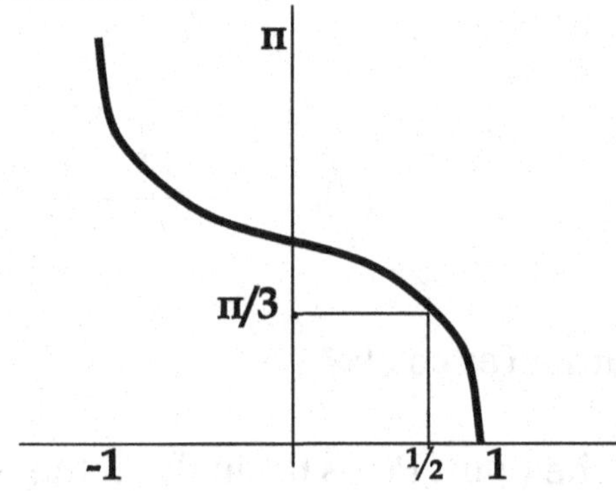

The log formula gives $\cos^{-1} ½ = i \ln(½ - i\sqrt{3}/2)$

$= i \ln(e^{-i\pi/3})$

$= i. -i\pi/3$

$= \pi/3$ as required

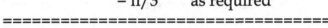

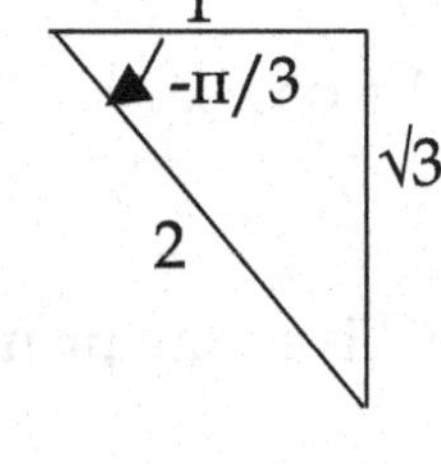

====================================

Exercise 7

[1] Check that the log formula gives $5\pi/6$ for $\cos^{-1}(-\sqrt{3}/2)$

[2] Check that the log formula gives $3\pi/4$ for $\cos^{-1}(-1/\sqrt{2})$

The log Formula for the Inverse Tangent

Let $\theta = \tan^{-1}m$, where θ is in the principal range $-\pi/2 \le \theta \le \pi/2$ so that $m=\tan\theta$ and $m=y/x$.
If $P(x,y)$ is the point $(\cos\theta, \sin\theta)$ on the unit circle

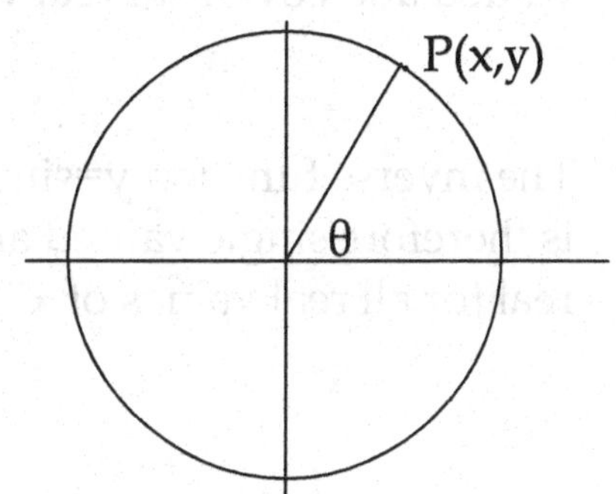

we have $e^{i\theta} = \cos\theta + i\sin\theta$

and $e^{-i\theta} = \cos\theta - i\sin\theta$

dividing
$$e^{2i\theta} = \frac{\cos\theta + i\sin\theta}{\cos\theta - i\sin\theta}$$

$$= \frac{1+im}{1-im} \qquad (\text{since } m = \sin\theta/\cos\theta)$$

hence
$$2i\theta = \ln\left[\frac{1+im}{1-im}\right]$$

$$\tan^{-1}m = \frac{1}{2i}\ln\left[\frac{1+im}{1-im}\right]$$

changing the variable we have

$$\mathbf{\tan^{-1}x = \frac{1}{2i}\ln\left[\frac{1+ix}{1-ix}\right]}$$

======================

Example

$$\tan^{-1}1 = \frac{1}{2i}\ln\left[\frac{1+i}{1-i}\right] = \frac{1}{2i}\left(\ln(1+i) - \ln(1-i)\right)$$

$$= \frac{1}{2i}\left(\ln(e^{i\pi/4}) - \ln(e^{-i\pi/4})\right) = (i\pi/4 + i\pi/4)/2i = \pi/4$$

==========================

The Inverse Shine Function ($\sinh^{-1}x$)

y=sinh x

For all real values of x, **y=sinh x** is single valued and covers all real values of y.

The inverse function $y=\sinh^{-1}x$ is therefore single valued and real for all real values of x.

The graph of $y=\sinh^{-1}x$

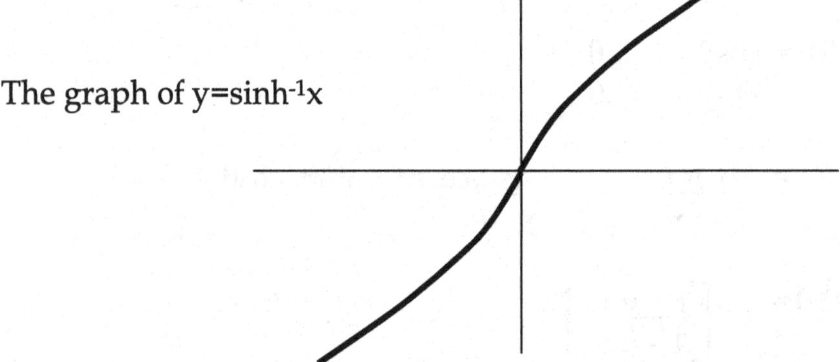

If $y=\sinh^{-1}x$ then $$x=\sinh y = \frac{e^y - e^{-y}}{2}$$

so that

$$e^{2y}-2xe^y - 1 = 0$$

solving for e^y gives

$$e^y = \frac{2x \pm\sqrt{(4x^2 + 4)}}{2}$$

$$e^y = x \pm\sqrt{(x^2 + 1)}$$

now e^y must be positive, for real x, so we have to take the positive sign, thus

$$e^y = x +\sqrt{(x^2 + 1)}$$

thus

$$y = \ln(x +\sqrt{(x^2 + 1)}\,)$$

therefore $$\mathbf{\sinh^{-1} x = \ln(\, x +\sqrt{(x^2 + 1)}\,)}$$

==========================

Alternatively we may deduce the same formula by using to the parametric coordinates that we found for the rectangular hyperbola (we here use the shorthand ch θ and sh θ for cosh θ and sinh θ):

Let P(x,y) be the point P(ch θ, sh θ) on $x^2 - y^2 = 1$

Then we have $\text{ch}\,\theta = x,\quad \text{sh}\,\theta = y,\quad x = \pm\sqrt{(1+y^2)}$

Further, we know that $\text{ch}\,\theta - \text{sh}\,\theta = e^{-\theta}$ (from the original definitions)

Therefore $\pm\sqrt{(1+y^2)} - y = e^{-\theta}$

So that $-\theta = \ln(\pm\sqrt{(1+y^2)} - y)$

Since θ is real, we cannot have the log of a negative number here (see graph of $y = \ln x$), and since $\sqrt{(1+y^2)}$ is always bigger than y, we have

$$-\text{sh}^{-1}y = \ln(\sqrt{(1+y^2)} - y)$$

or

$$\text{sh}^{-1}y = \ln(\sqrt{(1+y^2)} - y)^{-1}$$

Now $(\sqrt{(1+y^2)} - y)(\sqrt{(1+y^2)} + y) = (1+y^2) - y^2 = 1$

therefore $(\sqrt{(1+y^2)} - y)^{-1} = (\sqrt{(1+y^2)} + y)$

hence $\text{sh}^{-1}y = \ln(\sqrt{(1+y^2)} + y)$ for any real y

thus $$\mathbf{\sinh^{-1} x = \ln(x + \sqrt{(x^2+1)})}$$ ■

Exercise 8 Use the log formula to show that

[i] $\sinh^{-1}\left(\dfrac{e^2 - 1}{2e}\right) = 1$

[ii] $\sinh^{-1}\left(\dfrac{e^4 - 1}{2e^2}\right) = 1$

==

The Inverse Cosh ($\cosh^{-1}x$)

Since $y = \cosh x$ is symmetrical about the y axis, the inverse function $Y = \cosh^{-1}x$ is a two valued (Compare with $y = x^2$ where $x = \pm\sqrt{y}$).

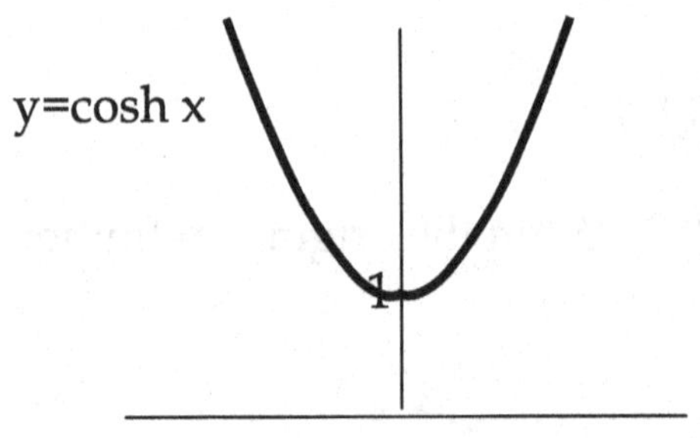

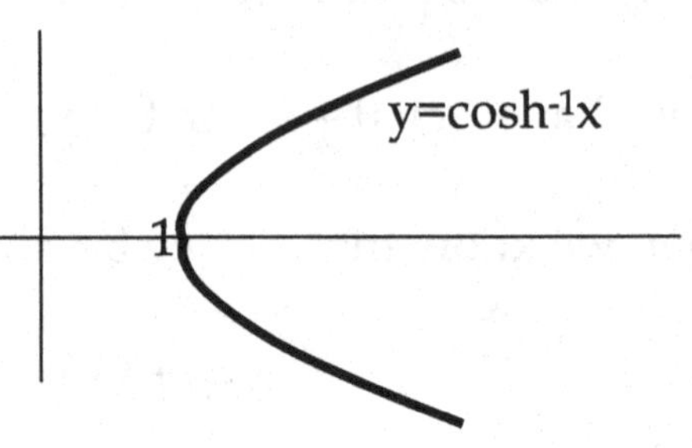

if $y = \cosh^{-1}x$ then $x = \cosh y = \dfrac{e^y + e^{-y}}{2}$

giving $$e^y + e^{-y} - 2x = 0$$

multiply by e^y $$e^{2y} - 2xe^y + 1 = 0$$

solving for e^y we find $$e^y = x \pm\sqrt{(x^2 - 1)}$$

giving $$y = \ln(x \pm\sqrt{(x^2 - 1)})$$

we note that $(x - \sqrt{(x^2 - 1)})(x + \sqrt{(x^2 - 1)}) = x^2 - (x^2 - 1) = 1$

so that $$(x - \sqrt{(x^2 - 1)}) = (x + \sqrt{(x^2 - 1)})^{-1}$$

and taking logs of both sides, we get

$$\ln(x - \sqrt{(x^2 - 1)}) = -\ln(x + \sqrt{(x^2 - 1)})$$

these are the two equal and opposite values for $\cosh^{-1}x$

thus $$\mathbf{\cosh^{-1}x = \ln(x \pm\sqrt{(x^2 - 1)})}$$

================================

Alternatively we may argue as follows:

Let P(x,y) be the point P(ch θ, sh θ) on $x^2 - y^2 = 1$

Then we have $\quad \text{ch}\,\theta = x, \quad \text{sh}\,\theta = y, \quad y = \pm\sqrt{x^2 - 1}$

Further, we know that $\quad \text{ch}\,\theta + \text{sh}\,\theta = e^\theta$ (from the original definitions)

Therefore $\quad x \pm \sqrt{x^2 - 1} = e^\theta$

so that

$$\theta = \ln(x \pm \sqrt{x^2 - 1})$$

$$\mathbf{\cosh^{-1}x = \ln(x \pm \sqrt{x^2 - 1})}$$

========================

The Inverse Hyperbolic Tangent ($\tanh^{-1}x$)

If $y = \tanh^{-1}x$ then $x = \tanh y$

$$x = \frac{e^y - e^{-y}}{e^y + e^{-y}}$$

giving

$$x(e^{2y} + 1) = e^{2y} - 1$$

thus

$$e^{2y} = \frac{1 + x}{1 - x}$$

so that

$$y = \tanh^{-1}x = \tfrac{1}{2}\ln\left[\frac{1 + x}{1 - x}\right]$$

We choose to define $\cosh^{-1}x$ using the positive value, so that we have the following single valued inverse hyperbolic functions for real values of x:

$$\cosh^{-1}x = \ln(x + \sqrt{x^2 - 1})$$

$$\sinh^{-1}x = \ln(x + \sqrt{x^2 + 1})$$

$$\tanh^{-1}x = \tfrac{1}{2}\ln\left[\frac{1 + x}{1 - x}\right]$$

Summary of Results

$$\sinh^{-1} x = \ln(x + \sqrt{(x^2 + 1)}) \quad \text{(valid for all real x)}$$

$$\cosh^{-1}x = \ln(x + \sqrt{(x^2 - 1)}) \quad \text{(for x >=1)}$$

$$\tanh^{-1}x = \tfrac{1}{2} \ln \left[\frac{1+x}{1-x}\right] \quad \text{(valid for -1<x<+1)}$$

$$\sin^{-1}x = i \ln (\sqrt{(1-x^2)} - i x) \quad \text{(valid for -1<x<+1)}$$

$$\cos^{-1}x = i \ln (x - i \sqrt{(1-x^2)}) \quad \text{(valid for -1<x<=1)}$$

$$\tan^{-1}x = \frac{1}{2i} \ln \left[\frac{1+ix}{1-ix}\right] \quad \text{(valid for all real x)}$$

==

Inverse sine from inverse shine:

Recall **$i.\sin\theta = \sinh i\theta$**

Let $ix = i \sin\theta = \sinh i\theta$ so that $\theta = \sin^{-1}x$ and $i\theta = \sinh^{-1} ix$

Then $$\sin^{-1}x = \frac{1}{i} \sinh^{-1} ix$$

$$= -i \ln(ix + \sqrt{(-x^2 + 1)})$$

$$= -i \ln(\sqrt{(1 - x^2)} + ix)$$

$$\mathbf{\sin^{-1}x = i \ln(\sqrt{(1 - x^2)} - ix)}$$

since $(\sqrt{(1 - x^2)} - ix) = (\sqrt{(1 - x^2)} + ix)^{-1}$

Exercise 9

[1] Check that the log formula gives $\pi/2$ for $\sin^{-1} 1$.

[2] Show that $\cosh^{-1} 2 = \ln(2 \pm \sqrt{3})$

and verify that $\cosh(\ln(2 \pm \sqrt{3})) = 2$

[3] Show that $\cosh^{-1} i = i\pi/2 + \ln(1 \pm \sqrt{2})$

Inverse cos from inverse cosh

Let $x = \cos\theta = \cosh i\theta$

Then $\theta = \cos^{-1}x$ and $i\theta = \cosh^{-1}x$

Thus $$\cos^{-1}x = \frac{1}{i}\cosh^{-1}x$$

$$= -i\ln(x + \sqrt{(x^2 - 1)})$$

$$\mathbf{\cos^{-1}x = i\ln(x - i\sqrt{(1 - x^2)})}$$

========================

The inverse tan from the inverse tanh

Let $x = \tan i\theta = i\tanh\theta$

Then $$i\theta = \tan^{-1}x = i\tanh^{-1}(x/i)$$

$$= \frac{i}{2}\ln\left[\frac{1 + x/i}{1 - x/i}\right]$$

$$\tan^{-1}x = \frac{i}{2}\ln\left[\frac{i + x}{i - x}\right]$$

======================================

Functions of Complex Numbers

The next three sections have been introduced in chapter 27, however, we recall them here again since they will complete the treatment of the elementary functions of complex numbers.

Complex Powers of Real Numbers

In the previous chapter we defined the exponential function e^z, where z is a complex number say $x+iy$, from the relation

$$e^z = e^{x+iy} = e^x \, . \, e^{iy} = e^x.(\cos y + i.\sin y)$$

The complex power of a real number a is then defined by

$$a^z = e^{z \ln a}$$

=====================================

Logs of Complex Numbers

Further, if $z = r(\cos\theta + i.\sin\theta) = r.e^{i\theta}$ with $-\pi < \theta \leq \pi$, we define the principal value of the log of the complex number z by

$$\log z = \ln r + i.\theta$$

We realize that there are other values for the angle θ that will give the same z since $\cos\theta + i.\sin\theta = \cos(\theta + 2n\pi) + i.\sin(\theta + 2n\pi)$ for any integer value of n.
Thus

$$z = r.e^{i\theta} = r.e^{i(\theta+2n\pi)}$$

We define the multi-valued Log by

$$\text{Log } z = \ln r + i.(\theta + 2n\pi)$$

The value of Log z for n=0 is called the principal value of Log z and will be denoted by log z.
Thus

$$\text{Log } z = \log z + 2n\pi i$$

(Many books define these the other way around and call log z the multi-valued log and call Log z the principal value. We follow the definitions given in the classic book "A Course of Pure Mathematics" by G. H. Hardy C.U.P.)

Complex Powers of Complex Numbers

Having defined the Log of a complex numbers, we use this to define a complex power of a complex number:

Suppose that **w** is the complex numbers $\mathbf{w = r\, e^{i\phi}}$
Then we define

$$\mathbf{z^w = e^{z.Log w}}$$

$$\mathbf{= e^{z.(\ln r + i(\phi + 2n\pi))}}$$

These functions have been treated in chapter 27, but we are now in a position to extend our definitions further so that the elementary function are also defined for complex numbers. The elementary functions are usually taken to be (i) polynomials, (ii) rational algebraic functions, (iii) exponential and log functions, (iv) trig functions and hyperbolic functions, (v) inverse trig functions and inverse hyperbolic functions.

Trig Functions of Complex Numbers

If **z** is any complex number, we define

$$\mathbf{\cos z = \frac{e^{iz} + e^{-iz}}{2}} \qquad \mathbf{\sin z = \frac{e^{iz} - e^{-iz}}{2i}}$$

$$\mathbf{\tan z = \frac{\sin z}{\cos z}} \qquad \mathbf{\cot z = \frac{\cos z}{\sin z}}$$

$$\mathbf{\sec z = \frac{1}{\cos z}} \qquad \mathbf{\mathrm{cosec}\, z = \frac{1}{\sin z}}$$

All the familiar formulae for trig functions of real variables hold for these definitions, for example

$$\cos^2 z + \sin^2 z = \left(\frac{e^{iz} + e^{-iz}}{2}\right)^2 + \left(\frac{e^{iz} - e^{-iz}}{2i}\right)^2$$

$$= \frac{e^{2iz} + 2 + e^{-2iz}}{4} + \frac{e^{2iz} - 2 + e^{-2iz}}{-4} = \frac{2}{4} + \frac{2}{4} = 1$$

Exercise 10
Show that the compound angle formula

$$\mathbf{\cos(z+w) = \cos z.\cos w - \sin z.\sin w}$$

holds for these definitions.

Hyperbolic Functions of Complex Numbers

If z is any complex variable, we define

$$\cosh z = \frac{e^z + e^{-z}}{2}$$

$$\sinh z = \frac{e^z - e^{-z}}{2}$$

$$\tanh z = \frac{\sinh z}{\cosh z}$$

$$\coth z = \frac{\cosh z}{\sinh z}, \quad \text{sech } z = \frac{1}{\cosh z}, \quad \text{cosech } z = \frac{1}{\sinh z}$$

Inverse Trig and Hyperbolic Functions of Complex variables

If **z = sin w** then **w = sin^{-1}z** is called the inverse sine of **z.** The other inverse circular functions are defined in a similar way.
If **z = sinh w** then **w = sinh^{-1}z** is called the inverse hyperbolic sine of **z.** The other inverse hyperbolic functions are similarly defined.

The inverse circular and inverse hyperbolic functions are all multiple valued functions and can be expressed in terms of log functions as shown above.

==============================

The elementary functions are thus defined for all positive, negative, real or complex numbers.

We recall here, the relations proved above, between the circular and hyperbolic functions for the real variable θ. **These relations hold also, for the complex variable z.** To see this, all we need to do is to change θ into z !

$$\sin iz = i \sinh z \qquad \cos iz = \cosh z \qquad \tan iz = i \tan z$$

$$\sinh iz = i \sin z \qquad \cosh iz = \cos z \qquad \tanh iz = i \tan z$$

Example 1 Prove that **$\cos(x + iy) = \cos x.\cosh y - i.\sin x.\sinh y$**

Proof

$$\cos(x + iy) = \cos x.\cos iy - \sin x.\sin iy$$

$$= \cos x.\cosh y - i.\sin x.\sinh y$$

Example 2 Prove that **$\sin(x + iy) =\sin x.\cosh y + i.\cos x.\sinh y$**

Proof

$$\sin(x + iy) = \sin x.\cos iy + \cos x.\sin iy$$

$$= \sin x.\cosh y + i.\cos x.\sinh y$$

Example 3 Solve $\cos z = i$

Solution

Let $z = x + iy$

then by example 1 above

$$\cos x.\cosh y \; - \; i.\sin x.\sinh y \; = \; i$$

$\therefore \quad \cos x.\cosh y = 0$ and $\sin x.\sinh y = -1$

now $\cosh y \geq 1$ so therefore $\quad \cos x = 0$

$\therefore \quad$ (i) $x = 2n\pi + \pi/2$

or $\quad$ (ii) $x = 2n\pi - \pi/2$

Case (i) $\quad x = 2n\pi + \pi/2$ gives $\quad \sinh y = -1$

$$\frac{e^y - e^{-y}}{2} = -1$$

$$e^{2y} + 2\,e^y \; -1 \; = \; 0$$

which gives $\quad (\,e^y \; +1\,)^2 \; = \; 2$

so that $\quad e^y \; = \; -1 \pm \sqrt{2}$

now e^y cannot be negative hence $e^y \; = \; \sqrt{2} - 1$

$\therefore \quad y = \ln(\,\sqrt{2} - 1\,)$

the first solution is therefore

$$z = 2n\pi + \pi/2 + i.\ln(\,\sqrt{2} - 1\,)$$

Case (ii) gives $\quad z = 2n\pi - \pi/2 + i.\ln(\,\sqrt{2} + 1\,)$

====================================

Example 4 Solve $\sin z = \sqrt{2}$

Solution Let $z = x + iy$

then by example 1 above

$$\sin x.\cosh y + i.\cos x.\sinh y = \sqrt{2}$$

$\therefore$ $\cos x.\sinh y = 0$ and $\sin x.\cosh y = \sqrt{2}$

If $\cos x.\sinh y = 0$ the either $\cos x = 0$ or $\sinh y = 0$
Now if $\sinh y = 0$ then $\cosh y = 1$ so that $\sin x = \sqrt{2}$ giving no solution since x is real.

Therefore, $\cos x = 0$ so that $x = 2n\pi + \pi/2$ or $x = 2n\pi - \pi/2$

Giving either $\sin x = 1$ and $\cosh y = \sqrt{2}$

or $\sin x = -1$ and $\cosh y = -\sqrt{2}$ but $\cosh y > 0$ therefore, the only solution is

$$\sin x = 1 \text{ and } \cosh y = \sqrt{2}$$

$$\frac{e^y + e^{-y}}{2} = \sqrt{2}$$

$$e^{2y} - 2\sqrt{2}\, e^y + 1 = 0$$

which gives $$(e^y - \sqrt{2})^2 + 1 = 2$$

so that $$e^y = \sqrt{2} \pm 1$$

Therefore the solution is $z = 2n\pi + \pi/2 + i.\ln(\sqrt{2} \pm 1)$

==============================

Exercise 10

Solve

[1] $\cos z = 4i$

[2] $\sin z = 4i$

[3] $\cos z = 6i$

[4] $\sin z = 6i$

[5] $\cos z = i/\sqrt{2}$

[6] $\sin z = i/\sqrt{2}$

[7] $\cos z = i/2$

[8] $\sin z = i/2$

Answers

Exercise 1

[1] $2x^2 - xy - 4x + 2y + 2 = 0$ is a hyperbola that touches the x axis at (1,0) and cuts the y axis at (0,-1). The asymptotes are x=2 and 2y=x.
There is a local maximum point at (1,0) and a local minimum at (3,8).

[2] $2x^2 + y^2 - 4x + 2y + 2 = 0$ is an ellipse with centre at (1,-1) with its axes parallel to the coordinate axes. The axis parallel to Ox is length 1 and the axis parallel to Oy is length 2. There is a local maximum at (2,0) where the ellipse touches the x axis and a local minimum at (2,-2).

[3] $2x^2 - y^2 - 4x + 2y + 2 = 0$ is a degenerate hyperbola that consists of the two straight lines $y-1 = \sqrt{2}\,x - \sqrt{2}$ and $y-1 = -\sqrt{2}\,x + \sqrt{2}$

[4] $x^2 - 2xy + y^2 - 2x + y + 1 = 0$ is a parabola that touches the x axis at (1,0). The axis of the parabola is $y = x + \frac{3}{4}$. The curve lies wholly in the first quadrant and passes through the points (¾ , ¼), (1 ¾, ¼), (1,1) and (3,1).

[5] $x^2 + y^2 - 2x + y + 1 = 0$ is a circle, centre (1, -½), radius ½ which touches the x axis at (1,0).

Exercise 5

[1] no they do not $\dfrac{e^x+e^{-x}}{2} = \dfrac{e^x - e^{-x}}{2}$ gives $e^{-x} = 0$ which is not possible

Exercise 6

$$\sin^{-1}\tfrac{1}{2} = i.\ln(\sqrt{(1 - \tfrac{1}{4})} - i/2)$$

$$= i.\ln(\sqrt{3}/2 - i/2)$$

$$= i.\ln(\cos(-\pi/6) + i.\sin(-\pi/6))$$

$$= i.\ln(e^{-i\pi/6})$$

$$= i.\,-i\pi/6 = \pi/6 \quad \blacksquare$$

Exercise 7

[1] $\sin^{-1}(-\sqrt{3}/2) = i.\ln(-\sqrt{3}/2 - i.\sqrt{(1 - 3/4)})$

$= i.\ln(-\sqrt{3}/2 - i/2)$

$= i.\ln(\cos(-5\pi/6) + i.\sin(-5\pi/6))$

$= i.\ln(e^{-i.5\pi/6})$

$= i.-i5\pi/6 = 5\pi/6$ ■

[2] $\cos^{-1}(-1/\sqrt{2}) = i.\ln(-1/\sqrt{2} - i.\sqrt{(1 - 1/2)})$

$= i.\ln(-1/\sqrt{2} - i/\sqrt{2})$

$= i.\ln(\cos(-3\pi/4) + i.\sin(-3\pi/4))$

$= i.\ln(e^{-i.3\pi/4})$

$= i.-i3\pi/4 = 3\pi/4$ ■

Exercise 11

[1] $z = 2n\pi + \pi/2 + i.\ln(\sqrt{5} + 2)$ or $2n\pi - \pi/2 + i.\ln(\sqrt{5} - 2)$

[2] $z = 2n\pi + i.\ln(\sqrt{5} + 2)$ or $(2n+1)\pi + i.\ln(\sqrt{5} - 2)$

[3] $z = 2n\pi + \pi/2 + i.\ln(\sqrt{10} + 3)$ or $2n\pi - \pi/2 + i.\ln(\sqrt{10} - 3)$

[4] $z = 2n\pi + i.\ln(\sqrt{10} + 3)$ or $(2n + 1)\pi + i.\ln(\sqrt{10} - 3)$

[5] $z = 2n\pi + \pi/2 + i.\ln \frac{1}{2}(\sqrt{6} + \sqrt{2})$ or $2n\pi - \pi/2 + i.\ln \frac{1}{2}(\sqrt{6} - \sqrt{2})$

[6] $z = 2n\pi + i.\ln \frac{1}{2}(\sqrt{6} + \sqrt{2})$ or $(2n + 1)\pi + i.\ln \frac{1}{2}(\sqrt{6} - \sqrt{2})$

[7] $z = 2n\pi + \pi/2 + i.\ln \frac{1}{2}(\sqrt{5} - 1)$ or $2n\pi - \pi/2 + i.\ln \frac{1}{2}(\sqrt{5} + 1)$

[8] $z = 2n\pi + i.\ln \frac{1}{2}(\sqrt{5} + 1)$ or $(2n + 1)\pi + i.\ln \frac{1}{2}(\sqrt{5} - 1)$

==================================

Index for Volume 2

================== ==================